Lecture Notes in Artificial Intelligence 10358

Subseries of Lecture Notes in Computer Science

More information about this series at http://www.springer.com/series/1244

Petra Perner (Ed.)

Machine Learning and Data Mining in Pattern Recognition

13th International Conference, MLDM 2017
New York, NY, USA, July 15–20, 2017
Proceedings

 Springer

Editor
Petra Perner
Institute of Computer Vision and Applied
  Computer Sciences
Leipzig, Sachsen
Germany

ISSN 0302-9743   ISSN 1611-3349   (electronic)
Lecture Notes in Artificial Intelligence
ISBN 978-3-319-62415-0   ISBN 978-3-319-62416-7   (eBook)
DOI 10.1007/978-3-319-62416-7

Library of Congress Control Number: 2017945280

LNCS Sublibrary: SL7 – Artificial Intelligence

Printed on acid-free paper

This Springer imprint is published by Springer Nature
The registered company is Springer International Publishing AG
The registered company address is: Gewerbestrasse 11, 6330 Cham, Switzerland

Preface

The 13th event of the International Conference on Machine Learning and Data Mining MLDM 2017 was held in New York (www.mldm.de) running under the umbrella of the World Congress "The Frontiers in Intelligent Data and Signal Analysis, DSA2017" (www.worldcongressdsa.com).

After the peer-review process, we accepted 35 high-quality papers for oral presentation. The topics range from theoretical topics for classification, clustering, association rule and pattern mining to specific data-mining methods for the different multimedia data types such as image mining, text mining, video mining, and Web mining. Extended versions of selected papers will appear in the international journal *Transactions on Machine Learning and Data Mining* (www.ibai-publishing.org/journal/mldm).

A tutorial on Data Mining, a tutorial on Case-Based Reasoning, a tutorial on Intelligent Image Interpretation and Computer Vision in Medicine, Biotechnology, Chemistry and Food Industry, and a tutorial on Standardization in Immunofluorescence were held before the conference.

We would like to thank all reviewers for their highly professional work and their effort in reviewing the papers. We would also thank the members of the Institute of Applied Computer Sciences, Leipzig, Germany (www.ibai-institut.de), who handled the conference as secretariat. We appreciate the help and understanding of the editorial staff at Springer, and in particular Alfred Hofmann, who supported the publication of these proceedings in the LNAI series.

Last, but not least, we wish to thank all the speakers and participants who contributed to the success of the conference. See you in 2018 in New York at the next World Congress (www.worldcongressdsa.com) on "The Frontiers in Intelligent Data and Signal Analysis, DSA2018," which combines under its roof the following three events: International Conferences on Machine Learning and Data Mining, MLDM, the Industrial Conference on Data Mining, ICDM, and the International Conference on Mass Data Analysis of Signals and Images in Medicine, Biotechnology, Chemistry and Food Industry, MDA.

July 2017   Petra Perner

Organization

Program Chair

Petra Perner   IBaI Leipzig, Germany

Program Committee

Sergey V. Ablameyko   Belarus State University, Belarus
Reneta Barneva   The State University of New York at Fredonia, USA
Patrick Bouthemy   Inria VISTA, France
Michelangelo Ceci   University of Bari, Italy
Xiaoqing Ding   Tsinghua University, China
Christoph F. Eick   University of Houston, USA
Ana Fred   Technical University of Lisbon, Portugal
Giorgio Giacinto   University of Cagliari, Italy
Makato Haraguchi   Hokkaido University Sapporo, Japan
Dimitrios A. Karras   Chalkis Institute of Technology, Greece
Adam Krzyzak   Concordia University, Montreal, Canada
Thang V. Pham   Intelligent Systems Lab Amsterdam (ISLA), The Netherlands
Linda Shapiro   University of Washington, USA
Shashank Shekhar   VMware Inc.
Francis E.H. Tay   National University of Singapore, Singapore
Alexander Ulanov   HP Labs, Russian Federation
Zeev Volkovich   ORT Braude College of Engineering, Israel
Patrick Wang   Northeastern University, USA

Additional Reviewers

Battista Biggio   University of Cagliari, Italy
Giorgio Fumera   University of Cagliari, Italy
Ahmad Pahlavan Tafti   Marshfield Clinic Research Institute, USA

Contents

An Information Retrieval Approach for Finding Dependent Subspaces of Multiple Views

Ziyuan Lin[1,2(✉)] and Jaakko Peltonen[1,2]

[1] Faculty of Natural Sciences, University of Tampere, Tampere, Finland
{ziyuan.lin,jaakko.peltonen}@uta.fi
[2] Department of Computer Science, Aalto University, Espoo, Finland

Abstract. Finding relationships between multiple views of data is essential both in exploratory analysis and as pre-processing for predictive tasks. A prominent approach is to apply variants of Canonical Correlation Analysis (CCA), a classical method seeking correlated components between views. The basic CCA is restricted to maximizing a simple dependency criterion, correlation, measured directly between data coordinates. We introduce a new method that finds dependent subspaces of views directly optimized for the data analysis task of *neighbor retrieval between multiple views*. We optimize mappings for each view such as linear transformations to maximize cross-view similarity between neighborhoods of data samples. The criterion arises directly from the well-defined retrieval task, detects nonlinear and local similarities, measures dependency of data relationships rather than only individual data coordinates, and is related to well understood measures of information retrieval quality. In experiments the proposed method outperforms alternatives in preserving cross-view neighborhood similarities, and yields insights into local dependencies between multiple views.

1 Introduction

Finding dependent subspaces across views (subspaces where some property of data is statistically related or similar across views) is a common data analysis need, where Canonical Correlation Analysis (CCA) [10] is a standard unsupervised tool. Preprocessing to find dependent subspaces is useful both for prediction and for analysis: in predictive tasks, such subspaces help if non-dependent parts of each view may arise from noise and distortions. In some data analysis tasks, finding the dependent subspaces may itself be the main goal; for example in bioinformatics domains dependency seeking projections have been used to identify relationships between different views of cell activity [11,25]; in signal processing a similar task could be identifying optimal filters for dependent signals of different nature, e.g., speech and the corresponding tongue movements of the speakers as in [32].

Methods like CCA maximize simple correlations between data point coordinate features across the projected subspaces. However, in many data domains

---

Ziyuan Lin and Jaakko Peltonen contributed equally to the paper.

P. Perner (Ed.): MLDM 2017, LNAI 10358, pp. 1–16, 2017.
DOI: 10.1007/978-3-319-62416-7_1

the coordinates may not be of main interest but rather the *data relationships that they reveal*. It is then of great interest to develop dependency seeking methods that directly focus on the data relationships. For example, consider a database of scientists, defined in one view by their level of interest in various research topics, and in another view by their level of interest in various hobbies. In a database like this, finding relationships of people is the common interest, e.g. to find nearest colleagues for a scientist, having the most similar (neighboring) research interests; or to find hobby partners having the most similar (neighboring) hobby interests; the question is then, can we predict the research colleagues from hobby partners or vice versa? Research topics and hobbies are very dissimilar views, and not all of their variation will be related, but we can try to find subspaces of research and hobby interests, so that research neighbors and hobby neighbors are as highly related as possible in those subspaces.

In this paper we propose a method that solves this task: we present a novel method for seeking dependent subspaces across multiple views, preserving neighborhood relationships of data. Our method directly maximizes the *between-view similarity of neighborhoods of data samples*, a natural measure for similarity of data relationships among the views. The method detects nonlinear and local dependencies, has strong invariance properties, is related to an information retrieval task of the analyst, and performs well in experiments.

Relating data items is one of the main elementary tasks in Shneiderman's taxonomy of tasks in visual data analysis [22]. Our method is optimized for finding related (neighboring) data items, formulated as optimizing an information retrieval task. Information retrieval of neighbors has been successful as a criterion for dimensionality reduction [3,19–21,27,34–36], here we use a similar approach to learn dependent subspaces of multiple views. Since our method directly serves the task of relating data items (across views) in Shneiderman's taxonomy, in this sense it arguably comes closer to needs of data analysts than maximizing some variant of coordinate correlation.

We find linear projections (linear subspaces) of views. Linear projections have advantages of simplicity and easy interpretability with respect to original data features. Even if projections are linear, the dependency criterion we optimize is flexible and detects nonlinear dependencies across views.

We present our method in Sect. 2, properties and extensions in Sect. 3, related work in Sect. 4, experiments in Sect. 5, and conclusions in Sect. 6.

2   Method: Dependent Neighborhoods of Views

Our method finds similar neighborhood relationships between views. We define the neighborhood relationships and then discuss how to measure their cross-view similarity. Instead of hard neighborhoods where two points are or are not neighbors, we use more realistic probabilistic neighborhoods.

Assume there are N_{data} input data items, where each item has paired features over multiple views $V = 1, \ldots, N_{\text{views}}$. We consider transformations of each view by a mapping f_V which is typically a dimensionality reducing transformation to a subspace of interest; in this paper, for simplicity and interpretability we use linear mappings $\mathbf{f}_V(\mathbf{x}_{i,V}) = \mathbf{W}_V \mathbf{x}_{i,V}$ where $\mathbf{W}_V \in \mathbb{R}^{dim^{\text{low}}(V) \times dim^{\text{orig}}(V)}$ are the to-be-optimized parameters of the mapping, and $dim^{\text{orig}}(V)$ and $dim^{\text{low}}(V)$ are the number of dimensions of V and its subspace respectively.

Probabilistic neighborhood between data items. Consider a particular view (feature set) V. The most simple representation of a neighborhood around some point i is the set of other points that are close enough to it, for example all points inside a fixed radius or a fixed number of nearest points. However, such neighborhoods would be overly simple: typically analysts would not make a hard binary decision whether some point should be considered a neighbor of i or not. Instead of hard neighborhoods where two points are or are not neighbors, we use more realistic probabilistic neighborhoods. Suppose the analyst is inspecting the other points around i, by looking at the points in some low-dimensional transformation of view V, and we ask the analyst to choose one of the other points j as an example neighbor (e.g., in order to pick out that point to inspect it next). Then each of the other points j has some probability that the analyst will pick that one, intuitively so that points close-by to i have large probability to get picked and points far-off from i have low probability. Following this principle, the local neighborhood of a data item i in any transformation $\mathbf{f}_V$ of view V can be represented by the conditional probability distribution $p_{i,V} = \{p_V(j|i; \mathbf{f}_V)\}$ over other data items $j \neq i$, where $p_V(j|i; \mathbf{f}_V)$ tells the probability that data item j is picked as a representative neighbor of i; that is, the probability that an analyst who inspected item i will next pick j for inspection. The $p_V(j|i; \mathbf{f}_V)$ can be defined in several ways, as a decreasing function of distance $d_V(i, j; \mathbf{f}_V)$ between features of data items i and j in view V. Here we define it by a simple exponential falloff with respect to squared distance of i and j, as

$$p_V(j|i; \mathbf{f}_V) = \frac{\exp(-d_V^2(i,j; \mathbf{f}_V)/\sigma_{i,V}^2)}{\sum_{k \neq i} \exp(-d_V^2(i,k; \mathbf{f}_V)/\sigma_{i,V}^2)} \tag{1}$$

where $\sigma_{i,V}$ sets the falloff rate around i in the view. We tried two simple ways to set the $\sigma_{i,V}$: one is as a fraction of maximal pairwise distance so

$$\sigma_{i,V} = \sigma_0 \cdot \max_{j,k} \|\mathbf{x}_{j,V} - \mathbf{x}_{k,V}\|, \tag{2}$$

or alternatively, set

$$\sigma_{i,V} = \sqrt{\frac{dim^{\text{low}}(V)}{dim^{\text{orig}}(V)}} \cdot \operatorname*{mean}_{j,l \in kNN(j)} \{\|\mathbf{x}_{l,V} - \mathbf{x}_{j,V}\|\} \tag{3}$$

i.e., calculate the average distance between $\mathbf{x}_{j,V}$ and its k-th nearest neighbor $\mathbf{x}_{l,V}$, then give the average a heuristic correction factor $\sqrt{dim^{\text{low}}(V)/dim^{\text{orig}}(V)}$ since the average distance is obtained from the original space yet $\sigma_{i,V}$ is used in a subspace. Other local choices to e.g. achieve a desired entropy are possible, see [27]. With linear mappings the probabilities become

$$p_V(j|i; \mathbf{f}_V) = \frac{\exp(-\|\mathbf{W}_V(\mathbf{x}_{i,V} - \mathbf{x}_{j,V})\|^2/\sigma_{i,V}^2)}{\sum_{k \neq i} \exp(-\|\mathbf{W}_V(\mathbf{x}_{i,V} - \mathbf{x}_{k,V})\|^2/\sigma_{i,V}^2)} \tag{4}$$

where the matrix $\mathbf{W}_V$ defines the subspace of interest for the view and also the distance metric within the subspace. Our method learns the mapping parameters $\mathbf{W}_V$ for each view.

2.1   Comparison of Neighborhoods Across Views

Neighborhoods represented as probability distributions can be compared by difference measures. We discuss two measures for different purposes, and their information retrieval interpretations.

Kullback-Leibler divergence. For two distributions $p = \{p(j)\}$ and $q = \{q(j)\}$, the Kullback-Leibler (KL) divergence is an information-theoretic asymmetric difference measure defined as

$$D_{KL}(p, q) = \sum_j p(j) \log \frac{p(j)}{q(j)} \; . \tag{5}$$

The KL divergence is nonnegative and zero if and only if $p = q$. Traditionally it is interpreted to measure the amount of extra coding length needed when coding examples with codes generated for distribution q when the samples actually come from distribution p. We treat views symmetrically and compute the symmetrized divergence $(D_{KL}(p, q) + D_{KL}(q, p))/2$.

KL divergence is related to an *information retrieval criterion*: $D_{KL}(p, q)$ is the cost of *misses in information retrieval of neighbors*, when neighbors are retrieved using retrieval distribution q but they actually follow distribution p. $D_{KL}(p, q)$ is also the cost of *false neighbors* when neighbors are retrieved using p but they actually follow q. The relationships were shown in [27] and used to compare a reduced-dimensional neighborhood to an original one; we use it in a novel way to compare neighborhoods across (transformed) views of data. The symmetrized divergence is the *total cost of both misses and false neighbors* when neighbors following the distribution in one transformed view are retrieved from the other transformed view with its neighbor distribution.

The value of the KL divergence can depend highly on differences between individual probabilities $p(j)$ and $q(j)$. A single missed neighbor can yield a high divergence value: for any index j if $p(j) > \epsilon$ for some $\epsilon > 0$, $D_{KL}(p, q) \rightarrow \infty$ as $q(j) \rightarrow 0$. In real-life multi-view data differences between views may be unavoidable, so we prefer a less strict measure focusing more on overall similarity of neighborhoods than severity of individual misses. We discuss such a measure below.

Angle cosine. A simple similarity measure between discrete distributions is the angle cosine between the distributions as vectors, that is,

$$\mathrm{Cos}(p, q) = \frac{\sum_j p(j) q(j)}{\sqrt{(\sum_j (p(j))^2)(\sum_j (q(j))^2)}},$$

which can be seen as the Pearson correlation coefficient between elements of p and q; it is thus a neighborhood correlation—a neighborhood based analogue of the coordinate correlation cost function of CCA.[1] The angle cosine is bounded above and below: it has highest value 1 if and only if $p = q$ and lowest value 0 if supports of p and q are nonoverlapping.

Similarity of neighborhoods by itself is not enough. The KL divergence and angle cosine (neighborhood correlation) measures only compare similarity of neighborhoods but not potential usefulness of the found subspaces. In high-dimensional data it is often possible to find subspaces where neighborhoods are trivially similar. For example, in data with sparse features it is often possible to find two dimensions where all data is reduced to a single value; in such dimensions neighborhood distributions would become uniform across all data since, hence any two such dimensions appear similar. To avoid discovering trivial similarities we wish to complement the measures of similarity between neighborhoods with terms favoring nontrivial (sparse) neighborhoods. A simple way to prefer sparse neighborhoods is to omit the normalization from neighborhood correlation, yielding

$$\text{Sim}(p, q) = \sum_j p(j)q(j) \tag{6}$$

which is the inner product between the vectors of neighborhood probabilities. Unlike $\text{Cos}(p, q)$, $\text{Sim}(p, q)$ favors sparse neighborhoods: it has highest value 1 if and only if $p = q$ and $p(j) = q(j) = 1$ for only one element j, and lowest value 0 if the supports of p and q are nonoverlapping.

The information retrieval interpretation is: $\text{Sim}(p, q)$ is a proportional count of true neighbors from p retrieved from q or vice versa. If p has K neighbors with near-uniform high probabilities $p(j) \approx 1/K$ and other neighbors have near-zero probabilities, and q has L neighbors with high probability $q(j) \approx 1/L$, then $\text{Sim}(p, q) \approx M/KL$ where M is the number of neighbors for which both p and q are high (retrieved true neighbors). Thus $\text{Sim}(p, q)$ rewards matching neighborhoods and favors sparse neighborhoods (small K and L). One advantage of this formulation is to avoid matching two neighborhoods that seem to match only because they are highly uninformative: for example if p and q are both uniform over all neighbors, they have the same probability values and would be "similar" in a naive comparison of probability values, but both are actually simply uninformative about the choice of neighbors. $\text{Sim}(p, q)$ would prefer a more sparse, more informative match, as desired.

2.2 Final Cost and Optimization Technique

We wish to evaluate similarity of neighborhoods between subspaces of each view, and optimize the subspaces to maximize the similarity, while favoring subspaces having sparse (informative) neighborhoods for data items. We then evaluate

---

[1] To make the connection exact, typically correlation is computed after substracting the mean from coordinates; for neighbor distributions of n data items, the mean neighborhood probability is the data-independent value $1/(n-1)^2$ which can be substracted from each sum term if an exact analogue to correlation is desired.

similarities as $\mathrm{Sim}(p_{i,V}, p_{i,U})$ where $p_{i,V} = \{p_V(j|i; f_V)\}$ is the neighborhood distribution around data item i in the dependent subspace of view V and f_V is the mapping (parameters) of the subspace, and $p_{i,U} = \{p_U(j|i; f_U)\}$ is the corresponding neighborhood distribution in the dependent subspace of view U having the mapping f_U. As the objective function for finding dependent projections, we sum the above over each pair of views (U, V) and over the neighborhoods of each data item i, yielding

$$C(f_1, \ldots, f_{N_{\mathrm{views}}}) = \sum_{V=1}^{N_{\mathrm{views}}} \sum_{U=1, U \neq V}^{N_{\mathrm{views}}} \sum_{i=1}^{N_{\mathrm{data}}} \sum_{j=1, j \neq i}^{N_{\mathrm{data}}} p_V(j|i; f_V) p_U(j|i; f_U) \qquad (7)$$

where, in the setting of linear mappings and neighborhoods with Gaussian falloffs, p_V is defined by (4) and is parameterized by the projection matrix W_V of the linear mapping.

Optimization. The function $C(f_1, \ldots, f_{N_{\mathrm{views}}})$ is a well-defined objective for dependent projections and can be maximized with respect to the projection matrices W_V of each view. We use gradient techniques for optimization, specifically limited memory Broyden-Fletcher-Goldfarb-Shanno (L-BFGS). Even with L-BFGS, (7) can be hard to optimize due to several local optima. To find a good local optimum, we optimize over L-BFGS rounds with a shrinking *added penalty* driving the objective away from the worst local optima during the first rounds; we use the optimum in each round as initialization of the next. For the penalty we use KL divergence based dissimilarity between neighborhoods, summed over neighborhoods of all data items i and view pairs (U, V), giving

$$C_{\mathrm{Penalty}}(f_1, \ldots, f_{N_{\mathrm{views}}}) = \frac{1}{2} \sum_{V=1}^{N_{\mathrm{views}}} \sum_{U=1, U \neq V}^{N_{\mathrm{views}}} \sum_{i=1}^{N_{\mathrm{data}}} (D_{KL}(p_{i,V}, p_{i,U}) + D_{KL}(p_{i,U}, p_{i,V}))$$

$$(8)$$

which is a function of all mapping parameters and can be optimized by L-BFGS; (8) penalizes severe misses (pairs (i, j) with nonzero neighborhood probability in one view but near-zero in another) driving the objective away from bad local optima. In practice KL divergence is too strict about misses; we use two remedies below.

Bounding KL divergence by neighbor distribution smoothing. To bound the KL divergence, one way is to give the neighbor distributions (1) a positive lower bound. In the spirit of the well-known Laplace smoothing in statistics, we revise the neighbor probabilities (1) as

$$p_V(j|i; f_V) = \frac{\exp(-d_V^2(i, j; \mathbf{f}_V)/\sigma_{i,V}^2) + \epsilon}{\sum_{k \neq i} \exp(-d_V^2(i, k; \mathbf{f}_V)/\sigma_{i,V}^2) + (N_{\mathrm{data}} - 1)\epsilon} \qquad (9)$$

where $\epsilon > 0$ is a small positive number. Without smoothing KL-divergence $D_{KL}(p_U, p_V)$ could become unbounded: if there is a pair (i, j) for which

$p_U(j|i; \mathbf{f}_U)$ is nonzero and $p_V(j|i; \mathbf{f}_V)$ tends to zero, then $D_{KL}(p_U, p_V)$ tends to infinity.

With smoothing, it is easy to see $p_V(j|i; \mathbf{f}_V) \geq \epsilon/((1 + \epsilon)(N_{\mathrm{data}} - 1))$ and thus the divergence has an upper bound: we have $D_{KL}(p_U, p_V) \leq \log(1/\min_j p_V(j|i; \mathbf{f}_V)) \leq \log((N_{\mathrm{data}} - 1)(1 + \epsilon)/\epsilon)$ where the first inequality arises because probabilities in p_U sum to one, and the second arises from the lower bound of probabilities in p_V. We set $\epsilon = 10^{-6}$ to give a reasonable upper bound for the KL-divergence. To keep notations simple, we still denote this smoothed neighbor distribution as $p_V(j|i; \mathbf{f}_V)$. To avoid over-complicated formulation and for consistency, we also use this version of neighbor probabilities in our objective function (7), even though the value of the objective is bounded by itself. We simply set $\epsilon = 10^{-6}$ which empirically works well.

Shrinking the penalty. Even with bounded KL divergence, optimization stages need different amounts of penalty. At end of optimization, nearly no penalty is preferred, as views may not fully agree even with the best mapping. We shrink the penalty during optimization; the objective becomes

$$C_{\mathrm{Total}}(f_1, \ldots, f_{N_{\mathrm{views}}}) = C(f_1, \ldots, f_{N_{\mathrm{views}}}) - \gamma\, C_{\mathrm{Penalty}}(f_1, \ldots, f_{N_{\mathrm{views}}}) \quad (10)$$

where γ controls the penalty. We initially set γ so the two parts of the objective function are equal for the initial mappings, $C(f_1, \ldots, f_{N_{\mathrm{views}}}) = \gamma\, C_{\mathrm{Penalty}}(f_1, \ldots, f_{N_{\mathrm{views}}})$, and multiply γ by a small factor at the start of each L-BFGS round to yield exponential shrinkage. The initial γ should be set as a tradeoff, sufficiently large to penalize bad local optima at the start of optimization while avoiding overpenalization of potential useful optima near the end of optimization. We set $\gamma = 0.9$ based on our internal experiments.

Time complexity. We calculate the neighbor distributions for all views, and optimize the objective function involving each pairs of views, thus the naive implementation takes $O(dN_{\mathrm{data}}^2 N_{\mathrm{views}}^2)$ time, with d the maximal dimensionality among views. Acceleration techniques [26,29,34] from neighbor embedding could be adopted to reduce time complexity of a single view from $O(N_{\mathrm{data}}^2)$ to $O(N_{\mathrm{data}} \log N_{\mathrm{data}})$ or even $O(N_{\mathrm{data}})$. But scalability is not our first concern in this paper, so we use the naive $O(N_{\mathrm{data}}^2)$ implementation for calculating the neighbor distributions for each view involved.

On the other hand, L-BFGS in practice has similar or better performance than BFGS, which has been shown to have a fast superlinear convergence rate as a quasi-Newton method, given the Hessian matrix is Lipschitz continuous near a minimizer [18].

3 Properties of the Method and Extensions

Information retrieval. Our objective measures success in a neighbor retrieval task of the analyst: we maximize count of retrieved true neighbors across views, and penalize by severity of misses.

Invariances. For any subspace of any view, (1) and (4) depend only on pairwise distances and are thus invariant to global translation, rotation, and mirroring of data in that subspace. The cost is then invariant to differences of global translation, rotation, and mirroring between views and finds view dependencies despite such differences. If in any subspace the data has isolated subsets (where data pairs from different subsets have zero neighbor probability) invariance holds for local translation/rotation/mirroring of the subsets as long as they preserve the isolation.

Dependency is measured between whole subspaces. Unlike CCA where each canonical component of one view has a particular correlated pair in the other view, we maximize dependency with respect to the entire subspaces (transformed representations) of each view, as neighborhoods of data depend on all coordinates within the dependent subspace. Our method thus takes into account within-view feature dependencies when measuring dependency. Moreover, dependent subspaces do not need to be same-dimensional, and in some views we can choose not to reduce dimensionality but to learn a metric (full-rank linear transformation).

Finding dependent neighborhoods between feature-based views and views external neighborhoods. In some domains, some data views may directly provide neighborhood relationships or similarities between data items, e.g., friendships in a social network, followerships in Twitter, or citations between scientific papers. Such relationships or similarities can be used in place of the feature-based neighborhood probabilities $p_V(j|i; f_V)$ above. This shows an interesting similarity to a method [19] used to find similarities of one view to an external neighborhood definition; our method contains this task as one special case.

Other falloffs. Exponential falloff in (1) and (4) can be replaced with other forms like t-distributed neighborhoods [16]. Such replacement preserves the invariances.

Other transformations. Our criterion is extensible to nonlinear transformations in future work; replace linear projections by another parametric form, e.g. neural networks, optimize (10) with respect to its parameters; the transformation can be chosen on a view-by-view basis. Optimization difficulty of transformations varies; the best form of nonlinear transformation is outside the paper scope.

4   Related Work

In general, multi-view learning [33] denotes learning models by leveraging multiple potentially dependent data views; such models could be built either for unsupervised tasks based on the features in the views or for supervised tasks that involve additional annotations like categories of samples (e.g. [9]). Multi-view learning often assumes paired data items across views, unlike multi-task learning (e.g. [5,6,14,15]) where usually only underlying statistical trends are shared across data sets. In this paper we concentrate on unsupervised multi-view learning, and our prediction tasks of interest are predicting neighbors across views.

The standard Canonical Correlation Analysis (CCA) [10] iteratively finds component pairs maximizing correlation between data points in the projected subspaces. Such correlation is a simple restricted measure of linear and global dependency. To measure dependency in a more flexible way and handle nonlinear local dependency, linear and nonlinear CCA variants have been proposed: Local CCA (LCCA) [31] seeks linear projections for local patches in both views that maximize correlation locally, and aligns local projections into a global nonlinear projection; its variant Linear Local CCA (LLCCA) finds a linear approximation for the global nonlinear projection; Locality Preserving CCA (LPCCA) [24] maximizes reweighted correlation between data coordinate differences in both views. In experiments we compare to the well known traditional CCA and LPCCA as an example of recent state of the art.

As a more general framework, Canonical Divergence Analysis [17] minimizes a general divergence between distributions of data coordinates in the projected subspace.

The methods mentioned above work on data coordinates in the original spaces. There are also nonlinear CCA variants [1,2,8,12,28,30] for detecting nonlinear dependency between multiple views. Although some variants above are locality-aware, they introduce locality from the original space before maximizing correlation or other similarity measures in the low-dimensional subspaces. Since locality in the original space may not reflect locality in the subspaces due to noise or distortions, such criteria may not be suited for finding local dependencies in subspaces.

The CODE method [7] creates an embedding of co-occurrence data of pairs of original categorical variables, mapping both variables into a shared space. Our method is not restricted to categorical inputs – its main applicability is to high-dimensional vectorial data, with several other advantages. In contrast to CODE, we find dependent subspaces (mappings) from multiple high-dimensional real-valued data views. Instead of restricting to a single mapping space we find several mappings, one from each view, which do not need to go into the same space; our output spaces can even be different dimensional for each view. Unlike CODE our method is not restricted to maximizing coordinate similarity: we only need to make neighborhoods similar which is more invariant to various transformations.

The above methods and several in [33] all maximize correlation or alternative dependency measures between data coordinates across views. As we pointed out, in many domains coordinates are not of main interest but rather the data relationships they reveal; we consider neighborhood relationships and our method directly finds subspaces having similar neighborhoods.

5 Experiments

We demonstrate our method on artificial data with multiple dependent groups between views, and three real data sets: a variant of MNIST digits [13], video data, and stock prices. In this paper we restrict our method to find linear sub-spaces, important in many applications for interpretability, and compare with two prominent linear subspace methods for multiple views, CCA and LPCCA. To our knowledge, no other information retrieval based approaches for finding linear subspaces is known up to the time when we did the experiment. Future work could compare methods for nonlinear mappings [33] to variants of ours for the same mapping; we do not focus on the mapping choice, and focus on showing the benefit or our neighborhood based objective.

On the artificial data set, we measure performance by correspondence between found projections and the ground truth. On the real data we use mean precision-mean recall curves, a natural performance measure for information retrieval tasks, and a measure for dependency as argued in Sect. 2.

Experiment on artificial data sets. We generate an artificial data set with 2 views with multiple dependent groups in each pair of corresponding dimensions as follows. Denote view $V(\in \{1,2\})$ as $X^{(V)} \in \mathbb{R}^{5 \times 1000}$, and its i-th dimension as $X_i^{(V)}$. For each i, we divide the 1000 data points in that dimension into 20 groups $\{g_{ij}\}_{j=1}^{20}$ with 50 data points each. For each g_{ij} and view V, we let $\hat{x}_{ijk}^{(V)} \triangleq F_{ij} m_{ij}^{(V)} + \epsilon_{ijk}$ $(1 \leq k \leq 50)$, with $m_{ij}^{(V)} \sim \mathcal{N}(0,5)$, $\epsilon_{ijk} \sim U[-0.5,0.5]$ and $F_{ij} \in \{-1,1\}$ a random variable allowing positive or negative correlation inside the group. We assemble $\hat{x}_{ijk}^{(V)}$ into $\hat{X}^{(V)} \in \mathbb{R}^{5 \times 1000}$, and randomly permute entries of $\hat{X}_i^{(1)}$ and $\hat{X}_i^{(2)}$ in the same way but differently for different i, to ensure cross-dimension independency. Lastly we perform a PCA between $\hat{X}_i^{(1)}$ and $\hat{X}_i^{(2)}$ for each i, to remove cross-dimension correlation. We assemble the resulting $X_i^{(V)}$ into $X^{(V)}$.

We use the neighborhood falloff as in (2) with $\sigma_0 = 0.05$, and pursue 2 transformations mapping from the 5D original space to a 1D latent space for each of the two views. Ground truth projections for both views will then be $W^{(i)} = (\delta_{ij})_{j=1}^5 \in \mathbb{R}^{1 \times 5}$. Results are in Fig. 1: compared with CCA, our method

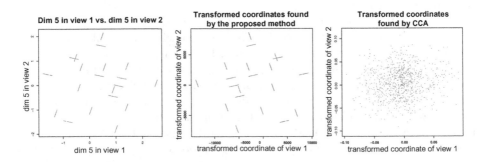

Fig. 1. Result for artificial data with dependent groups. Left to right: one of the ground truths; 1D subspace pair recovered by our method; 1D subspace pair recovered by CCA. Our method recovers the dependency between views in the 5th dimension despite mirroring and scale differences; CCA fails to do so.

Table 1. Means and standard deviations of the correspondence measure (11); our method outperforms CCA and LPCCA, recovering the dependency in all artificial data sets.

	Our method	CCA	LPCCA
Mean	1.00	0.51	0.51
Std	0.00	0.043	0.043

successfully finds one of the ground truth transformations (e.g., the 5th one), despite mirroring and scale, recovering the between-view dependency.

We measure performance by correspondence between the found projections and the ground truth transformation: let $W_1, W_2 \in \mathbb{R}^{1\times 5}$ be projections found by a method, define

$$\mathrm{Corr}(W_1, W_2) = \max_i \frac{1}{2}\left(\frac{|W^{(i)}W_1^{\mathrm{T}}|}{\|W_1\|_2} + \frac{|W^{(i)}W_2^{\mathrm{T}}|}{\|W_2\|_2} \right) \tag{11}$$

as the correspondence score. High score means good match between the found projections and ground truth. We repeat the experiment calculating correspondence on 20 artificial data sets generated in the same way. Table 1 summarizes the statistics. Our method outperforms CCA and LPCCA (with $k = 5$), finding the dependency on all 20 data sets.

Experiment on real data sets with two views. We show our method helps match neighbors between the subspaces of two views after transformation. We use the following 3 data sets.

MNIST handwritten digit database (MNIST). MNIST contains 70000 grayscale hand-written digit images of size 28×28. We create a training set and a testing set with 2000 images each. In the training set, we randomly choose 200 images from each digit to balance the distribution, while the testing set is another 2000 random images without balancing. We apply nonlinear dimensionality algorithm on both the left half and the right half of the images to create the two views to simulate a scenario where views have nonlinear dependency; we use Neighbor Retrieval Visualizer (NeRV) [27] embedding to 5 dimensions with $\lambda = 0.1$ and $\lambda = 0.9$ respectively. The experiment is repeated 17 times, covering 68000 images.

Image patches from video (Video). We take image patches from a subset of frames in a video (db.tt/rcAS5tII). Starting from frame 50, we take 5 consecutive frames as a short clip at every 50 frames until frame 5200, then create two views from image patches in two fixed rectangles in those frames, $rect_1 = [800, 900] \times [250, 350]$ and $rect_2 = [1820, 1920] \times [800, 900]$. We measure 5-fold cross-validation performance after randomly permuting the clips.

Stock prices (Stock), from Kaggle competition "Winton stock market challenge" (goo.gl/eqdhKK). It contains prices of a stock at different times. We split the time series in the given training set into two halves, and let view 1 be the amplitudes from the Fourier transform results of the first half, and view 2 be the phases from the Fourier transform results of the second half.

For each data set we seek a pair of transformations onto 2D subspaces for the views. We measure performance by mean precision-mean recall curves of neighbor retrieval between (1) the two subspaces from the transformations, and (2) one of the original views and the subspace from the transformation for the other view. The better the performance is, the more to the top (better mean precision) and right (better mean recall) the curve will be in the figure. We set the neighborhood falloff as in (3) with $k = 5$, and the number of neighbors in the ground truth as 5 for MNIST and Stock, 4 for Video, and let the number of retrieved neighbors vary from 1 to 10 as we focus on the performance of the matching for the nearest neighbors. We compare the performance with CCA and LPCCA. Figure 2 (column 1–3) shows the results.

We now show our method can find **dependent subspaces for multiple (more than two) views**. The task can be particularly essential to bioinformatics since it is shown that "no single inference method performs optimally across all datasets" [4]. In this experiment we use *Cell-Cycle* data with five views. The views are from different measurements of cell cycle-regulated gene expression for the same set of 5670 genes [23]. We preprocess data as in [25] with an extra feature normalization step. We seek five two-dimensional subspaces from the five views, comparing to the PCA baseline with 2 components. We again use mean precision-mean recalls curves as the performance measure, additionally average the curves across the 10 view pairs or view-transformed coordinate pairs, besides averaging over the five repetitions in 5-fold cross-validation. Figure reffig:realspsdata (column 4) shows we outperform the baseline.

Finding subspaces with different dimensions. We show our method can find dependent subspaces with different dimensions. We create three two-dimensional Lissajous curves $L_k(t) = (\cos \sqrt{2k-1}t + 2\pi(k-1)/3, \cos \sqrt{2k}t + 2\pi(k-1)/3)$, $k = 1, 2, 3$. We create the first view $X^{(1)} \in \mathbb{R}^{6 \times 1000}$ as $X^{(1)}_{1,1:1000} = (0, \cdots, 999)$ and $X^{(1)}_{d \geq 2, 1:1000} \overset{i.i.d}{\sim} \mathcal{N}(0, 1)$, and the second view $X^{(2)} \in \mathbb{R}^{6 \times 1000}$ as the concatenation of the coordinates in the Lissajous curves. We seek a one-dimensional subspace from $X^{(1)}$, and a two-dimensional subspace from $X^{(2)}$; the aim is to find the nonlinear dependency between one-dimensional timestamps, and a two-dimensional representation for the three trajectories summarizing the two-dimensional movements of the three points along Lissajous curves. Figure 3 shows the Lissajous curves, found subspaces, and optimized projection pair. Our method successfully finds the informative feature in $X^{(1)}$, and keeps transformed coordinates of $X^{(2)}$ smooth, with roughly the same amount of "quick turns" as in original Lissajous curves. The magnitudes in the optimized projections also suggest they capture the correct dependency.

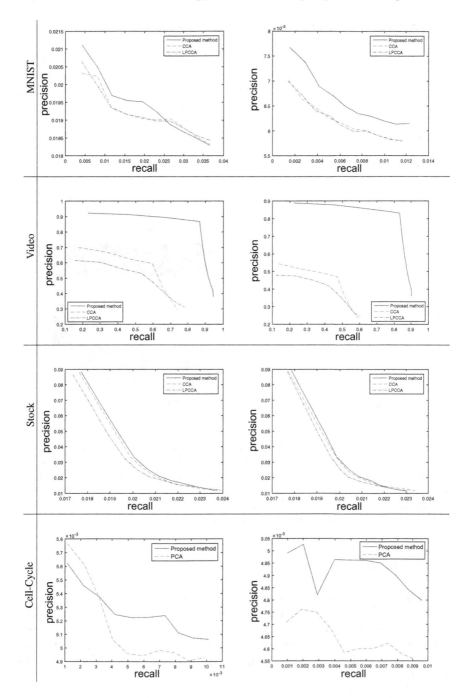

Fig. 2. Mean precision-mean recalls curves from different real data on the test sets. **Left column**: view 1 as the ground truth; **right column**: subspace from view 1 as the ground truth. We can see curves from our method are to the top and/or right of the curves from other methods in most parts of all sub-figures, meaning our method achieves better precision and recall on average. Row 1–3: our method outperforms CCA and LPCCA; Row 4: our method outperforms PCA.

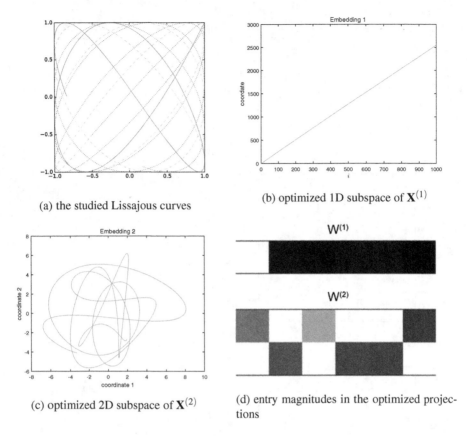

(a) the studied Lissajous curves

(b) optimized 1D subspace of $\mathbf{X}^{(1)}$

(c) optimized 2D subspace of $\mathbf{X}^{(2)}$

(d) entry magnitudes in the optimized projections

Fig. 3. Lissajous curves (a) and the found subspaces from our method. (b)–(d) show we find the correct dependency: (b): perfect linear correlation shows the time dimension was found. (c): the number of "quick turns" (14 in total) in the smooth curves roughly matches that in the original curves. (d): projection weights, darker color means smaller magnitude; high magnitude of $\mathbf{W}^{(1)}$'s first entry and the complementary pattern in $\mathbf{W}^{(2)}$ suggest we capture the dependency correctly.

6   Conclusions

We presented a novel method for seeking dependent subspaces across multiple views, preserving neighborhood relationships of data. It has strong invariance properties, detects nonlinear dependencies, is related to an information retrieval task of the analyst, and performs well in experiments. Potential future work includes further evaluation across a range of data domains, dimensionalities, and case studies.

Acknowledgments. We acknowledge the computational resources provided by the Aalto Science-IT project. Authors belong to the Finnish CoE in Computational Inference Research COIN. The work was supported in part by TEKES (Re:Know project).

The work was also supported in part by the Academy of Finland, decision numbers 252845, 256233, and 295694.

References

1. Andrew, G., Arora, R., Livescu, K., Bilmes, J.: Deep canonical correlation analysis. In: Proceedings of ICML (2013)
2. Bach, F.R., Jordan, M.I.: Kernel independent component analysis. J. Mach. Learn. Res. **3**, 1–48 (2003)
3. Bunte, K., Järvisalo, M., Berg, J., Myllymäki, P., Peltonen, J., Kaski, S.: Optimal neighborhood preserving visualization by maximum satisfiability. In: Proceedings of AAAI (2014)
4. Ceci, M., Pio, G., Kuzmanovski, V., Deroski, S.: Semi-supervised multi-view learning for gene network reconstruction. PLOS ONE **10**(12), 1–27 (2015)
5. Faisal, A., Gillberg, J., Leen, G., Peltonen, J.: Transfer learning using a nonparametric sparse topic model. Neurocomputing **112**, 124–137 (2013)
6. Faisal, A., Gillberg, J., Peltonen, J., Leen, G., Kaski, S.: Sparse nonparametric topic model for transfer learning. In: Proceedings of ESANN (2012)
7. Globerson, A., Chechik, G., Pereira, F., Tishby, N.: Euclidean embedding of co-occurrence data. J. Mach. Learn. Res. **8**, 2265–2295 (2007)
8. Hodosh, M., Young, P., Hockenmaier, J.: Framing image description as a ranking task: data, models and evaluation metrics. J. Artif. Intell. Res. **47**(1), 853–899 (2013)
9. Honkela, A., Peltonen, J., Topa, H., Charapitsa, I., Matarese, F., Grote, K., Stunnenberg, H., Reid, G., Lawrence, N., Rattray, M.: Genome-wide modeling of transcription kinetics reveals patterns of RNA production delays. Proc. Natl. Acad. Sci. **112**(42), 13115–13120 (2015)
10. Hotelling, H.: Relations between two sets of variates. Biometrika **28**(3–4), 321–377 (1936)
11. Klami, A., Virtanen, S., Kaski, S.: Bayesian canonical correlation analysis. J. Mach. Learn. Res. **14**, 965–1003 (2013)
12. Lai, P., Fyfe, C.: Kernel and nonlinear canonical correlation analysis. Int. J. Neural Syst. **10**(5), 365–377 (2000)
13. LeCun, Y., Cortes, C.: MNIST handwritten digit database (2010)
14. Leen, G., Peltonen, J., Kaski, S.: Focused multi-task learning using Gaussian processes. In: Gunopulos, D., Hofmann, T., Malerba, D., Vazirgiannis, M. (eds.) ECML PKDD 2011. LNCS, vol. 6912, pp. 310–325. Springer, Heidelberg (2011). doi:10.1007/978-3-642-23783-6_20
15. Leen, G., Peltonen, J., Kaski, S.: Focused multi-task learning in a Gaussian process framework. Mach. Learn. **89**(1–2), 157–182 (2012)
16. van der Maaten, L., Hinton, G.: Visualizing data using t-SNE. J. Mach. Learn. Res. **9**, 2579–2605 (2008)
17. Nguyen, H., Vreeken, J.: Canonical divergence analysis. CoRR abs/1510.08370 (2015)
18. Nocedal, J., Wright, S.J.: Numerical Optimization. Springer, New York (1999)
19. Peltonen, J.: Visualization by linear projections as information retrieval. In: Príncipe, J.C., Miikkulainen, R. (eds.) WSOM 2009. LNCS, vol. 5629, pp. 237–245. Springer, Heidelberg (2009). doi:10.1007/978-3-642-02397-2_27
20. Peltonen, J., Kaski, S.: Generative modeling for maximizing precision and recall in information visualization. In: Proceedings of AISTATS, pp. 579–587 (2011)

21. Peltonen, J., Lin, Z.: Information retrieval approach to meta-visualization. Mach. Learn. **99**(2), 189–229 (2015)
22. Shneiderman, B.: The eyes have it: a task by data type taxonomy for information visualizations. In: Proceedings of IEEE Symposium on Visual Languages, pp. 336–343. IEEE Computer Society Press (1996)
23. Spellman, P., Sherlock, G., Zhang, M., Iyer, V., Anders, K., Eisen, M., Brown, P., Botstein, D., Futcher, B.: Comprehensive identification of cell cycle-regulated genes of the yeast saccharomyces cerevisiae by microarray hybridization. Mol. Biol. Cell **9**(12), 3273–3297 (1998)
24. Sun, T., Chen, S.: Locality preserving CCA with applications to data visualization and pose estimation. Image Vis. Comput. **25**(5), 531–543 (2007)
25. Tripathi, A., Klami, A., Kaski, S.: Simple integrative preprocessing preserves what is shared in data sources. BMC Bioinform. **9**, 111 (2008)
26. Van Der Maaten, L.: Accelerating t-SNE using tree-based algorithms. J. Mach. Learn. Res. **15**(1), 3221–3245 (2014)
27. Venna, J., Peltonen, J., Nybo, K., Aidos, H., Kaski, S.: Information retrieval perspective to nonlinear dimensionality reduction for data visualization. J. Mach. Learn. Res. **11**, 451–490 (2010)
28. Verbeek, J.J., Roweis, S.T., Vlassis, N.A.: Non-linear CCA and PCA by alignment of local models. In: Proceedings of NIPS, pp. 297–304. MIT Press (2003)
29. Vladymyrov, M., Carreira-Perpinán, M.A.: Linear-time training of nonlinear low-dimensional embeddings. In: Proceedings of AISTATS, vol. 33 (2014)
30. Wang, W., Arora, R., Livescu, K., Bilmes, J.: On deep multi-view representation learning. In: Proceedings of ICML (2015)
31. Wei, L., Xu, F.: Local CCA alignment and its applications. Neurocomputing **89**, 78–88 (2012)
32. Westbury, J.R.: X-ray microbeam speech production database user's handbook. Waisman Center on Mental Retardation & Human Development, University of Wisconsin, 1.0 edn., June 1994
33. Xu, C., Tao, D., Xu, C.: A survey on multi-view learning. CoRR abs/1304.5634 (2013)
34. Yang, Z., Peltonen, J., Kaski, S.: Scalable optimization of neighbor embedding for visualization. In: Proceedings of ICML, pp. 127–135 (2013)
35. Yang, Z., Peltonen, J., Kaski, S.: Optimization equivalence of divergences improves neighbor embedding. In: Proceedings of ICML, pp. 460–468 (2014)
36. Yang, Z., Peltonen, J., Kaski, S.: Majorization-minimization for manifold embedding. In: Proceedings of AISTATS, pp. 1088–1097 (2015)

Predicting Target Events in Industrial Domains

Julio Borges[1]([✉]), Martin A. Neumann[1], Christian Bauer[2], Yong Ding[1],
Till Riedel[1], and Michael Beigl[1]

[1] TECO, Karlsruhe Institute of Technology (KIT), Karlsruhe, Germany
{julio.borges,martin.neumann,yong.ding,till.riedel,michael.beigl}@kit.edu
[2] TRUMPF Werkzeugmaschinen GmbH + Co. KG, Ditzingen, Germany
Christian.Bauer@de.TRUMPF.com

Abstract. In industrial environments, machine faults have a high
impact on productivity due to the high costs it can cause. Machine generated event logs are a abundant source of information for understanding
the causes and events that led to a critical event in the machine. In this
work, we present a Sequence-Mining based technique to automatically
extract sequential patterns of events from machine log data for understanding and predicting machine critical events. By experiments using
real data with millions of log entries from over 150 industrial computer
numerical control (CNC) cutting machines, we show how our technique
can be successfully used for understanding the root causes of certain
critical events, as well as for building monitors for predicting them long
before they happen, outperforming existing techniques.

1 Introduction

In today's industry, the success of manufacturing companies highly depends
on reliability and quality of their machines and products for their production
process. Unexpected machine failures in production processes are bounded to
high repair costs and production delays [1]. Therefore, understanding and predicting critical situations before they occur can be a valuable source for avoiding
unexpected breakdowns and saving costs associated to the failure.

Log-files keep a record on the flow of states and activities performed by
the machine presenting therefore an important source of information on how a
machine "behave" prior to a critical situation [2]. Understanding what are the
causes that lead to specific critical situations in a machine can help engineering
and maintenance teams to build a problem diagnosis and repair the machine. In
many industries, this task is still bounded to high manual efforts, as the maintenance staff must go through the data in order to evaluate what are the possible
causes for critical events, demanding expert domain knowledge of the system.
Automatically diagnosing and even predicting when critical events are about to
occur can thus represent a significant advantage in the industry for both **critical
event diagnosis**, i.e., understanding what caused a critical event, to **critical
event prediction**, monitoring and alarming when a critical event is about to
occur even before it happens, enabling pre-intrusion to avoid system problems.

© Springer International Publishing AG 2017
P. Perner (Ed.): MLDM 2017, LNAI 10358, pp. 17–31, 2017.
DOI: 10.1007/978-3-319-62416-7_2

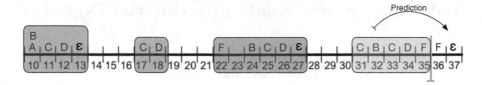

Fig. 1. Event-Based Prediction based on Sequence Mining. The sequential rule "$\langle B, C, D \rangle \Rightarrow \varepsilon$" describes the sequence of 3 events that always appear in sequence before the target (critical) event (ε). This information can be used for monitoring and predicting interesting target events in advance.

This paper focuses on the latter. This requires some short-term anticipation of upcoming critical event based on evaluation of the running state of the system.

The critical event prediction approach to be introduced by this paper is based on the batch offline analysis of logged error events. Frequent patterns of sequences of events that lead to errors are mined from the logs used to construct a monitor which can posteriorly observe events during runtime. Upcoming target critical events are then predicted by monitoring the events that have occurred recently before present time (cf. Fig. 1). Experiments on industrial data of industrial computer numerical control (CNC) cutting machines have shown superior prediction performance in comparison with the most well-known event-based prediction algorithms in that area.

The rest of this paper is organized as follows. Section 2 presents related work in the area of event based predictive maintenance. Section 3 describes the underlying collected dataset used for our experiments in this work. Section 4 presents the necessary data preparation steps for our algorithm. Section 5 presents our proposed algorithm which is evaluated and compared against the state of the art in Sect. 6. Section 7 then concludes the paper.

2   Related Work

In light of *Industry 4.0*, intelligent analysis of machine generated data specially for the task of predictive maintenance has gained increased attention recently.

Salfner et al. published a broad survey on the application of log analysis for what they call short-term **online failure prediction**. The goal is to predict the occurrence of failures during runtime based on the current system state [3]. They proposed in [2] an online failure prediction technique based on log files by using hidden semi-Markov predictor (HSMM) and appropriate pre-processing of the data. Two HSMMs are trained from previously recorded log data: one for failure and one for non-target sequences. Online failure prediction is then accomplished by computing likelihood of the observed error sequence for both models and by applying Bayes decision theory to classify the sequence (and hence the current system status) as failure-prone or not. They evaluated their approach on commercial telecommunication system data reporting failure prediction F-Measure up to 0.66.

Vilalta et al. proposed a prediction technique by applying itemset mining on a set of events preceding a target event on computer network data [4]. The dataset is partitioned in failure and non-failure event sets w.r.t. a target event. Events preceding the target event in a time window constitute the failure event sets. Vilalta et al. then applies association rule mining for extracting frequent and valid association rules on the partitioned dataset. In this process, the failure event sets are checked against frequency (support test) and validity (confidence). In contrast to sequence mining techniques (as this work is based upon), the event sets are unordered, so the chronological sequence of events plays no role in the training process. Consequently, our work considers the chronological order of occurring events. Vilalta et al. reported that depending on the target event, the prediction accuracy of their approach can significantly vary. Meanwhile they observed that for an event they had a false negative error rate of only around 4.5%, for another target event the error rate was so high as 83% with the same algorithm parametrization. We have made similar observations which will be discussed in Sect. 6.2.

3   Data Description

The log datasets used in this study derives from a real production scenario of a industrial computer numerical control (CNC) cutting machine system from the company TRUMPF GmbH + Co. KG, one of the world's biggest providers of machine tools. Overall, we collected big log datasets from 154 unique machines. Altogether, there are more than 4 million logged events in the datasets.

A **log dataset** is a collection of **events** recorded by various applications/components running on the machine (see Table 1). An event in this setup consists mainly of an unique and representative event code indicating an entering state or action being performed by the machine (ex.: Machine door opened), a timestamp indicating when the event occurred and an event severity class (Info, Warning, Error, etc.). They can have a systematic nature or be caused by an human intervention on the machine. Events reflect the developers' original idea about what are the valuable states to report.

Table 1. A piece of a real log file (with renamed event codes and messages for Blind Review) from a computer numerical control (CNC) cutting machine

Timestamp	Eventcode	Severity	Message
12/22/2014 07:24:23	300	Info	Data transfer 12345
12/22/2014 07:57:49	600	Warning	Check XX level
12/22/2014 08:01:28	900	Error	Error in YY system

In our application, we have altogether more than 1400 unique event codes from different categories. Although our approach can be used to predict any arbitrary target event, we focus particurlaly on predicting events which signalize critical machine events and situations.

4   Data Modeling

The first challenge that arises when applying Sequence Mining is how to proper model the raw machine log data into a sequence representation for the sequence miner to find patterns on. This section describes the necessary steps to transform the raw event logs into proper temporal event sequences used as input for the model to be proposed by this paper - while comparing to other modeling methods presented in the literature.

4.1   Static Time Window

Time window based approaches are techniques that group all events logs preceding a target event within a given time interval (the window size, cf. [4]). All grouped events within the time window are called a **target sequence**. i.e., a target sequence is a positive case containing the target critical event. **Non-Target sequences** denote sequences that have occurred between target sequences: starting at the beginning of the sequence, non-overlapping time windows of size W that do not intersect the set of time windows preceding target events (target sequences) are considered negative cases.

A big drawback of such strategies for data modeling is caused by the use of grids (partitioning the data in equidistant time intervals). In general, grid-based approaches heavily depend on the positioning of the grids: sequential patterns may be missed if they are inadequately oriented or shaped. Therefore, alternative data modeling strategies are desired.

4.2   Dynamic Time-Window: The Clustering Strategy

The clustering/tupling strategy for grouping events in a log file has been first presented by `Tsao et al.` in [5] and further discussed and analyzed by `Hansen et al.` in [6]. Tupling basically refers to grouping (*clustering*) events that occur within a pre-defined time interval in the log. The intuition underlying this approach is that two entries in the log, if related, are likely to occur near in time [6]. Consequently, if their inter-arrival time distance is below a predetermined threshold Δt (called the *coalescence window*), they are placed in the same group (called *tuple*). `Lal and Choi` have shown that the tupling method can also be used for clustering events that are related to the same target category in time [7]. They have reported that the clusters can successfully gather bursts of events in the machine logs. This is formalized in Definition 1.

Definition 1. *Cluster Based Sequence:* *given are n timestamped, chronologically ordered log entries $e_1, ..., e_n$. Let t_i stands for the time of occurrence of event e_i and Δt be a user specified threshold parameter. Then a sequence S in a sequence database SDB is represented by $S = \{\langle e_i, ..., e_k \rangle | i, k \in [1, n], i < k : t_{j+1} - t_j < \Delta t, j \in [i, k-1]\}$.*

After this step (which is independent of the target critical event to be predicted), a label definition of a **target** and **non-target** sequence (cf. Sect. 4.2) must be provided in order to specify which events in the sequences of the database will serve as possible patterns for pattern recognition techniques to act upon. Given an target critical event ε, the set of target sequences (positive cases) is a subset of the sequence database SDB which contains the target critical event ε. i.e., the set of target sequences for a target critical event $\varepsilon = \{S \in SDB | \varepsilon \in S\}$. Only events which appears prior to the appearance of the target event in the target sequence are relevant to be analyzed as the cause of possible critical situations This is defined in Definition 2 and shown in Fig. 2.

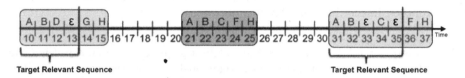

Fig. 2. For a given target critical target event ε, target and non-target clusters are depicted, highlighting target relevant sequences.

Definition 2. *Target Relevant Sequences*[1]: *A sequence set SDB_ε of Target Relevant Sequences for a target critical event ε is a subset of subsequences from a sequence database SDB defined as:*

$$SDB_\varepsilon = \{S' | S' \sqsubseteq S, S \in SDB\}, S' = \langle e_i, .., e_j \rangle, e_j = \varepsilon, \nexists k > j : e_k = \varepsilon \wedge e_k \in S.$$

Concluding this section, static time windows are sensitive to window resolution and position. Clustering based-approaches on the other side are position-independent and only depend on the resolution (coalescence window). Since critical events are rather rare, our approach just searches for patterns within target relevant sequences, dramatically reducing the search space.

5   Approach

Sequence Mining is a data mining technique that focuses on finding statistically relevant patterns in a sequence form for a given data set. We propose an approach to extract sequential patterns from temporal data to predict the occurrence of target events, such as critical situations logged by a machine. The first step in order to apply sequence mining to log data is transformation of the log into a proper sequence representation for the sequence miner to find patterns on, as discussed in the last section. In this section, we propose a prediction technique based on transforming the event prediction problem into a search for all frequent and valid sequences preceding a target event. Patterns are then integrated into a rule-based model for prediction.

---

[1] We use the notation '$\sqsubseteq$' to denote subsequence and '$\subset$' to denote subset relations.

5.1   Mining Sequential Patterns

A sequence in our context denotes a chronological ordered list of itemsets (each item is a machine event). The support (frequency) of a sequence S_j in a sequence database SDB is defined as the portion of sequences $S \in SDB$, such that $S_j \sqsubseteq S$. Let min_{sup} be a threshold on the support value: a sequence S_j is deemed frequent iff $support(S_j) > min_{sup}$. The problem of mining sequential patterns is to discover all frequent sequences. For our implementation in this work, we utilize the algorithm SPADE [8]. SPADE is an efficient algorithm that decomposes the original problem of finding frequent patterns into smaller subproblems which can be solved independently using efficient lattice search techniques and join operations. Nonetheless, several algorithms featuring different properties have been proposed in the literature for this task [9].

Mining Sequential Rules. Once the frequent sequences are known, they can be used to obtain rules that describe the relationship between different sequence events. We are particularly interested if it is possible to infer the probability of appearance of a target critical event (ε) given an observed previous sequence of events. To solve this kind of problem, using sequence mining by mining frequent sequences alone is very limited. For example, consider the pattern $S = \langle X, \varepsilon \rangle$, which means that it is possible that ε appears frequently after an observed sequence X. In this case, we talk about a *sequential rule*, denoted by $X \Rightarrow \varepsilon$, derived from $\langle X, \varepsilon \rangle$.

Definition 3. *ε-Rule: a ε-sequential-rule, for a target event ε, is a sequence in the form $R = \langle X, \varepsilon \rangle$, whereby X is a sequence in R preceding the target event ε. In this case we write: $R = \langle X, \varepsilon \rangle = (\langle X \rangle \Rightarrow \varepsilon) = (X \Rightarrow \varepsilon)$.*

The challenge is: there may be also many cases where X is not followed by ε in SDB. For prediction, we need a measurement of the conditional probability, that if X occurs, ε will also occur afterwards. The measure is the **confidence** defined as:

$$confidence_{SDB}(X \Rightarrow \varepsilon) = P(\varepsilon \mid X) = \frac{support_{SDB}(\langle X, \varepsilon \rangle)}{support_{SDB}(\langle X \rangle)} \qquad (1)$$

The higher the confidence, the higher the probability, that ε occurs **after** X. Given a user-specified minimum confidence (min_{conf}), sequential mining algorithms extract all rules which fulfills the condition, i.e., those that comply to the minimum desired confidence. Note that due to the apriori property, for a frequent sequence $S = \langle X, \varepsilon \rangle$, all the subsequences of X are also frequent, also those which do not contain ε. For this reason, they are filtered out of the output from the sequence miner. We call this result set the ε-Rules Data Base RDB_ε (cf. Algorithm 1). Sequence mining can thus be leveraged for creating sequential rules delivering the probability of occurrence of an critical situation in the machine given a previously observed sequence.

5.2   Filtering the Noise

Machinery logs often contain events that are unrelated to a specific critical target event, but due to different processes logging in the machine simultaneously, the target event may be interlaced with unrelated events. Applying Sequence Mining techniques for the task of predicting future critical situations can deliver many sequential patterns from which many of them· may be redundant and have no stronger predictive power. The apriori property states: if a sequence $S = \langle X, \varepsilon \rangle$ is frequent w.r.t. min_{sup}, so is every sequence S' with $S' \sqsubseteq S$. The Sequence Mining algorithm will thus output all subsequences of X for every closed sequence S, as a possible explanation to ε if they pass the confidence test.

Consequently, some post-processing noise remotion techniques become necessary in order to remove noisy, unpredictive and redundant patterns which reduce the size of the returned rule set. In the following, some definitions will be presented which will be necessary for presenting the post-processing noise remotion strategy.

Definition 4. *Confidence Gain:* *given two sequential rules R_i, $R_j \in RDB_\varepsilon$ with $R_i \sqsubset R_j$. The confidence gain of R_j w.r.t. R_i is defined as:*

$$confidence_{gain}(R_j, R_i) = confidence_{SDB}(R_j) - confidence_{SDB}(R_i) \qquad (2)$$

With the confidence gain, we measure the increase in predictive information gained by augmenting the left-hand of a rule R_i to an augmented rule R_j. We identify a sequential rule as redundant in RDB_ε if it delivers no gain in confidence. Formally:

Definition 5. *Redundancy-Free Rule:* *given a sequential rule $R_j \in RDB_\varepsilon$. The rule R_j is deemed redundancy-free iff:*

$$\forall R_i \sqsubset R_j, R_i \in RDB_\varepsilon : confidence_{gain}(R_j, R_i) > 0 \qquad (3)$$

*Otherwise it is deemed **redundant**.*

If a rule is not redundancy-free, it is deemed redundant and it is filtered out as noise in our approach. Consequently, a rule R_j is deemed redundant, if there exists another rule R_i with: $R_i \sqsubset R_j$ and $confidence(R_i) = confidence(R_j)$. Due to the anti-monotonicity property of the support constraint, R_i is at least as frequent as R_j [8]. Consequently, we eliminate all redundant sequences that are **not** more predictive than any of their proper subsequences.

The rationale is that, sequential patterns can be augmented with noisy events which do not contribute to a increase in confidence gain at predicting the target event ε. Take for example $R_1 := \langle e_1, e_2, e_3, n \rangle \Rightarrow \varepsilon$ as a rule with high confidence and with a noise event n. In this case, the event n does not contribute to a better predicability when compared to its subsequence rule $R_2 := \langle e_1, e_2, e_3 \rangle \Rightarrow \varepsilon$. And in this case the prediction lead time of R_2 is per definition better that the prediction lead time of R_1, as n comes after e_3 as last element in the rule. R_1 is thus a redundant rule and can be removed from the result set.

In case R_j is a redundancy-free rule, we keep R_j and eliminate all those R_i, with $R_i \sqsubset R_j$. I.e., we eliminate all sequences, which are less predictive than any of its proper super-sequences. This leads to the concept of a *redundancy-free sequence set*:

Definition 6. Redundancy-Free Sequence Set: *given a sequence set* $\mathcal{KB} \subseteq RDB_\varepsilon$, $\mathcal{KB}$ *is a Redundancy-Free Sequence Set if:*

$$\forall R \in \mathcal{KB} : R \text{ is redundancy-free} \tag{4}$$

We still must guarantee, that we output all possible redundancy-free sequential rules from RDB_ε. For this, we introduce the concept of *concept-covering sequence set*. Intuitively, a set of rules $\mathcal{KB}$ is concept-covering if adding any new rule $\in RDB_\varepsilon$ into $\mathcal{KB}$ always results in redundancy.

Definition 7. Concept-Covering Sequence Set: *given a sequence set* $\mathcal{KB} \subseteq RDB_\varepsilon$, $\mathcal{KB}$ *is concept-covering if:*

$$\forall \overline{R} \in RDB_\varepsilon \backslash \mathcal{KB} : \overline{R} \text{ is not redundancy-free} \tag{5}$$

In our definition $\mathcal{KB}$ must be redundancy-free and concept covering. We output $\mathcal{KB}$ as *knowledge base* instead of RDB_ε, which contains just a small set of high confident and non-redundant rules. The knowledge base $\mathcal{KB}$ is guaranteed to contain only rules which do not contain noisy events w.r.t. Definition 4.

5.3   Building a Sequence-Based Predictive Model

The resulting redundancy-free and concept-covering set of rules $\mathcal{KB}$ (Definition 7) can be leveraged to build the so called predictive **monitor**.

Definition 8. Monitor: *a function that takes as input a sequence S and outputs a predicted event if any of the rules in a knowledge base* $\mathcal{KB}$ *matches the input sequence:*

$$Monitor(S) := \{\varepsilon \mid R \in \mathcal{KB}, R = (X \Rightarrow \varepsilon) : X \sqsubseteq S\} \tag{6}$$

Please note that due to the filtering of redundant rules (Definition 5) the monitor guarantees that no sequences which may be subsequences from each other can match an input sequence, avoiding redundant matches.

Let's take as example a monitor M_R with a single rule $R := \langle A, B \rangle \Rightarrow E$. The monitor must make the decision (prediction) to fire or not an alarm while observing the sequence of logged events of the machine for the target critical event. This is a typical binary decision case where each monitor's decision can fall into the following 4 categories:

- **TP - True-Positive (Hit):** an alert/alarm occurs and the target critical event is observed in the given sequence after the alert. The sequences $\langle A, B, C, E \rangle$ and $\langle E, A, B, C, E \rangle$ for example are counted as hit for M_R.

Algorithm 1. <u>S</u>equence <u>B</u>ased <u>F</u>ailure <u>P</u>redictor (SBFP)

```
 1: procedure SBFP(Target Event ε, Input Data Logs, min_sup, min_conf, Δt)
 2:     SDB ← TemporalCluster(Logs, Δt)                    ▷ Cluster log files w.r.t. Definition 1
 3:     SDB_ε ← FRS(ε, SDB)               ▷ Extract Failure Relevant Sequences w.r.t. Definition 2
 4:     FS ← SequenceMiner(SDB_ε, min_sup)                ▷ Sequence Miner, e.g. SPADE
 5:     RDB_ε ← SET()
 6:     for all S ∈ FS do                               ▷ Filter valid ε-Rules w.r.t. Definition 3
 7:         if S is ε-Rule AND confidence_SDB(S) ≥ min_conf then
 8:             RDB_ε.add(S)
 9:         end if
10:     end for
11:     KB ← redundancy-free and concept covering subset from RDB_ε w.r.t. Definitions 6 and 7
12:     return Monitor(KB)              ▷ Build Monitor with Knowledge Base KB w.r.t. Definition 8
13: end procedure
```

- **FP - False-Positive:** false alarms. E.g. $\langle A, B, C \rangle$ and $\langle E, A, B, C \rangle$ for M_R.
- **TN - True-Negative:** no alarm is raised and the observed log sequence does not contain the target event. E.g. $\langle B, A, C \rangle$ for M_R.
- **FN - False-Negative (Miss):** no alarm has been raised but the target event was observed in the sequence. E.g. $\langle A, C, E \rangle$ for M_R.

Based on these premises, an intensive evaluation of the presented rule-based predictive model will be performed and discussed in the next section. The technique is summarized in Algorithm 1.

Computational Efficiency. Looking for all frequent eventsets is in the worst case exponential in the number of single event types. Note that the Algorithm 1 (line 4) only scans SDB_ε and not the sequence database SDB itself in search for sequential patterns. Since the number of sequences containing target events is much lower than the total number of sequences, we scan a much smaller projection of SDB in initial steps guaranteeing inexpensive computational efficiency in both memory and time in first steps. However, the complexity of the approach is highly dependent on the complexity of the sequence mining algorithm chosen in this step. In our case, we deploy the SPADE algorithm, which features a linear scalability w.r.t. the number of sequences [10]. Finally, the line 7 of the algorithm filter out all frequent sequences which do not contain the target event ε or are invalid w.r.t. minimum confidence min_{conf}.

6   Evaluation

We carry out extensive experiments with the proposed predictive model, leveraging real world log datasets from industrial computer numerical control (CNC) cutting machine systems (cf. Sect. 3). The performance of the proposed approach is then compared to the state-of-the-art method in event prediction proposed by Vilalta et al. [4]. As a reminder, Vilalta et al. proposed a method based on time window modeling of the log data (cf. Sect. 4.1), where they apply temporal itemset mining to extract association rules for predicting target critical events. Please note that the underlying assumption of the method proposed by Vilalta et al. is that the chronological sequence in which the events appear prior to a critical event is not significant. Just the set of logged events, independent of the sequence in which they appear in the log, are leveraged by Vilalta et al. for their prediction model. We claim that the set of events alone is not as powerful as the actual chronological order of the events in the set (Sequence Mining) for our use-case.

We evaluate the prediction performance of a Monitor in a holdout setting using binary classification performance metrics such as the precision and the recall. The point of partition of the dataset is set to the 50%th-index of the target critical event in the log: log events to the left of this point are used for training and in the right for testing, assuring that both partitions contain the exact same occurence number of the target event in it and that the training data lies on the past of the test data. The precision of a monitor is the percentage of times the monitor signals a critical event, and the critical event actually occurs (i.e., the ratio of correct critical situation signals to the total number of critical situation signals). The recall of a monitor is the percentage of critical events signaled prior to their occurrence. We leverage the F_β-*Measure* which balances the precision and the recall, as a target function to be optimized (maximized):

$$F_\beta\text{-}Measure = (1 + \beta^2) \cdot \left(\frac{Precision \cdot Recall}{(\beta^2 \cdot Precision) + Recall}\right) \tag{7}$$

The F_1-Measure is the most widespread metric for comparing or evaluating classifiers. Two other commonly used F measures are the F_2-Measure, which weights recall higher than precision, and the $F_{0.5}$-Measure, which puts more emphasis on precision than recall. We argue, that for the use-case in question (predicting machine critical events), the $F_{0.5}$-Measure is more relevant, as the precision of the monitor/predictor is more important than the recall, i.e., misses are in our scenario not so bad as false alarms. This argument lies on the assumption that producing false alarms may generate *reaction costs*, so we prefer more conservative and confident models in opposite to more general ones. For this reason, we set the $F_{0.5}$-Measure as the main evaluation metric to be used in this work.

Given a target critical event, we are interested in answering following questions: How does the proposed approach perform at predicting the target critical event? Which parametrization leads to best prediction performance? Which prediction lead time can we expect for predicting the target event in our context?

These questions will be answered on top of experiments. Domain experts ranked the 13 most relevant target critical events to be analyzed in this work. We select the machine log dataset where the most important ranked target event most often occurs for a detailed evaluation in Sect. 6.1. The other remaining 12 target events are also be analyzed and evaluated across varying machine datasets in Sect. 6.2.

6.1   Parametrization and Prediction Performance

We optimize the 3 parameters of the proposed approach, namely the coalescence window (Δt), the min_{sup} and min_{conf} parameters through grid search [11] for the $F_{0.5}$-Measure as scorer function. In grid search the set of trials is formed by assembling every possible combination of values and the best parametrization w.r.t. scorer function is returned. The overall best parameter space is identified and used for training the end model. Our grid search yells as result the parameters $\Delta t = 450\,s$, $min_{sup} = 10\%$ and $min_{conf} = 70\%$ as the overall best parametrization of the model which optimizes the $F_{0.5}$-Measure.

Naturally, the minimum desired support and confidence parameters (min_{sup} & min_{conf}) have a big impact on the prediction performance. The use of a low support can allow rare rules to pass the frequency test, but this comes at the cost of increasing the complexity of the search space generating excessively many rules. For our scenario, setting $min_{sup} = 0.1$ turned out to be optimal since mining more rules (by setting a lower min_{sup}) did not the enrich the knowledge base significantly (i.e., most new rules are redundant below this value w.r.t. confidence gain).

Figure 3 shows how the prediction performance behave in dependence of the confidence parameter. The minimum desired confidence (min_{conf}) is a trade-off between precision and recall. Rules with high confidence make predictions only with strong evidence so they lead to overall higher precision, but they often also lead to low recall rates.

Last but not least, the coalescence window also impact the prediction performance as well as the prediction lead time of the proposed algorithm. As can be seen on the Fig. 4 (left), all other parameters being equal, the prediction performance increases together with larger coalescence windows up to a certain point.

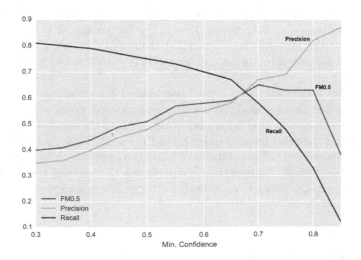

Fig. 3. Variation of prediction metrics depending on the minimum desired confidence parameter.

Larger coalescence windows enable us to capture more information preceding target events. But just until to the point where the number of collisions[2] causes several sequences to be merged, loosing contrast for separating the events which are related to the target critical event and those which are not.

Figure 4 (right) shows the distribution of the prediction lead time for the tested dataset. Prediction models which are based on modeling the log data with a static time window strategy are bounded to a maximum prediction lead time of the size of the window itself, which is not our case. Such a data modeling strategy enables a flexible short-term prediction lead time, which can be by orders of magnitude longer than the coalescence window parameter itself (up to 40 min), as shown in Fig. 4 (right).

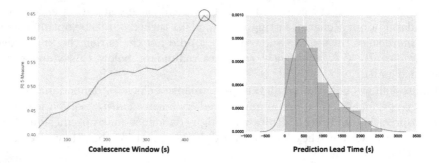

Fig. 4. Left: Variation of prediction performance ($F_{0.5}$-Measure) in dependence of the coalescence window. **Right:** Distribution of Monitor's prediction lead time with a kernel density estimator for the optimized model.

Table 2 shows the prediction results for the overall best parametrization ($\Delta t = 450$ s, $min_{sup} = 0.1$, $min_{conf} = 0.7$). By means of comparison, the performance of the state-of-the-art method from Vilalta et al. is also presented. We set its time window parameter to be $W = 660$ s, which optimizes the $F_{0.5}$-Measure.

In total, 89 sequential rules have been mined for creating the predictive monitor evaluated in this work. Without the proposed noisy reduction steps, the maximum achievable prediction performance ($F_{0.5}$-Measure) would fall from 0.65 to 0.62 and the total number of extracted rules would lie by the thousands, which can easily exceed the capabilities of a human operator to identify interesting results. Additionally, our method outperforms the state-of-the-art by both precision and recall, leading to an overall better prediction performance.

---

[2] Collision occurs when unrelated events are combined into the same group (cluster). The contrary of collision is also denoted "truncation", referring to related events being separated into different groups when the coalescence window is too small.

Table 2. Comparison of prediction performance results between the method proposed by `Vilalta et al.` and the proposed SBFP. Both methods have been parameterized to optimize the $F_{0.5}$-Measure.

Algorithm	TP	TN	FP	FN	Precision	Recall	FM05
SBFP	130	792	65	94	0.67	0.58	0.65
Vilaltas'	124	1300	80	109	0.6	0.53	0.59

Number of observations differ due to the different data modeling strategies in both algorithms.

6.2   Performance Across Varying Target Events and Datasets

Up to this point, we discussed the main properties and influencing factors of our proposed prediction model for a single target event in a single machine log dataset. It remains to discuss the prediction performance of the approach beyond an unique example. This is the goal of this section.

Naturally, within the scope and capabilities of this work, we can just provide detailed description of the proper parametrization and prediction performance of the proposed approach for selected cases. Nevertheless, in order to provide a broader (but rough) overview of the capabilities of the proposed approach, we perform the following experiment: after the labeling/ranking from domain experts (cf. Sect. 6) we take the remaining 12 target critical events to be predicted, and select the machine log datasets in which the selected target event appears more frequently than a given threshold (in our case in at least 20 distinct sequences). We then run our approach onto those eligible datasets without parameter tuning. Therefore, we can not guarantee that the best solution is found for each run, but we can provide an overview on how the prediction performance roughly varies for a target event across different log datasets from different machines for a given parametrization. The results of this experiment is shown in Fig. 5. For each target event (E1–E12), we show the distribution of the prediction performance when running the algorithm on different datasets in form of Tukey boxplots.

For each run, the prediction performance ($F_{0.5}$-Measure) of the algorithm for the given target event and machine dataset is shown as a point on the boxplots of Fig. 5. As shown in the figure, different events can be predicted to varying degrees. Events $E6$ and $E9$ are for example generally less predictable than most other events with the given parameters. Even for a particular event type, the prediction performance varies significantly according to the machine log dataset used for training and validation of the model. Results for event $E10$ shows a great midspread (IQR) while for other events, like event $E8$, the results dispersion is by orders of magnitude lower. This complies with previous observations of similar work in the literate [4] (cf. Sect. 2). Considering that, the success of the algorithm is contingent on the proper parametrization and existence of sequential pattern of events preceding the target events in the different datasets.

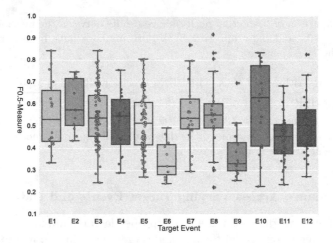

Fig. 5. Variation of prediction performance for 12 target event types without parametrization tuning. For each target event, the algorithm was trained and tested individually across several distinct machine log datasets. A point on the graph depicts the prediction performance ($F_{0.5}$-Measure) w.r.t. the given target event for an individual machine log dataset.

7 Conclusion

In this work, we analyzed log files generated by industrial computer numerical control (CNC) cutting machines from the company TRUMPF GmbH + Co. KG, with the goal of predicting critical events in machinery log data. We presented a Sequence-Mining based technique for mining sequential patterns of events from log data. Sequence-Mining constitutes an extension to related approaches which so far have just leveraged itemset and association rules mining, which did not consider strict chronological order of events for issuing predictions. We show how our approach with this additional property requires more evidence for issuing positive predictions than methods based just on event sets, leading to a more precise performance outperforming existing techniques.

Furthermore, we presented contributions in how machinery log data can be proper modeled into sequence definitions prior to its usage in sequence mining as well as a information gain based noise reduction technique that guarantees that the detected patterns are related to the target critical event, reducing the result set and increasing prediction's performance.

Our approach is suitable for short-term predictions, achieving prediction lead times of up to 40 min in advance for a specific critical event.

Together with domain experts, we extensively evaluated our approach and its parametrization using real data from over 150 machines containing millions of log entries and discuss the benefit of such techniques for the industry.

References

1. Raman, S., De Silva, C.W.: Sensor-fault tolerant condition monitoring of an industrial machine. Int. J. Inf. Acquis. **9**(01), 1350001 (2013)
2. Salfner, F., Tschirpke, S.: Error log processing for accurate failure prediction. In: WASL (2008)
3. Salfner, F., Lenk, M., Malek, M.: A survey of online failure prediction methods. ACM Comput. Surv. **42**(3), 1–42 (2010)
4. Vilalta, R., Ma, S.: Predicting rare events in temporal domains. In: Proceedings of the 2002 IEEE International Conference on Data Mining, 2002. ICDM 2003, pp. 474–481. IEEE (2002)
5. Tsao, M.M., Siewiorek, D.P.: Trend analysis on system error files. In: Proceedings of the 13th International Symposium on Fault-Tolerant Computing, Milano, vol. 116119 (1983)
6. Hansen, J.P., Siewiorek, D.P.: Models for time coalescence in event logs. In: Twenty-Second International Symposium on Fault-Tolerant Computing, FTCS-22. Digest of Papers, pp. 221–227. IEEE (1992)
7. Lal, R., Choi, G.: Error and failure analysis of a unix server. In: Proceedings of the Third IEEE International on High-Assurance Systems Engineering Symposium, pp. 232–239. IEEE (1998)
8. Zaki, M.: SPADE: an efficient algorithm for mining frequent sequences. Mach. Learn. **42**(1–2), 31–60 (2001)
9. Zhao, Q., Bhowmick, S.S.: Sequential pattern mining: a survey, ITechnical report CAIS Nayang Technological University Singapore, pp. 1–26 (2003)
10. Slimani, T., Lazzez, A.: Sequential mining: patterns and algorithms analysis. arXiv preprint arXiv:1311.0350 (2013)
11. Bergstra, J., Bengio, Y.: Random search for hyper-parameter optimization. J. Mach. Learn. Res. **13**, 281–305 (2012)

Importance of Recommendation Policy Space in Addressing Click Sparsity in Personalized Advertisement Display

Sougata Chaudhuri[1(✉)], Georgios Theocharous[2],
and Mohammad Ghavamzadeh[2]

[1] University of Michigan, Ann Arbor, USA
sougata@umich.edu
[2] Adobe Research, San Jose, USA
{theochar,ghavamza}@adobe.com

Abstract. We study a specific case of personalized advertisement recommendation (PAR) problem, which consist of a user visiting a system (website) and the system displaying one of K ads to the user. The system uses an internal ad recommendation policy to map the user's profile (context) to one of the ads. The user either clicks or ignores the ad and correspondingly, the system updates its recommendation policy. The policy space in large scale PAR systems are generally based on classifiers. A practical problem in PAR is extreme click sparsity, due to very few users actually clicking on ads. We systematically study the drawback of using classifier-based policies, in face of extreme click sparsity. We then suggest an alternate policy, based on rankers, learnt by optimizing the Area Under the Curve (AUC) ranking loss, which can significantly alleviate the problem of click sparsity. We create deterministic and stochastic policy spaces and conduct extensive experiments on public and proprietary datasets to illustrate the improvement in click-through-rate (CTR) obtained by using the ranker-based policy over classifier-based policy.

1 Introduction

Large scale personalized advertisement recommendation (PAR) system is intrinsic to many major tech companies like Google, Yahoo, Facebook and others. The particular PAR setting we study here consists of a policy that displays one of the K possible ads/offers, when a user visits the system. The user's profile is represented as a context vector, consisting of relevant information like demographics, geo-location, frequency of visits, etc. Depending on whether user clicks on the ad, the system gets a reward of value 1, which in practice translates to dollar revenue. The policy is (continuously) updated from historical data, which consist of tuples {*user context*, *displayed ad*, *reward*}. We will, in this paper, concern ourselves with PAR systems that are geared towards maximizing total number of clicks.

The plethora of papers written on various forms of PAR problem makes it impossible to provide an exhaustive list. Interested readers may refer to a comprehensive paper by a team of researchers in Google [17] and references therein.

© Springer International Publishing AG 2017
P. Perner (Ed.): MLDM 2017, LNAI 10358, pp. 32–46, 2017.
DOI: 10.1007/978-3-319-62416-7_3

While the techniques in different papers differ in their details, the majority of them employ an algorithmic framework that is focused on efficient learning of recommendation policies, where the policies are essentially some form of multi-class classifier. For example, one important class of algorithms learn a classifier per ad from the batch of data [11,17,18] and convert it into a policy. Other, more theoretically sophisticated online algorithms, essentially learn a cost-sensitive multi-class classifier by updating after every round of user-system interaction [1,10]. Both kinds of classifier based policies usually display the ad with the highest estimated expected reward to the user; a practice referred to as *exploitation*. Due to the prevalent *cold start* problem, where there is little feedback on a new or relatively unexplored ad, some *exploration* strategy is combined with exploitation, where a random ad, not recommended by the policy, is displayed to get feedback. The explorations techniques can be explicit like ϵ-greedy [13,21] or implicit Bayesian type sampling from the posterior distribution maintained on classifier parameters [6].

A fundamental problem in PAR systems stems from *click sparsity*, i.e., very few users actually ever click on online ads and this lack of positive feedback makes it difficult to learn good classifier based recommendation policies. Our main objective is to systematically study this important practical issue and provide an alternate solution to partly alleviate this problem. We list our contributions:

- We systematically study the negative effect of click sparsity on classifier based policies.
- We suggest a simple ranker-based policy to overcome the click sparsity problem. The rankers are learnt by optimizing the *Area Under Curve* (AUC) ranking loss via use of *stochastic gradient descent* (SGD) [20] technique, leading to a highly scalable algorithm. The rankers are then combined to create a recommendation policy.
- We conduct extensive experiments to illustrate the improvement provided by our suggested method over both linear and ensemble classifier-based policies for the PAR problem. Our first set of experiments compare *deterministic policies* on publicly available classification datasets, that are converted to bandit datasets following standard techniques. Our second set of experiments compare *stochastic policies* on three proprietary bandit datasets, for which we employ a high confidence offline evaluation technique.

2   Effect of Click Sparsity on Classifier Based Policies

We provide a formal description of the general algorithmic framework that is suited to our particular PAR setting, and then discuss the problem that arises due to click sparsity.

Let $\mathcal{X} \subseteq \mathbb{R}^d$ and $[K] = \{1, 2, \ldots, K\}$ denote the user context space and K different ads/arms. Π is the space of policies such that for any $\pi \in \Pi$, $\pi : \mathcal{X} \mapsto [K]$. Note that our setting does not have context on ads. The algorithmic framework have the following steps:

- At each round t, the context vector x_t is revealed, i.e., a user visits the system.
- The system selects ad a_t according to the current policy $\pi_t \in \Pi$ (exploitation strategy). Optionally, an exploration strategy is sometimes added, creating a distribution $p_t(\cdot)$ over the ads and a_t is drawn from $p_t(\cdot)$. Policy π_t and distribution p_t are sometimes used synonymously by considering π_t to be a stochastic policy.
- Reward r_{t,a_t} is revealed and the new policy $\pi_{t+1} \in \Pi$ is computed, using information $\{x_i, a_i, r_{i,a_i}\}_{i \le t}$. We *emphasize* that the system does not get to know $r_{t,a'}, \forall a' \ne a_t$.

Traditionally, the algorithmic framework is known as *contextual bandits* [15], due to the fact that system only gets to see the reward on the ad displayed and not the other ads. Assuming the user-system interaction happens over T rounds, the objective of the system is to maximize its click-through-rate (**CTR**):

$$\frac{\sum_{t=1}^{T} r_{t,a_t}}{T}. \tag{1}$$

Any specific algorithm following the framework, in course of updating its policy at every round, tries to converge to the oracle optimal policy π^*, where $\pi^* = \underset{\pi \in \Pi}{\text{argmax}} \sum_{t=1}^{T} r_{t,\pi(x_t)}$. Crucially, the algorithms are only concerned with converging as fast as possible to the best policy π^* in the policy space Π and do not take into account the nature of the policy space. *Hence, if the policy space itself is poorly designed, then the algorithm will have very low cumulative reward, regardless of its sophistication.*

Policy space Π considered for large scale PAR systems are based on classifiers. They can be tuples of binary classifiers, with one classifier for each ad, or global cost-sensitive multi-class classifier, depending on the nature of the algorithm. Since clicks on the ads are rare and small improvement in click-through rate can lead to significant reward, it is vital for the policy space to have good policies that can identify the correct ads for the rare users who are highly likely to click on them. Extreme click sparsity makes it *practically challenging* to design a classifier-based policy space. We discuss the issue in details for binary and multi-class classifiers.

Binary Classifier Based Policies: The algorithmic framework preceding Eq. 1 is presented as an online framework, with policy updated at end of every round. Usually, in large scale industrial PAR systems, it is highly impractical to update policies continuously, due to thousands of users visiting a system in a small time frame. Thus, policy update happens, i.e. new policy is learnt, after intervals of time, using the bandit data produced from the interaction between the current policy and users, collected in batch. It is convenient and practical to learn a binary classifier per ad in such a setting. To explain the process concisely, we note that the bandit data consists of tuples of the form $\{x, a, r_a\}$. For each ad a, the users who had not clicked on the ad ($r_a = 0$) would be considered as negative examples and the users who had clicked on the ad ($r_a = 1$) would be considered as positive examples, creating a binary training set for ad a. The K binary classifiers are converted into a recommendation policy using a "one-vs-all" method [19]. *Thus, each policy in policy space Π can be considered to be a tuple of K binary classifiers.*

A number of research publications show that researchers consider binary linear classifiers, that are learnt by optimizing the logistic loss [18], while ensemble classifiers, like random forests, are also becoming popular [13]. We note that the majority of the papers that learn a logistic linear classifier focus on feature selection [11], novel regularizations to tackle high-dimensional context vectors [17], or propose clever combinations of logistic classifiers [2].

Click sparsity poses difficulty in designing accurate binary classifiers in the following way: each ad a would have received very few clicks as compared to the number of users who did not click on the ad. A binary classifier learnt in such setting will almost always predict that its corresponding ad will not be clicked by a *future user*, failing to identify the rare, but very important, users who are likely to click on the ad. This is colloquially referred to as "class imbalance problem" in binary classification [12]. Due to the extreme nature of the imbalance problem, tricks like under-sampling of negative examples [7] might not be very useful (additionally, aggressive under-sampling can lead to some important features within the negative set to be discarded). Sophisticated techniques like cost-sensitive SVMs generally require prior knowledge of class importance, which is not available in the setting.

Cost Sensitive Multi-class Classifier Based Policies. Another type of policy space consist of cost-sensitive multi-class classifiers [1, 10, 15]. They can be cost-sensitive multi-class svms [5], multi-class logistic classifiers or filter trees [3]. Click sparsity poses slightly different kind of problem in a policy space of such classifiers.

Cost sensitive multi-class classifier works as follows: assume a context-reward vector pair (x, r) is generated as described in the PAR setting. The classifier will try to select a class (ad) a such that the reward r_a is maximum among all choices of $r_{a'}$, $\forall\ a' \in [K]$ (we consider reward maximizing classifiers, instead of cost minimizing classifiers). Unlike in traditional multi-class classification, where one entry of r is 1 and all other entries are 0; in cost sensitive classification, r can have any combination of 1 and 0. Now consider the reward vectors r_ts generated over T rounds. A poor quality classifier π^p, which fails to identify the correct ad for most users x_t, will have very low average reward, i.e., $\frac{\sum_{t=1}^{T} r_{t,\pi^p(x_t)}}{T} \sim O(\epsilon)$, with $\epsilon \sim 0$. From the model perspective, extreme click sparsity translates to almost all reward vectors r_t being $\mathbf{0}$ vector. Thus, even a very good classifier π^g, which can identify the correct ad a_t for almost all users, will have very low average reward, i.e., $\frac{\sum_{t=1}^{T} r_{t,\pi^g(x_t)}}{T} \sim O(\delta)$, with $\delta \sim 0$. From practical perspective, it is difficult to distinguish between the performance of a good and poor classifier, in face of extreme sparsity, and thus, cost sensitive multi-class classifiers are not ideal policies for algorithms addressing the PAR problem.

3   Ranker Based Policy

We propose a ranking-based alternative to learning a classifier per ad that is capable of overcoming the click sparsity problem. We learn a ranker per ad by optimizing the Area Under the Curve (AUC) loss, and use a ranking score normalization technique to create a policy, mapping context to ad.

Ranker Learning Technique. For an ad a, let $S^+ = \{x : r_a = 1\}$ and $S^- = \{x : r_a = 0\}$ be the set of positive and negative instances, respectively. Let f_w be a linear ranking function parameterized by $w \in \mathbb{R}^d$, i.e., $f_w(x) = w \cdot x$ (inner product). AUC-based loss (AUCL) is a ranking loss that is minimized when positive instances get higher scores than negative instances, i.e., the positive instances are ranked higher than the negatives when instances are sorted in descending order of their scores [9]. Formally, we define empirical AUCL for function $f_w(\cdot)$:

$$\text{AUCL} = \frac{1}{|S^+||S^-|} \sum_{x^+ \in S^+} \sum_{x^- \in S^-} \mathbb{1}(\underbrace{f_w(x^+) - f_w(x^-)}_{t} < 0).$$

Direct optimization of AUCL is a NP-hard problem, since AUCL is sum of discontinuous indicator functions. To make the objective function computationally tractable, the indicator functions are replaced by a continuous, convex surrogate $\ell(t)$. Examples include hinge $\ell(t) = [1 - t]_+$ and logistic $\ell(t) = \log(1 + exp(-t))$ surrogates. Thus, the final objective function to optimize is

$$L(w) = \frac{1}{|S^+||S^-|} \sum_{x^+ \in S^+} \sum_{x^- \in S^-} \ell(\underbrace{f_w(x^+) - f_w(x^-)}_{t}). \tag{2}$$

Advantage of Learning a Ranker over a Classifier: A classifiers is learnt by optimizing a surrogate (hinge, logistic) of $0 - 1$ loss (accuracy loss) and is penalized for incorrect class prediction. For highly imbalanced dataset, a biased classifier gets high accuracy by always predicting the dominant class. However, the concept of class imbalance does not apply to a ranker learnt by optimizing AUCL. Irrespective of the actual number of positive and negative instances in the training set, AUCL is only concerned with the relative position of positive and negative instances. As an example, consider a highly imbalanced sample with 1 positive instance and 999 negative instances. A classifier, which predicts all negatives will suffer almost no loss (99.9% accuracy), while if a ranker ranks the positive instance below most of the negative instances, it will suffer a very large loss. Thus, the classifier will remain biased after training and fail to identify the valuable but rare positive instances. However, the ranker will ideally be trained to rank positive instances above the negative instances.

We note that AUC is a popular measure to evaluate a classifier on an imbalanced dataset and this has led to directly learning a classifier by optimizing AUC during training [4]. *However, our objective is not to evaluate AUC per class during testing.* We explicitly use the loss to learn a ranker that overcomes the imbalance problem and then create a context to ad mapping policy. We also note that due to the lack of context on ads, we are focused on policies which learn ranker/classifier per ad, instead of joint modeling of users and ads.

Constructing Policy from Rankers: Similar to learning a classifier per ad, a separate ranking function $f_{w_a}(\cdot)$ is learnt for each ad a from the bandit batch data. Then a threshold score s_a is learnt for each ad a by maximizing *F-score* measure on a validation set. That is, score of each (positive and negative) instance corresponding to a is calculated and the F-scores corresponding to different thresholds are compared. The threshold that gives the maximum F-score value is assigned to s_a. For a new user x, the combined policy π works as follows:

$$\pi(x) = \underset{a \in [K]}{\mathrm{argmax}} \; (f_{w_a}(x) - s_a). \tag{3}$$

Thus, π maps x to ad a with maximum "normalized score". This shifting negates the inherent scoring bias that might exist for each ranking function. That is, a ranking function for an action $a \in [K]$ might learn to score all instances (both positive and negative) higher than a ranking function for an action $b \in [K]$. Therefore, for a new instance x, ranking function for a will always give a higher score than the ranking function for b, if no normalization is done, leading to incorrect predictions.

SGD based Optimization Procedure: The objective function (2) is a convex function and can be efficiently optimized by standard convex optimization techniques. One memory-related computational issue associated with AUCL is that it pairs every positive and negative instance, effectively squaring the training set size. *Stochastic gradient descent* (SGD) procedure [20] is a suitable optimization technique to overcome the computational burden. At every step of SGD, a positive and a negative instance are randomly selected from the separate positive and negative sample set, followed by a gradient descent step on objective (2). This makes the training procedure memory-efficient and mimics full gradient descent optimization on the entire loss. We also note that the rankers for the ads can be trained in parallel and any regularizer like $\|w\|_1$ and $\|w\|_2$ can be added to (2), to introduce sparsity or avoid overfitting. Lastly, powerful non-linear kernel ranking functions can be learnt, but at the cost of memory efficiency, and the rankers can be learnt from streaming data as well [22].

4   Competing Policies and Evaluation Techniques

To support our hypothesis that ranker based policies address the click-sparsity problem better than classifier based policies, we set up two sets of experiments. We

- compared deterministic policies (only "exploitation") on full information (classification) datasets.
- compared stochastic policies ("exploitation + exploration") on bandit datasets, with a specific offline evaluation technique.

Both of our experiments were designed for policies constructed from classifiers/rankers per ad. The classifiers considered were *logistic linear* and *Random-Forest* classifiers and ranker considered was the *AUC optimized linear ranker*.

Deterministic Policies: Policies from the trained classifiers were constructed using the "one-vs-all" technique, i.e., for a new user x, the ad with the maximum score according to the classifiers was predicted. For the policy constructed from rankers, the ad with the maximum normalized score according to the rankers was predicted, using Eq. 3. Deterministic policies are "exploit only" policies.

Stochastic Policies: Stochastic policies were constructed from deterministic policies by adding an ϵ-greedy exploration technique on top. Briefly, let one of the stochastic policies be denoted by π_e and let $\epsilon \in [0, 1]$. For a context x in the test set, let a be the offer with maximum score according to underlying deterministic policy. Then, π_e selects a w.p. $1 - \epsilon$ (i.e. $\pi_e(a|x) = 1 - \epsilon$), and selects any other ad a' uniformly at random (i.e., $\pi_e(a'|x) = \frac{\epsilon}{K-1}$, where K is the total number of offers). Thus, π_e is a probability distribution over the offers. Stochastic policies are "exploit+ explore" policies.

4.1   Evaluation on Full Information Classification Data

Benchmark bandit data are usually hard to obtain in public domains. So, we compared the deterministic policies on benchmark K-class classification data, converted to K-class bandit data, using the technique in [14]. Briefly, the standard conversion technique is as follows: A K-class dataset is randomly split into training set X_{train} and test set X_{test} (in our experiments, we used 70–30 split). *Only the labeled training set is converted into bandit data.* Let $\{x, a\}$ be an instance and the corresponding class in the training set. A class $a' \in [K]$ is selected uniformly at random. If $a = a'$, a reward of 1 is assigned to x; otherwise, a reward of 0 is assigned. The new bandit instance is of the form $\{x, a', 1\}$ or $\{x, a', 0\}$, and the true class a is hidden. The bandit data is then divided into K separate binary class training sets, as detailed in the section "Binary classifier based policies".

Evaluation Technique: We compared the *deterministic policies* by calculating the CTR of each policy. For a policy π, CTR on a test set of cardinality n is measured as: $\frac{1}{n} \sum_{(x,a) \in \text{test set}} \mathbb{1}(\pi(x) = a)$. Note that we can calculate the true CTR of a policy π because the correct class a for an instance x is known in the test set.

4.2   Evaluation on Bandit Information Data

Bandit datasets have both training and test sets in bandit form, and datasets we use in the paper are industry proprietary in nature.

Evaluation Technique: We compared the *stochastic policies* on bandit datasets. Comparison of policies on bandit test set comes with the following unique challenge: for a stochastic policy π and a test context x, the expected reward is:

$$\rho(\pi|x) = \mathbb{E}_{a\sim\pi(\cdot|x)} r_a = \sum_a r_a \pi(a|x) \tag{4}$$

The true CTR of π is the average of expected reward over entire test set. *Since the bandit form of test data does not give any information about rewards for offers which were not displayed, it is not possible to calculate the expected reward!*

We evaluated the policies using a particular offline bandit policy evaluation technique. There exist various such evaluation techniques in the literature, with adequate discussion about the process [16]. We used one of the importance weighted techniques similar to one in Theocharous et al. [21]. The reason was that we could give high confidence lower bound on the performance of the policies. Mathematical details of the technique follows.

The bandit test data was logged from the interaction between users and a fully random policy π_u, over an interaction window. The random policy produced the following distribution over offers: $\pi_u(a|x) = \frac{1}{K}, \forall a \in [K]$. For an instance $\{x, a', r_{a'}\}$ in the test set, the importance weighted reward of evaluation policy π is computed as:

$$\hat{\rho}(\pi|x, a') = r_{a'} \frac{\pi(a'|x)}{\pi_u(a'|x)}. \tag{5}$$

The importance weighted reward is an unbiased estimator of the true expected reward of π, i.e.,

$$\mathbb{E}_{a'\sim\pi_u(\cdot|x)} \hat{\rho}(\pi|x, a') = \rho(\pi|x).$$

Let the cardinality of the test set be n. The **importance weighted CTR of π** is defined as

$$\frac{1}{n} \sum_{(x,a)\in\text{test set}} r_a \frac{\pi(a|x)}{\pi_u(a|x)} \tag{6}$$

Since (x, r) are assumed to be generated i.i.d., the importance weighted CTR is an unbiased estimator of the true CTR of π. Moreover, it is possible to construct a *t-test* based lower confidence bound on the expected reward, using the unbiased estimator, as follows: let $X_i = r_{a_i} \frac{\pi(a_i|x_i)}{\pi_u(a_i|x_i)}$, $\hat{X} = \frac{1}{n}\sum_{i=1}^{n} X_i$, and $\sigma = \sqrt{\frac{1}{n-1}\sum_{i=1}^{n}(X_i - \hat{X})^2}$. Then, $\hat{X} = $ importance weighted CTR and

$$\hat{X} - \frac{\sigma}{\sqrt{n}} t_{1-\delta, n-1} \tag{7}$$

is a $1 - \delta$ **lower confidence bound** on the true CTR. Thus, during evaluation, we plotted the importance weighted CTR and lower confidence bounds for the competing policies.

We provide a pseudo-algorithm to summarize the policy construction and evaluation process:

Algorithmic Framework 1. Policy Construction and Evaluation

Construction of rankers and classifiers:
    Input: for each ad a, S_a^+ = positive instances, S_a^- = negative instances.
    Linear rankers $f_{w_a}^R()$ and threshold score s_a: learn by optimizing Eq. 2, using SGD
    (optionally add ℓ_1 or ℓ_2 regulzarizer to objective in Eq. 2).
    Linear classifiers $f_{w_a}^C()$: learn by optimizing hinge/logistic loss
    (optionally add ℓ_1 or ℓ_2 regulzarizer to the loss).
    Random forest classifier $RF^C()$: learn using any standard random forest package.

Construction of deterministic policy: for a test context x
    Ranker based policy: $\pi(x) = \underset{a \in [K]}{\text{argmax}} \ (f_{w_a}^R(x) - s_a)$ (Eq. 3).
    Classifier based policy: $\pi(x) = \underset{a \in [K]}{\text{argmax}} \ f_{w_a}^C(x)$ (or $\underset{a \in [K]}{\text{argmax}} \ RF_{w_a}^C(x)$).

Construction of stochastic policy (with greedy parameter ϵ): for a test context x, and a
deterministic policy π for which $\pi(x) = a$,
$\pi_e(a|x) = 1 - \epsilon$ and $\pi_e(a'|x) = \frac{\epsilon}{K-1}$. (Stochastic policies, Sect. 4).

Evaluation:
    Deterministic policy: CTR of $\pi = \frac{1}{n} \sum_{(x,a) \in \text{test set}} \mathbb{1}(\pi(x) = a)$. Deterministic policies are
evaluated on full information multi-class classification datasets.
    Stochastic policy: Importance weighted CTR of $\pi_e = \frac{1}{n} \sum_{(x,a) \in \text{test set}} r_a \frac{\pi(a|x)}{\pi_u(a|x)}$ (Eq. 6)
and lower confidence bound $\hat{X} - \frac{\sigma}{\sqrt{n}} t_{1-\delta, n-1}$ (Eq. 7). Stochastic policies are evaluated
on bandit datasets.

Table 1. All multi-class datasets were obtained from UCI repository (https://archive.
ics.uci.edu/ml/). Five different datasets were selected. In the table, size represents the
number of examples in the complete dataset. Features indicate the dimension of the
instances. Avg. % positive gives the number of positive instances per class, divided by
the total number of instances for that class, in the bandit training set, averaged over
all classes, expressed as percentage. The lower the value, the more is the imbalance per
class during training.

	OptDigits	Isolet	Letter	PenDigits	Movementlibras
Size	5620	7797	20000	10992	360
Features	64	617	16	16	91
Classes	10	26	26	10	15
Avg. % positive	10	4	4	10	7

5 Empirical Results

We detail the parameters and results of our experiments.

Linear Classifiers and Ranker: For each ad a, a linear classifier was learnt by
optimizing the logistic surrogate, while a linear ranker was learnt by optimizing

the objective function (2), with $\ell(\cdot)$ being the logistic surrogate. Since we did not have the problem of sparse high-dimensional features in our datasets, we added an ℓ_2 regularizer instead of ℓ_1 regularizer. We applied SGD with 100,000 iterations; varied the parameter λ of the ℓ_2 regularizer in the set $\{0.01, 0.1, 1, 10\}$ and recorded the best result.

Ensemble Classifiers: We learnt a RandomForest classifier for each ad a. The RandomForests were composed of 200 trees, both for computational feasibility and for the more theoretical reason outlined in [13].

5.1   Comparison of Deterministic Policies

The multi-class datasets are detailed in Table 1. To compare the deterministic policies, we conducted two sets of experiments; one without under-sampling of negative classes during training (i.e., no class balancing) and another with heavy under-sampling of negative classes (artificial class balancing). Training and testing were repeated 10 times for each dataset to account for the randomness introduced during conversion of classification training data to bandit training data, and the average accuracy over the runs are reported. Figure 1 top and bottom show performance of various policies learnt without and with under-sampling during training, respectively. Under-sampling was done to make positive:negative ratio as 1:2 for every class (this basically means that Avg % positive was 33%). *The ratio of 1:2 generally gave the best results.*

Observations

- With heavy under-sampling, the performance of classifier-based policies improve significantly during training. Ranker-based policy is not affected, validating that class imbalance does not affect ranking loss.
- The linear ranker-based policy performs uniformly better than the linear classifier-based policy, with or without under-sampling. This shows that restricting to same class of functions (linear), rankers handles class-imbalance much better than classifiers.
- The linear ranker-based policy does better than more complex RandomForest (RF) based policy, when no under-sampling is done during training, and is competitive when under-sampling is done.
- Complex classifiers like RFs are relatively robust to moderate class imbalance. However, as we will see in real datasets, when class imbalance is extreme, gain from using a ranker-based policy becomes prominent. Moreover, growing big forests may be infeasible due to memory constraints.

5.2   Comparison of Stochastic Policies

The next domain that we consider is digital marketing, as described previously, using data from the websites of three large companies. We refer to these three data sets as DM1, DM2, and DM3.

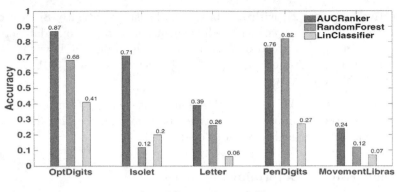

(a) Without under-sampling

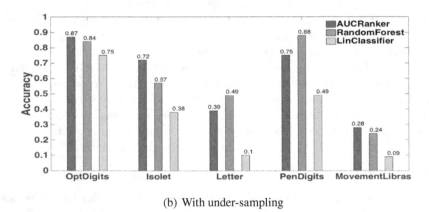

(b) With under-sampling

Fig. 1. Comparison of classification accuracy of various policies for different datasets.

Datasets: All the datasets were collected from campaigns run by major institutions. When a user visited the campaign website, she was either assigned to a targeted policy or a purely random policy. The targeted policy was some specific ad serving policy, particular to the campaign. The data was collected in the form $\{x, a, r_a\}$, where x denotes the user context, a denotes the offer displayed, and $r_a \in \{0, 1\}$ denotes the reward received. We trained our competing policies on data collected from the targeted policy and testing was done on the data collected from the random policy. We focused on the top-5 offers by number of impressions in the training set. Table 2 provides information about the training sets collected from the DM1 and DM2 campaigns. DM3's campaign has similar training set as DM2. *As can be clearly observed, each offer in the DM2's campaign suffers from extreme click sparsity.*

Feature Selection: We used a feature selection strategy to select around 20% of the users' features, as some of the features were of poor quality and led to difficulty in learning. We used the *information gain* criteria to select features [8].

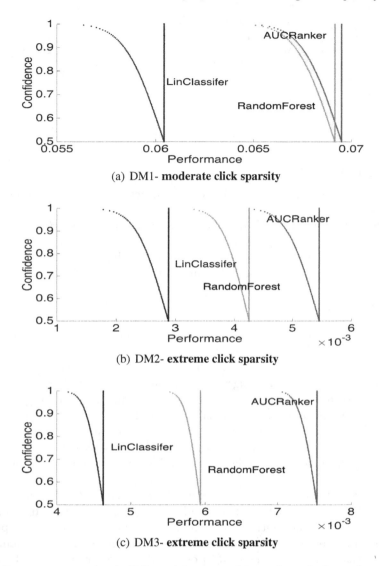

(a) DM1- **moderate click sparsity**

(b) DM2- **extreme click sparsity**

(c) DM3- **extreme click sparsity**

Fig. 2. Importance weighted CTR and lower confidence bounds for policies. Ranking based approach **gains 15–25% in importance weighted CTR** over RF based approach on data with **extreme click sparsity**. Classifiers were trained after **class balancing**.

Results: Figures 2(a), (b), and (c) show the results of our experiments. We used heavy under-sampling of negative examples, at the ratio 1:1 for positive:negative examples per offer, while training the classifiers. During evaluation, $\epsilon = 0.2$ was used as exploration parameter. Taking $\epsilon < 0.2$, meaning more exploitation, did not yield better results.

Table 2. Training set statistics

Domain	Offer	Impressions	Avg % positive (clicks/impressions)	Domain	Offer	Impressions	Avg % positive (clicks/impressions)
DM1	1	36164	8.1	DM2	1	37750	0.17
	2	37944	8.2		2	38254	0.40
	3	30871	7.8		3	182191	0.45
	4	32765	7.7		4	168789	0.30
	5	20719	5.5		5	17291	0.23

Observations:

- The ranker-based policy generally performed better than the classifier-based policies.
- For the DM2 campaign, *where the click sparsity problem is extremely severe, it can be stated with high confidence that the ranker-based policy performed significantly better than classifier based policies.* This shows that the ranker-based policy can handle class imbalance better than the classifiers.

6   Conclusion

In this paper, we analyzed the importance of recommendation policy space for a particular setting of PAR problem. In this setting, there is context on users but no context on ads. We discussed the drawback of classifier-based policies due to extreme click sparsity. We suggested learning a ranker per ad to overcome imbalance, and showed how to convert it into a recommendation policy. We conducted extensive experiments on classification datasets, as well as proprietary bandit datasets, to demonstrate the advantage of our suggested ranker-based policy over classifier-based policies. For the real datasets, we compared the competing policies using a high-confidence lower-bound on their importance weighted CTRs.

An exciting future direction of work involves development of online ranking algorithms for the PAR problem. Instead of considering context free ads, we would like to consider ads with context and learn a joint ranking model, updated in an online manner, after every interaction between user and PAR system. The intuition is that when the true CTR of every ad is very low, it might be easier to rank the ads according to possibility of being clicked, rather than predict the correct ad for a new user. The difficulty lies in the fact that the PAR system can display only one ad to the user, which makes it non-trivial to learn a ranker which would rank the ads at every round. We plan to combine techniques from contextual bandits, learning to rank and partial monitoring to develop scalable ranking algorithms with theoretical guarantees.

References

1. Agarwal, A., Hsu, D., Kale, S., Langford, J., Li, L., Schapire, R.: Taming the monster: a fast and simple algorithm for contextual bandits. In: Proceedings of the 31st International Conference on Machine Learning, pp. 1638–1646 (2014)

2. Agarwal, D., Gabrilovich, E., Hall, R., Josifovski, V., Khanna, R.: Translating relevance scores to probabilities for contextual advertising. In: Proceedings of the 18th ACM Conference on Information and Knowledge Management, pp. 1899–1902. ACM (2009)
3. Beygelzimer, A., Langford, J., Ravikumar, P.: Multiclass classification with filter trees. Preprint, 2 June 2007
4. Calders, T., Jaroszewicz, S.: Efficient AUC optimization for classification. In: Kok, J.N., Koronacki, J., Lopez de Mantaras, R., Matwin, S., Mladenič, D., Skowron, A. (eds.) PKDD 2007. LNCS (LNAI), vol. 4702, pp. 42–53. Springer, Heidelberg (2007). doi:10.1007/978-3-540-74976-9_8
5. Cao, P., Zhao, D., Zaiane, O.: An optimized cost-sensitive SVM for imbalanced data learning. In: Pei, J., Tseng, V.S., Cao, L., Motoda, H., Xu, G. (eds.) PAKDD 2013. LNCS, vol. 7819, pp. 280–292. Springer, Heidelberg (2013). doi:10.1007/978-3-642-37456-2_24
6. Chapelle, O., Li, L.: An empirical evaluation of thompson sampling. In: Advances in Neural Information Processing Systems, pp. 2249–2257 (2011)
7. Chawla, N., Japkowicz, N., Kotcz, A.: Editorial: special issue on learning from imbalanced data sets. ACM SIGKDD Explor. Newsl. 6(1), 1–6 (2004)
8. Cheng, T., Wang, Y., Bryant, S.: FSelector: a ruby gem for feature selection. Bioinformatics 28(21), 2851–2852 (2012)
9. Cortes, C., Mohri, M.: AUC optimization vs. error rate minimization. Adv. Neural Inf. Proces. Syst. 16(16), 313–320 (2004)
10. Dudik, M., Hsu, D., Kale, S., Karampatziakis, N., Langford, J., Reyzin, L., Zhang, T.: Efficient optimal learning for contextual bandits. In: Proceedings of the 27th Conference on Uncertainty in Artificial Intelligence (2011)
11. He, X., et al.: Practical lessons from predicting clicks on ads at facebook. In: Proceedings of 20th ACM SIGKDD Conference on Knowledge Discovery and Data Mining, pp. 1–9. ACM (2014)
12. Japkowicz, N., Stephen, S.: The class imbalance problem: a systematic study. Intell. Data Anal. 6(5), 429–449 (2002)
13. Koh, E., Gupta, N.: An empirical evaluation of ensemble decision trees to improve personalization on advertisement. In: Proceedings of KDD 14 Second Workshop on User Engagement Optimization (2014)
14. Langford, J., Li, L., Dudik, M.: Doubly robust policy evaluation and learning. In: Proceedings of the 28th International Conference on Machine Learning, pp. 1097–1104 (2011)
15. Langford, J., Zhang, T.: The epoch-greedy algorithm for multi-armed bandits with side information. In: Advances in Neural Information Processing Systems, pp. 817–824 (2008)
16. Li, L., Chu, W., Langford, J., Wang, X.: Unbiased offline evaluation of contextual-bandit-based news article recommendation algorithms. In: Proceedings of the fourth ACM International Conference on Web Search and Data Mining, pp. 297–306. ACM (2011)
17. McMahan, H., et al.: Ad click prediction: a view from the trenches. In: Proceedings of the 19th ACM SIGKDD International Conference on Knowledge Discovery and Data Mining, pp. 1222–1230. ACM (2013)
18. Richardson, M., Dominowska, E., Ragno, R.: Predicting clicks: estimating the click-through rate for new ads. In: Proceedings of the 16th International Conference on World Wide Web, pp. 521–530. ACM (2007)
19. Rifkin, R., Klautau, A.: In defense of one-vs-all classification. J. Mach. Learn. Res. 5, 101–141 (2004)

20. Shamir, O., Zhang, T.: Stochastic gradient descent for non-smooth optimization: convergence results and optimal averaging schemes. In: Proceedings of the 30th International Conference on Machine Learning, pp. 71–79 (2013)
21. Theocharous, G., Thomas, P., Ghavamzadeh, M.: Ad recommendation systems for life-time value optimization. In: Proceedings of the 24th International Conference on World Wide Web Companion, pp. 1305–1310 (2015)
22. Zhao, P., Jin, R., Yang, T., Hoi, S.C.: Online AUC maximization. In: Proceedings of the 28th International Conference on Machine Learning (ICML-11), pp. 233–240 (2011)

Global Flow and Temporal-Shape Descriptors for Human Action Recognition from 3D Reconstruction Data

Georgios Th. Papadopoulos$^{(\boxtimes)}$ and Petros Daras

Information Technologies Institute,
Centre for Research and Technology Hellas,
57001 Thermi, Thessaloniki, Greece
papad@iti.gr

Abstract. In this paper, global-level view-invariant descriptors for human action recognition using 3D reconstruction data are proposed. 3D reconstruction techniques are employed for addressing two of the most challenging issues related to human action recognition in the general case, namely view-variance and the presence of (self-) occlusions. Initially, a set of calibrated Kinect sensors are employed for producing a 3D reconstruction of the performing subjects. Subsequently, a 3D flow field is estimated for every captured frame. For performing action recognition, a novel global 3D flow descriptor is introduced, which achieves to efficiently encode the global motion characteristics in a compact way, while also incorporating spatial distribution related information. Additionally, a new global temporal-shape descriptor that extends the notion of 3D shape descriptions for action recognition, by including temporal information, is also proposed. The latter descriptor efficiently addresses the inherent problems of temporal alignment and compact representation, while also being robust in the presence of noise. Experimental results using public datasets demonstrate the efficiency of the proposed approach.

Keywords: Action recognition · 3D reconstruction · 3D flow · 3D shape

1 Introduction

Efficiently and accurately recognizing human actions has emerged as one of the most challenging and active areas of research in the computer vision field over the past decades [1,11,20]. This is mainly due to the very wide set of possible applications with great commercialization potentials that can benefit from the resulting accomplishments, such as surveillance, security, human computer interaction, smart houses, helping the elderly/disabled, gaming, e-learning, to name a few. For achieving robust recognition performance, the typical requirements for rotation, translation and scale invariance need to be incorporated. Additional significant challenges need also to be sufficiently addressed, like the differences

© Springer International Publishing AG 2017
P. Perner (Ed.): MLDM 2017, LNAI 10358, pp. 47–62, 2017.
DOI: 10.1007/978-3-319-62416-7_4

in the appearance of the subjects, the human silhouette features, the execution of the same actions, etc. Despite the fact that human action recognition constitutes the central point of focus for multiple research groups/projects and that numerous approaches have already been proposed, significant obstacles towards fully addressing the problem in the general case still remain.

Two of the most significant challenges in human action recognition in the general case (i.e. in unconstrained environments) that current state-of-art algorithms face are view-variance and the presence of (self-) occlusions. In order to simultaneously handle both challenges in a satisfactory way, 3D reconstruction information is used in this work. This choice is further dictated by the recent technological breakthrough, which has resulted in the introduction of portable, affordable, high-quality and accurate motion capturing devices to the market; these devices have already gained tremendous acceptance in several research and daily-life application fields.

In this paper, global-level view-invariant descriptors for human action recognition using 3D reconstruction data are proposed. 3D reconstruction techniques are employed in this work for addressing two of the most significant challenges in human action recognition in the general case, namely view-variance (i.e. when the same action is observed from different viewpoints) and the presence of (self-) occlusions (i.e. when for a given point of view a body-part of an individual conceals an other body-part of the same or an other subject). In the first step, a 3D reconstruction of the performing subjects is generated using a set of calibrated Kinect sensors. Subsequently, a 3D flow field is estimated for every captured frame. A novel global 3D flow descriptor is proposed for performing action recognition. Among the advantages of this descriptor is that it efficiently encodes the global motion characteristics in a compact way, while also incorporating spatial distribution related information. Additionally, a new global temporal-shape descriptor that extends the notion of 3D shape descriptions for action recognition, by including temporal information, is also introduced. The latter descriptor efficiently addresses the challenging problems of temporal alignment and compact representation, while also being robust in the presence of noise (as opposed to similar tracking-based methods of the literature). Experimental results as well as comparative evaluation using datasets from the Huawei/3DLife 3D human reconstruction and action recognition Grand Challenge demonstrate the efficiency of the proposed approach.

The remainder of the paper is organized as follows: Previous work is reviewed in Sect. 2. Section 3 describes the 3D information processing. The descriptor extraction procedure is detailed in Sect. 4. Section 5 outlines the adopted action recognition scheme. Experimental results are presented in Sect. 6 and conclusions are drawn in Sect. 7.

2   Previous Work on 3D Action Recognition

The recent introduction of accurate motion capturing devices, with the Microsoft Kinect being the most popular one, has given great boost in human action recognition tasks and has decisively contributed in shifting the research focus towards the analysis in 3D space. This is mainly due to the wealth of information present in the captured stream, where the estimated 3D depth maps facilitate in overcoming typical barriers (e.g. scale estimation, presence of occlusions, etc.) of traditional visual analysis on the 2D plane and hence significantly extending the recognition capabilities. The great majority of the methods that belong to this category typically exploit human skeleton-tracking or surface (normal vectors) information, which is readily available by applying widely-used open-source software (e.g. OpenNI API[1], Kinect SDK[2], etc.). In [24], a depth similarity feature is proposed for describing the local 3D cuboid around a point of interest with an adaptable supporting size. Cheng et al. [4] introduce a descriptor of depth information, which depicts the structural relations of spatio-temporal points within action volumes, making use of the distance information in the depth data. Additionally, Wang et al. [21] introduce the so-called semi-local random occupancy pattern (ROP) features, which employ a sampling scheme that explores a large sampling space. In [22], an actionlet ensemble model is learnt to represent each action and to capture the intra-class variance. Moreover, Xia et al. [25] utilize histograms of 3D joint locations (HOJ3D) as a compact representation of human postures. The spherical angles between selected joints, along with the respective angular velocities, are calculated in [15].

2.1 Flow Descriptors

Although numerous approaches to 3D action recognition have already been proposed, they mainly focus on exploiting human skeleton-tracking or surface (normal vectors) information. Therefore, more elaborate information sources, like 3D flow, have not received the same attention yet. The latter is mainly due to the increased computational complexity that inherently 3D flow estimation involves, since its processing includes an additional disparity estimation problem. However, methods that emphasize on reducing the required computational complexity, by adopting several optimization techniques (hardware, algorithmic, GPU implementation), have achieved processing rates up to 20 Hz [12,17]. Consequently, these recent advances have paved the way for introducing action recognition methods that make use of 3D flow information.

Regarding methods that utilize 3D flow information for recognizing human actions, Holte et al. [9] introduce a local 3D motion descriptor; specifically, an optical flow histogram (HOF3D) is estimated, taking into account the 4D spatio-temporal neighborhood of a point-of-interest. In [12], a 3D grid-based flow descriptor is presented, in the context of a real-time human action recognition system. Additionally, histograms of 3D optical flow are also used in [26], along with other descriptions (spatio-temporal interesting points, depth data, body posture). Gori et al. [8] build a frame-level 3D Histogram of Flow (3D-HOF), as part of an incremental method for 3D arm-hand behaviour modelling and recognition. In [16], a local-level 3D flow descriptor is introduced, which

---

[1] http://structure.io/openni.
[2] http://www.microsoft.com/en-us/kinectforwindows/.

among others incorporates spatial and surface information in the flow representation and efficiently handles the problem of defining 3D orientation at every local neighborhood. Furthermore, Fanello et al. [6] present an approach to simultaneous on-line video segmentation and recognition of actions, using histograms of 3D flow.

Although some works have recently been proposed for action recognition using 3D flow information, most of them rely on relatively simple local/global histogram- or grid-based representations. Therefore, significant challenges in 3D flow processing/representation still remain partially addressed or even unexplored, like incorporation of spatial information, view-invariance, introduction of a compact global representation, etc.

2.2 Shape Descriptors

Concerning the exploitation of 3D shape information for action recognition purposes, the overpowering majority of the literature methods refers to the temporal extension of the corresponding 2D spatial analysis (i.e. analysis in the $xy + t$ 3D space), which is typically initiated by e.g. concatenating the binary segmentation masks or outer contours of the examined object in subsequent frames. Consequently, analysis in the 'actual' xyz 3D space (or equivalently analysis in the $xyz + t$ 4D space, if the time dimension is taken into account) is currently avoided. In particular, Weinland et al. [23] introduce the so called Motion History Volumes (MHV), as a free-viewpoint representation for human actions, and use Fourier transforms in cylindrical coordinates around the vertical axis for efficiently performing alignment and comparison. In [7], human actions are regarded as three-dimensional shapes induced by the silhouettes in the space-time volume and properties of the solution to the Poisson equation are utilized to extract features, such as local space-time saliency, action dynamics, shape structure and orientation. Additionally, Efros et al. [5] present a motion descriptor based on optical flow measurements in a spatio-temporal volume for each stabilized human figure and an associated similarity measure.

Towards the goal of performing shape analysis for action recognition in the above-mentioned $xyz + t$ 4D space, Huang et al. [10] present time-filtered and shape-flow descriptors for assessing the similarity of 3D video sequences of people with unknown temporal correspondence. In [2], an approach to non-sequential alignment of unstructured mesh sequences that is based on a shape similarity tree is detailed, which allows alignment across multiple sequences of different motions, reduces drift in sequential alignment and is robust to rapid non-rigid motions. Additionally, Yamasaki et al. [27] present a similar motion search and retrieval system for 3D video based on a modified shape distribution algorithm. The problem of 3D shape representation, which is formulated using Extremal Human Curve (EHC) descriptors extracted from the body surface, and shape similarity in human video sequences is the focus of the work in [18].

Despite the fact that some works on temporal-shape descriptions have already been proposed, their main limitation is that they include in their analysis the problem of the temporal alignment of action sequences (typically using common techniques, like e.g. dynamic programming, Dynamic Time Warping, etc.). The latter often has devastating effects in the presence of noise or leads to cumulative errors in case of misalignment occurrences. To this end, a methodology that would alleviate from the burden of the inherent problem of temporal alignment when performing temporal-shape analysis, while maintaining a compact action representation, would be beneficial.

3   3D Information Processing

The 3D information processing step is initiated by the application of a 3D reconstruction algorithm, which makes use of a set of calibrated Kinect sensors. Output of this algorithm is a uniform voxel grid $VG_t = \{v_t(x_g, y_g, z_g) : x_g \in [1, X_g], y_g \in [1, Y_g], z_g \in [1, Z_g]\}$, where each voxel corresponds to a cuboid region in the real 3D space with edge length equal to $10\,\mathrm{mm}$ and t denotes the currently examined frame. It is considered that $v_t(x_g, y_g, z_g) = 1$ if $v_t(x_g, y_g, z_g)$ belongs to the subject's surface and $v_t(x_g, y_g, z_g) = 0$ otherwise.

For 3D flow estimation, an approach similar to the one described in [9] is followed, where pixel correspondences (obtained by the application of 2D flow estimation algorithms) are converted to voxel correspondences. Output of this procedure is the computation of a flow field $\bar{\mathbf{F}}_t^{3D}(x_g, y_g, z_g)$ for every voxel grid VG_t.

4   Descriptor Extraction

4.1   Global Flow Descriptor

For extracting a discriminative global 3D flow descriptor, the following challenges need to be addressed: (a) the difficulty in introducing a consistent orientation definition for different action instances, in order to produce comparable low-level descriptions, and (b) the incorporation of spatial distribution information in a compact way, while maintaining 3D rotation invariance.

The fundamental problem of orientation definition is addressed in this work by assuming a vertical direction consideration. The latter selection is justified by the fact that the angle of the principal axis of the 3D human silhouette with the vertical direction typically does not exhibit significant deviations among different instances of a given action. Subsequently, the descriptor extraction procedure is initiated by estimating a vertically aligned minimum bounding cylinder of all $v_t(x_g, y_g, z_g)$ for which a flow vector $\bar{\mathbf{F}}_t^{3D}(x_g, y_g, z_g)$ is estimated for all frames t that comprise the examined action. The center of the cylinder (i.e. the central point of its axis) is denoted $v_{cg}(x_{cg}, y_{cg}, z_{cg})$, while its radius is represented by ζ. Additionally, the upper and lower cylinder boundaries are denoted y_{max}^c and y_{min}^c, respectively. Then, a set of con-centric ring-shaped areas are defined, according to the following expressions:

$$B_{\kappa,\lambda} = \begin{cases} (\lambda - 1)\gamma \leq \xi \leq \lambda\gamma \\ y_{min}^c + (\kappa - 1)\delta \leq y_g \leq y_{min}^c + \kappa\delta \\ \xi = \sqrt{(x_g - x_{cg})^2 + (z_g - z_{cg})^2} \end{cases} \tag{1}$$

where $\kappa \in [1, K]$, $\lambda \in [1, \Lambda]$, $\gamma = \zeta/\Lambda$ and $\delta = (y_{max}^c - y_{min}^c)/K$. For describing the flow information in every $B_{\kappa,\lambda}$ region, a loose representation is required that will render the respective descriptor robust to differences in the appearance of the subjects and the presence of noise. To this end, a histogram-based representation is adopted. In particular, for every $B_{\kappa,\lambda}$ area, a 2D angle histogram is estimated, taking into account all flow vectors $\bar{\mathbf{F}}_t^{3D}(x_g, y_g, z_g)$ during the whole duration of the examined action that correspond to voxels $v_t(x_g, y_g, z_g)$ that lie in that spatial area. More specifically, for each of the aforementioned $\bar{\mathbf{F}}_t^{3D}(x_g, y_g, z_g)$, the following two angles are calculated:

$$\psi = \tan^{-1}(\frac{z_g - z_{cg}}{x_g - x_{cg}}) - \tan^{-1}(\frac{\bar{\mathbf{F}}_{z,t}^{3D}(x_g, y_g, z_g)}{\bar{\mathbf{F}}_{x,t}^{3D}(x_g, y_g, z_g)})$$

$$o = \cos^{-1}(\frac{\langle(0, 1, 0), \bar{\mathbf{F}}_t^{3D}(x_g, y_g, z_g)\rangle}{\|\bar{\mathbf{F}}_t^{3D}(x_g, y_g, z_g)\|}) \tag{2}$$

where $\bar{\mathbf{F}}_{x,t}^{3D}(x_g, y_g, z_g)$ and $\bar{\mathbf{F}}_{z,t}^{3D}(x_g, y_g, z_g)$ are the x- and z-component of the flow vector $\bar{\mathbf{F}}_t^{3D}(x_g, y_g, z_g)$, respectively. $\psi \in [-\pi, \pi]$ corresponds to the angle between the horizontal projection of $\bar{\mathbf{F}}_t^{3D}(x_g, y_g, z_g)$ and the projection of the vector connecting the cylindrical center (x_{cg}, y_{cg}, z_{cg}) with the examined voxel position (x_g, y_g, z_g) on the horizontal xz plane, while $o \in [0, \pi]$ corresponds to the angle of $\bar{\mathbf{F}}_t^{3D}(x_g, y_g, z_g)$ with the vertical axis. Then, the above-mentioned 2D histogram for area $B_{\kappa,\lambda}$ is computed by partitioning the value ranges of ψ and o into b_ψ and b_o equal-width non-overlapping bins, respectively. During the calculations, $\|\bar{\mathbf{F}}_t^{3D}(x_g, y_g, z_g)\|$ is aggregated to the appropriate histogram bin. The global flow descriptor is computed by concatenating the estimated angle histograms of all $B_{\kappa,\lambda}$ areas, while it is subsequently $L1$ normalized for rendering the descriptor robust to the difference in the speed with which every action is executed. From the definitions of the ring-shaped areas $B_{\kappa,\lambda}$ and angle ψ, it can be justified that the proposed global 3D flow descriptor satisfies the requirement for rotation invariance, while it also incorporates spatial distribution related information in the flow representation. In this work, the following parameter values were selected after experimentation: $\Lambda = 4$, $K = 4$, $b_\psi = 6$ and $b_o = 3$. An example of ring-shaped $B_{\kappa,\lambda}$ areas formation for a 'push away' action instance is given in Fig. 1.

4.2 Global Shape Descriptor

As described in Sect. 2.2, current temporal-shape techniques include in their analysis the problem of the temporal alignment of the action sequences, which has devastating effects in the presence of noise or leads to cumulative errors in case of misalignment occurrences. To this end, a temporal-shape descriptor

Fig. 1. Example of ring-shaped areas $B_{\kappa,\lambda}$ formation for $\kappa = 3$ and $\lambda \in [1,4]$ for a 'push-away' action instance.

that encodes the dominant shape variations and avoids the need for exact action sequence alignment, while maintaining a compact shape representation, is proposed in this section.

The biggest challenge in using the temporal dimension for realizing 3D shape-based action recognition is the temporal alignment of different action executions, which is often misleading and causes devastating aggregated errors. Additionally, this alignment is more likely to lead to mismatches if high-dimensional vector representations need to be used, which is the case of 3D shape-based analysis. For overcoming these obstacles, a frequency domain analysis is followed in this work for identifying and modeling the dominant shape characteristics and their variation in time. In this way, the temporal sequence of the action constituent postures is captured, although this is not a strict temporal alignment of the respective action frames. In particular, for every frame t that belongs to the examined action segment an individual global 3D shape descriptor $\mathbf{q}_t$ is extracted. More specifically, for every frame t a composite voxel grid VG_t^{co} is computed, by superimposing all VG_t from the beginning of the action segment until frame t and estimating their outer surface. $\mathbf{q}_t$ is then computed by estimating a 3D shape descriptor for VG_t^{co}. Using VG_t^{co}, instead of VG_t, for descriptor extraction was experimentally shown to lead to better temporal action dynamics encoding, as it will be demonstrated in the experimental evaluation. Indicative examples of VG_t^{co} estimation for different human actions are depicted in Fig. 2.

For producing a compact temporal-shape descriptor, the descriptor vector sequence $\mathbf{q}_t$ is initially adjusted to a predefined length H forming sequence $\overline{\mathbf{q}}_h$, using linear interpolation; the latter accounts for action sequences that typically consist of a different number of frames. $H = 20$ based on experimentation. Subsequently, 1D frequency domain analysis is applied to each of the value sequences $\overline{\mathbf{q}}_{s,h}$ that are formed by considering the s-th element of $\overline{\mathbf{q}}_h$ each time. For frequency domain analysis, the Discrete Cosine Transform (DCT) is applied to $\overline{\mathbf{q}}_{s,h}$, as follows:

$$fc_s(\beta) = \sum_{h=1}^{H} \overline{\mathbf{q}}_{s,h} \cos \frac{\pi}{H}[(h-1) + \frac{1}{2}(\beta-1)] \tag{3}$$

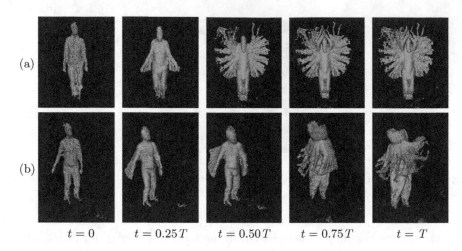

$t = 0$ $t = 0.25\,T$ $t = 0.50\,T$ $t = 0.75\,T$ $t = T$

Fig. 2. Indicative examples of composite voxel grid VG_t^{co} estimation for actions: (a) jumping-jacks and (b) tennis-forehand. T denotes the overall duration of the action.

where $fc_s(\beta)$ are the estimated DCT coefficients and $\beta \in [1, H]$. The reason for using the DCT transform is twofold: (a) its simple form requires relatively reduced calculations, and (b) it is a frequency domain transform that receives as input a real sequence and its output is also a real set of values. Other common frequency analysis methods (e.g. Fourier transform) were also evaluated; however, they did not lead to increased performance compared with the one received when using DCT. Out of the $H\ fc_s(\beta)$ coefficients, only the first P are considered, since the remaining ones were experimentally shown to correspond mainly to noise or did not add to the discriminative power of the formed descriptor. The P selected coefficients for each $\overline{\mathbf{q}}_{s,h}$ are concatenated in a single vector that constitutes the proposed global 3D temporal-shape descriptor. It must be noted that modeling the correlations between different $\overline{\mathbf{q}}_{s,h}$ sequences during the descriptor extraction procedure led to inferior recognition performance, mainly due to overfitting occurrences.

Although the proposed 3D temporal-shape descriptor extraction methodology is independent of the particular 3D static shape descriptor to be used, in this work the 'shape distribution' descriptor [14] (3D distance histogram) was utilized; this was experimentally shown to lead to better overall action recognition performance than other common shape descriptors. In [10], description and comparative evaluation of different static 3D shape descriptors for action recognition are given.

5   Action Recognition

After extracting the global 3D flow/shape descriptors for every examined human action (as detailed in Sect. 4), each descriptor is L1-normalized for incorporating invariance with respect to the execution speed of different instances of the same action from the same or different subjects. Action recognition is then realized using multi-class Support Vector Machines (SVMs).

6   Experimental Results

In this section, experimental results from the application of the proposed approach to the Huawei/3DLife[3] datasets for 3D human reconstruction and action recognition, which were used in the ACM Multimedia 2013 'Multimedia Grand Challenge' and are among the most comprehensive and broad ones in the literature, are presented. In particular, the first (dataset D_1) and the second (dataset D_2) sessions of the first dataset are used, which provide RGB-plus-depth video streams from five and two Kinect sensors, respectively. For dataset D_2, the data stream from only the frontal Kinect was utilized. D_1 and D_2 include captures of 17 and 14 human subjects, respectively, and each action is performed at least 5 times by every individual. Out of the available 22 supported actions, the following set of 17 dynamic ones were considered for the experimental evaluation: $E = \{e_g, \ g \in [1, G]\} \equiv \{$Hand waving, Knocking the door, Clapping, Throwing, Punching, Push away, Jumping jacks, Lunges, Squats, Punching and kicking, Weight lifting, Golf drive, Golf chip, Golf putt, Tennis forehand, Tennis backhand, Walking on the treadmill$\}$. The remaining 5 discarded actions (namely 'Arms folded', 'T-Pose', 'Hands on the hips', 'T-Pose with bent arms' and 'Forward arms raise') correspond to static ones that can be easily detected using a simple representation. Performance evaluation was realized following the 'leave-one-out' methodology, where in every iteration one subject was used for performance measurement and the remaining ones were used for training.

In Fig. 3, quantitative results in terms of the estimated recognition rates and overall accuracy are given for the proposed global flow and shape descriptors. From the presented results, it can be seen that both descriptors achieve high recognition rates in both datasets; namely, the flow (shape) descriptor exhibits recognition rates equal to 81.27% and 78.99% (76.53% and 69.83%) in D_1 and D_2, respectively. From these results, it can be observed that the global flow descriptor outperforms the respective shape one in both utilized datasets; this is mainly due to the more detailed and discriminative information contained in the estimated 3D flow fields. Due to the latter factor, the flow descriptor is advantageous for actions that incorporate more fine-grained body/body-part movements (e.g. 'Hand waving', 'Knocking the door', 'Punching and kicking' and 'Weight lifting'). On the other hand, the cases that the shape descriptor is better involve body movements with more extensive and distinctive whole body postures (e.g. actions 'Clapping' and 'Squats').

In order to investigate the behavior of the proposed global temporal-shape descriptor, comparison with the following benchmarks is performed: (a) global static shape descriptor: A static shape descriptor (the 'shape distribution' descriptor described in Sect. 4.2) is extracted for the composite voxel grid VG_t^{co}

---

[3] http://mmv.eecs.qmul.ac.uk/mmgc2013/.

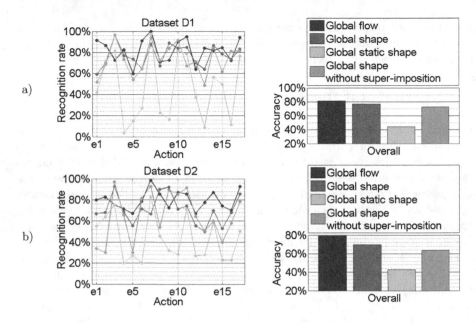

Fig. 3. Action recognition results for (a) D_1 and (b) D_2 datasets.

for $t = T$, i.e. when all constituent voxel grids VG_t of an action are superimposed. This can be considered as the counterpart of the respective volumetric descriptions for the 2D analysis case, i.e. methods that estimate a 3D volumetric shape of the examined action from the 2D video sequence and subsequently estimating a 3D shape descriptor of the generated volume (like in [7,23]). (b) variant of the proposed temporal-shape descriptor, where voxel grids VG_t are used instead of the composite ones VG_t^{co} during the descriptor extraction procedure. From the results presented in Fig. 3, it can be seen that the proposed temporal-shape descriptor significantly outperforms the static one in both datasets. This fact highlights the significant added value of incorporating temporal information in the global 3D representation. Additionally, it can be observed that the use of the composite voxel grids VG_t^{co} is advantageous compared with when using the voxel grids VG_t. The latter implies that superimposing information from multiple frames during the descriptor extraction procedure can lead to more discriminative shape representations.

6.1 Parameter Selection

In order to apply and evaluate the performance of the proposed descriptors, particular values inevitably need to be selected for the defined parameters. In this section, quantitative evaluation results are given for the most crucial parameters, aiming at shading light on the behavior of the respective descriptors. It must be noted that in the followings experimental results are given only for D_1, while similar behavior of the proposed descriptors has been observed in D_2.

Table 1. Global flow descriptor parameter selection

Parameters	Accuracy
$K = 3$, $\Lambda = 3$, $b_\psi = 6$, $b_o = 3$	75.44%
$K = 4$, $\Lambda = 4$, $b_\psi = 6$, $b_o = 3$	**81.27%**
$K = 5$, $\Lambda = 5$, $b_\psi = 6$, $b_o = 3$	80.88%
$K = 4$, $\Lambda = 4$, $b_\psi = 6$, $b_o = 3$	**81.27%**
$K = 4$, $\Lambda = 4$, $b_\psi = 4$, $b_o = 3$	79.97%
$K = 4$, $\Lambda = 4$, $b_\psi = 6$, $b_o = 3$	**81.27%**
$K = 4$, $\Lambda = 4$, $b_\psi = 6$, $b_o = 6$	77.22%

In particular, the descriptor behavior for different values of the following parameters, along with justification where particular values were selected, is investigated:

- Parameters K, Λ, b_ψ, b_o: K and Λ control the partitioning of the longitudinal and the polar axis, when defining the ring-shaped areas $B_{\kappa,\lambda}$ (Sect. 4.1), respectively. Additionally, b_ψ and b_o define the number of bins of the histograms calculated with respect to angles ψ and o (Sect. 4.1), respectively. In Table 1, action recognition results from the application of the proposed global flow descriptor for different sets of values of the aforementioned parameters are given. From the first group of experimental results, it can be seen that the ring-shape partitioning using $K = 4$ and $\Lambda = 4$ leads to the best overall performance. Additionally, the second group of experiments shows that using more bins in the histogram representation with respect to angle ψ, which corresponds to the angle between the horizontal projection of $\bar{\mathbf{F}}_t^{3D}(x_g, y_g, z_g)$ and the projection of the vector connecting the cylindrical center (x_{cg}, y_{cg}, z_{cg}) with the examined voxel position (x_g, y_g, z_g) on the horizontal xz plane, is advantageous. On the other hand, using a decreased number of bins in the histogram representation with respect to angle o, which corresponds to the angle of $\bar{\mathbf{F}}_t^{3D}(x_g, y_g, z_g)$ with the vertical axis, leads to increased performance (third group of experiments).
- Parameter H: This adjusts the length of the shape descriptor vector sequence $\bar{\mathbf{q}}_h$ (Sect. 4.2). In the current implementation, H was set equal to 20, which is close to the average action segment duration in frames in the employed datasets.
- Parameter P: This defines the number of selected DCT coefficients to be used in the produced global shape representation (Sect. 4.2). The performance obtained by the application of the proposed temporal-shape descriptor for different values of P is given for both datasets in Table 2. From the presented results, it can be seen that the best performance is achieved when only relatively few frequency coefficients are used; these are shown to be adequate for accomplishing a good balance between capturing sufficient temporal information and maintaining the dimensionality of the overall descriptor low.

Table 2. Temporal-shape descriptor parameter selection

Dataset	Parameter P			
	5	10	15	20
D_1	**76.53%**	71.68%	68.11%	66.64%
D_2	**69.83%**	66.12%	61.64%	57.82%

6.2 Comparative Evaluation

Comparative evaluation results of the proposed descriptors (and their combination) with similar literature approaches are reported in this section. In particular, in Fig. 4, quantitative results in terms of the estimated recognition rates and overall accuracy are given for the following cases: (i) The HOF3D descriptor with 'vertical rotation' [9], which is a local histogram-based 3D flow descriptor that does not incorporate spatial information. (ii) The local 3D flow descriptor of [16], which is again a local histogram-based 3D flow descriptor; however, it incorporates spatial and surface information in the flow representation. (iii) The proposed global flow descriptor. (iv) The LC-LSF local 3D shape descriptor of [13], which employs a set of local statistical features for describing non-rigid 3D models. (v) The global 3D temporal-shape descriptor of [10], where a self-similarity matrix is computed for every action (by means of static shape descriptor extraction for every frame) and subsequently a temporal-shape descriptor is estimated by applying a time filter to the calculated matrix. (vi) The proposed global shape descriptor. (vii) The overall proposed approach, which combines the proposed global flow and shape descriptors, by means of simple concatenation in a single feature vector. (viii) The skeleton-tracking-based methods of [3,15,19], which estimate human posture representations at every frame, making use of the detected human joints (for the methods of [3,19], only the reported overall classification accuracy in D_2 is provided in Fig. 4).

From the presented results, it can be seen that the proposed global flow descriptor performs significantly better than the local flow ones of [9,16]. This is mainly due to the inefficiency of the local descriptors in fully addressing the problem of defining a consistent orientation, during the analysis in local 3D neighborhoods for descriptor extraction. The aforementioned difference in performance is more pronounced and clear in D_1, i.e. the most challenging dataset due to the relatively increased presence of noise in the provided depth maps, which is in turn mainly due the increased interference among the higher number of Kinect sensors used in D_1. Similar observations can be made for the case of the proposed global shape descriptor, which performs equally or significantly better than the local 3D shape descriptor of [13] for the same reasons described above. It worths noticing that the proposed global shape descriptor outperforms the local flow ones by a large margin, despite the fact that 3D flow is a more discriminative information source. Additionally, the proposed temporal-shape descriptor is also shown to outperform the temporal-shape method

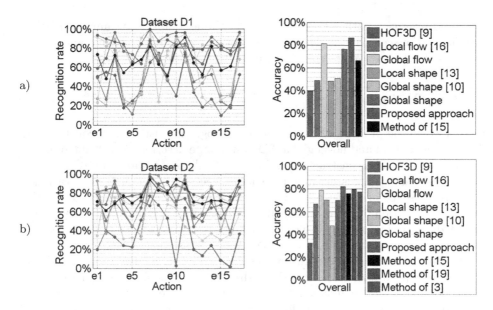

Fig. 4. Comparative evaluation results for (a) D_1 and (b) D_2 datasets.

of [10]. This denotes the increased efficiency of the frequency domain analysis on top of the per-frame extracted shape descriptors in capturing and modeling the human action dynamics, compared with the case of estimating the self-similarity matrix of the same descriptors and applying time filtering techniques. Moreover, the overall proposed approach (which consists of simple concatenation of the proposed global flow and shape descriptors in a single feature vector) achieves increased performance, compared with the cases of using each individual descriptor alone. The latter demonstrates the complementarity of the introduced descriptors and dictates that a robust human action recognition framework should incorporate multiple information sources for accomplishing increased recognition performance. Furthermore, it is shown that the proposed approach outperforms the skeleton-tracking-based methods of [3,15,19]. This difference in performance demonstrates the superiority of the proposed method in capturing the action dynamics more efficiently, mainly due to the combination of multiple and complementary information sources (i.e. flow/shape information). On the other hand, the methods of [3,15,19] suffer from the limitations of the employed skeleton tracker, which is reported not to perform well in case of fast movements and in the presence of noise in the captured depth maps. Another interesting observation is that the performance difference between the proposed approach and the method of [15] is higher in D_1 (i.e. the dataset with increased presence of noise) than in D_2; this again highlights the robustness in the presence of noise of the proposed method, while the method of [15] is significantly affected by the limitations of the employed skeleton tracker (as described above). It needs to be highlighted that an additional advantageous characteristic of the proposed approach is that it does not make the assumption of human(s) being present in the scene, i.e. it does not use domain-specific knowledge and it can be applied with any other type of object being captured.

7   Conclusions

In this work, the problem of human action recognition using 3D reconstruction data was examined and novel global 3D flow/shape descriptors were introduced. Exploitation of 3D reconstruction techniques facilitates towards addressing two of the most challenging issues in human action recognition, namely view-variance and the presence of (self-) occlusions. This choice is further endorsed by the recent introduction of low-cost, portable, high-quality and accurate motion capturing devices. The proposed global 3D flow descriptor efficiently encodes the global motion characteristics in a compact way. It was observed that this descriptor led to the best action recognition results. Additionally, the proposed global temporal-shape descriptor efficiently addresses the inherent problems of temporal alignment and compact representation. The proposed descriptors were experimentally shown to outperform similar methods of the literature, using publicly available datasets for the evaluation. Moreover, the comparative evaluation of the overall proposed approach (concatenation of the introduced global 3D flow/shape descriptors) was shown to outperform action recognition methods that rely on human skeleton-tracking methodologies.

Acknowledgment. The work presented in this paper was supported by the European Commission under contract H2020-700367 DANTE.

References

1. Borges, P.V.K., Conci, N., Cavallaro, A.: Video-based human behavior understanding: a survey. IEEE Trans. Circuits Syst. Video Technol. **23**(11), 1993–2008 (2013)
2. Budd, C., Huang, P., Klaudiny, M., Hilton, A.: Global non-rigid alignment of surface sequences. Int. J. Comput. Vis. **102**(1–3), 256–270 (2013)
3. Cai, X., Zhou, W., Wu, L., Luo, J., Li, H.: Effective active skeleton representation for low latency human action recognition. IEEE Trans. Multimedia **18**(2), 141–154 (2016)
4. Cheng, Z., Qin, L., Ye, Y., Huang, Q., Tian, Q.: Human daily action analysis with multi-view and color-depth data. In: Fusiello, A., Murino, V., Cucchiara, R. (eds.) ECCV 2012. LNCS, vol. 7584, pp. 52–61. Springer, Heidelberg (2012). doi:10.1007/978-3-642-33868-7_6
5. Efros, A.A., Berg, A.C., Mori, G., Malik, J.: Recognizing action at a distance. In: Proceedings of the Ninth IEEE International Conference on Computer Vision, pp. 726–733. IEEE (2003)
6. Fanello, S.R., Gori, I., Metta, G., Odone, F.: Keep it simple and sparse: real-time action recognition. J. Mach. Learn. Res. **14**(1), 2617–2640 (2013)
7. Gorelick, L., Blank, M., Shechtman, E., Irani, M., Basri, R.: Actions as space-time shapes. IEEE Trans. Pattern Anal. Mach. Intell. **29**(12), 2247–2253 (2007)

8. Gori, I., Fanello, S.R., Odone, F., Metta, G.: A compositional approach for 3D arm-hand action recognition. In: 2013 IEEE Workshop on Robot Vision (WORV), pp. 126–131. IEEE (2013)
9. Holte, M.B., Chakraborty, B., Gonzalez, J., Moeslund, T.B.: A local 3-D motion descriptor for multi-view human action recognition from 4-D spatio-temporal inter-est points. IEEE J. Sel. Top. Sig. Process. **6**(5), 553–565 (2012)
10. Huang, P., Hilton, A., Starck, J.: Shape similarity for 3D video sequences of people. Int. J. Comput. Vis. **89**(2–3), 362–381 (2010)
11. Ji, X., Liu, H.: Advances in view-invariant human motion analysis: a review. IEEE Trans. Syst. Man. Cybern. Part C Appl. Rev. **40**(1), 13–24 (2010)
12. Munaro, M., Ballin, G., Michieletto, S., Menegatti, E.: 3D flow estimation for human action recognition from colored point clouds. Biologically Inspired Cogn. Architectures **5**, 42–51 (2013)
13. Ohkita, Y., Ohishi, Y., Furuya, T., Ohbuchi, R.: Non-rigid 3D model retrieval using set of local statistical features. In: 2012 IEEE International Conference on Multimedia and Expo Workshops (ICMEW), pp. 593–598. IEEE (2012)
14. Osada, R., Funkhouser, T., Chazelle, B., Dobkin, D.: Shape distributions. ACM Trans. Graph. (TOG) **21**(4), 807–832 (2002)
15. Papadopoulos, G.T., Axenopoulos, A., Daras, P.: Real-time skeleton-tracking-based human action recognition using kinect data. In: International Conference on MultiMedia Modeling, pp. 473–483 (2014)
16. Papadopoulos, G.T., Daras, P.: Local descriptions for human action recognition from 3d reconstruction data. In: IEEE International Conference on Image Process-ing (ICIP 2014), pp. 2814–2818, November 2014
17. Sizintsev, M., Wildes, R.P.: Spatiotemporal stereo and scene flow via stequel matching. IEEE Trans. Pattern Anal. Mach. Intell. **34**(6), 1206–1219 (2012)
18. Slama, R., Wannous, H., Daoudi, M.: 3D human motion analysis framework for shape similarity and retrieval. Image Vis. Comput. **32**(2), 131–154 (2014)
19. Sun, L., Aizawa, K.: Action recognition using invariant features under unexampled viewing conditions. In: Proceedings of the 21st ACM International Conference on Multimedia, pp. 389–392. ACM (2013)
20. Turaga, P., Chellappa, R., Subrahmanian, V.S., Udrea, O.: Machine recognition of human activities: a survey. IEEE Trans. Circuits Syst. Video Technol. **18**(11), 1473–1488 (2008)
21. Wang, J., Liu, Z., Chorowski, J., Chen, Z., Wu, Y.: Robust 3D action recognition with random occupancy patterns. In: Fitzgibbon, A., Lazebnik, S., Perona, P., Sato, Y., Schmid, C. (eds.) ECCV 2012. LNCS, pp. 872–885. Springer, Heidelberg (2012). doi:10.1007/978-3-642-33709-3_62
22. Wang, J., Liu, Z., Wu, Y., Yuan, J.: Mining actionlet ensemble for action recog-nition with depth cameras. In: 2012 IEEE Conference on Computer Vision and Pattern Recognition (CVPR), pp. 1290–1297. IEEE (2012)
23. Weinland, D., Ronfard, R., Boyer, E.: Free viewpoint action recognition using motion history volumes. Comput. Vis. Image Underst. **104**(2), 249–257 (2006)
24. Xia, L., Aggarwal, J.: Spatio-temporal depth cuboid similarity feature for activity recognition using depth camera. In: 2013 IEEE Conference on Computer Vision and Pattern Recognition (CVPR), pp. 2834–2841. IEEE (2013)
25. Xia, L., Chen, C.-C., Aggarwal, J.: View invariant human action recognition using histograms of 3D joints. In: 2012 IEEE Computer Society Conference on Computer Vision and Pattern Recognition Workshops (CVPRW), pp. 20–27. IEEE (2012)

26. Xia, L., Gori, I., Aggarwal, J., Ryoo, M.: Robot-centric activity recognition from first-person rgb-d videos (2015)
27. Yamasaki, T., Aizawa, K.: Motion segmentation and retrieval for 3D video based on modified shape distribution. EURASIP J. Appl. Sig. Process. **2007**(1), 211–211 (2007)

Reverse Engineering Gene Regulatory Networks Using Sampling and Boosting Techniques

Turki Turki[1,2](✉) and Jason T.L. Wang[2](✉)

[1] Department of Computer Science, King Abdulaziz University,
P.O. Box 80221, Jeddah 21589, Saudi Arabia
tturki@kau.edu.sa
[2] Bioinformatics Program and Department of Computer Science, New Jersey
Institute of Technology, University Heights, Newark, NJ 07102, USA
{ttt2,wangj}@njit.edu

Abstract. Reverse engineering gene regulatory networks (GRNs), also known as network inference, refers to the process of reconstructing GRNs from gene expression data. Biologists model a GRN as a directed graph in which nodes represent genes and links show regulatory relationships between the genes. By predicting the links to infer a GRN, biologists can gain a better understanding of regulatory circuits and functional elements in cells. Existing supervised GRN inference methods work by building a feature-based classifier from gene expression data and using the classifier to predict the links in GRNs. Observing that GRNs are sparse graphs with few links between nodes, we propose here to use under-sampling, over-sampling and boosting techniques to enhance the prediction performance. Experimental results on different datasets demonstrate the good performance of the proposed approach and its superiority over the existing methods.

Keywords: Graph mining · Sampling methods · Boosting techniques · Supervised learning · Applications in biology and medicine

1 Introduction

Gene regulation is a series of processes that control gene expression and its extent. The connections among genes and their regulatory molecules, usually transcription factors, and a descriptive model of such connections, are known as gene regulatory networks (GRNs). Elucidating GRNs is crucial to understand the inner workings of the cell and the interactions among genes. Furthermore, GRNs could be the basis to infer more complex networks, encompassing gene, protein, and metabolic spaces, as well as the entangled and often over-looked signaling pathways that interconnect them [2,3,12,13,16].

Existing GRN inference methods can be broadly categorized into two groups: unsupervised and supervised [23]. Unsupervised methods infer GRNs based solely on gene expression data. The accuracy of these methods is usually low. By contrast, supervised methods use machine learning algorithms and training data

© Springer International Publishing AG 2017
P. Perner (Ed.): MLDM 2017, LNAI 10358, pp. 63–77, 2017.
DOI: 10.1007/978-3-319-62416-7_5

to achieve higher accuracy. These methods work as follows. We represent a GRN by a directed graph in which each node is a gene or transcription factor, and a directed link or edge from node A to node B indicates that gene A regulates the expression of gene B. The training data contains a partially known network with known present edges and absent edges between nodes. These known present edges are positive training examples, and the known absent edges are negative training examples. We train a machine learning algorithm using the training data and apply the trained model to predict the remaining unknown edges in the network. With the predicted present and absent edges, we are able to infer or construct a complete GRN.

GRNs are always sparse graphs. The ratio between the number of gene interactions (i.e., edges or links) and the number of genes (i.e., nodes) falls between 1.5 and 2.75 regardless of the differences in phylogeny, phenotypic complexity, life history, and the total number of genes in an organism [14]. Thus, all GRNs have relatively few present edges and a lot of absent edges. This means there are few positive examples and a lot of negative examples when modeling the GRN inference problem as the link prediction problem described above. This poses an imbalanced classification problem in which the positive class (i.e., the minority class) is much smaller than the negative class (i.e., the majority class). However, existing supervised GRN inference methods [9,17] do not take into consideration the imbalanced datasets, and hence their performance is unsatisfactory.

In this paper, we present a new approach to supervised GRN inference. We tackle the imbalanced classification problem by using sampling techniques, including under-sampling and over-sampling, to obtain a balanced training set. This balanced training set, containing the same number of positive and negative training examples, is used to train a machine learning algorithm to make predictions. Furthermore, we develop several boosting techniques to enhance the prediction performance. As our experimental results show later, this new approach outperforms the existing supervised GRN inference methods [9,17].

The rest of this paper is organized as follows. Section 2 transforms the GRN inference problem to a link prediction problem in which we infer a GRN by predicting the links in the GRN. We then present several sampling and boosting techniques for enhancing the link prediction performance. Section 3 reports experimental results, comparing our approach with the existing methods [9,17]. Section 4 concludes the paper and points out some directions for future research.

2   Methods

2.1   Problem Statement

We are given n genes where each gene has p expression values. The gene expression profile of these n genes is denoted by $G \subseteq R^{n \times p}$, which contains n rows, each row corresponding to a gene, and p columns, each column corresponding to an expression value [9,17,18,20,22]. In addition, we are given known regulatory relationships or links among some genes. Suppose these known regulatory relationships are stored in a matrix $X \subseteq R^{m \times 3}$, which forms the training dataset.

X contains m rows, where each row shows a known regulatory relationship between two genes, and three columns. The first column shows a transcription factor (TF). The second column shows a target gene. The third column shows the label, which is $+1$ if the TF is known to regulate the expression of the target gene or -1 if the TF is known not to regulate the expression of the target gene. The matrix X represents a partially observed or known gene regulatory network for the n genes. If the label of a row in X is $+1$, then the TF in that row regulates the expression of the target gene in that row, and hence that row represents a directed link or edge of the network. That row is a positive training example. If the label of a row in X is -1, then there is no link between the corresponding TF and target gene in that row. That row is a negative training example. The positive and negative training examples in X are used to train a machine learning or classification algorithm. There are much more negative training examples than positive training examples in X.

The test dataset contains ordered pairs of genes (g_1, g_2) where the regulatory relationship between g_1 and g_2 is unknown. Given a test example, i.e., an ordered pair of genes (g_1, g_2) in the test dataset, the goal of *link prediction* is to use the trained classifier to predict the label of the test example. The predicted label is either $+1$ (i.e., a directed link is predicted to be present from g_1 to g_2) or -1 (i.e., a directed link is predicted to be absent from g_1 to g_2). Here, the present link means g_1 (a transcription factor) regulates the expression of g_2 (a target gene) whereas the absent link means g_1 does not regulate the expression of g_2.

2.2 Feature Vector Construction

To perform training and prediction, we construct a feature matrix $D \subseteq R^{q \times 2p}$ with q feature vectors based on the gene expression profile G. Let g_1 and g_2 be two genes. Let $g_1^1, g_1^2, ..., g_1^p$ be the gene expression values of g_1 and $g_2^1, g_2^2, ..., g_2^p$ be the gene expression values of g_2. The feature vector of the ordered pair of genes (g_1, g_2), denoted D_d, is stored in the feature matrix D and constructed by concatenating their gene expression values as follows:

$$D_d = (g_1^1, g_1^2, ..., g_1^p, g_2^1, g_2^2, ..., g_2^p) \qquad (1)$$

Thus, the ordered pair of genes (g_1, g_2) corresponds to a point in $2p$-dimensional space. Each training and test example is represented by a $2p$-dimensional feature vector. For a positive training example, the label of its feature vector is $+1$. For a negative training example, the label of its feature vector is -1. For a test example, the label of its feature vector is unknown and to be predicted. This feature vector construction method has been widely used by existing supervised GRN inference methods [9,17,18,20,22].

2.3 Under-Sampling

Given is a training dataset X that is the union of two disjoint subsets X_+ and X_-. X_+ is the minority class, containing positive training examples (i.e., known present links). X_- is the majority class, containing negative training examples (i.e., known absent links). X_+ is much smaller than X_-. Our under-sampling method works as follows [10,11,15]. It samples a random subset $X^s_- \subseteq X_-$ such that the size of X^s_- is equal to the size of X_+ (i.e., $|X^s_-| = |X_+|$). Thus, $X^s = X_+ \cup X^s_-$ forms a balanced dataset. We then use X^s to train a machine learning algorithm. The trained model will be used to predict the labels of test examples.

2.4   Over-Sampling

Our over-sampling method is based on SMOTE [4,6,10]. Given is the training dataset $X = X_+ \cup X_-$ as described above. Our over-sampling method creates a new dataset X_{++} that contains all examples in X_+ and many synthetic examples generated as follows. For each example $x_i \in X_+$, we select h-nearest neighbors of x_i, where

$$h = \left\lfloor \frac{1.1 \times |X_-|}{|X_+|} \right\rfloor \tag{2}$$

Here, Euclidean distances are calculated to find the h-nearest neighbors. Denote these h-nearest neighbors as x_r, $1 \leq r \leq h$.

A new synthetic example x_{new} along the line between x_i and x_r, $1 \leq r \leq h$, is created as follows:

$$x_{new} = x_i + (x_r - x_i) \times \delta \tag{3}$$

where $\delta \in (0, 1)$ is a random number. We add x_{new} to X_{++}. We continue generating and adding such synthetic examples to X_{++} until X_{++} is larger than X_-. Then a random subset $X^s_{++} \subseteq X_{++}$ is selected such that the size of X^s_{++} is equal to the size of X_- (i.e., $|X^s_{++}| = |X_-|$). Thus, $X^s = X^s_{++} \cup X_-$ forms a balanced dataset. We then use X^s to train a machine learning algorithm. The trained model will be used to predict the labels of test examples.

2.5   Boosting

We further improve the performance of our link prediction algorithms through boosting. Boosting algorithms such as AdaBoost [8,21,25,26], described below, have been used in various domains with great success. We use a weighted decision tree [1] as the base learning algorithm and create a strong classifier through an iterative procedure as follows. Let X be the set of training examples $\{x_1, x_2, ..., x_m\}$. The label associated with example x_i is y_i such that

$$y_i = \begin{cases} +1 \text{ if } x_i \text{ is a positive example (i.e., present link)} \\ -1 \text{ if } x_i \text{ is a negative example (i.e., absent link)} \end{cases} \tag{4}$$

Initially, in iteration 1, each example is assigned an equal weight, i.e., $W_1(x_i) = \frac{1}{m}$, $1 \leq i \leq m$. In iteration k, $1 \leq k \leq K$, AdaBoost generates a base learner (i.e., model) H_k by calling the base learning algorithm on the training set X with weights W_k. Then H_k is used to classify each training example x_i as either

+1 (i.e., x_i is a predicted present link) or -1 (i.e., x_i is a predicted absent link). That is,

$$H_k(x_i) = \begin{cases} +1 \text{ if } H_k \text{ classifies } x_i \text{ as a positive example (i.e., present link)} \\ -1 \text{ if } H_k \text{ classifies } x_i \text{ as a negative example (i.e., absent link)} \end{cases} \quad (5)$$

Let $E_k = \{x_i | H_k(x_i) \neq y_i\}$. The error ε_k of H_k is:

$$\varepsilon_k = \sum_{x_i \in E_k} W_k(x_i) \quad (6)$$

The weight α_k of H_k is:

$$\alpha_k = \frac{1}{2} \ln\left(\frac{1 - \varepsilon_k}{\varepsilon_k}\right) \quad (7)$$

AdaBoost then updates the weight of each training example x_i, $1 \leq i \leq m$, as follows:

$$W_{k+1}(x_i) = \begin{cases} \frac{W_k(x_i)}{Z_k} \times e^{-\alpha_k} \text{ if } H_k(x_i) = y_i \\ \frac{W_k(x_i)}{Z_k} \times e^{\alpha_k} \quad \text{if } H_k(x_i) \neq y_i \end{cases}$$
$$= \frac{W_k(x_i) \exp(-\alpha_k y_i H_k(x_i))}{Z_k} \quad (8)$$

where Z_k is a normalization factor chosen so that W_{k+1} is normally distributed. The weights of incorrectly classified examples will increase in iteration $k + 1$. Then, in iteration $k + 1$, AdaBoost generates a base learner H_{k+1} by calling the base learning algorithm again on the training set X with weights W_{k+1}. Such a process is repeated K times. Using this technique, each weak classifier H_{k+1} should have greater accuracy than its predecessor H_k. The final, strong classifier H is derived by combining the votes of the weighted weak classifiers H_k, $1 \leq k \leq K$, where the weight α_k of a weak classifier H_k is calculated as shown in Eq. (7).

Specifically, given an unlabeled test example $\hat{x}$, $H(\hat{x})$ is calculated as follows:

$$H(\hat{x}) = sign\left(\sum_{k=1}^{K} \alpha_k H_k(\hat{x})\right) \quad (9)$$

The *sign* function indicates that if the sum of the results of the weighted K weak classifiers is greater than or equal to zero, then H classifies $\hat{x}$ as $+1$ (i.e., $\hat{x}$ is a predicted present link); otherwise H classifies $\hat{x}$ as -1 (i.e., $\hat{x}$ is a predicted absent link).

We propose to extend AdaBoost by modifying Eq. (8) to obtain the following variants:
Boost I:

$$W_{k+1}(x_i) = \begin{cases} \frac{e^{-\alpha_k}}{Z_k} \text{ if } H_k(x_i) = y_i \\ \frac{e^{\alpha_k}}{Z_k} \quad \text{if } H_k(x_i) \neq y_i \end{cases}$$
$$= \frac{\exp(-\alpha_k y_i H_k(x_i))}{Z_k} \quad (10)$$

Boost II:

$$W_{k+1}(x_i) = \frac{\exp\left(-\left(\sum_{j=1}^{k} \alpha_j H_j(x_i)\right) y_i\right)}{Z_k} \tag{11}$$

Boost III:

$$W_{k+1}(x_i) = \begin{cases} \frac{C \times e^{\alpha_k}}{Z_k} & \text{if } H_k(x_i) = -1 \text{ and } y_i = +1 \\ \frac{e^{-\alpha_k}}{Z_k} & \text{if } H_k(x_i) = +1 \text{ and } y_i = +1 \\ \frac{e^{\alpha_k}}{Z_k} & \text{if } H_k(x_i) = +1 \text{ and } y_i = -1 \\ \frac{e^{-\alpha_k}}{Z_k} & \text{if } H_k(x_i) = -1 \text{ and } y_i = -1 \end{cases} \tag{12}$$

where C is the number of examples in the majority class (i.e., negative class) divided by the number of examples in the minority class (i.e., positive class) in the training set.

Each one of the above variants is taken as a new boosting technique. Boost I is a simplified version of AdaBoost. For each training example x_i, $1 \le i \le m$, Boost I does not consider $W_k(x_i)$ when calculating $W_{k+1}(x_i)$. Boost II is an accumulative version of AdaBoost. It considers all weak classifiers H_j, $1 \le j \le k$, obtained in the previous k iterations when calculating the weight $W_{k+1}(x_i)$. Specifically, Boost II will increase the weight of a training example x_i in iteration $k+1$ if the majority of the weak classifiers obtained in the previous k iterations incorrectly classify x_i. Boost III can be regarded as a cost-sensitive boosting technique. For an imbalanced dataset, positive examples (i.e., those with labels of $+1$) are much fewer than negative examples (i.e., those with labels of -1). Our objective here is to improve the classification performance on the minority (i.e., positive) class. Hence we introduce the cost C, giving more weights to misclassified examples in the minority class where the examples are classified as negative though they have labels of $+1$. For a training example x_i that is correctly classified in iteration k, we decrease its weight in iteration $k+1$ so that the next classifier H_{k+1} pays less attention to x_i while focusing more on the other examples that are incorrectly classified in iteration k.

2.6   The Proposed Approach

Figure 1 presents an overview of our approach. In (A), we are given a training set containing imbalanced labeled links. These labeled links include few positive examples (i.e., known present links with labels of $+1$) and a lot of negative examples (i.e., known absent links with labels of -1). In addition, we are given a test set in which each test example is an unlabeled ordered gene pair. We construct feature vectors for both training examples and test examples as described in Sect. 2.2. In (B), we apply a sampling technique, either under-sampling as described in Sect. 2.3 or over-sampling as described in Sect. 2.4, to the training set to obtain a balanced training set. In (C), we apply a boosting technique as described in Sect. 2.5 to the balanced training set to learn K models (weak classifiers). These models predict the labels of the test examples. In (D), we take the weighted majority vote from the weak classifiers as shown in Equation (9) to make final predictions of the labels of the test examples. A test example is a predicted present link if its predicted label is +1; a test example is a predicted absent link if its predicted label is −1.

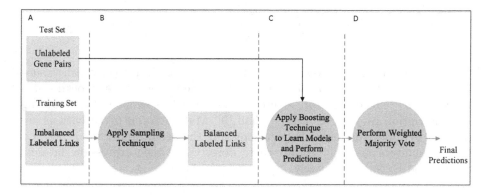

Fig. 1. The proposed approach for link prediction in gene regulatory networks.

3   Experiments and Results

3.1   Datasets

We used GeneNetWeaver [19] to generate the datasets related to yeast and E. coli. GeneNetWeaver has been widely used to generate benchmark datasets for evaluating GRN inference tools. Specifically we built four different networks for each organism where the networks contained 50, 100, 150, 200 genes (or nodes) respectively. Table 1 presents details of the yeast networks, showing the number of nodes (edges, respectively) in each network. The present edges or links in a network form positive examples. The absent edges or links in a network form negative examples. Table 2 presents details of the E. coli networks.

Table 1. Yeast networks used in the experiments.

Network	Directed	#Nodes	#Edges	#Positive examples	#Negative examples
Yeast 50	Yes	50	63	63	2387
Yeast 100	Yes	100	281	281	9619
Yeast 150	Yes	150	333	333	22017
Yeast 200	Yes	200	517	517	39283

For each network, we generated three files of gene expression data. These files were labeled as knockouts, knockdowns and multifactorial, respectively.

Table 2. E. coli networks used in the experiments.

Network	Directed	#Nodes	#Edges	#Positive examples	#Negative examples
E. coli 50	Yes	50	68	68	2382
E. coli 100	Yes	100	177	177	9723
E. coli 150	Yes	150	270	270	22080
E. coli 200	Yes	200	415	415	39385

A knockout is a technique to deactivate the expression of a gene, which is simulated by setting the transcription rate of this gene to zero [9]. A knockdown is a technique to reduce the expression of a gene, which is simulated by reducing the transcription rate of this gene by half [9]. Multifactorial perturbations are simulated by randomly increasing or decreasing the activation of the genes in a network simultaneously [9]. Totally there were twelve gene expression datasets for yeast and E. coli respectively.

3.2 Experimental Methodology

We compared our proposed approach with existing supervised GRN inference methods [9,17]. The existing methods employ support vector machines (SVM) and use the same feature vector construction method as ours (cf. Sect. 2.2); however, they lack sampling and boosting techniques. Table 3 lists the abbreviations of the fifteen algorithms that we evaluated and compared in this study where twelve algorithms are boosting-related and three algorithms are SVM-related.

The performance of each algorithm was evaluated through 10-fold cross validation. The positive examples (negative examples, respectively) were evenly distributed to the ten folds. When testing a fold, the Area Under the ROC Curve (AUC) of an algorithm was calculated where the AUC is defined as

$$\text{AUC} = \frac{1}{2} \times \left(\frac{\text{TP}}{\text{TP} + \text{FN}} + \frac{\text{TN}}{\text{TN} + \text{FP}} \right) \tag{13}$$

Here TP (FP, TN, FN, respectively) denotes the number of true positives (false positives, true negatives, false negatives, respectively) for the test set. A true positive (true negative, respectively) is a predicted present link (a predicted absent link, respectively) that is indeed a known present link (a known absent link, respectively). A false positive (false negative, respectively) is a predicted present link (a predicted absent link, respectively) that is in fact a known absent link (a known present link, respectively). For each algorithm, the mean AUC, denoted MAUC, over the ten folds was computed and recorded. The higher MAUC an algorithm has, the better performance that algorithm achieves.

3.3 Experimental Results

We first conducted experiments to evaluate the performance of SVM with different kernel functions, including the linear kernel, polynomial kernel of degree 2,

Table 3. Abbreviations of the fifteen algorithms studied in this paper.

Abbreviation	Algorithm
AdaBoost	AdaBoost technique
AdaBoost+U	AdaBoost with under-sampling technique
AdaBoost+O	AdaBoost with over-sampling technique
Boost I	Boost I technique
Boost I+U	Boost I with under-sampling technique
Boost I+O	Boost I with over-sampling technique
Boost II	Boost II technique
Boost II+U	Boost II with under-sampling technique
Boost II+O	Boost II with over-sampling technique
Boost III	Boost III technique
Boost III+U	Boost III with under-sampling technique
Boost III+O	Boost III with over-sampling technique
SVM	SVM technique
SVM+U	SVM with under-sampling technique
SVM+O	SVM with over-sampling technique

Gaussian kernel, and sigmoid kernel. It was observed that the Gaussian kernel performed the best. In subsequent experiments, we fixed the SVM kernel at the Gaussian kernel.

Figure 2 shows the AMAUC values of the three SVM-related algorithms, namely SVM, SVM+U, SVM+O, on the twelve yeast datasets used in the experiments. For each algorithm, the AMAUC was calculated by taking the average of the MAUC values the algorithm received over the twelve yeast datasets. It can be seen from Fig. 2 that SVM+U performed better than SVM+O and SVM. Figure 3 shows the AMAUC values of the twelve boosting-related algorithms on the twelve yeast datasets used in the experiments. It can be seen from Fig. 3 that Boost III+U performed the best on the twelve yeast datasets.

Table 4 shows the MAUC values of Boost III+U and SVM+U, and compares them with the existing approaches using SVM only [9,17] on the twelve yeast datasets. For each dataset, the algorithm with the best performance (i.e., the highest MAUC) is shown in bold. It can be seen from Table 4 that Boost III+U has the best overall performance on the yeast datasets, and beats the existing approaches using SVM only [9,17].

Figure 4 shows the AMAUC values of the three SVM-related algorithms, namely SVM, SVM+U, SVM+O, on the twelve E. coli datasets used in the experiments. It can be seen that SVM+O outperformed SVM+U and SVM. Figure 5 shows the AMAUC values of the twelve boosting-related algorithms on the twelve E. coli datasets used in the experiments. It can be seen from Fig. 5 that Boost II+U performed the best among the twelve boosting-related algorithms.

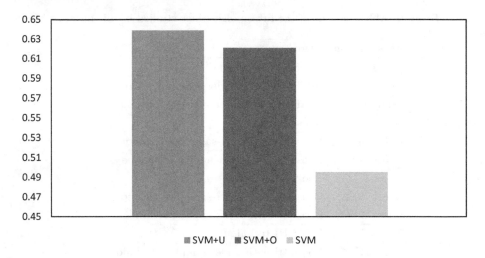

Fig. 2. AMAUC values of three SVM-related algorithms on twelve yeast datasets.

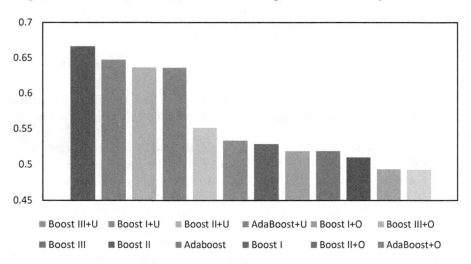

Fig. 3. AMAUC values of twelve boosting-related algorithms on twelve yeast datasets.

Table 5 shows the MAUC values of Boost II+U and SVM+O, and compares them with the existing approaches using SVM only [9,17] on the twelve E. coli datasets. It can be seen from Table 5 that Boost II+U has the best overall performance on the E. coli datasets, and beats the existing approaches using SVM only [9,17].

To summarize, one of our proposed boosting methods coupled with the under-sampling technique achieves the best performance among all the fifteen algorithms studied in this paper on the yeast and E. coli datasets respectively. For the yeast datasets, this proposed boosting method is Boost III. For the E. coli

Table 4. MAUC values of three algorithms on twelve yeast datasets.

Dataset	Boost III+U	SVM+U	SVM
Yeast 50 knockouts	0.709	**0.802**	0.534
Yeast 50 knockdowns	0.664	**0.790**	0.529
Yeast 50 multifactorial	0.679	**0.736**	0.526
Yeast 100 knockouts	**0.729**	0.638	0.487
Yeast 100 knockdowns	**0.706**	0.690	0.489
Yeast 100 multifactorial	**0.701**	0.639	0.472
Yeast 150 knockouts	**0.633**	0.529	0.486
Yeast 150 knockdowns	**0.673**	0.519	0.494
Yeast 150 multifactorial	**0.680**	0.538	0.474
Yeast 200 knockouts	0.599	**0.622**	0.490
Yeast 200 knockdowns	0.612	**0.623**	0.490
Yeast 200 multifactorial	**0.608**	0.540	0.471
AMAUC	**0.666**	0.638	0.495

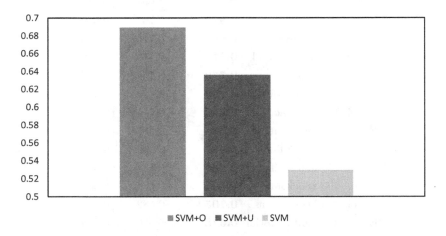

Fig. 4. AMAUC values of three SVM-related algorithms on twelve E. coli datasets.

datasets, this proposed boosting method is Boost II. Both boosting methods coupled with the under-sampling technique are superior to the existing approaches using SVM only [9,17].

Our boosting techniques are based on a weighted decision tree [1]. We have combined the boosting techniques with other machine learning algorithms including random forests [5], SVM with the linear kernel, SVM with the sigmoid kernel, SVM with the Gaussian kernel, and SVM with the polynomial kernel of degree 2. However, the performance of these other machine learning algorithms is inferior to the performance of the weighted decision tree used in this paper.

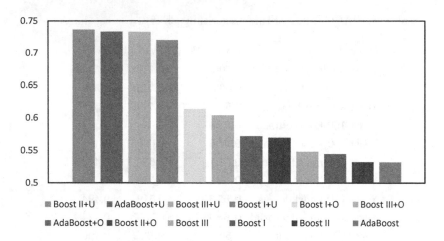

Fig. 5. AMAUC values of twelve boosting-related algorithms on twelve E. coli datasets.

Table 5. MAUC values of three algorithms on twelve E. coli datasets.

Dataset	Boost II+U	SVM+O	SVM
E. coli 50 knockouts	**0.872**	0.767	0.669
E. coli 50 knockdowns	**0.866**	0.776	0.505
E. coli 50 multifactorial	**0.879**	0.819	0.706
E. coli 100 knockouts	**0.770**	0.708	0.493
E. coli 100 knockdowns	**0.758**	0.712	0.492
E. coli 100 multifactorial	**0.750**	0.711	0.490
E. coli 150 knockouts	**0.625**	0.567	0.498
E. coli 150 knockdowns	**0.597**	0.596	0.500
E. coli 150 multifactorial	**0.636**	0.584	0.508
E. coli 200 knockouts	**0.702**	0.683	0.504
E. coli 200 knockdowns	**0.705**	0.682	0.495
E. coli 200 multifactorial	**0.678**	0.666	0.495
AMAUC	**0.736**	0.689	0.529

As a consequence, the results from the other machine learning algorithms are not reported here.

To evaluate the effectiveness of the proposed sampling and boosting techniques, we have also tested and compared the following four algorithms: (i) the weighted decision tree without boosting and sampling techniques; (ii) the weighted decision tree with boosting techniques only; (iii) the weighted decision tree with sampling techniques only; and (iv) the weighted decision tree with both boosting and sampling techniques. The results from the yeast and E. coli datasets are similar. For example, for the yeast datasets, the AMAUC value for the weighted decision tree without boosting and sampling techniques is 0.45. This is lower than the existing approaches using SVM only (with AMAUC being 0.495 as shown in Table 4). When the weighted decision tree is coupled with only the Boost III technique, its AMAUC value is 0.53. When the weighted decision tree is coupled with only the under-sampling technique, its AMAUC value is 0.61. When the weighted decision tree is coupled with both Boost III and under-sampling techniques, its AMAUC value is 0.666 as shown in Table 4, which is much higher than the AMAUC value of 0.45 achieved by the weighted decision tree without boosting and sampling techniques.

4   Conclusion and Future Work

In this paper we present a new approach to gene network inference through regulatory link prediction. Our approach uses a weighted decision tree as the base learning algorithm coupled with sampling and boosting techniques to improve prediction performance. Experimental results demonstrated the superiority of the proposed approach over existing methods [9,17], and the effectiveness of our sampling and boosting techniques.

Moving forward, we are extending the techniques described here to miRNA-mediated regulatory networks. In addition to their importance as regulatory elements in gene expression, the capacity of miRNAs to be transported from cell to cell implicates them in a panoply of pathophysiological processes that include antiviral defense, tumorigenesis, lipometabolism and glucose metabolism [25,26]. This role in disease complicates our understanding of translational regulation via endogenous miRNAs. In addition, miRNAs seem to be present in different types of foods with potential implications on human health and disease. Understanding the biogenesis, transport and mechanisms of action of miRNAs on their target genes would result in possible therapies. Thus, reverse engineering of miRNA-mediated regulatory networks would contribute to human health improvement and disease treatment. We are exploring new ways for inferring such networks. We are also investigating other algorithms such as guided regularized random forests (GRRF) [7,24] and compare them with the weighted decision tree employed here.

References

1. Alhammady, H., Ramamohanarao, K.: Using emerging patterns to construct weighted decision trees. IEEE Trans. Knowl. Data Eng. **18**(7), 865–876 (2006)
2. Altaf-Ul-Amin, M., Afendi, F.M., Kiboi, S.K., Kanaya, S.: Systems biology in the context of big data and networks. In: BioMed Research International 2014 (2014)
3. Altaf-Ul-Amin, M., Katsuragi, T., Sato, T., Kanaya, S.: A glimpse to background and characteristics of major molecular biological networks. In: BioMed Research International 2015 (2015)
4. Batista, G.E., Prati, R.C., Monard, M.C.: A study of the behavior of several methods for balancing machine learning training data. ACM SIGKDD Explor. Newslett. **6**(1), 20–29 (2004)

5. Breiman, L., Friedman, J., Stone, C.J., Olshen, R.A.: Classification and Regression Trees. CRC Press, New York (1984)
6. Bunkhumpornpat, C., Sinapiromsaran, K., Lursinsap, C.: Safe-Level-SMOTE: safe-level-synthetic minority over-sampling technique for handling the class imbalanced problem. In: Theeramunkong, T., Kijsirikul, B., Cercone, N., Ho, T.-B. (eds.) PAKDD 2009. LNCS, vol. 5476, pp. 475–482. Springer, Heidelberg (2009). doi:10. 1007/978-3-642-01307-2_43
7. Deng, H., Runger, G.: Gene selection with guided regularized random forest. Pattern Recogn. **46**(12), 3483–3489 (2013)
8. Freund, Y., Schapire, R.E.: A desicion-theoretic generalization of on-line learning and an application to boosting. In: Vitányi, P. (ed.) EuroCOLT 1995. LNCS, vol. 904, pp. 23–37. Springer, Heidelberg (1995). doi:10.1007/3-540-59119-2_166
9. Gillani, Z., Akash, M., Rahaman, M., Chen, M.: CompareSVM: supervised, support vector machine (SVM) inference of gene regularity networks. BMC Bioinformatics **15**, 395 (2014). http://dx.doi.org/10.1186/s12859-014-0395-x
10. He, H., Garcia, E.A.: Learning from imbalanced data. IEEE Trans. Knowl. Data Eng. **21**(9), 1263–1284 (2009)
11. He, H., Ma, Y.: Imbalanced Learning: Foundations, Algorithms, and Applications. Wiley, Piscataway (2013)
12. Ideker, T., Krogan, N.J.: Differential network biology. Mol. Syst. Biol. **8**(1), 565 (2012)
13. Kibinge, N., Ono, N., Horie, M., Sato, T., Sugiura, T., Altaf-Ul-Amin, M., Saito, A., Kanaya, S.: Integrated pathway-based transcription regulation network mining and visualization based on gene expression profiles. J. Biomed. Inform. **61**, 194–202 (2016)
14. Leclerc, R.D.: Survival of the sparsest: robust gene networks are parsimonious. Mol. Syst. Biol. **4**(1), 213 (2008)
15. Liu, X.Y., Wu, J., Zhou, Z.H.: Exploratory undersampling for class-imbalance learning. IEEE Trans. Syst. Man Cybern. Part B (Cybern.) **39**(2), 539–550 (2009)
16. Margolin, A.A., Wang, K., Lim, W.K., Kustagi, M., Nemenman, I., Califano, A.: Reverse engineering cellular networks. Nat. Protoc. **1**(2), 662–671 (2006)
17. Mordelet, F., Vert, J.P.: SIRENE: supervised inference of regulatory networks. Bioinformatics **24**(16), i76–i82 (2008). http://bioinformatics.oxfordjournals. org/content/24/16/i76.abstract
18. Patel, N., Wang, J.T.L.: Semi-supervised prediction of gene regulatory networks using machine learning algorithms. J. Biosci. **40**(4), 731–740 (2015). http://dx.doi.org/10.1007/s12038-015-9558-9
19. Schaffter, T., Marbach, D., Floreano, D.: GeneNetWeaver: in silico benchmark generation and performance profiling of network inference methods. Bioinformatics **27**(16), 2263–2270 (2011)
20. Turki, T., Bassett, W., Wang, J.T.L.: A learning framework to improve unsupervised gene network inference. In: Perner, P. (ed.) MLDM 2016. LNCS (LNAI), vol. 9729, pp. 28–42. Springer, Cham (2016). doi:10.1007/978-3-319-41920-6_3
21. Turki, T., Ihsan, M., Turki, N., Zhang, J., Roshan, U., Wei, Z.: Top-k parametrized boost. In: Prasath, R., O'Reilly, P., Kathirvalavakumar, T. (eds.) MIKE 2014. LNCS (LNAI), vol. 8891, pp. 91–98. Springer, Cham (2014). doi:10.1007/ 978-3-319-13817-6_10
22. Turki, T., Wang, J.T.L.: A new approach to link prediction in gene regulatory networks. In: Jackowski, K., Burduk, R., Walkowiak, K., Woźniak, M., Yin, H. (eds.) IDEAL 2015. LNCS, vol. 9375, pp. 404–415. Springer, Cham (2015). doi:10. 1007/978-3-319-24834-9_47

23. Turki, T., Wang, J.T.L., Rajikhan, I.: Inferring gene regulatory networks by combining supervised and unsupervised methods. In: 15th IEEE International Conference on Machine Learning and Applications, ICMLA 2016, Anaheim, CA, USA, 18–20 December 2016

24. Wang, S., Yao, X.: Relationships between diversity of classification ensembles and single-class performance measures. IEEE Trans. Knowl. Data Eng. **25**(1), 206–219 (2013). http://dx.doi.org/10.1109/TKDE.2011.207

25. Zhong, L., Wang, J.T.L., Wen, D., Aris, V., Soteropoulos, P., Shapiro, B.A.: Effective classification of microRNA precursors using feature mining and AdaBoost algorithms. OMICS: J. Integr. Biol. **17**(9), 486–493 (2013)

26. Zhong, L., Wang, J.T.L., Wen, D., Shapiro, B.A.: Pre-miRNA classification via combinatorial feature mining and boosting. In: 2012 IEEE International Conference on Bioinformatics and Biomedicine, BIBM 2012, Philadelphia, PA, USA, 4–7 October 2012, Proceedings, pp. 1–4 (2012). http://doi.ieeecomputersociety.org/10.1109/BIBM.2012.6392700

Detecting Large Concept Extensions
for Conceptual Analysis

Louis Chartrand[1], Jackie C.K. Cheung[2], and Mohamed Bouguessa[1(✉)]

[1] Department of Computer Science,
University of Quebec at Montreal,
Montreal, QC, Canada
bouguessa.mohamed@uqam.ca
[2] School of Computer Science,
McGill University, Montreal, QC, Canada

Abstract. When performing a conceptual analysis of a concept, philosophers are interested in all forms of expression of a concept in a text—be it direct or indirect, explicit or implicit. In this paper, we experiment with topic-based methods of automating the detection of concept expressions in order to facilitate philosophical conceptual analysis. We propose six methods based on LDA, and evaluate them on a new corpus of court decision that we had annotated by experts and non-experts. Our results indicate that these methods can yield important improvements over the keyword heuristic, which is often used as a concept detection heuristic in many contexts. While more work remains to be done, this indicates that detecting concepts through topics can serve as a general-purpose method for at least some forms of concept expression that are not captured using naive keyword approaches.

Keywords: Concept mining · Topic models · Conceptual analysis

1 Conceptual Analysis as a Computational Linguistics Problem

Conceptual analysis in philosophy can refer, in a technical sense, to the discovery of *a priori* knowledge in the concepts we share [4,12,14]. For instance, philosophers will say that "male sibling" is a proper analysis of the concept BROTHER, because it decomposes its meaning into two other concepts: a brother is nothing more and nothing less than a male sibling. Doing so allows us to make explicit knowledge that is *a priori*, or in other words, knowledge that is not empirical, that can be acquired without observation: for instance, the knowledge that a brother is always a male. In a broader sense, it can refer to the philosophical methods we use to uncover the meaning and use of a concept in order to clarify or improve it [6,10]. Given philosophy's focus on conceptual clarity, the latter has been ubiquitous in practice. These methods usually seek to make explicit key features of the concept under scrutiny, in order to construct an account of

© Springer International Publishing AG 2017
P. Perner (Ed.): MLDM 2017, LNAI 10358, pp. 78–90, 2017.
DOI: 10.1007/978-3-319-62416-7_6

it; be it a formalized representation that can be expressed in terms of necessary and sufficient conditions, or a more intuitive and pragmatic account of it.

Among the empirical sources upon which conceptual analysis relies, textual data is one of the most important. While armchair philosophy (which relies on thought experiments and intuitions) helps one give a better account of his or her own concepts, contact with texts provides an essential perspective. As a result, philosophers often build *corpora*, i.e. databases of texts that likely use or express a particular concept that is undergoing analysis.

In philosophy as elsewhere, corpora need to be broad enough to cover all the types of usages of the concept under scrutiny, lest the analysis fails to be exhaustive. In other disciplines of social science and humanities, the necessity of grounding analysis in corpora has lead researchers to harness text mining and natural language processing to improve their interpretations of textual data. Philosophy, however, has remained untouched by those developments, save for a few projects [3,15].

One important obstacle to the adoption of those methods in philosophy lies in the lack of proper concept models for conceptual analysis. Keyword approaches to identifying concepts can run into ambiguity problems, like polysemy and synonymy. Furthermore, they can only detect explicit concepts, whereas passages where a concept is latent are bound to also interest the analyst. Latent concept approaches, such as latent semantic analysis (LSA) [5] or latent dirichlet allocation (LDA) [1], can work to alleviate ambiguity problems and detect latent semantic expressions, but the dimensions they generate ("concepts" in LSA, "topics" in LDA) are thematic, not conceptual. While concepts typically refer to abstract entities or entities in the world, themes, topics and other thematic units are discursive: they can only describe features and regularities in the text.

The problem we address in this paper is that of retrieving textual segments which are relevant to philosophical conceptual analysis. Considering that conceptual analysis is interested in the entire set of textual segments where a queried concept is present in any form, the task at hand is to detect segments whose discourse expresses, implicitly or explicitly, a queried concept. Our concept detection problem distinguishes itself from traditional information retrieval problems in that the aim is to retrieve text segments where the queried concept is *present*, rather than text segments that are *relevant* to the queried concept. In the context of a relevance search, the inquirer will look for the minimum number of documents that can give the maximum amount of generic information about the queried concept; for instance, a web search for "brother" will likely return dictionary definitions and the Wikipedia entry for this word. In the context of a presence query, the inquirer will look for all of the documents where the queried concept is present, thus enabling a more subtle understanding of the concept in all its shades. A search for the presence of the concept BROTHER might thus return texts in genetics or inheritance law as well as implicit evocations of brotherly love in a play. On the other hand, this concept detection problem also differs from entity recognition or traditional concept mining, as the concept does not need to be associated with a word or an expression. While these problems focus on one particular way a concept can be expressed, conceptual analysis will be interested in any kind of expression of a concept, be it direct or indirect, explicit or implicit.

As such, in Sect. 2, we clarify what counts as concept expression for the sake of conceptual analysis, and we distinguish it from other similar notions. In Sect. 3, we describe methods to detect a queried concept's expression in textual segments from a corpus. In Sect. 4, we present how these methods were implemented and tested, including how an annotated corpus was built, and in Sect. 5, results are laid out. Finally, in Sect. 6, results are discussed, in a bid to shed light on the underlying assumptions of the methods employed.

2   Concept Detection

While conceptual analysis can take many forms, it can always be enhanced by taking empirical data into account. Philosophers who set out to make a concept's meaning explicit through its analysis typically already possess the said concept, and can thus rely on their own intuitions to inform their analysis. However, their analysis can be improved, both in terms of quality and in validity, by being compared with other sources. This explains, for instance, the appeal of experimental philosophy, which has developed in the last 15 years as a way of testing philosophical intuitions using the tools of cognitive and social psychology [13]. However, these inquiries have their limits: the intuitions they aim to capture are restricted to a specific time and scope, as they are provoked in an artificial setting. Textual corpora give us the opportunity to study concepts in a more natural setting, and in broader populations, or in populations which are hard to reach via conventional participant recruitment schemes (experts, authors from past centuries, etc.).

In order to use data from textual corpora, philosophers now have access to the methods and techniques of computer-assisted analysis of textual data. Those methods and techniques are both numerous and diverse, but there are some common characteristics. For instance, they typically involve various steps, which, together, form treatment chains [7,15]: textual data are preprocessed (cleaning, lemmatisation, etc.) and transformed into suitable representations (e.g. vector-space model); then, specific treatment tasks are performed, and finally their output is analysed and interpreted. Furthermore, concepts must be identified in the text, in order to extract their associations to other features that can be found in textual data, such as words, themes and other concepts.

One way of identifying a concept in the text is to identify textual segments in which it is expressed. This expression can take many forms: it can be a word that explicitly refers to a concept in a very wide variety of contexts ("moose" for MOOSE), a description ("massive North American deer"), or embedded in an anaphoric reference ("the animal that crossed the street", "*its* habitat"). It can also be expressed in such a way that it is not tied to any specific linguistic expression. For instance, it can appear in the background knowledge that is essential in understanding a sentence (for instance, in talking about property damage that only a moose could have done), or in relation to the ontological hierarchy (for instance, the concept MOOSE can be expressed when talking about a particular individual moose, or when talking about cervidae).

Our objective here is to test methods of identifying such expressions in textual segments. In other words, our goal is to detect, within a corpus, which passages are susceptible to inform our understanding of how the concept is expressed in a corpus. As such, concept detection can be seen as a useful step in a wide variety of computer-assisted conceptual analysis methods. For instance, it can act as a way of reducing the study corpus (i.e., the corpus on which a concepual analysis is based) to make it more digestible to a human reader, or it can signal that the semantic content of the segments where the queried concept is detected is likely to be related to the concept, and thus enable new ways of representing it.

Because conceptual analysis is focused on a concept's expression in discourse, concept detection is interested in its presence in discourse. This can mean that the concept is explicitly present, and that it can be matched to a word or an expression, but this presence can also be found in other ways. It can be present in the postulates of the argumentation, without which the passage would be impossible to understand. (For instance, talk of incarceration takes on a very different meaning if we lack the concept of sentencing for a crime). It can be a hypernym to an explicit hyponym, if its properties, expressed to the hyponym, are important enough to the discourse content that we can identify the hypernym as a relevant contributor to the proposition. It can be present in the theme that's being expanded in the passage. It can be referenced using a metaphor or an anaphor. To synthesize, this criterion can be proposed: *a concept is expressed in a textual segment if and only if possession of a concept is necessary to understand the content of the segment.*

Concept detection is similar to other popular problems and projects that have been developed within NLP. However, important distinctions justify our treating it as a different kind of problem.

For instance, concept detection differs from information retrieval (IR) in that presence, rather than relevance, is what we are looking for. For instance, while IR might be interested in giving priority to text segments where a queried concept is central, this is of little importance to a conceptual analysis, as salient and less salient expressions of a concept are likely to give different yet equally important dimensions of a concept of interest. Conversely, while IR is interested in relevance of a document to a concept even if it is absent, such a rating is meaningless if one is only looking for presence or absence.

It also differs from other tasks which are geared towards presence detection, such as named-entity recognition or coreference resolution. While a concept can be present because a word or expression directly refers to it, it is not absent because no such expression exists in a sentence or another textual unit. In other words, a concept can be present in a text segment even if no single word or expression refers to it. It can be present in virtue of being part of the necessary background knowledge that is retrieved by the reader to make sense of what she or he is reading. Concept detection, as we mean it, should detect both direct and indirect presence of concepts.

3   LDA Methods for Detecting Concepts

The presence of a concept as described in Sect. 2 can therefore be expressed in various ways: direct explicit reference, anaphorical or metaphorical reference, implicit argumentative or narrative structures, etc. In order to detect these different types of presence, one may expect that we should fragment the task of concept detection into more specific tasks attuned to specific types of presence. In other words, we could detect concept presence by running various algorithms of named-entity recognition or extraction, coreference resolution, topic models, etc. Each of these would detect a specific way in which a concept can become present in a text, and we would rule that a concept is present in a text segment if it has been detected with any of the methods employed. However, not only is such an approach potentially very time consuming, it makes it very hard to have a constant concept representation: these various algorithms will accept different types of representations of the queried concept, and as such, it will be hard to guarantee that they are all looking for the same concept.

One way around this problem is to hypothesize that while these various expressions of a concept are expressed in different ways, they may be conditioned in similar ways by latent variables. We suppose, in this way, that topics—i.e. underlying discursive and narrative constructions which structure a text, cf. [1]—are such latent variables that condition the expression of words and concepts alike. For instance, if the topic "family dinner" is present in a text excerpt, it makes it likely for words such as "table", "mother", "brother" to be present, and unlikely for words such as "clouds" or "mitochondria" to be present; and in a similar fashion, concepts such as FOOD, BROTHER and MOTHER are likely to be expressed and concepts such as ORGANELLE and CLOUD are likely absent.

Therefore, given a concept expressed as a word that is typically associated with it, we can find topics in which it is expressed, and use those topics to find the textual segments where it is likely to be present.

We implement this approach using two different algorithms for learning an LDA model, one that is based on Hoffman's online learning algorithm [11] and one that is based on Griffiths & Steyvers's Gibbs sampler [8].

3.1   Online Learning

Hoffman's algorithm [11] is an online variational Bayes algorithm for the LDA. As such, it relies on the generative model that was introduced by Blei [1].

Blei's model uses this generative process, which assumes a corpus D of M documents each of length N_i:

1. Choose $\theta_i \sim \text{Dirichlet}(\alpha)$, where $i \in \{1 \dots M\}$, the topic distribution for document i
2. Choose $\phi_k \sim \text{Dirichlet}(\beta)$, where $k \in \{1 \dots K\}$, the word distribution for topic k

3. For each of the word positions i, j, where $j \in \{1, \ldots, N_i\}$, and $i \in \{1, \ldots, M\}$:
   (a) Choose a topic $z_{i,j} \sim \text{Multinomial}(\theta_i)$.
   (b) Choose a word $w_{i,j} \sim \text{Multinomial}(\varphi_{z_{i,j}})$.

Here, α and β are parameters of the Dirichlet prior on the per-document topic distributions and on the per-topic word distribution respectively; θ_i is the topic distribution for document i; and ϕ_k is the word distribution for topic k.

Through online stochastic optimization, the online LDA algorithm learns θ (the topic distributions for each document) and ϕ (the word distribution for each topic). Thus, it is possible to know which topics are likely to be found in each document, and which words are likely to be found for each topic.

Using this information and given a queried concept represented as a word, we can use ϕ to find the topics for which it is among the most important words, relatively, and then use θ to find the documents in which these topics have a non-negligible presence. We thus have a set of documents which are likely to contain the queried concept.

3.2 Gibbs Sampling-LDA

While it uses the same LDA model, Griffiths & Steyvers's algorithm [8] operates very differently. Rather than estimating θ and ϕ, it learns instead the posterior distribution over the assignments of words to topics $P(z \mid w)$, and it does so with the help of Gibbs sampling, thus assigning topics to each word. After a certain number of sampling iterations (the "burn-in"), these assignments are a good indicator of there being a relationship between word and topic, and between topic and document. From them, we can pick the topics have been assigned to a given word in its various instanciations, and retrieve the documents to which these topics have been assigned. Furthermore, when necessary, ϕ and θ can be calculated from the assignments.

3.3 Concept Presence in Topics

We assume that the presence of a concept in a topic is indicated by the presence of a word typically associated with the concept in question. Therefore a topics association with a word is indicative of its association with the corresponding concept. The LDA model explicitly links words to topics, but in a graded way: each word is associated with each topic to a certain degree. From this information, we can use various heuristics to rule whether a concept is involved in a topic or not.

In this study, we tested these heuristics:

Most Likely: The queried concept is associated to the topic which makes its corresponding word most likely to occur.
Highest Rank: The queried concept is associated with the topic in which its corresponding word has the highest rank on the topic's list of most likely words.

Top 30 Rank: The queried concept is associated with the topics in which its corresponding word is among the top 30 words on the topic's list of most likely words.

Concrete Assignment: In the Gibbs Sampling method, individual words are assigned to topics, and word likelihood given a topic is calculated from these assignments. We can thus say that a word is involved in a topic if there is at least one assignment of this topic to this word in the corpus.

Using these heuristics and an LDA model (learned using either Hoffman's or Griffiths & Steyvers's method), we can determine for a given concept the topics in which it is involved.

Depending on the learning method, we can then determine which textual segments are associated to a given topic. On one hand, in Hoffman's method, when a topic is assigned to a segment, there will be a non-zero probability that any given word in the segment is associated with the topic in question. On the other hand, when learning the LDA model using Gibbs Sampling, we'll consider that a topic is associated to a textual segment if there is at least one word of this segment that is associated with the topic in question.

Thus, from a given concept, we can retrieve the segments in which the concept is likely expressed by retrieving the textual segments that are associated to the topics which are associated to the queried concept.

4   Experimentation

4.1   Corpus

Algorithms were tested on a French-language corpus of 5,229 decisions from the *Cour d'appel du Québec* (Quebec Court of Appeal), the highest judicial court in Quebec. Much like philosophical discussions, arguments in juridical texts, and in decisions in particular, are well-developed, and nuances are important, so we can expect concepts to be explained thoroughly and employed with precision. However, there is much more homogeneity in style and vocabulary, and this style and vocabulary are more familiar to the broader public than in typical philosophical works, which facilitates annotation. Thus, court decisions are likely to afford complex conceptual analyses, but lack the difficulties that come with the idiosyncrasies of individual philosophical texts.

Court decisions were divided into paragraphs, yielding 198,675 textual segments, which were then broken down into words and lemmatized using TreeTagger [17]. Only verbs, adjectives, nouns and adverbs were kept, and stopwords were removed.

In order to provide a gold standard against which we could evaluate the performances of the chosen algorithms, annotations were collected using Crowd-Flower[1].

In a first "tagging" step, French-speaking participants were given a textual segment and were instructed to write down five concepts which are expressed in

---

[1] http://www.crowdflower.com.

the segment—more specifically, the criterion mentioned in the instructions was that the concept must contribute to the discourse (in French: "*propos*") expressed in the segment. 25 participants annotated 105 segments in this way, yielding 405 segment annotations for a total of 3,240 segment-concept associations.

Data obtained from this first step can tell us that a concept is present in a segment, but we can never infer its absence from it, as its absence from the annotations could simply mean that the annotator chose to write down five other concepts and had no more place for another one. Therefore, it was necessary to add another step to assess absence.

In the second "rating" step, participants were given a segment and six concepts (from the pool of concepts produced in the tagging step), and were instructed to rate each concept's degree of presence or absence from 1 (absent) to 4 (present). The degree of presence is meant to give options to the participant to mark a concept as present, but to a lesser degree, if, say, it is not particularly salient, or if lack of context gives way to some doubt as to whether it really is present. Using this strategy, we can get participants to mark the absence of a concept (degree 1 of the scale) in a way that is intuitive even if one has not properly understood the instructions. For our purposes, we assume that CrowdFlower participants mark a concept as absent when they give it a rating of 1, and as present (even if minimally) if they make any other choice. After removing low-quality annotations, we get 104 segments annotated by 37 participants, for a total of 5,256.

In order to ensure that annotations by CrowdFlower participants reflect a genuine understanding of the text, we also recruited legal experts to make similar taggings and judgments and to compare annotations. While the first task was the same for the experts, the second was slightly different in that there were only two options, and in that they were given oral and written instructions to only mark as absent concepts which were definitely absent. This is because the contact we had with these participants made it possible to ensure that instructions were well understood: we did not need to add options to reinforce the idea that a concept is only absent when it is completely and undoubtedly absent. In total, 5 experts tagged 82 text segments in the tagging step, producing a total of 361 tag-segment pairs, and 4 experts rated concepts on 58 segments in the rating step, producing a total of 412 tag-segment pairs.

As Table 1 shows, the distribution is skewed towards presence, which makes Cohen's κ a poor choice of metric [9]. Gwet's AC1 coefficient [9] was used instead, and it revealed that CrowdFlower participants and legal experts have moderate but above-chance agreement, with a coefficient of 0.30 and p-value of less than 0.05 (indicating that there is less than 5% chance that this above-chance agreement is due to random factors)[2]. As the confusion matrix of Table 1 shows, the error mostly comes from the fact that CrowdFlower participants seem much more likely to mark concepts as absent than legal experts.

---

[2] The scenario on the tagging step does not fit any of the common inter-annotator agreement metrics. Firstly, a single item is given five values for the same property. Secondly, in our annotations, absence of annotation does not mean absence of concept; the converse would have been a common assumption in inter-annotator metrics.

Table 1. Contingency table of the CrowdFlower ratings against the legal experts' ratings for the rating step.

		Legal experts	
		Present	Absent
CrowdFlower participants	Present	32	2
	Absent	24	4

4.2 Algorithms

Both LDA algorithms were implemented as described in the previous section. For the online LDA, we have used the implementation that is part of Gensim [16], and for the Gibbs sampler-LDA, we have adapted and optimized code from Mathieu Blondel [2]. In both cases, we used $k = 150$ topics as parameter, because observing the semantic coherence of the most probable words in each topic (as indicated by ϕ_z) suggests that greater values for k yield topics that seem less coherent and less interpretable overall. For the Gibbs sampler-LDA, we did a burn-in of 150 iterations.

The baseline chosen was the keyword heuristic: a concept is marked as present in a segment if the segment contains the word that represents it, and absent if it does not.

Each method was successively applied to our corpus, using, as queries, items from a set of concept-representing words that were both used in annotations from the rating steps and found in the corpus lexicon. In total, this set numbers 229 concepts for the legal experts' annotations and 808 concepts for the CrowdFlower annotations. Among these, 170 terms are found in both sets of annotations.

5 Results

Results from the application of the baseline and our methods on all concepts were compared to the gold standards obtained from the rating step using overall precision, recall, and F1-score. They are illustrated in Table 2.

Table 2. Performance for each method, calculated using data from the rating task.

		CrowdFlower			Experts		
		Recall	Precision	F1	Recall	Precision	F1
	Keyword	0.03	0.56	0.07	0.01	**1.00**	0.04
Online LDA	Most Likely	0.06	0.63	0.13	0.03	0.67	0.07
	Highest Rank	0.07	0.51	0.16	0.03	0.50	0.07
	Top 30 Rank	**0.18**	0.60	**0.32**	**0.15**	0.61	**0.29**
Gibbs Sampling- LDA	Most likely	0.05	0.55	0.12	0.05	0.50	0.13
	Highest Rank	0.00	0.60	0.01	0.03	0.50	0.07
	Top 30 Rank	0.01	0.64	0.03	0.01	0.25	0.04
	Concrete Assign	0.08	**0.65**	0.19	0.12	0.53	0.25

Apart from the Gibbs Sampling-LDA/Highest Rank method, all of the proposed methods improved on the baseline, except for the ones using word rankings among the Gibbs Sampling-LDA methods. This is due in particular to improvements in recall. This is to be expected, as the keyword only targets one way in which a concept can be expressed, and thus appears to be overly conservative.

Table 3. Reuse rate in annotation tasks.

	CrowdFlower	Legal experts
Tagging task (step 1)	0.35	0.10
Rating task (step 2)	0.75	0.24

Among the Gibbs Sampling-LDA methods, Concrete Assignment fares significantly better, but the best overall, both in recall and F1-score, is the Online LDA/Top 30 Rank. On this, experts and non-experts are in agreement.

6 Discussion

These results seem to validate this study's main hypothesis, that is, LDA methods can improve on the keyword heuristic when it comes to detection of concept expression.

This said, recall remains under 20%, indicating that topic models are still insufficient to detect all forms of expression of a concept. As such, while it is a clear improvement on the keyword heuristic, it would seem to contradict our hypothesis that topic models can be used to detect all sorts of concept expressions.

6.1 Quality of Annotations

While experts' and non-experts' annotations are mostly in agreement, there are important discrepancies. Experts' annotations systematically give better scores to Gibbs Sampling methods, and lower scores to Online LDA methods, than non-experts'. For instance, while the Online LDA/Top 30 Rank method beats the Gibbs Sampling/Assignment method by 0.11 in F1-scores using Crowd-Flower annotations, this difference shrinks to 0.4 when using experts' annotations. These discrepancies, however, can be traced to a difference in types of heuristics employed in the tagging step: CrowdFlower participants are more likely to employ words from the excerpt as annotations (i.e. using the concept BROTHER when the word "brother" is present *verbatim* in the text segment), which favors the baseline.

In order to give evidence for this claim, we calculated the propensity of a participant to mark as present a tag that is also a word in the text segment.

Specifically, we estimated the reuse rate[3] as depicted by Table 3.[4]). As it turns out, in the initial tagging step, CrowdFlower participants are more than three times more likely to write down a word that is present in the text. As participants in the rating step are only rating tags entered by people of the same group, this translates into a similar ratio in the rating task. However, as it seems that in the rating step, participants are less likely to mark as present a word which is not specifically in the text, reuse rate is inflated for both participant groups. As a result, a large majority of one-word annotations by CrowdFlower participants are already in the text, while the reverse is still true of expert annotations.

Thus, when we discriminate between tags that are present in the textual segment and those that are not, we get a much clearer picture (Table 4). In the first case, the best heuristic is still the baseline, with Online LDA methods offering much better results than Gibbs Sampling-LDA methods. But in the second, the baseline is unusable, and while F1-scores of Online LDA methods drop by more than half, Gibbs Sampling-LDA methods stay the same or improve. Having fewer annotations where the concept's keyword is in the textual segment will penalize the Online LDA methods, but not the Gibbs Sampling-LDA ones.

As such, this discrepancy should not count as evidence against the hypothesis that CrowdFlower annotations are invalidated by their discrepancies with experts' annotations. However, it suggests that future annotations should control for the ratio of present and absent words in the rating step. Furthermore, it would be useful to test participants of a same group on the same textual segment/concept pair in order to compare in-group inter-annotator agreement with between-group inter-annotator agreement.

Table 4. F1-scores against CrowdFlower annotations for each method, based on presence or absence of the queried concept keyword in the textual segment.

		Keyword in segment	Keyword absent
	Keyword	**0.72**	0.00
Online LDA	Most Likely	0.21	0.13
	Highest Rank	0.48	0.14
	Top 30 Rank	0.67	**0.30**
Gibbs Sampling- LDA	Most Likely	0.00	0.12
	Highest Rank	0.00	0.01
	Top 30 Rank	0.00	0.03
	Concrete Assignment	0.21	0.19

---

[3] The reuse rate here is simply the number of tags which are a word in the text segment divided by the total number of tags that are words. Multi-word expressions were excluded because detecting whether they are in the text or not would be complicated.

[4] Experts' annotations were ignored because there were too few annotation instances where the queried concept's keyword was in the textual segment, and, as a result, values for the "Keyword in segment" condition were uninformative.

6.2   Improving on Topic Model Methods

In any case, while it does not solve the problem of retrieving all the textual segments where a concept is expressed, the Online LDA/Top 30 Rank method makes important headway towards a more satisfactory solution. It improves on the keyword heuristic's F1-score by 0.25 (both when experts' and non-experts' annotations are used as gold standard), and, as such, constitutes a clear improvement and a much better indicator of concept presence.

Improvements could be reached by associating different approaches to concept detection, when we know that some methods do better than others in specific contexts. For example, the keyword heuristic does slightly better than Online LDA/Top 30 Rank when the queried concept's keyword is present in the text, so it could be used in these situations, while the former method could be used in other cases. In fact, this produces a minor improvement (F1-score of 0.33 with the CrowdFlower gold standard, as compared to 0.32 for pure Online LDA/Top 30 Ranks). We can hope that including other methods for other means of expressing a concept can contribute to further improvements.

7   Conclusion

In this paper, we expressed the problem of concept detection for the purpose of philosophical conceptual analysis, and sought LDA-based methods to address it. In order to evaluate them, we devised an annotation protocol and had experts and non-experts annotate a corpus.

Our results suggest that LDA-based methods and the Online LDA/Top 30 Rank method in particular, can yield important improvements over the keyword heuristic that is currently used as a concept detection heuristic in many contexts. Despite important improvement, it remains a high-precision, low-recall method. However, while more work remains to be done, this indicates that detecting concepts through topics can serve as a general-purpose method for at least some forms of concept expression that are not captured using naive keyword approaches.

As such, we suggest that further research should try to integrate other methods of detecting concept presence in textual data that focus on other means of expressing concepts in texts and discourse.

Acknowledgments. This work is supported by research grants from the Natural Sciences and Engineering Research Council of Canada (NSERC) and from the Social Sciences and Humanities Research Council of Canada (SSHRC).

References

1. Blei, D.M., Ng, A.Y., Jordan, M.I.: Latent dirichlet allocation. J. Mach. Learn. Res. **3**, 993–1022 (2003)
2. Blondel, M.: Latent Dirichlet Allocation in Python (2010). https://gist.github.com/mblondel/542786

3. Braddon-Mitchell, D., Nola, R.: Introducing the Canberra plan. In: Braddon-Mitchell, D., Nola, R. (eds.) Conceptual Analysis and Philosophical Naturalism, pp. 1–20. MIT Press (2009)
4. Chalmers, D.J., Jackson, F.: Conceptual analysis and reductive explanation. Philos. Rev. **110**(3), 315–361 (2001)
5. Deerwester, S., Dumais, S.T., Furnas, G.W., Landauer, T.K., Harshman, R.: Indexing by latent semantic analysis. J. Am. Soc. Inf. Sci. **41**(6), 391 (1990)
6. Dutilh Novaes, C., Reck, E.: Carnapian explication, formalisms as cognitive tools, and the paradox of adequate formalization. Synthese **194**(1), 195–215 (2017)
7. Fayyad, U., Piatetsky-Shapiro, G., Smyth, P.: From data mining to knowledge discovery in databases. AI Magazine **17**(3), 37 (1996)
8. Griffiths, T.L., Steyvers, M.: Finding scientific topics. Proc. Natl. Acad. Sci. **101**(suppl 1), 5228–5235 (2004)
9. Gwet, K.L.: Computing inter-rater reliability and its variance in the presence of high agreement. Br. J. Math. Stat. Psychol. **61**(1), 29–48 (2008)
10. Haslanger, S.: Resisting Reality: Social Construction and Social Critique. Oxford University Press, Oxford (2012)
11. Hoffman, M., Bach, F.R., Blei, D.M.: Online learning for latent Dirichlet allocation. In: Advances in Neural Information Processing Systems, pp. 856–864 (2010)
12. Jackson, F.: From Metaphysics to Ethics: A Defence of Conceptual Analysis. Oxford University Press, New York (1998)
13. Knobe, J., Nichols, S.: An experimental philosophy manifesto. In: Knobe, J., Nichols, S. (eds.) Experimental philosophy, pp. 3–14. Oxford University Press (2008)
14. Laurence, S., Margolis, E.: Concepts and conceptual analysis. Philos. Phenomenological Res. **67**(2), 253–282 (2003)
15. Meunier, J.G., Biskri, I., Forest, D.: Classification and categorization in computer assisted reading and analysis of texts. In: Lefebvre, C., Cohen, H. (eds.) Handbook of Categorization in Cognitive Science, pp. 955–978. Elsevier (2005)
16. Řehůřek, R., Sojka, P.: software framework for topic modelling with large corpora. In: Proceedings of LREC 2010 workshop New Challenges for NLP Frameworks, pp. 46–50. University of Malta, Valletta, Malta (2010). http://www.fi.muni.cz/usr/sojka/presentations/lrec2010-poster-rehurek-sojka.pdf
17. Schmid, H.: Probabilistic part-of-speech tagging using decision trees. In: Proceedings of the International Conference on New Methods in Language Processing, vol. 12, pp. 44–49 (1994)

Qualitative and Descriptive Topic Extraction from Movie Reviews Using LDA

Christophe Dupuy[1,2(✉)], Francis Bach[1,3], and Christophe Diot[2]

[1] INRIA, Paris, France
christophe.dupuy@inria.fr
[2] Technicolor, Issy-les-Moulineaux, France
[3] ENS, Paris, France

Abstract. Crowdsourced review services such as IMDB or Rotten Tomatoes provide numerical ratings and raw text reviews to help users decide what movie to watch. As the amount of reviews per movie can be very large, selecting a movie among a catalog can be a tedious task. This problem can be addressed by providing the user the most relevant reviews or building automatic reviews summaries. We take a different approach by predicting personalized movie description from text reviews using Latent Dirichlet Allocation (LDA) based topic models. Our models extract distinct qualitative and descriptive topics by combining text reviews and movie ratings in a joint probabilistic model. We evaluate our models on an IMDB dataset and illustrate its performance through comparison of topics.

1 Introduction

Crowdsourced review services allow any user to assess contents (e.g., movies, restaurants, hotels) through both numerical rating and free form text. They exploit the ratings to suggest contents to users while making raw reviews available for users to get a precise opinion on a particular content. As it is tedious to extract useful information among thousands of reviews, it is more intuitive and less time consuming for users to access a personalized summary of the other users' opinions on each content. In this regard, many existing implementations either manually label each content—e.g., with the genre—or select "useful" reviews to read in order to make a quick opinion on the particular content.

These existing approaches suffer from several limitations. First, the textual information —labels of contents or "useful" reviews— provided by the system is not personalized and may not be adapted to every user. The aspects decribed in the selected "useful" reviews may not be decisive in the opinion of every user and the labels of contents may be too generic to make a decision. The labelling of the movies is also a cumbersome human task.

In our approach, we combine both text reviews and numerical ratings to automatically predict the words a user would employ to assess an unseen content. We leverage topic models to separate descriptive information and qualitative information in distinct topics. This distinction between descriptive and qualitative

P. Perner (Ed.): MLDM 2017, LNAI 10358, pp. 91–106, 2017.
DOI: 10.1007/978-3-319-62416-7_7

information is crucial as it enables to decompose the reviews in (1) the aspects evaluated in the reviews and (2) the opinion attached to these aspects. Without this distinction, it is impossible to automatically identify the decisive aspects in the opinion of the user. We extend LDA (latent Dirichlet allocation) model [2] to include ratings along with texts.

Our contributions are (1) extensions of LDA in order to separate descriptive and qualitative topics in a corpus of text reviews using movie ratings and (2) their theoretical evaluation and comparison to state of the art [12,13] for the task of word prediction. As a by-product (3) we show that profiling users from reviews is not possible with this method because of the diversity of movies reviewed and opinions expressed by users. Note that this work can easily be generalized to other types of content such as restaurants, hotels and products. We have indeed applied our models on a restaurant dataset with similar observations and performance.

2   Topic Extraction with LDA

We define a corpus as a list of documents and a document (e.g., a single user review) as a list of words. In this section we present the methodology we use to pre-process raw text reviews. We then describe two types of topic models. The first type of model only applies to text while the second type of model applies to text and ratings. Each review is generated by an user on a movie. We apply each type of model to three different corpora, depending on wether a document is (1) a single review, (2) the concatenation of reviews related to a single movie or (3) the concatenation of reviews written by a single user. These three different processes with the two types of models lead us to six different models.

2.1   Pre-processing Reviews

Many words in reviews are not relevant for our purpose as they convey neither qualitative nor descriptive information (e.g., stop words), or because they appear too frequently in reviews and are too generic (e.g., *movie, film, scene* in movie reviews).

We first remove the stop words using the NLTK toolbox [1] and words appearing in more than 20% of the reviews (for instance, *movie* appears in 80% of the reviews). We choose 20% as a threshold because it only filters out very frequent words. We then select from the remaining words the 10,000 most frequent words in the database. We observe that in our IMDB dataset of 97,000 reviews (we use a subset of the dataset decribed by [5] that spreads over 5,900 movies and 2,400 users), after filtering, each word appears in at least 10 reviews, which means only words that appear in less than 10 reviews — i.e., 0.01% of the dataset — have been pruned.

2.2   LDA and Extensions

Let D be the number of documents of a corpus $\mathcal{W} = \{\mathbf{w}^1, \ldots, \mathbf{w}^D\}$, $V = 10,000$ the number of words in the vocabulary and K the number of latent topics in

Table 1. Top 5 topics extracted with LDA-O and LDA-C, $K = 128$ (ordered by importance $\hat{\theta}_k = \frac{1}{D} \sum_d \theta_k^d$).

LDA-O, 1 doc = 1 review					LDA-C, 1 doc = 1 movie				
Comedy	Best	Action	Horror	Love	Comedy	Horror	Action	Love	Animation
Funny	Great	Comic	Thriller	Family	Funny	Original	Die	Son	Animated
Laugh	Love	Great	Best	Romantic	Laugh	Remake	Hard	Best	Toy
Fun	Old	Best	Killer	Comedy	Joke	Gore	Bad	Mother	Voice
Love	Totally	Fun	Great	Beautiful	Fun	Scary	Hero	Comedy	Pixar
Great	Award	X-man	Kill	Best	Hilarious	Dead	John	Beautiful	Child
Best	Funny	Kill	Bad	Young	Great	Scare	Fight	Young	Great
Hilarious	Benjamin	Fight	Dark	Romance	Sex	Scream	Bruce	Brother	Disney

the corpus. Each topic ϕ^k corresponds to a distribution on the V words. For each document d, LDA infers a discrete distribution θ^d over the K topics. In practice, the inference may be done using variational EM [2], Gibbs sampling [6] algorithm or online learning [7]. LDA is a generative model applied to a corpus of text documents which assumes that the nth word w_n^d of the dth document is generated as follows:

- Choose $\theta^d \sim$ Dirichlet(α)
- For each word $w_n^d \in \mathbf{w}^d$:
  - Choose a topic $z_n^d \sim$ Mult(θ^d)
  - Choose a word $w_n^d \sim$ Mult($\phi^{z_n^d}$)

In a recommendation setting, each review refers to a unique (user, movie) tuple. One could use this information to learn specific topics of (movies, users) pairs and get a more precise representation of the reviews. In our baseline model, noted LDA-O, we run LDA on a corpus where each document is a single review (i.e., we consider that reviews are independently generated).

We then aggregate our reviews accounting for the fact that multiple reviews either belong to a movie or to a user. We build user profiles in LDA by constraining the topic document distribution θ to be the same for all the reviews written by the same user. It is equivalent to considering a new corpus where each document is the concatenation of all the reviews written by a single user. We refer to this model as LDA-U and we refer to the topic document distributions θ^u inferred with LDA-U as "topic user distributions", where $u \in \{1, \ldots, N_u\}$ denotes users. The same aggregation is done to profile movies with the corpus where a document is the aggregation of all the reviews belonging to the same movie. We refer to this model as LDA-C and we refer to the topic document distributions θ^c as "topic movie distributions", where $c \in \{1, \ldots, N_c\}$ denotes movies.

Inferred Topics. Topic consistency is an empirical notion that we use in the rest of the paper. A strongly consistent topic has all its top words belonging to the same lexical field. This notion can be applied to any list of words such as a document or a corpus.

Parameter K influences the consistency of the descriptive topics. Ideally, we would like each topic to represent a movie *feature*, e.g. actors, genres, or sequels.

After multiple experiments with values of K between 8 and 260, we observed that for $K \leq 30$, the descriptive topics mix several features. On the contrary, for $K \geq 150$, each feature is spread over multiple topics and topic consistency decreases. In Table 1, $K = 128$ is a good compromise [1].

The first 5 topics inferred with the models LDA-O and LDA-C proposed above are presented in Table 1. As the topics extracted with LDA-U are not consistent and due to a lack of space, they are not presented. We observe that topics in LDA-O and LDA-C are very consistent around a genre, an actor, a director or a sequel. The main difference between these two methods is that there are more outliers in LDA-O topics than in LDA-C topics. These outliers are generic words appearing frequently enough in the corpus to influence the topics. These words are not frequent enough to be removed during the filtering (e.g., *best*, *action*, *review*). In LDA-C, the aggregation of reviews related to the same movie lowers the impact of these outliers.

The corpus used in LDA-C is consistent because reviews of the same movie share a common vocabulary, each document of this corpus brings out words associated to the movie and lowers the influence of noisy words in the topics. We also observe that topics obtained with our models are mostly descriptive and that very few qualitative information is "lost" in the middle of the topics, making it difficult to tell a user whether or not she would like a movie.

2.3   Inclusion of Ratings in the Inference Process

Given the lack of qualitative in the top words of the topics, we introduce a method to extract qualitative topics (i.e., words with a positive or negative connotation) using numerical ratings in addition to text reviews. Given a text review and the corresponding rating, we expect user's opinion to be conveyed in the text and summarized in the rating. Ideally, positive (resp. negative) words should be more likely to appear in a high (resp. low) rated review. In order to keep the model simple, we reduce first numerical ratings (initially between 1 and 10) to binary ratings $\{-1, +1\}$. Using these new ratings, we infer a positive (resp. negative) topic from $+1$ (resp. -1) reviews in a new generative model.

Rating Reduction. We use a standard rating prediction technique [9] to extract users and movie features given ratings. We denote by D the number of reviews, r^d the rating of the dth review, related to user $u^d \in \{1, \ldots, N_u\}$ and movie $c^d \in \{1, \ldots, N_c\}$ with $d \in \{1, \ldots, D\}$. We first model the ratings to be generated as a sum of a user factor and a movie factor (ANOVA, [3]):

$$r^d = \mu + a_{u^d} + b_{c^d} + \epsilon, \ \ d \in \{1, \ldots, D\},$$

---

[1] For restaurants, the number of features is smaller and the optimal value of K is around 60.

with $\epsilon \sim \mathcal{N}(0, \sigma^2)$. We optimize (a_u), (b_c) and μ as the solution of a penalized (in L^2 norm) least-squares problem.

The training ratings are reduced to -1 if the following residual is negative and to $+1$ if the residual is positive:

$$\hat{r}^d = \begin{cases} +1 & \text{if } r^d - (a_{u^d} + \mu) \geq 0 \\ -1 & \text{otherwise} \end{cases}$$

In the model, μ represents the average rating, b_c the average deviation from μ of ratings on movie c, a_u the average deviation from μ of ratings from user u. We consider here that a review is positive $(\hat{r}^d = +1)$ if user u^d rated movie c^d with a higher score than her average rating $(a_{u^d} + \mu)$. The reduced rating $\hat{r}^d$ represents the binary user opinion.

For any test review t, the user opinion $\hat{r}^t$ is unknown. This quantity is then a random variable with:

$$\mathbf{Pr}[\hat{r}^t = +1] = \mathbf{Pr}(b_{c^t} + \epsilon \geq 0 | a, b)$$
$$\mathbf{Pr}[\hat{r}^t = -1] = 1 - \mathbf{Pr}[\hat{r}^t = +1]$$

During prediction of the review t, we use these probabilities to extract a mixture of positive and negative words.

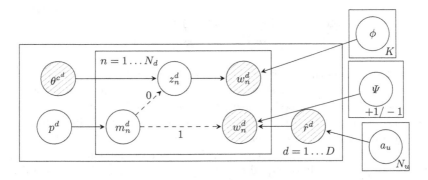

Fig. 1. Graphical representation of the model **LDA-R-C** including reduced ratings applied on D documents. White nodes represent hidden variables and colored nodes represent observed variables. The observed rating r^d is not reported for the sake of clarity.

Qualitative Topics Inference. We use these reduced ratings $(\hat{r}^d)$ to infer qualitative topics. We consider a corpus where each document is a single review. We denote by D the number of documents, by $c^d \in \{1, \ldots, N_c\}$ and $\hat{r}^d \in \{-1, +1\}$ respectively the movie and reduced rating of the dth document, $d = 1, \ldots, D$. K is the number of descriptive topics, i.e., the number of topics extracted with **LDA-C**. Given K, the corresponding topic movie distributions $\theta^c, c \in \{1, \ldots, N_c\}$ inferred with **LDA-C** on the D reviews and the reduced ratings $(\hat{r}^d)$, we extract a positive topic and a negative topic with the following generative model. For each word w_n^d of the dth document $\mathbf{w}^d$:

- Draw $m_n^d \sim$ Bernoulli(p^d)
- If m_n^d is a success ($m_n^d = 1$):
  - Choose $w_n^d \sim$ Multinomial$(\Psi^{\hat{r}^d})$, with $\hat{r}^d \in \{+1, -1\}$
- Otherwise ($m_n^d = 0$):
  - Choose a topic $z_n^d \sim$ Multinomial(θ^{c^d})
  - Choose $w_n^d \sim$ Multinomial$(\phi^{z_n^d})$.

In other words, if m_n^d is a success, w_n^d is qualitative otherwise w_n^d is descriptive. We still infer K descriptive topics with this method, represented by $\{\phi^1, \ldots, \phi^K\}$. The main difference with LDA-C is that we only infer ϕ, Ψ parameters, knowing topic document distributions θ. We also learn the proportion of qualitative/descriptive words in each document p^d, embedded with a Dirichlet prior. In the following, we refer to this model as LDA-R-C. The graphical representation of LDA-R-C is presented Fig. 1.

We build two additional models by replacing in LDA-R-C the topic movie distributions θ^c by the topic document distributions $\theta^d, d = 1 \ldots D$ — model LDA-R — or by the topic user distributions $\theta^u, u = 1 \ldots N_u$ — model LDA-R-U.

For the three models, the inference is done with a variational EM algorithm adapted from LDA inference [2]—see Appendix A for complete derivations. We also run a full EM for learning at the same time both topics (parameters ϕ, Ψ) and topic distribution (parameter θ). As the algorithm is very sensitive to initialization, we observe that random initialization gives poor results while initialization with parameters resulting from LDA-O leads to a steady state. Indeed, LDA-O returns a local miminum of the full algorithm.

Inferred Topics. As the qualitative topics extracted with LDA-R-U are not consistent and due to a lack of space, they are not presented. The qualitative topics inferred with LDA-R and LDA-R-C are presented in Table 2. In both LDA-R and LDA-R-C the average proportion $\frac{1}{D}\sum_d p^d$ of qualitative words in documents is 10%. Some qualitative words appear in the top words of the opposed topic (e.g., *bad* comes up in the positive topic of LDA-R). In our model, each training document d is assigned to only one qualitative topic — Ψ^{+1} or Ψ^{-1} — depending on the corresponding rating $\hat{r}^d$. Consequently, if a positive word frequently appears in low rated reviews — e.g., in negation phrases — it is likely to be a top negative word and vice versa. Some neutral words also come up in these qualitative topics (e.g., *suppose*). These words appear in reviews of different types of movies and are then in the tail of descriptive topics. They are still used frequently enough to have an influence on the qualitative topics.

3   Evaluation

In this section we evaluate our model for the review prediction task. Using our models, a review prediction is a discrete distribution over the V words of the vocabulary given a set of training reviews $\mathcal{W}$ and a test couple (u^t, c^t) of user, movie. The evaluation of such prediction is done by splitting the review dataset in a training set and a testing set. The parameters of the model are learned on the training set and evaluated on the test set.

Table 2. Positive (left column) and negative (right column) topic inferred with LDA-R and LDA-R-C, $K = 32$ topics.

LDA-R		LDA-R-C	
Great	Bad	Great	Bad
Best	Worst	Bad	Great
Nice	Boring	Love	Boring
Bad	Waste	Best	Worst
Entertaining	Stupid	Original	Stupid
Original	Terrible	Entertaining	Totally
Love	Suppose	Definitely	Waste

Methodology. Given a test review $\mathbf{w}$, the best predictor maximizes the likelihood $\mathbf{P}(\mathbf{w}|\mathcal{W})$ of the test review, $\mathbf{P}$ depending on the model. We then use the log-perplexity measure defined by $LP = -\log \mathbf{P}(\mathbf{w}|\mathcal{W})$ to evaluate our models. The log-perplexity is a theoretical measure of the quality of the model for the word prediction task; it is not an indicator of user satisfaction. As shown by [4], the perplexity is not suited to measure user satisfaction. However, perplexity measures the precision of the prediction, which is what we need in order to compare our models to the state of the art.

For LDA-O, the likelihood is intractable to compute. We approximate $\mathbf{P}(\mathbf{w}|\mathcal{W})$ with the "left-to-right" evaluation algorithm [16] applied to each test document. This algorithm is a mix of particle filtering and Gibbs sampling, easily adjustable to other graphical models. For LDA-C, as the topic distributions are learned for each movie (resp. user), the likelihood of a new review is computed for movies (resp. users) seen in the training corpus through pointwise estimation. Finally, we adapt the "left-to-right" algorithm to approximate the likelihood of LDA-R and LDA-R-C for each test document.

We compare our models to two existing approaches. In the model proposed by [12], authors use words and ratings to predict ratings by learning a mapping function between LDA parameters and rating matrix factorization parameters. We refer to this model as HFT. In the model proposed by [13], authors incorporate scores directly in LDA in order to predict the score of a new review given the words used. We refer to this model as SLDA. These two methods will be discussed in related works. Both HFT and SLDA use documents and ratings associated to each document to infer similar parameters than LDA — i.e., topics ϕ and topic proportions θ. While the application presented by [13] and [12] is rating prediction, we can use the parameters ϕ and θ inferred with these methods to predict words of new reviews.

We could not compare to the method described by [11] because of the lack of information available to implement the method accurately.

The review dataset was collected from IMDB, with a catalogue of 5,900 movies, 2,400 users and 97,000 reviews — a subset of the data described by [5].

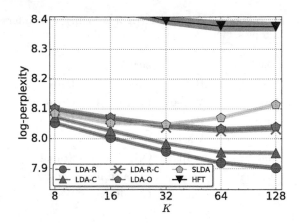

Fig. 2. Average log-perplexity per word of the presented models on test reviews.

We apply our models to 10 random splits of train and test sets, representing respectively 90% and 10%. Each movie and each user of a testing set is seen at least once in the corresponding train reviews.

Figure 2 present the average log-perplexity measures per word on the test sets for LDA-O, LDA-C, LDA-R, LDA-R-C, HFT and SLDA. Given the poor topic consistency observed with LDA-U and LDA-R-U and their poor performance, we do not display perplexity for these models.

Evaluation. The number of topics K is the unique input of LDA and its extensions. It represents the dimension of the latent topic document distribution θ. The Perplexity on test sets still decreases at $K = 128$ for LDA-C and LDA-R, which means we do not reach overfitting with $K \leq 128$.

LDA-O overfits the training reviews as it infers one distribution over topics (θ) by review. This method does not catch the structure of the vocabulary in the reviews as all reviews are processed separately. The predictions with LDA-O are then polluted by frequent words which appear in the inferred topics and lower the quality of the predictions. Conversely, the aggregation of reviews in LDA-C improves the consistency of the topics and generates better movie profiles (through topics movie distribution θ^c). LDA-C outperforms LDA-O.

By comparing the performance of LDA-O and LDA-R, we can observe that including reduced ratings in LDA-O improves perplexity for all values of K. Indeed, for all K, descriptive topics of LDA-R contain noisy words. The representation of the qualitative words in the training set results in an important decrease of perplexity compared with LDA-O.

The consistency of qualitative topics in LDA-R makes predictions outperform LDA-R-C. Predictions with LDA-R-C are also beaten by predictions with LDA-C. The consistency of descriptive topics of LDA-C is affected when including ratings.

With HFT the topic-document distribution θ is mapped to matrix factorization parameters. The objective being optimized can be expressed as a sum: $f =$ (rating prediction error) $-$ (corpus likelihood). The quality of word prediction is deteriorated as the objective is not directly the corpus likelihood. As a result, HFT is the worst predictor.

SLDA overfits at $K = 32$ as it infers one distribution over topics (θ) by review. This model includes ratings in LDA to infer a score along each topic. As a result, SLDA extracts descriptive and qualitative topics but they may not be automatically differentiated (score associated with a descriptive topic may be as high as the score associated with a positive/negative topic). SLDA is able to extract qualitative and descriptive information from the reviews but is outperformed by LDA-C and LDA-R for all values of K.

Even if the best predictors are LDA-R and LDA-C, the most practical model is LDA-R-C for three main reasons. Performances of LDA-R-C — in terms of perplexity — are comparable to the best models. This model is the only model combining strongly consistent descriptive topics with consistent qualitative topics. LDA-R suffers from polluted descriptive topics while LDA-C does not infer any qualitative information. The profiles of movies θ^c extracted with LDA-R-C are useful to compute a distance between movies — e.g., with the Kullback-Leibler divergence. As a result, one could recommend movies to a cold start user (i.e., user with an insufficient number of reviews) with LDA-R-C.

4   Empirical Discussion

In this section we compare and discuss side-by-side the topics produced by the methods discussed in the paper. We only present the 8 most significant topics out of 100 topics inferred with LDA-R-C and SLDA in Table 3. As the topics extracted with HFT are less consistent than with LDA-R-C and SLDA and due to lack of space, the topics of HFT are presented in Appendix B. The comparison of the topics is a premise to the analysis of words prediction. Indeed, for any LDA based algorithm, the probability of occurence of word w is given by $\sum_k \theta_k^\top \phi_w^k$, where θ represents the topic proportions of the current review and ϕ the inferred topic matrix. A topic vector ϕ^k may be seen as the probabilities of occurence of words in a review generated from a single topic i.e., $\theta_k = 1$ and $\forall j \neq k, \theta_j = 0$.

Note that this section is an empirical discussion and that it does not replace a formal evaluation with real users that is planned for future works due to the complexity of the task (methodologically and business-wise).

Both LDA-R-C and SLDA extract qualitative topics. The descriptive topics of the two methods are consistent around genres, sequels, actors or directors. We observe that the two methods extract similar topics (Table 3, topics T1 to T5 extracted with LDA-R-C are similar to topics T3 to T7 extracted with SLDA). From topics extracted with LDA-R-C, T1 is consistent around *comedy*, T2 around *horror*, T3 around *animation*, T4 around *family*, T5 around *science-fiction*. We can see the same consistency for topics T3 to T7 with SLDA. We observe that topics extracted with LDA-R-C share many top words with SLDA. Indeed, the inference schemes of LDA-R-C and SLDA are close (i.e., variational EM), leading to similar topics.

The difference between LDA-R-C and SLDA is that with LDA-R-C we obtain two distinct sets of words. A first set of words that illustrates the rating of

Table 3. 8 topics extracted with `LDA-R-C` and `SLDA`, $K = 100$ and the associated score for `SLDA` (see [13] for details).

LDA-R-C, 1 doc = 1 movie							
Qualitative		Descriptive					
Q1	Q2	T1	T2	T3	T4	T5	T6
Great	Bad	Funny	Horror	Animation	Child	Alien	Action
Love	Waste	Comedy	Scary	Disney	Family	Space	Car
Best	Boring	Laugh	House	Voice	Father	Earth	Jones
Beautiful	Worst	Fun	Scare	Animated	Son	Sci-fi	Fast
Perfect	Stupid	Great	Creepy	Toy	Mother	Science	Bad
Excellent	Money	Hilarious	Suspense	Pixar	Daughter	Planet	Agent
Far	Suppose	Joke	Ghost	Adult	Kevin	Fiction	Fun
Enjoy	Long	Star	Gore	Child	Boy	Ship	Diesel
SLDA							
T1	T2	T3	T4	T5	T6	T7	T8
(1.01)	(−1.59)	(0.73)	(0.89)	(0.91)	(0.80)	(0.70)	(1.20)
Enjoy	Bad	Funny	Horror	Voice	Family	Alien	Classic
Surprise	Worst	Comedy	Scary	Animation	Father	Space	Era
Entertaining	Terrible	Laugh	Scare	Disney	Mother	Earth	Noir
Fun	Waste	Joke	House	Animated	Son	Science	Today
Nice	Horrible	Hilarious	Creepy	Toy	Child	Planet	Early
Pretty	Awful	Fun	Atmosphere	Child	Daughter	Fiction	Silent
Great	Worse	Parody	Ghost	Adult	Young	Sci-fi	Modern
Enjoyable	Stupid	Satire	Haunt	Pixar	Parent	Ship	Kane

the users — corresponding to topics Q1 and Q2 in Table 3 — and a second set of words that the user would employ to describe a movie — topics T1 to T6 in Table 3. There is no such distinction in `SLDA` as the output is a single list of topics associated with a score. We notice that most of the extracted topics with `SLDA` are descriptive — e.g., topics T3 to T8 in Table 3 — and qualitative topics are associated to high scores (in absolute value) — e.g., topics T1 and T2 in Table 3. However, some descriptive topics are associated with high scores — e.g., topic T8 in Table 3 — which makes difficult the distinction between qualitative and descriptive topics. As our model `LDA-R-C` extracts two sets of topics, there is no possible confusion between descriptive and qualitative topics. `SLDA` extracts a single list of topics and it is difficult nay impossible to automatically classify the topics as a qualitative or a descritpive.

5   Related Work

Several techniques have been proposed to extract information from raw text data. LDA [2] is a probabilistic model that infers hidden topics given a text corpus where each document of the corpus can then be represented as topic probabilities. The assumption behind LDA is that each document is generated from a mixture of topics and the model infers the hidden topics and the topic proportions of each document. For increasing consistency of inferred topics, a regularized version of LDA is proposed by [14]. This regularized version puts structured priors on the hidden topics. The parameters of the prior are pre-computed and consist in a "covariance" matrix which captures the short-range dependencies between words. This matrix has a regularization effect on the topics. The authors compare different priors. We increase the consistency of topics even further by aggregating reviews in LDA. [15] present a LDA based model with two types of topics; this model infers global topics that contain the different types of movie being reviewed, while the local topics extract the specific aspects of the movies. Instead, we chose to add qualitative information in the LDA topics, similar to the model proposed by [10] where a sentiment label is inferred for each document. The difference with our method is that we leverage the ratings found in movie reviews datasets to extract qualitative and descriptive information in separate topics.

In the rating prediction area, [12] propose a transformation between LDA parameters and collaborative filtering parameters. Numerical ratings and words are processed in two separated models. Instead, we combine scores and words in the same LDA model. The model proposed by [13] infers parameters from both a text corpus and numerical ratings. The model is then able to predict a score given a new text. We use a similar approach to predict directly a list of words instead of numerical ratings. The model proposed by [11] combines LDA with matrix factorization to predict ratings. The main difference with our approach is that we distinguish qualitative and descriptive words in the topics while the topics inferred by [11] mix qualitative and descriptive words. Their model is also suited for rating prediction while we focus on word prediction.

6   Conclusion

We have proposed six LDA-based models for word prediction from crowd sourced reviews and ratings. We show on an IMDB dataset that our LDA-R-C model combining movie profiling and ratings performs slightly better than the state of the art. It builds a set of descriptive topics that convey the features of movies — e.g., genres, actors, directors — and contain the words the user would employ to describe a movie. It also builds a set of qualitative topics that convey the opinion of users about movies and contain the words that influence — positively or negatively — the final ratings of users.

While studying our LDA-U model, we came to the conclusion that it is difficult to build a user profile as each user writes reviews about very different movies expressing very different opinions.

For now, our models only extract two qualitative topics (positive and negative). We plan to build models that would extract a wider range of qualitative topics by reducing the observed ratings to a wider range of values than $\{+1, -1\}$.

The review prediction is currently based on single words preciction, which is not intuitive for users. We plan to predict readable sentences which would facilitate evaluation by users with A/B testing. Implemented with readable reviews or tags, our model could be integrated in a recommender to provide a personalized opinion summary to users for each content.

A   Variational derivation of LDA-R-C

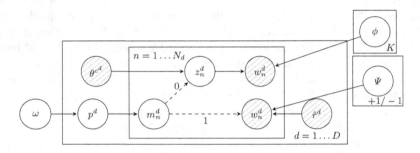

Fig. 3. Graphical representation of the model LDA-R-C including reduced ratings applied on D documents. White nodes represent hidden variables and colored nodes represent observed variables. The observed rating r^d is not reported for the sake of clarity. (Color figure online)

In this section, we provide the full variational derivation of our model LDA-R-C presented Fig. 3. Our objective is to maximize the likelihood of the observed corpus of documents $\mathcal{W} = w^1, \ldots, W^D$:

$$p(\mathcal{W}|\omega, \{\theta^{c^d}\}_d, \eta, \{\hat{r}^d\}_d) = \prod_{d=1}^{D} p(w^d|\omega, \theta^{c^d}, \eta, \hat{r}^d).$$

As this likelihood is intractable to compute, we maximize an approximation of the likelihood $\mathcal{L}(q) = \sum_{d=1}^{D} \mathcal{L}^d(q)$ over a variational family of distributions. Following [8], we have for any $w^d \in \mathcal{W}$:

$$\log p(w^d|\omega, \theta^{c^d}, \eta, \hat{r}^d) \geq \mathbb{E}_q[\log p(w^d, z^d, m^d, p^d, \phi, \Psi|\omega, \theta^{c^d}, \eta, \hat{r}^d)] - \mathbb{E}_q[\log q(z^d, m^d, p^d, \phi, \Psi)] \equiv \mathcal{L}^d(q),$$

where q represents the variational model. We choose the variational model q to be in the meanfield variational family:

$$q(z^d, m^d, p^d, \phi, \Psi) = q(\phi|\lambda)q(\Psi|\Lambda)q(p^d|\pi^d) \prod_{n=1}^{N} q(z_n^d|\alpha_n^d)q(m_n^d|\mu_n^d),$$

with, $\forall d = 1, \ldots, D$:

- $q(\phi^k|\lambda^k) \sim \text{Dirichlet}(\lambda^k)$ with $\lambda^k \in \mathbb{R}^V$ and $k = 1, \ldots, K$,
- $q(\Psi^s|\Lambda^s) \sim \text{Dirichlet}(\Lambda^s)$ with $\Lambda^s \in \mathbb{R}^V$ and $s \in \{+1, -1\}$,
- $q(p^d|\pi^d) \sim \text{Dirichlet}(\pi^d)$ with $\pi^d \in \mathbb{R}^2$,
- $q(z_n^d|\alpha_n^d) \sim \text{Multinomial}(\alpha_n^d)$ with $\alpha_n^d \in \mathbb{R}^K$, $\sum_k \alpha_{n,k}^d = 1$ and $n = 1, \ldots, N$,
- $q(m_n^d|\mu_n^d) \sim \text{Multinomial}(\mu_n^d)$ with $\mu_n^d \in \mathbb{R}^2$, $\mu_{n,1}^d + \mu_{n,2}^d = 1$ and $n = 1, \ldots, N$.

We also have:

$$p(w^d, z^d, m^d, p^d, \phi, \Psi|\omega, \theta^{c^d}, \eta, \hat{r}^d) = p(\phi|\eta)p(\Psi|\eta)p(p^d|\omega) \prod_{n=1}^{N} p(w_n^d|\phi, \Psi, z_n^d, r^d, m_n^d)p(z_n^d|\theta^{c^d})p(m_n^d|p^d),$$

with, $\forall d = 1, \ldots, D$:

- $p(\phi^k|\eta) \sim \text{Dirichlet}(\eta\mathbf{1})$ with $\eta \in \mathbb{R}$ and $k = 1, \ldots, K$,
- $p(\Psi^s|\eta) \sim \text{Dirichlet}(\eta\mathbf{1})$ with $\eta \in \mathbb{R}$ and $s \in \{+1, -1\}$,
- $p(p^d|\omega) \sim \text{Dirichlet}(\omega)$ with $\omega \in \mathbb{R}^2$,
- $p(z_n^d|\theta^{c^d}) \sim \text{Multinomial}(\theta^{c^d})$ with $\theta^{c^d} \in \mathbb{R}^K$, $\sum_k \theta_k^{c^d} = 1$ and $n = 1, \ldots, N$,
- $p(m_n^d|p^d) \sim \text{Multinomial}(p^d)$ with $p^d \in \mathbb{R}^2$, $p_1^d + p_2^d = 1$ and $n = 1, \ldots, N$,
- $p(w_n^d|\phi, \Psi, z_n^d, r^d, m_n^d) \sim \text{Multinomial}\left(\phi^{z_n^d}\mathbf{1}[m_n^d = 0] + \Psi^{r^d}\mathbf{1}[m_n^d = 1]\right)$.

We then maximize $\mathcal{L}(q)$ by iteratively maximizing $\mathcal{L}(q)$ with respect to variational parameters $\lambda, \Lambda, \pi, \alpha, \mu$ (E-step) then maximizing $\mathcal{L}(q)$ with respect to hyperparameters ω, η (M-step).

A.1   Variational E-step

For the E-step, we maximize $\mathcal{L}(q)$ with respect to variational parameters $\lambda, \Lambda, \pi, \alpha, \mu$ by alternatively setting the gradient of $\mathcal{L}(q)$ with respect to each paramater to zero. It gives the following updates for the variational parameters, for $n = 1, \ldots, N$; $k = 1, \ldots, K$; $i = 1, 2$ and $s \in \{-1, +1\}$:

$$
\begin{cases}
\alpha_{n,k}^d \propto \theta_k^{c^d} \exp\left[\mu_{n,1}^d\left(\psi(\lambda_{w_n^d}^k) - \psi(\sum_j \lambda_j^k)\right)\right], \\
\pi_i^d = \omega_i + \sum_{n=1}^{N} \mu_{n,i}^d, \\
\mu_{n,1}^d \propto \exp\left[\psi(\pi_1^d) + \sum_{k=1}^{K} \psi(\lambda_{w_n^d}^k) - \psi(\sum_j \lambda_j^k)\right], \\
\mu_{n,2}^d \propto \exp\left[\psi(\pi_2^d) + \sum_{s \in \{-1,+1\}} \psi(\Lambda_{w_n^d}^s) - \psi(\sum_j \Lambda_j^s)\right], \\
\\
\lambda_v^k = \eta + \sum_{d=1}^{D} \sum_{n=1}^{N_d} \mu_{n,1}^d \alpha_{n,k}^d \mathbf{1}[w_n^d = v], \\
\Lambda_v^s = \eta + \sum_{d:r^d=s} \sum_{n=1}^{N_d} \mu_{n,2}^d \mathbf{1}[w_n^d = v].
\end{cases}
$$

ψ is the digamma function: $\psi(x) = \frac{d}{dx} \ln \Gamma(x)$.

A.2   Variational M-step

For the M-step, we maximize $\mathcal{L}(q)$ with respect to the hyperparameters ω, η. We use the Newton method for each parameter, using the same scheme than in LDA [2]. We have the following derivatives for ω:

$$
\begin{cases}
\frac{\partial}{\partial \omega_i}\mathcal{L}(q) & = D\left(\psi(\sum_j \omega_j) - \psi(\omega_i)\right) + \sum_{d=1}^D \left(\psi(\pi_i^d), -\psi(\sum_j \pi_j^d)\right) \\
\frac{\partial^2}{\partial \omega_i \partial \omega_j}\mathcal{L}(q) & = D\psi'(\sum_l \omega_l) - \mathbf{1}[i=j]D\psi'(\omega_i).
\end{cases}
$$

We have the following derivatives for η

$$
\begin{cases}
\frac{\partial}{\partial \eta}\mathcal{L}(q) & = (K+2)V\left(\psi(V\eta) - \psi(\eta)\right) + \sum_{v=1}^V \left(\sum_{k=1}^K \psi(\lambda_v^k) + \sum_{s=\{-1,+1\}} \psi(\Lambda_v^s)\right) \\
& \quad -V\left(\sum_{k=1}^K \psi(\sum_{v=1}^V \lambda_v^k) + \sum_{s=\{-1,+1\}} \psi(\sum_{v=1}^V \Lambda_v^s)\right), \\
\frac{\partial^2}{\partial \eta^2}\mathcal{L}(q) & = (K+2)V\left(V\psi'(V\eta) - \psi'(\eta)\right).
\end{cases}
$$

We maximize $\mathcal{L}(q)$ with respect to ω by doing iterations of Newton steps until convergence:

$$
\omega^{(t+1)} = \omega^{(t)} - H^{-1}\nabla_{\omega^{(t)}}\mathcal{L}(q),
$$

where H is the Hessian $H = \nabla^2_{\omega^{(t)}}\mathcal{L}(q)$. We then maximize $\mathcal{L}(q)$ with respect to η by again doing iterations of Newton steps until convergence:

$$
\eta^{(t+1)} = \eta^{(t)} - \left[\frac{\partial^2}{(\partial \eta^{(t)})^2}\mathcal{L}(q)\right]^{-1}\left(\frac{\partial}{\partial \eta^{(t)}}\mathcal{L}(q)\right).
$$

B   Topics Extracted with HFT [12]

In this section, we present the 8 most significant topics out of 100 topics inferred with LDA-R-C, SLDA and HFT in Table 4. Both LDA-R-C and SLDA extract qualitative topics, while we could not extract qualitative topics with HFT. The descriptive topics of the three methods are consistent around genres, sequels, actors or directors and the three methods extract similar topics. We observe that topics extracted with LDA-R-C share more top words with SLDA than with HFT. For instance, in Table 3, 6 out of 10 top words of topic T1 obtained with LDA-R-C also appear in the top of SLDA's topic T3. In the same way, topics T2 to T5 extracted with LDA-R-C are respectively closer to topics T4 to T7 extracted with SLDA than topics T4 to T7 extracted with HFT.

In HFT, the parameters of LDA are linked to rating prediction parameters. As a result, the top words of the topics are still centered around generic genres, sequels, actors, directors but also contain words related to specific movies. For instance, in Table 3, the topic T4 extracted with HFT is centered around *comedy* and contains the words *sandler, ferrell* which are specific actor names and *wedding* which is a specific part of a plot. In the topic T5 extracted with

HFT, centered around *animation movies*, we find the words *wall-e, nemo* which are specific titles and *costner* which is an actor name. The top words in both LDA-R-C and SLDA topics are more generic, leading to better predictions. Indeed, it is more likely that a review about a *comedy* movie contains *funny* than *wedding*, as only few comedy movies are related to a wedding.

Table 4. 8 topics extracted with LDA-R-C, SLDA and HFT, $K = 100$ and the associated score for SLDA (see [13] for details).

LDA-R-C, 1 doc = 1 movie

Qualitative		Descriptive					
Q1	Q2	T1	T2	T3	T4	T5	T6
Great	Bad	Funny	Horror	Animation	Child	Alien	Action
Love	Waste	Comedy	Scary	Disney	Family	Space	Car
Best	Boring	Laugh	House	Voice	Father	Earth	Jones
Beautiful	Worst	Fun	Scare	Animated	Son	Sci-fi	Fast
Perfect	Stupid	Great	Creepy	Toy	Mother	Science	Bad
Excellent	Money	Hilarious	Suspense	Pixar	Daughter	Planet	Agent
Far	Suppose	Joke	Ghost	Adult	Kevin	Fiction	Fun
Enjoy	Long	Star	Gore	Child	Boy	Ship	Diesel
Long	Awful	Love	Night	Fun	Young	Ape	Rock
Wonderful	Pretty	Brooks	Remake	Princess	Town	Crew	Chase

SLDA

T1	T2	T3	T4	T5	T6	T7	T8
(1.01)	(−1.59)	(0.73)	(0.89)	(0.91)	(0.80)	(0.70)	(1.20)
Enjoy	Bad	Funny	Horror	Voice	Family	Alien	Classic
Surprise	Worst	Comedy	Scary	Animation	Father	Space	Era
Entertaining	Terrible	Laugh	Scare	Disney	Mother	Earth	Noir
Fun	Waste	Joke	House	Animated	Son	Science	Today
Nice	Horrible	Hilarious	Creepy	Toy	Child	Planet	Early
Pretty	Awful	Fun	Atmosphere	Child	Daughter	Fiction	Silent
Great	Worse	Parody	Ghost	Adult	Young	Sci-fi	Modern
Enjoyable	Stupid	Satire	Haunt	Pixar	Parent	Ship	Kane
Interesting	Boring	Gag	Gore	Cartoon	Boy	Predator	Simple
Definitely	Crap	Silly	Disturbing	Age	Brother	Scientist	Citizen

HFT

T1	T2	T3	T4	T5	T6	T7	T8
Action	Vampire	Comedy	Horror	Animation	Father	Sci-fi	Hulk
Franchise	Dracula	Funny	Halloween	Disney	Son	Science	Fox
Installment	Twilight	Sandler	Slasher	Pixar	Fanning	Space	Banner
Diesel	Blade	Hilarious	Scary	Animated	Dakota	Mars	Car
Explosion	Beckinsale	Ferrell	Eli	Wall-e	Dad	Spaceship	Bana
Sequel	Helsing	Laugh	Scare	Costner	Precious	Planet	Ross
Cgi	Underworld	Wedding	House	Nemo	Boy	Earth	Wax
Stunt	Jacob	Joke	Myers	Chicken	Mother	Robot	Racing
Vin	Bella	Gag	Creepy	Toy	Bike	Scientist	Eric
Fun	Edward	Comedic	Carrie	Dreamworks	Parent	Alien	Norton

References

1. Bird, S., Klein, E., Loper, E.: Natural language processing with Python. O'Reilly Media, Inc. (2009)
2. Blei, D.M., Ng, A.Y., Jordan, M.I.: Latent Dirichlet allocation. JMLR **3**, 993–1022 (2003)
3. Casella, G., Berger, R.L.: Statistical Inference, vol. 2. Duxbury Pacific Grove, CA (2002)
4. Chang, J., Gerrish, S., Wang, C., Boyd-Graber, J.L., Blei, D.M.: Reading tea leaves: How humans interpret topic models. In: Advance in Neural Information Processing Systems (2009)
5. Diao, Q., Qiu, M., Wu, C.Y., Smola, A.J., Jiang, J., Wang, C.: Jointly modeling aspects, ratings and sentiments for movie recommendation (JMARS). In: Proceeding of ACM SIGKDD (2014)
6. Griffiths, T.L., Steyvers, M.: Finding scientific topics. Proc. Nat. Acad. Sci. **101**(suppl 1), 5228–5235 (2004)
7. Hoffman, M.D., Blei, D.M., Bach, F.R.: Online learning for latent Dirichlet allocation. In: Advance in Neural Information Processing Systems (2010)
8. Hoffman, M.D., Blei, D.M., Wang, C., Paisley, J.: Stochastic variational inference. JMLR **14**(1), 1303–1347 (2013)
9. Koren, Y., Bell, R., Volinsky, C.: Matrix factorization techniques for recommender systems. Computer **42**(8), 30–37 (2009)
10. Lin, C., He, Y.: Joint sentiment/topic model for sentiment analysis. In: Proceeding of the 26th International Conference on Machine Learning, pp. 375–384 (2009)
11. Ling, G., Lyu, M.R., King, I.: Ratings meet reviews, a combined approach to recommend. In: Proceedings of the 8th ACM Conference on Recommender Systems, pp. 105–112. ACM (2014)
12. McAuley, J.J., Leskovec, J.: Hidden factors and hidden topics: understanding rating dimensions with review text. In: Proceedings of the 7th ACM Conference on Recommender Systems (2013)
13. Mcauliffe, J., Blei, D.: Supervised topic models. In: Advance in Neural Information Processing Systems, pp. 121–128 (2008)
14. Newman, D., Bonilla, E., Buntine, W.: Improving topic coherence with regularized topic models. In: Advance in Neural Information Processing Systems, pp. 496–504 (2011)
15. Titov, I., McDonald, R.: Modeling online reviews with multi-grain topic models. In: Proceedings of the 17th International Conference on World Wide Web, pp. 111–120. ACM (2008)
16. Wallach, H.M., Murray, I., Salakhutdinov, R., Mimno, D.: Evaluation methods for topic models. In: Proceedings of the 26th International Conference on Machine Learning(2009)

Towards an Efficient Method of Modeling "Next Best Action" for Digital Buyer's Journey in B2B

Anit Bhandari[1(✉)], Kiran Rama[1], Nandini Seth[2], Nishant Niranjan[1],
Parag Chitalia[1], and Stig Berg[1]

[1] VMware Inc., Palo Alto, USA
{anitb,rki,nniranjan,pchitalia,stigb}@vmware.com
[2] Indian Institute of Management – Bangalore, Bengaluru, India
nandini.seth15@iimb.ernet.in

Abstract. The rise of Digital B2B Marketing has presented us with new opportunities and challenges as compared to traditional e-commerce. B2B setup is different from B2C setup in many ways. Along with the contrasting buying entity (company vs. individual), there are dissimilarities in order size (few dollars in e-commerce vs. up to several thousands of dollars in B2B), buying cycle (few days in B2C vs. 6–18 months in B2B) and most importantly a presence of multiple decision makers (individual or family vs. an entire company). Due to easy availability of the data and bargained complexities, most of the existing literature has been set in the B2C framework and there are not many examples in the B2B context. We present a unique approach to model next likely action of B2B customers by observing a sequence of digital actions. In this paper, we propose a unique two-step approach to model next likely action using a novel ensemble method that aims to predict the best digital asset to target customers as a next action. The paper provides a unique approach to translate the propensity model at an email address level into a segment that can target a group of email addresses. In the first step, we identify the high propensity customers for a given asset using traditional and advanced multinomial classification techniques and use non-negative least squares to stack rank different assets based on the output for ensemble model. In the second step, we perform a penalized regression to reduce the number of coefficients and obtain the satisfactory segment variables. Using real world digital marketing campaign data, we further show that the proposed method outperforms the traditional classification methods.

Keywords: Multi-class classification · Literature survey · Ensemble · Regression · Digital · Robustness · Non-negativity constraint · B2B · Next-best action

1 Introduction

VMware (VMW) is a virtualization, end user computing and cloud company with annual revenues of USD 6,571 million (as of 2015) and a market capital of USD 34 BB as of 2017. VMware sells products in the Software Defined Data Center (vSphere,

© Springer International Publishing AG 2017
P. Perner (Ed.): MLDM 2017, LNAI 10358, pp. 107–116, 2017.
DOI: 10.1007/978-3-319-62416-7_8

VSAN, NSX for computing, storage & network virtualization respectively), end user computing (Air-watch, Horizon, and Fusion/Workstation) and cloud. The company exclusively caters to business customers – i.e. B2B.

VMW is characterized by 100% digital supply chain which means that all products are downloadable from the website (www.vmware.com). The company also promotes them online. Different individuals across companies worldwide visit the site to familiarize themselves with the products before making a purchase decision. Along with the overview of the product, there are various customer- interaction digital assets that are show to the VMW audience including and not limited to:

– Hands-on Labs (HoL): Here the visitor can evaluate a VMW product before making a decision to buy. HoLs provide a virtual environment where a visitor/email id can acquaint himself with the product by using it first-hand.
– Eval: Here the visitor can download a version of the software for his/her personal use.
– Whitepapers: These downloadable papers cover a wide range of topics related to the product - such as usage, features, vs. competition summaries, Gartner research reports etc.
– Seminar & Webinar: Here the visitor can register for a seminar or a webinar.

The digital buyer journey goes through 4 stages (Awareness → Interest → Trial → Action) which VMW would like to personalize. Of the digital assets available on the website, HoL is more of an Action, Eval is more of a Trial, and Whitepaper is more of Awareness whereas Seminar/Webinar are expressions of interest. To personalize the buyer's journey, it is imperative to identify the appropriate digital assets that need to be pressed to the consumers and an optimal ordering for the same. An example for digital assets is shown in Fig. 1.

Traditionally, the models built for B2B interactions focus all their marketing effort assuming the company to be a unified entity. While this is a prudent assumption to design a "Propensity to buy" model, it is not appropriate for the "propensity to respond" model we are trying to propose. A company as a united entity may have a propensity to buy a product but the diverse individuals within the company will have varying propensities to respond. There are numerous individuals involved in the decision making process who go through various phases of nurture program before they complete a purchase. The traditional response models attempt to predict the likelihood of response for a marketing effort around a particular digital asset. However, these models do not take into account the past consumption event of the individual with the particular asset. The method presented here eliminates this drawback and work towards targeting customers more effectively.

1.1   Objective

In consultation with the Digital Marketing team, the Advanced Analytics & Data Sciences team came up with the following objectives:

Fig. 1. Digital assets example – case study download, webinar, hands-on-lab, eval (in clockwise)

- Determine the right order of digital assets to display to an individual email id
- Since the website would like to target groups of email ids, come up with a set of segment rules to identify top individuals for a digital asset and to target them with personalization on the website

2 Literature Review

2.1 Digital Marketing Literature

Digital Marketing is the process of fostering customer relationships through online activities to expedite the exchange of ideas, products, and services that satisfy the twofold objectives of the customer as well as the seller [1]. Lately, the use of internet to expand marketing efforts has intensified. With the internet serving as a computer facilitated marketplace in which customers and sellers can access each other, they can easily execute trade functions like sales, distribution and marketing [2]. Customers go through a compound decision making process before they make their final consumption. As McKinsey [3] rightly pointed out, "consumers don't want to feel subjected to the hard sell — they expect marketers to engage them, not dictate to them." As a result, traditional marketing strategies have reformed drastically leading to additional customer value, improved targeted marketing & escalated company profits [4]. Due to the comparative novelty and an ever-growing exigency, digital marketing is an exciting area for research – not just academically but in the industry. Academicians and practitioners have emphasized on the significance of digital marketing to deal with marketing mixes, which include global accessibility, convenience in updating, real-time information services, interactive communications features, and unique customization and personalized capabilities [5]. Plenty of attention has been focused on the tremendous opportunities digital marketing presents, with very little attention on the real challenges companies are facing with analyzing and interpreting digital. These challenges are discussed at length in the recent paper by Leeflang et al. [8]. A number of books have been published in the recent years concentrating on B2B digital marketing (see [6, 7]). Even though these books are wide-ranging, there is still a great deal of understanding that is required to formulate models on digital data, especially in a B2B framework.

2.2 Machine Learning Models Literature

In the past decade, striving attempts have been made to design frameworks and models to derive value out of digital data using multi-class classification methods like logistic regression, random forests, gradient boosting etc. Random forest and gradient boosting models have been extensively used on digitally generated data in analytical Customer Relationship Management (CRM) to develop churn-prediction models [9–11, 22]. Random forest has been lucratively used in answering other marketing questions like predicting subject line open rates for targeted emails [12], privacy preserving data mining [13], defect prediction [14], building different attribution models [15, 16] etc. One of the principal application of gradient boosting in digital marketing has been in obtaining better recommender systems by improved unified search system [23] and adaptive advertising [24]. Various application of Regularized Logistic Regression can be found in the marketing literature such as for better targeting of display ads [19], multi-touch attribution model [20], analyzing the effectiveness of an ad [21] etc. Irrespective of the sizeable literature research done on digital marketing, very little has been said about identification and targeting of customers in a B2B framework. In an antecedent to the work present here, ensemble decision tree method has been used in a B2C setup to improve personalized advertising and behavioral targeting [18]. Ensemble methods use numerous learning algorithms to achieve improved predictive capabilities than what could be obtained from any of the featuring algorithms. Ensemble methods are popular for superior performance than other algorithms in most cases [17]. A current study [25] presents an ensemble method approach to developing a more accurate and implementable recommender system for B2B clients using collaborative filtering and gradient boosting.

3 Solution Framework

3.1 Integrating Online and Offline Features

The dataset for the model was created at an email address level using Pivotal Greenplum and Hadoop. In Greenplum which is an enterprise data warehouse, we created dataset with all historic bookings information. Similarly in the Hadoop environment from the click stream data, we created dataset containing individual's online behavior. We then combined both these datasets to obtain the single view of an individual's behavior across offline and online features. We ended up with 800 + explanatory features and one target variable comprising of 5 classes. Figure 2 presents a block diagram declaring all the features that were used to build the model. We used time based cross validation to examine the performance of the created models. Time Based

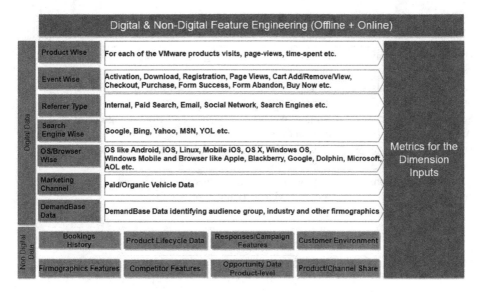

Fig. 2. Feature bucket diagram

Cross Validation is the right technique in digital marketing scenarios where we try to predict the future based on what we observe in the past. Such models lead to generalization and limit overfitting. To form the training set, we took the information for explanatory variables from FY13Q1 (01/01/13) to FY15Q2 (06/30/15) and the target variable from FY15Q3 (07/01/15) to FY15Q4 (12/31/15). Similarly for validation of the model, we took explanatory variables from FY13Q1 (01/01/13) to FY15Q4 (12/31/15) and the target variable was from FY16Q1 (01/01/16) to FY16Q2 (06/30/16).

3.2 Multiple Classification Methods

Since we had to find the best possible asset for better targeting of an email address, the target variable was assigned based on the order of action taken among the assets: HoL, Eval, Seminar/Webinar, Download or other events. Table 1 shows how the target variable was defined based on order of action.

Once the target variable was defined, we observed that the multi-class variable had high sparsity. We ran algorithms which could handle highly imbalanced data and at the same time can solve for a multinomial classification problem. Table 2 shows the numbers of positives in each of the target class for training and validation datasets.

Once the target variable was defined, we executed multiple algorithms designed to solve a multinomial classification problem.

(a) Since the data was highly imbalanced, the first approach we undertook was random forest [26, 28] with an under-sample method. The random forest method relies on an autonomous, pseudo-random procedure to select a small number of dimensions from a larger feature space [27] - in our case is 800 + features. We

Table 1. Target variable definition

Criterion	Details	Target
Hands-on-Lab < Any Action	Hands – on – Lab was registered before doing one or more actions	5
Any Action < Hands – on – Lab	Hands – on – Lab was registered after doing one or more actions	4
Seminar/Webinar < Any Action	Seminar/Webinar was signed up before doing one or more actions	3
Download < Any Action	Any download event occurred before doing one or more actions	2
All remaining actions	Any other action done other than download, seminar/webinar, Hands – on – Lab	1

Table 2. Target variable sparsity in datasets

Models	Population	Target 5 [y = 5]	Target 4 [y = 4]	Target 3 [y = 3]	Target 2 [y = 2]	Target 1 [y = 1]	Target 0 [y = 0]
Training	1,468,690	3,275	897	54	112	2,891	1,461,461
Validation	1,876,064	4,150	1,074	52	125	3,014	1,867,649

performed this for multiple iterations with different number of trees and under-samples of target class.

(b) The next attempted model was the L2 regularized logistic regression using LIBLINEAR: A library for Large Linear Classification [29]. It has been shown that LIBLINEAR is very efficient on large sparse data [30]. We tried L2 logistic implementation for different costs and arrived at the best cost to obtain the maximum AUC. We also ran an iteration of LIBLINEAR with assigned weights for each class in the model.

(c) The final model we tried was the extreme gradient boosting (xgboost) [31, 32] with two different approaches - optimizing eta, depth and number of trees. In the first trial we considered a higher depth of 9, with eta = 0.2 and no of rounds of trees = 200 which we call big_xg_boost and in the second trial, we decrease the depth to 7, eta = 0.5 and no of rounds = 15, called small_xg_boost.

We observed the AUC for all these five models for the multi-class target to compare their performances. Table 3 lists the AUC obtained for these models.

3.3   Proposed Ensemble Method Using Outputs of Traditional Methods

Now that we got the results from each of the multi-class classification models - To improve the AUC of the output, we propose an ensemble using the output of the models that were ran in the previous iteration along with enforcing a non-negativity

Table 3. Model performance measure based on AUC for traditional methods

Models	Target 5	Target 4	Target 3	Target 2	Target 1
Random forest under-sample	0.78	0.84	0.83	0.85	0.89
L2 regularized logistic regression	0.62	0.68	0.72	0.72	0.75
Weighted L2 regularized logistic regression	0.73	0.80	0.66	0.78	0.74
XGBoost (Big)	0.72	0.75	0.74	0.72	0.85
XGBoost (Small)	0.78	0.85	0.88	0.87	0.89

Table 4. Model Performance measure based on AUC including model ensemble results

Models	Target 5	Target 4	Target 3	Target 2	Target 1
Random forest under-sample	0.78	0.84	0.83	0.85	0.89
L2 regularized logistic regression	0.62	0.68	0.72	0.72	0.75
Weighted L2 regularized logistic regression	0.73	0.80	0.66	0.78	0.74
XGBoost (Big)	0.72	0.75	0.74	0.72	0.85
XGBoost (Small)	0.78	0.85	0.88	0.87	0.89
Ensemble method	0.82	0.87	0.90	0.89	0.91

constraint [33] in predicting digital action. In this ensemble method, for each target class we took the probability output from random forest under-sample, L2 regularized logistic regression, weighted L2 regularized logistic regression, XGBoost implementation as inputs into Lawson-Hanson NNLS implementation of non-negative least squares method in R [34] which results in the ensemble of models and is better performing than existing traditional methods. Table 4 shows how the ensemble method performs better then each of the individual models in each of the target class.

4 Conclusion

4.1 Results

As presented in Table 4, we observe that the ensemble technique presented in this paper performs better than the other ensemble methods – i.e. the ones obtained by combining identical classification methods. Intuitively, this improvement can be credited to the combination of the regularized logic regression method and the gradient boosting method. While the regularization takes care of the sparseness, boosting tends to increase the predictability of the model.

4.2 Uniqueness of the Approach

In varied ways, the current research is innovative. The novelty of the research lies in the following particulars.

(i)   The model attempts to study the digital behavior of B2B customers which has been inadequately explored in both academia and industry as of yet. In a B2B scenario, the decision maker is not an individual (as in the B2C case). It is possible that a targeted company is highly diverse in its structure and needs. In addition to that, each individual who is a part of the decision-making process might respond differently to the target asset leading to increased complexity in the model.

(ii) Disparate to most existing models, the current research presents the use of an ensemble of various dissimilar multi-class classification models (random forest, L2 regularized logistic regression- weighted and non-weighted and XGBoost-small and big). We demonstrate the superiority of this ensemble of models approach in mining highly imbalanced digital datasets in B2B which is forward-looking and highly sophisticated as compared to the existing literature.

(iii) The model goes beyond conventional propensity to buy models and contributes by providing a Propensity to Respond model. This alteration leads to tremendous managerial implication which have been provided below in Sect. 4.3.

(iv) The model also extends in identifying high propensity to respond individuals by rules which can be implemented to do website personalization and give an engaging experience to the individual. This is particularly useful in the web analytics world where many organizations might not have the tools to enable targeting at an individual email id or cookie level. Translating the propensity model output into segments allows organizations the ability to do 1:M personalization where M is the number of segments in cases where the technology stack of the organization is not equipped for 1:1 marketing.

4.3 Managerial Implication

The ability to predict a customer's response to a digital action has remarkable value. Along with identifying probable customers who will choose a particular action; we can determine an order in which digital actions should be pressed to targeted customers. A predictive model based on the digital behavior and historic data of the existing customers can help identify potential sales leads and enhance customer experience by improved personalized marketing. The output of the model can be directly used to create a list of probable leads for varied marketing channels, re-targeting and social targeting. Further, the derived rules can be used to target customers who are most likely to consume specific digital content. This leads to enhanced segmentation and better understanding of the customized behavior for each segment. This in turn indicates to better identification of potential buyers or raw leads. These refined ways of identification and targeting advance to an increased efficiency due to a considerable reduction in cost and unnecessary marketing efforts.

4.4 Future Extensions

The method presented here can be used as a foundational framework to design the Next Best Digital Action (NBDA) using which B2B company should target prospects. Given the sparse literature around B2B, this work can be a basis for a system for the B2B digital buyer journey. Future research areas could include design of new metrics to analyze the digital buyer journey and development of models to optimize this metric directly. The multinomial problem could be framed as a recommendation problem as well and compared vs. existing models. A very interesting research area will be development of an algorithm to generate discriminating segments of users from the propensity model output that are business-user discernible and can be used by digital business sites that do not have the ability for individual email id/user level targeting.

References

1. Imber, J., Toffler, B.A.: Dictionary of marketing terms. Barron's snippet (2008)
2. Farhoomand, A.F., Lovelock, P.: Global e-Commerce: Text and Cases Plus Instructor's Manual (2001)
3. Edelman, D.C.: Four ways to get more value from digital marketing. McKinsey Q. 6 (2010)
4. Strauss, J.: E-Marketing. Routledge, New York (2016)
5. Kian Chong, W., Shafaghi, M., Woollaston, C., Lui, V.: B2B e-marketplace: an e-marketing framework for B2B commerce. Market. Intell. Plan. 28(3), 310–329 (2010)
6. Miller, M.: B2B digital marketing: using the web to market directly to businesses. Que Publishing (2012)
7. Järvinen, J., Tollinen, A., Karjaluoto, H., Jayawardhena, C.: Digital and social media marketing usage in B2B industrial section. Market. Manage. J. 22(2) (2012)
8. Leeflang, P.S., Verhoef, P.C., Dahlström, P., Freundt, T.: Challenges and solutions for marketing in a digital era. Eur. Manage. J. 32(1), 1–12 (2014)
9. Burez, J., Van den Poel, D.: CRM at a pay-tv company: using analytical models to reduce customer attrition by targeted marketing for subscription services. Expert Syst. Appl. 32(2), 277–288 (2007)
10. Nafis, S., Makhtar, M., Awang, M.K., Rahman, M.N.A., Deris, M.M.: Feature selections and classification model for customer churn. J. Theor. Appl. Inf. Technol. 75(3) (2015)
11. Xie, Y., Li, X., Ngai, E.W.T., Ying, W.: Customer churn prediction using improved balanced random forests. Expert Syst. Appl. 36(3), 5445–5449 (2009)
12. Balakrishnan, R., Parekh, R.: Learning to predict subject-line opens for large-scale email marketing. In: 2014 IEEE International Conference on Big Data (Big Data), pp. 579–584. IEEE (2014)
13. Szűcs, G.: Decision trees and random forest for privacy-preserving data mining. In: Xu, L. (ed.) Research and Development in E-Business through Service-Oriented Solutions, pp. 71–90. IGI Global, Hershey (2013)
14. Pushpavathi, T.P., Suma, V., Ramaswamy, V.: Defect prediction in software projects-using genetic algorithm based fuzzy c-means clustering and random forest classifier. Int. J. Sci. Eng. Res. 5(9) (2014)
15. Sinha, R., Saini, S., Anadhavelu, N.: Estimating the incremental effects of interactions for marketing attribution. In: 2014 International Conference on Behavior, Economic and Social Computing (BESC), pp. 1–6. IEEE (2014)

16. Yadagiri, M.M., Saini, S.K., Sinha, R.: A non-parametric approach to the multi-channel attribution problem. In: Wang, J., et al. (eds.) WISE 2015. LNCS, vol. 9418, pp. 338–352. Springer, Cham (2012). doi:10.1007/978-3-319-26190-4_23

17. Opitz, D., Maclin, R.: Popular ensemble methods: an empirical study. J. Artif. Intell. Res. **11**, 169–198 (1999)

18. Koh, E., Gupta, N.: An empirical evaluation of ensemble decision trees to improve personalization on advertisement. In: Proceedings of KDD 14 Second Workshop on User Engagement Optimization (2014)

19. Perlich, C., Dalessandro, B., Raeder, T., Stitelman, O., Provost, F.: Machine learning for targeted display advertising: transfer learning in action. Mach. Learn. **95**(1), 103–127 (2014)

20. Shao, X., Li, L.: Data-driven multi-touch attribution models. In: Proceedings of the 17th ACM SIGKDD International Conference on Knowledge Discovery and Data Mining, pp. 258–264. ACM (2011)

21. Farahat, A., Shanahan, J.: Econometric analysis and digital marketing: how to measure the effectiveness of an ad. In: Proceedings of the Sixth ACM International Conference on Web Search and Data Mining, p. 785. ACM (2013)

22. Lu, N., Lin, H., Lu, J., Zhang, G.: A customer churn prediction model in telecom industry using boosting. IEEE Trans. Ind. Inf. **10**(2), 1659–1665 (2014)

23. Wang, J., Zhang, Y., Chen, T.: Unified recommendation and search in e-commerce. In: Hou, Y., Nie, J.-Y., Sun, L., Wang, B., Zhang, P. (eds.) AIRS 2012. LNCS, vol. 7675, pp. 296–305. Springer, Heidelberg (2012). doi:10.1007/978-3-642-35341-3_25

24. Addicam, S., Balkan, S., Baydogan, M.: Adaptive advertisement recommender systems for digital signage (2015)

25. Zhang, W., Enders, T., Li, D.: GreedyBoost: an accurate, efficient and flexible ensemble method for B2B recommendations. In: Proceedings of the 50th Hawaii International Conference on System Sciences (2017)

26. Ho, T.K.: Random decision forests (PDF). In: Proceedings of the 3rd International Conference on Document Analysis and Recognition, Montreal, QC, 14–16 August 1995, pp. 278–282 (1995)

27. Ho, T.K.: The random subspace method for constructing decision forests (PDF). IEEE Trans. Pattern Anal. Mach. Intell. **20**(8), 832–844 (1998). doi:10.1109/34.709601

28. Breiman, L.: Random forests. Mach. Learn. **45**(1), 5–32 (2001)

29. Chang, C.-C., Lin, C.-J.: LIBSVM: a library for support vector machines (2001). http://www.csie.ntu.edu.tw/~cjlin/libsvm

30. Fan, R.-E., Chang, K.-W., Hsieh, C.-J., Wang, X.-R., Lin, C.-J.: Liblinear: a library for large linear classification. J. Mach. Learn. Res. **9**, 1871–1874 (2008)

31. Friedman, J.: Greedy function approximation: a gradient boosting machine. Ann. Stat. **29**(5), 1189–1232 (1999). Reitz Lecture

32. Friedman, J., Hastie, T., Tibshirani, R.: Additive logistic regression: a statistical view of boosting (with discussion and a rejoinder by the authors). Ann. Statist. **28**(2), 337–407 (2000). doi:10.1214/aos/1016218223

33. Chen, D., Plemmons, R.J.: Nonnegativity constraints in numerical analysis. Symposium on the Birth of Numerical Analysis (2009). CiteSeerX:10.1.1.157.9203

34. https://cran.r-project.org/web/packages/nnls/nnls.pdf

Detecting Relative Anomaly

Richard Neuberg[1,2(✉)] and Yixin Shi[2]

[1] Columbia University, New York City, USA
rn2325@columbia.edu
[2] Google, Mountain View, USA

Abstract. System states that are anomalous from the perspective of a domain expert occur with high density in some anomaly detection problems. The performance of commonly used unsupervised anomaly detection methods may suffer in that setting, because they use density as a proxy for anomaly. We propose a novel concept for anomaly detection, called *relative anomaly detection*. It is tailored to be robust towards anomalies that have high density, by taking into account their location relative to the most typical observations. The approaches we develop are computationally feasible even for large data sets, and they allow real-time detection. We illustrate using data sets of potential scraping attempts and Wi-Fi channel utilization, both from Google.

Keywords: Anomaly detection · Unsupervised learning

1 Introduction

Multivariate anomaly detection may be categorized broadly into supervised and unsupervised detection. In supervised anomaly detection, training data are labeled by domain experts as normal or anomalous, and a model is trained to classify future observations. In unsupervised anomaly detection, which is the focus of this article, labels are not known, because labeling is too difficult or costly. The goal is to approximately recover the missing expert judgements using empirical characteristics of the data. The data themselves typically first undergo a feature selection and feature engineering process to devise informative covariates. An unsupervised model can be evaluated by comparing its predictions with actual domain expert labels. Potential applications include intrusion detection, fraud detection and process control.

Density is commonly chosen as the target criterion for unsupervised anomaly detection. The population definition of anomalous observations then is $\{\mathbf{x} : f(\mathbf{x}) < \lambda\}$, where f is the data generating density, and λ is a user-selected threshold. Methods that exactly or approximately fall under this paradigm are density estimators and the closely related nearest neighbor approaches, besides many others; for a review on commonly used anomaly detection methods, see [5].

However, density may not align well with expert judgements in some applications. For example, scraping (the automated collection of information from websites) may occur frequently, but it nevertheless constitutes anomalous user

© Springer International Publishing AG 2017
P. Perner (Ed.): MLDM 2017, LNAI 10358, pp. 117–131, 2017.
DOI: 10.1007/978-3-319-62416-7_9

behavior. The performance of common approaches to unsupervised anomaly detection may suffer in the presence of such frequently occurring anomalies.

We propose a framework which we call relative anomaly detection to better handle cases where anomalies may have high density. We use the term *relative* to emphasize that in this framework the anomaly of an observation is determined by taking into account not only its own location and that of neighboring observations, but also the location of the most typical system states. The underlying assumption in relative anomaly detection is that large clusters of high-density system states are indeed normal from an expert's perspective, and that observations that are far from these most typical system states are anomalous. Such anomalies may have high density.

The rest of this paper is organized as follows. In Sect. 2, we discuss the approach to anomaly detection of [10], which is closely related to the PageRank algorithm [11]. We discuss the similarity graph of the observations in the training data set. We show connections with other approaches to anomaly detection, and discuss their shortcomings in the presence of anomalies that have high density. In Sect. 3, we introduce two novel *relative* anomaly detection approaches. In Sect. 4, we compare our *relative* approaches with the density-targeting approach of [10], using data sets of potential scraping attempts and Wi-Fi usage from Google. We conclude in Sect. 5.

2   Many Approaches to Anomaly Detection Target Density Criterion

Density is not necessarily an appropriate criterion for anomaly detection, especially not when anomalies occur with high density. We show in this section that the anomaly detection approach of [10], which is similar to the PageRank method [11], approximately targets the density criterion. Other commonly used methods for anomaly detection, such as kernel density estimation, the nearest neighbor approach, the one-class SVM [14] and also clustering approaches approximately target the density criterion, too. The CBLOF anomaly detection method of [8] is based on both density and distance, but it requires the labeling of clusters, which is difficult without extensive domain knowledge.

2.1   Similarity Graph

We begin by introducing the similarity graph of examples in a data set, which will also serve as a basis for the relative anomaly detection approaches we develop in Sect. 3. The relationship between unlabeled observations in a data set may be described through a weighted similarity graph. Observations form the nodes of the graph, and the weight of an edge expresses the similarity between two observations. Two observations $\mathbf{x}_i$ and $\mathbf{x}_j$ are typically considered similar when their distance is small. However, non-monotonic transformations can be useful with time series data, to take into account periodic behavior of the underlying system; for a reference on such transformations, see [13, Chap. 4]. A common monotonic transformation from distance $d(\mathbf{x}_i, \mathbf{x}_j)$ to similarity $s(\mathbf{x}_i, \mathbf{x}_j)$ uses the kernel function

$$s(\mathbf{x}_i, \mathbf{x}_j) = \exp(-d(\mathbf{x}_i, \mathbf{x}_j)^2/\gamma), \tag{1}$$

which is symmetric in its arguments. The parameter γ controls the degree of localization, meaning how far one observation can lie from another observation for the two to still be considered similar. When $\gamma = \infty$, all observations are equally similar to $\mathbf{x}_i$, and when $\gamma \downarrow 0$ only $\mathbf{x}_i$ is similar to itself. More localization is needed when the data come from a complicated distribution. The resulting matrix of similarities, $\mathbf{S}$, holds the edge weights in the similarity graph. In methods that apply the "kernel trick," such as the support vector machine and kernel principal components analysis, such a similarity matrix is called the kernel matrix.

Common choices for the distance between two real data points, $d(\mathbf{x}_i, \mathbf{x}_j)$, are Euclidean ($L_2$) and Manhattan ($L_1$) distance. Both of these distance measures assume that each dimension of the data has been appropriately normalized. Euclidean distance has the advantage of being rotation invariant, and the order of the resulting distances typically remains meaningful even in high dimension [18]. Furthermore, data points often approximately lie in a lower-dimensional subspace; then Euclidean distance calculations are effectively carried out in the lower-dimensional subspace. For very high-dimensional problems, Manhattan distance may be preferred over Euclidean distance [1]. However, if the data truly cover the high-dimensional space, that means that the system components are barely correlated, even after feature selection and feature engineering. Then a multivariate anomaly analysis may add only little value as compared to running separate univariate analyses. If variables are measured on a nominal or ordinal scale, they may be converted into numerical data using dummy variables, or specialized distance measures for that scale level can be used; for a reference, see [7, Chap. 14].

2.2   A Random Walk Approach, Density Estimation and Distance-Based Anomaly Detection All Target the Density Criterion

The approach of [10] proposes to take a random walk on the similarity graph, and to label an observation as anomalous when the stationary probability of the random walk at that observation is low. For cases where the similarity matrix is not irreducible and aperiodic, random restarts are introduced in the random walk, like it was proposed as part of the PageRank algorithm [11]. In the case that the similarity matrix is irreducible and aperiodic, which we will assume in to following to keep technical discussions at a minimum, the matrix of transition probabilities in the graph is simply the similarity matrix normalized by row,

$$\mathbf{P} = [\mathrm{diag}(\mathbf{S1})]^{-1}\mathbf{S}, \tag{2}$$

where $\mathbf{1}$ is a column vector of ones. The vector of stationary (unnormalized) probabilities, $\mathbf{p}$, follows from the stationarity condition $\mathbf{P}^\mathsf{T}\mathbf{p} = \mathbf{p}$ as the dominant left-eigenvector of $\mathbf{P}$ by the Perron–Frobenius theorem; see [9] for a reference.

We now show that the approach of [10] is closely related to both a density-based and a distance-based approach to anomaly detection, and therefore approximately targets the density criterion. To see the connection with density-based anomaly detection, consider the case when $\mathbf{S}$ is symmetric; then the dominant left-eigenvector of $\mathbf{P} = [\mathrm{diag}(\mathbf{S1})]^{-1}\mathbf{S}$ is, up to scaling, $\mathbf{S1}$. This follows from plugging in $\mathbf{S1}$ for $\mathbf{p}$ in $\mathbf{P}^\mathsf{T}\mathbf{p} = \mathbf{p}$, and using that $\mathbf{P}^\mathsf{T} = \mathbf{S}^\mathsf{T}[\mathrm{diag}(\mathbf{S1})]^{-1}$, with $\mathbf{S}^\mathsf{T} = \mathbf{S}$, which yields the true statement $\mathbf{S}[\mathrm{diag}(\mathbf{S1})]^{-1}\mathbf{S1} = \mathbf{S1}$. We see that the stationary probability at observation $\mathbf{x}_i$ is proportional to its (weighted) vertex degree $\mathrm{VD}(\mathbf{x}_i)$ in the similarity graph, where

$$\mathrm{VD}(\mathbf{x}_i) := (\mathbf{S1})_i = \sum_{j=1}^{n} s(\mathbf{x}_i, \mathbf{x}_j). \tag{3}$$

Expression (3) is proportional to a kernel density estimate with Gaussian kernel, whose kernel covariance matrix is diagonal with all diagonal elements equaling $\gamma/2$. As a density estimate, $\mathrm{VD}(\mathbf{x}_i)$ is typically mis-specified, because the kernel matrix is not tuned to fit the particular data generating process. This may actually be desired in anomaly detection problems where a low density observation close to a very typical system state does not make for an interesting anomaly. However, the close connection with kernel density estimation suggests that if anomalous system states occur too frequently, they may not be labeled correctly as anomalies, even if they are for from the most typical system states.

To also see the connection with distance-based anomaly detection, consider a directed k nearest neighbor graph instead of a fully connected similarity graph. Here the (i, j)th element of $\mathbf{S}$ takes value $s(\mathbf{x}_i, \mathbf{x}_j)$ if $\mathbf{x}_j$ is in the set $\mathcal{N}_k(\mathbf{x}_i)$, which contains the k nearest neighbors of $\mathbf{x}_i$, and it is zero otherwise. The resulting similarity matrix is a sparse approximation of the full similarity matrix. The additional tuning parameter k controls the degree of localization. Localization via the k nearest neighbor graph is also used in spectral clustering, manifold learning, and local multidimensional scaling; for a reference, see [7, Chap. 14]. Consider a linear expansion of the radial kernel function, defined in Eq. (1), around some distance level $v > 0$. Then $\mathrm{VD}(\mathbf{x}_i)$ is approximately an affine decreasing function of the average distance to the k nearest neighbors:

$$\mathrm{VD}^{\mathrm{approx}}(\mathbf{x}_i) = k \exp(-v^2/\gamma)(1 + 2v^2/\gamma) \tag{4}$$
$$- 2(v \exp(-v^2/\gamma))/\gamma \sum_{j:\mathbf{x}_j \in \mathcal{N}_k(\mathbf{x}_i)} d(\mathbf{x}_i, \mathbf{x}_j).$$

This effectively eliminates the dependency on the kernel parameter γ, because γ does not influence the ordering of the anomaly scores. Using the average distance to the k nearest neighbors as a measure of anomaly was suggested in both [2,6]. However, for relative anomalies, the average distance to the k nearest neighbors can be small, and what is an anomalous system state may not be considered anomalous by the anomaly detection model.

2.3   Other Anomaly Detection Methods Also Target Density or Have Issues Related to Clustering

Another approach to anomaly detection is the one-class SVM [14]. This method labels data as anomalous if they are dissimilar to data that the method was trained on. We see that the one-class SVM uses density as its target criterion, which means that it cannot be used with training data that contain a considerable amount of anomalous observations, and that it cannot be used for exploratory analyses. A closely related approach is relative novelty detection [15].

Clustering approaches are also sometimes used for anomaly detection. While these may perform satisfactorily in certain applications, our main concern with clustering approaches is that it is difficult or impossible to determine in the first place whether a cluster is normal or not. Furthermore, especially in out-of-sample applications, an observation that is very distant from all previously seen data may be assigned to a normal cluster and therefore be judged normal, even though it is very atypical.

The CBLOF method [8] is a clustering-based approach to handling anomalies that occur with high density. It takes into account both the density of an example's closest cluster and the distance of the example to the cluster. While this is an improvement over pure density-based anomaly detection approaches in the presence of high-density anomalies, this approach also has several issues: firstly, no clear guidance exists for choosing the number of clusters, and, secondly, the choice of when to call a cluster anomalous depends only on the cluster itself, and not the location of the cluster relative to the other data points. This means that extensive domain-knowledge is required when training this model, and that it is difficult to use it as an exploratory tool.

3   Detecting Relative Anomaly

Approaches to unsupervised anomaly detection that target the density criterion may not perform well in the presence of frequently occurring anomalies, as discussed in the previous sections. We now introduce two anomaly detection models that take into account the location of the most typical observations when determining how anomalous a new observation is. Both of these methods have the advantages that they do not depend on a clustering solution and a decision of when to call a cluster anomalous and that they provide a natural quantitative ordering of the data points in terms of their degree of anomaly (unlike, for example, approaches based on clustering). We also investigate relationships and differences with other approaches to anomaly detection, especially the approach of [10], which we discussed in Sect. 2.2.

3.1   Popularity Approach

We propose to consider a "random walk" between nodes based on the *unnormalized* similarity matrix $\mathbf{S}$ — instead of the transition probability matrix $\mathbf{P}$ considered in Sect. 2.2. From

$$\mathbf{S} = \mathrm{diag}(\mathbf{S1})\mathbf{P}, \tag{5}$$

we see that the similarity $[\mathbf{S}]_{ij}$ between two nodes $\mathbf{x}_i$ and $\mathbf{x}_j$ factors into the transition probability $[\mathbf{P}]_{ij}$ and the vertex degree of $\mathbf{x}_i$. This has the effect that the random walk weakens when transitioning through nodes whose vertex degree is medium or small, and that it strengthens when passing through nodes of high vertex degree. We label an observation $\mathbf{x}_i$ as anomalous if its *relative anomaly*,

$$\mathrm{RA}(\mathbf{x}_i) := -(\mathbf{s})_i, \tag{6}$$

is small, where $\mathbf{s}$ is the dominant left-eigenvector of $\mathbf{S}$. This eigenvector is unique with all elements positive by the Perron–Frobenius theorem.

We can gain further insight into this algorithm via a connection with the network analysis literature. [3] considers a network of persons, where each person rates each other person as popular or not. Their goal is to determine an overall popularity score for each person, based on the pairwise ratings. They suggest that a measure of overall popularity of person i should depend not only on how many people in the network deem that person to be popular, but also whether those people are themselves popular. This leads to the eigenproblem $\sum_j [\mathbf{S}]_{ij} v_j = \lambda v_i$, where v_i is the overall popularity of person i, and $[\mathbf{S}]_{ij}$ takes value one if person i considers person j popular. A person is labeled as overall popular when its entry in the dominant left-eigenvector of the adjacency matrix, called the eigenvector centrality, is large. We see that by measuring anomaly using (6) instead of (3), how anomalous an observation is depends not only on how many other observations are close, but also on whether these other observations themselves have close neighbors. As a result, high vertex degree observations that are sufficiently far from many other observations in the similarity graph will be labeled anomalous. Asymptotically, the leading eigenvector of a kernel matrix converges to the leading eigenfunction, φ, in the following eigenproblem [16]:

$$\int s(\mathbf{x}, \mathbf{y}) f(\mathbf{x}) \varphi(\mathbf{x}) \mathrm{d}\mathbf{x} = \delta \varphi(\mathbf{y}). \tag{7}$$

Here f is the data density, and δ is the eigenvalue that corresponds to φ. We see that, asymptotically, the popularity $\varphi(\mathbf{y})$ of an observation $\mathbf{y}$ is high if values $\mathbf{x}$ that are close to $\mathbf{y}$ have high density and are popular themselves. Here the size of the surrounding of $\mathbf{y}$ is determined by the choice of s.

The power method can be used to find the dominant left-eigenvector of $\mathbf{S}$. This iterative method starts from a random initialization, $\mathbf{s}_0$, and then follows the recurrence relation $\mathbf{s}_{t+1} = \mathbf{S}\mathbf{s}_t / \|\mathbf{S}\mathbf{s}_t\|_2$. The convergence is geometric, with ratio $|\lambda_2/\lambda_1|$, where λ_1 and λ_2 denote the first and second dominant eigenvalue of $\mathbf{S}$, respectively. We find that the error $\|\mathbf{S}\mathbf{s}_t - \mathbf{s}_t^\mathsf{T}\mathbf{S}\mathbf{s}_t\mathbf{s}_t\|_2$ typically becomes small after just a few iterations. This computation is highly parallelizable.

This algorithm scales linearly in the number of input dimensions, since only distances between data points are required as inputs. However, computations are typically convenient only for datasets whose size does not exceed a few thousands of examples because of the similarity matrix involved, whose size increases quadratically in the number of examples.

We find that typically more than half of the smallest elements of the kernel matrix can be set to a small constant—allowing sparse matrix computations and a hence a speed-up of more than two—without changing the rank order of the relative anomaly values. Furthermore, for high-dimensional problems, we can obtain good starting values for the power iteration as follows. [12] show that $\mathbf{S}$ can be approximated by $\boldsymbol{\Phi}^\mathsf{T}\boldsymbol{\Phi}$, where $\boldsymbol{\Phi}$ is a draw of random Fourier features calculated from the original data. If we choose only a small number of Fourier features as compared to the sample size, then $\operatorname{rank}(\boldsymbol{\Phi}^\mathsf{T}\boldsymbol{\Phi}) \ll \operatorname{rank}(\mathbf{S})$, and we can cheaply find an approximation to the leading eigenvector of $\mathbf{S}$ as $\boldsymbol{\Phi}^\mathsf{T}\operatorname{lev}(\boldsymbol{\Phi}\boldsymbol{\Phi}^\mathsf{T})$. Here $\operatorname{lev}(\boldsymbol{\Phi}\boldsymbol{\Phi}^\mathsf{T})$ denotes the leading eigenvector of $\boldsymbol{\Phi}\boldsymbol{\Phi}^\mathsf{T}$; it can again be found using the power iteration. In our experiments, this approach reduces the run time until the leading eigenvector of $\mathbf{S}$ is found to one fourth.

It is computationally expensive to retrain the model with every new observation. Furthermore, it may not even be desired to update the model in the presence of every new observation, because that new observation may come from a different, anomalous data generating process. We propose to instead determine the relative anomaly of a new observation with respect to the observations in the training data set as follows. Recall that the left-eigenproblem of $\mathbf{S}$ is $\lambda\mathbf{s} = \mathbf{S}^\mathsf{T}\mathbf{s}$, from which we see that $(\mathbf{s})_i = (\mathbf{S}^\mathsf{T}\mathbf{s})_i/\mathbf{s}^\mathsf{T}\mathbf{S}\mathbf{s}$. We can use this relation to predict the relative anomaly of a new observation $\bullet$, based solely on training data, as

$$\widehat{\mathrm{RA}}(\bullet) = -\frac{(s(\bullet, \mathbf{x}_1), \ldots, s(\bullet, \mathbf{x}_n))\mathbf{s}}{\mathbf{s}^\mathsf{T}\mathbf{S}\mathbf{s}}. \tag{8}$$

This can be viewed as an application of the Nyström method to approximate the leading eigenvector of the extended kernel matrix; for a reference on the Nyström method, see [17]. As a result, real-time detection is rapid.

3.2 Shortest Path Approach

We also propose an approach to relative anomaly detection based on highest similarity paths. The idea is to first identify those observations that can be considered very typical, and then to label an observation as anomalous if it is difficult to reach it from any of the typical observations. Here we interpret an element $[\mathbf{S}]_{ij}$ as a "connectivity" value between nodes $\mathbf{x}_i$ and $\mathbf{x}_j$. We use the following two-step approach:

1. Consider those observations for which the vertex degree is higher than that of $(1-q) \cdot 100\%$ of the observations in the training data set as highly normal. For each observation $\bullet$, we can express this as $\hat{F}_{\mathrm{VD}}(\bullet) > q$, using the empirical cumulative distribution function of vertex degrees in the training data set, $\hat{F}_{\mathrm{VD}}$. Note that by choosing the kernel bandwidth large enough we can smooth out local peaks in the data density, such that indeed the observations with highest vertex degrees can be considered normal.
2. Now, for each observation $\bullet$ that is not considered highly normal, find the length of the best-connected path from it to any of the observations deemed normal:

$$\max_{l\,:\,1-\hat{F}_{\mathrm{VD}}(\mathbf{x}_l)\leq q} \; \max_{\left\{\substack{\text{paths from} \\ \bullet \text{ to } \mathbf{x}_l}\right\}} \; \prod_{(i,j):\text{ is edge in path}} s_{ij}. \tag{9}$$

Alternatively, solve the equivalent shortest path problem

$$\mathrm{RA}_q(\bullet) := \min_{l\,:\,1-\hat{F}_{\mathrm{VD}}(\mathbf{x}_l)\leq q} \; \min_{\left\{\substack{\text{paths from} \\ \bullet \text{ to } \mathbf{x}_l}\right\}} \; \sum_{(i,j):\text{ is edge in path}} -\ln s_{ij}. \tag{10}$$

Then label $\bullet$ as anomalous if $\mathrm{RA}_q(\bullet)$ is large. $\mathrm{RA}_q(\bullet) = 1$ if $\bullet$ is one of the $q \cdot 100\%$ of observations which are considered most normal, and $\mathrm{RA}_q(\bullet) > 1$ otherwise. This shortest path problem can be solved more efficiently when considering a sparsified version of $\mathbf{S}$, for example by applying a directed k nearest neighbor truncation.

An advantage of this approach is that the tuning parameter q allows controlling the number of data points considered typical. Several central regions of the data may emerge for a larger value of q. A disadvantage is the higher computational complexity of the shortest path problem, which may however be reduced through subsampling.

We can gain further insight into this approach when used with the kernel function in (1). Then the path length in (10) becomes

$$\sum_{(i,j):\text{ is edge in path}} -\ln\exp(-d(\mathbf{x}_i,\mathbf{x}_j)^2/\gamma) \;\propto\; \sum_{(i,j):\text{ is edge in path}} d(\mathbf{x}_i,\mathbf{x}_j)^2. \tag{11}$$

We see that the squared distance between two observations discourages large jumps, and thereby paths through high density regions are encouraged. While the tuning parameter γ does not influence the comparison between two path lengths, since it is only a multiplicative constant, it influences the calculation of $\hat{F}_{\mathrm{VD}}$ in (10). A larger value for γ means that the bandwidth in the vertex degree estimator is higher, thereby smoothing the density more, which can be used to smear away small clusters of frequently occurring anomalies.

This algorithm, like the popularity approach, scales linearly with the number of input dimensions, because it is based on distances between data points only. However, shortest path solvers are typically fast only up to a few thousand data points. Real-time detection requires calculating only a single shortest path and is therefore very fast.

3.3   Normalization

A relative anomaly measure RA can be transformed into a degree of anomaly in $(0, 1)$ for each observation $\bullet$ using the empirical distribution function $\hat{F}$ of directed anomalies in the training data:

$$\mathrm{DORA}(\bullet) := \hat{F}(\mathrm{RA}(\bullet)) \in (0, 1). \tag{12}$$

3.4   Determining Largest Univariate Deviations

Once an anomalous state $\mathbf{x}_{\text{anomalous}}$ is identified, we can determine which univariate features deviate most from what is normal as follows:

1. Find that normal observation in the data set which is closest to the anomalous observation:

$$\mathbf{x}_{\text{closest}}(\mathbf{x}_{\text{anomalous}}) \quad = \quad \underset{\mathbf{x}_i : \text{DORA}(\mathbf{x}_i) < p}{\arg\min} \quad d(\mathbf{x}_i, \mathbf{x}_{\text{anomalous}}).$$

   The threshold $p \in (0, 1)$ determines how large the anomaly of $\mathbf{x}_{\text{closest}}$ may be to still be considered normal. Here it may be useful to use the L_1 distance to judge discrepancy, because the suggested change will be large in a few dimensions, unlike it is the case with L_2 distance, which will suggest smaller changes in many dimensions.
2. Calculate $\mathbf{x}_{\text{anomalous}} - \mathbf{x}_{\text{closest}}(\mathbf{x}_{\text{anomalous}})$; the largest elements of this vector difference show which univariate components need to be altered for the system to revert to a normal state.

4   Application

We compare the relative anomaly detection approaches, introduced in Sect. 3, to the vertex degree anomaly detection approach discussed in Sect. 2.2, which is a representative example of an anomaly detection method that targets the density criterion. We do not consider clustering-based approaches because they rely so strongly on the choice of the number of clusters, and because of the difficulty associated with specifying whether a cluster is normal or not, both discussed in Sect. 2.3.

We use two data sets from Google, of 1,000 data points each. For confidentiality reasons, our explanations of what these data show will have to be rather short. We pre-process each covariate using the Box–Cox transform [4],

$$x \mapsto \begin{cases} \frac{(x+\delta)^\lambda - 1}{\lambda}, & \text{if } \lambda \neq 0, \\ \ln(x + \delta), & \text{if } \lambda = 0, \end{cases} \tag{13}$$

to reduce skew and normalize kurtosis; special cases of this transform are the logarithmic and square-root transforms. We find the parameters (δ, λ) as those maximizing the normal log-likelihood of the data. We then standardize the data and form a fully connected similarity graph using the radial basis kernel.

4.1   Potential Scraping Data

The first data set contains information about potential scraping attempts. Scraping is the automated collection of information from websites. The two covariates are experimental features that measure aspects of user behavior for each access log.

We choose the hyperparameter γ by noting that it equals twice the variance parameter in a Gaussian distribution. We set $\gamma = 0.5$ so that the kernel "standard deviation" equals 0.5, which allows for good localization of the signal, given that user behavior 1 and 2 both roughly have a range of ten.

In Fig. 1a we show the anomaly detection results using the vertex degree approach of Sect. 2.2, which targets the density criterion. Here and in the following, the lighter the shade of gray is, the higher the respective region's detected degree of anomaly. The top twenty % detected anomalies are emphasized. However, domain experts have identified that the observations in the diffuse cluster on the right exhibit behavior that is typical of scrapers. As a result, there are false positives surrounding the very high density area around $(-1, 0)$, and the observations around $(5, -2)$ and $(6, 4)$ are false negatives.

The results for the popularity approach to relative anomaly detection, introduced in Sect. 3.1, are shown in Fig. 1b. We set $\gamma = 0.2$, because we find that the relative anomaly approach generally requires less smoothing than the vertex degree approach. The results are not very sensitive to the exact choice of γ; in contrast, lowering γ in the vertex degree approach would result in a significant increase in the number of false positives and false negatives. There are no false positives or false negatives, as compared with the expert judgement.

It is extremely labor-intensive—potentially even impossible—to assess with certainty whether an individual data point is or is not a scraper. Hence it may be desired to only label users as scrapers if we are highly certain. The detected level of relative anomaly in Fig. 1b tends to increase while moving away from the high density area on the left. Increasing the threshold of relative anomaly above which a user is labeled as a scraper will have the desired result that only observations on the far right—whose behavior is most different from what is typical—are labeled as anomalous. In contrast, the vertex degree approach will continue labeling observations in the low density area close to the cluster of normal users as anomalous.

In Fig. 2 we show how the empirical cumulative distribution of relative anomalies may be useful for determining the threshold above which an observation is labeled an anomaly. For a clearer presentation, we transformed the relative anomaly values according to $\bullet \mapsto -\ln(-\bullet)$. The top 20% of observations have much higher relative anomaly values than the other observations. This approach to setting the anomaly threshold tuning parameter may be particularly useful in higher-dimensional problems, where a visual inspection is otherwise difficult.

We see that the largest univariate deviation (compare Sect. 3.4) is in user behavior 1, making observations anomalous if they have a large values for that covariate.

We also apply the shortest path approach from Sect. 3.2 to the scraping data set. In Fig. 3 we see that, compared with the approach of Sect. 3, the shortest path approach using $q = 0.5$ yields sharper bounds around the group of normal observations, which may be desired in some applications. In-sample, the classification outcomes are identical.

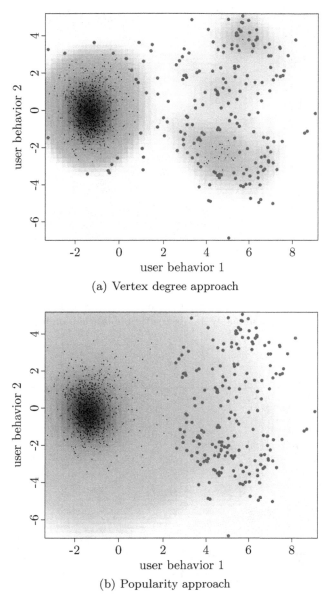

(a) Vertex degree approach

(b) Popularity approach

Fig. 1. The vertex degree approach labels low-density observation in the left cluster of normal observations as anomalous (marked in red), and mistakes some observations in the diffuse right cluster of scrapers as normal. The popularity approach correctly detects the left cluster of normal observations as normal, and labels the diffuse right cluster of scrapers as anomalous. In each figure the top 20% detected anomalies are highlighted. Lighter shades of gray correspond to higher anomaly. (Color figure online)

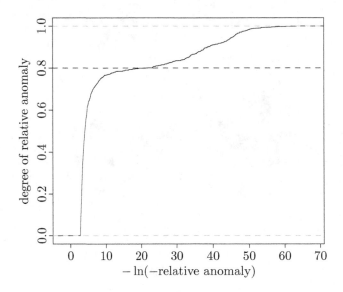

Fig. 2. The empirical distribution of relative anomalies can assist with deciding above which threshold of relative anomaly an observation is labeled anomalous

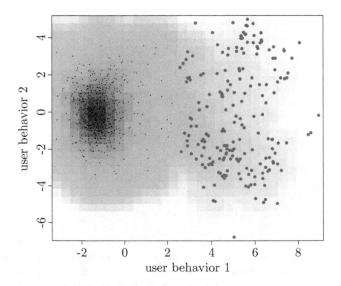

Fig. 3. The shortest path approach correctly detects the left cluster of normal observations as normal, and labels the diffuse right cluster of scrapers as anomalous

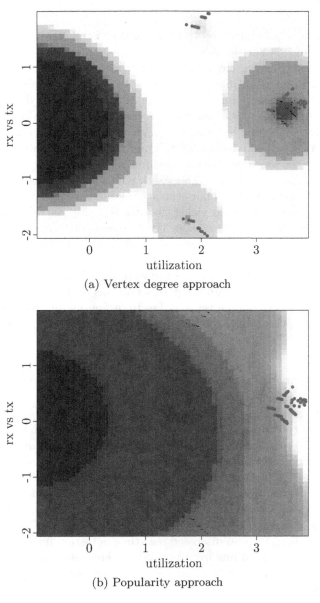

(a) Vertex degree approach

(b) Popularity approach

Fig. 4. The vertex degree approach labels the two clusters on top and bottom as anomalous, even though these correspond to medium overall Wi-Fi usage, as measured by the first covariate; most of the elements of the heavy-usage cluster on the right are labeled as normal, because heavy usage occurs relatively frequently; note that a medium-usage configuration at $(2, 0)$ would falsely be considered extremely anomalous. The popularity approach correctly labels only the heavy-usage cluster on the far right as anomalous; note that a medium-usage configuration at $(2, 0)$ would correctly be considered normal. In each figure the top 13% detected anomalies are highlighted.

4.2   Wi-Fi Usage Data

Our second data set contains observations on Wi-Fi channel utilization reported for wireless transmissions at different access points within a specific location in a corporate networking environment. The instantaneous channel utilization at each access point is an indication of how busy the transmission channel is, and whether the access point should change transition to a different channel. Detecting channel utilization anomalies is critical for identifying access points with low performance due to consistent high utilization. The data set contains two covariates for a Wi-Fi access point. The first covariate is measure of overall utilization, and the second covariate measures utilization of rx versus tx. 72% of the data points cluster at value $(-0.89, 0.04)$, which corresponds to no utilization. According to domain experts, high utilization states are anomalous.

The vertex degree approach yields the results in Fig. 4a, where again we set $\gamma = 0.5$. We see that the two smaller clusters around $(1.7, -1.5)$ and $(2, 1.8)$, as well as the few data points around $(0.4, -0.2)$, are jointly labeled as the top thirteen % anomalies.

In Fig. 4b we show the results for the approach from Sect. 3.1, again using $\gamma = 0.2$. Here the cluster of high usage observations on the far right is correctly labeled as anomalous — because it is far from the many observations at the left of the figure. The results for the shortest path approach from Sect. 3.2 are similar.

5   Conclusion

Unsupervised approaches to anomaly detection are commonly used because labeling data is too costly or difficult. Many common approaches for unsupervised anomaly detection target a density criterion. This means that their performance deteriorates when anomalies have high density, as for example in the case of scraping. We proposed a novel concept, relative anomaly detection, that is more robust to such frequently occurring anomalies. It is tailored to be robust towards anomalies that have high density by taking into account their location relative to the most typical observations. We presented two novel algorithms under this paradigm. We also discussed real-time detection for new observations, and how univariate deviations from normal system behavior can be identified. We illustrated these approaches using data on potential scraping and Wi-Fi usage from Google.

Acknowledgments. We thank Mitch Trott, Phil Keller and Robbie Haertel of Google as well as Lauren Hannah of Columbia University for many helpful comments, and furthermore Dave Peters and Taghrid Samak of Google for granting us access to their data sets.

References

1. Aggarwal, C.C., Hinneburg, A., Keim, D.A.: On the surprising behavior of distance metrics in high dimensional space. In: Bussche, J., Vianu, V. (eds.) ICDT 2001. LNCS, vol. 1973, pp. 420–434. Springer, Heidelberg (2001). doi:10.1007/3-540-44503-X_27
2. Angiulli, F., Basta, S., Pizzuti, C.: Distance-based detection and prediction of outliers. IEEE Trans. Knowl. Data Eng. **18**(2), 145–160 (2006)
3. Bonacich, P.: Factoring and weighting approaches to status scores and clique identification. J. Math. Sociol. **2**(1), 113–120 (1972)
4. Box, G.E.P., Cox, D.R.: An analysis of transformations. J. R. Stat. Soc. Ser. B (Methodological) **26**(2), 211–252 (1964)
5. Chandola, V., Banerjee, A., Kumar, V.: Anomaly detection: a survey. ACM Comput. Surv. (CSUR) **41**(3), 15 (2009)
6. Eskin, E., Arnold, A., Prerau, M., Portnoy, L., Stolfo, S.: A geometric framework for unsupervised anomaly detection. In: Barbará, D., Jajodia, S. (eds.) Applications of Data Mining in Computer Security. Advances in Information Security, vol. 6, pp. 77–101. Springer, Heidelberg (2002)
7. Hastie, T., Tibshirani, R., Friedman, J.: The Elements of Statistical Learning: Data Mining, Inference, and Prediction. Springer, New York (2009)
8. He, Z., Xiaofei, X., Deng, S.: Discovering cluster-based local outliers. Pattern Recognit. Lett. **24**(9), 1641–1650 (2003)
9. Isaacson, D.L., Madsen, R.W.: Markov Chains, Theory and Applications, vol. 4. Wiley, New York (1976)
10. Moonesinghe, H.D.K., Tan, P.-N.: Outlier detection using random walks. In: 18th IEEE International Conference on Tools with Artificial Intelligence (ICTAI 2006), pp. 532–539. IEEE (2006)
11. Page, L., Brin, S., Motwani, R., Winograd, T.: The pagerank citation ranking: bringing order to the web (1999)
12. Rahimi, A., Recht, B.: Random features for large-scale kernel machines. In: Advances in Neural Information Processing Systems, pp. 1177–1184 (2007)
13. Rasmussen, C.E., Williams, C.K.I.: Gaussian processes for machine learning (2006)
14. Schölkopf, B., Williamson, R.C., Smola, A.J., Shawe-Taylor, J., Platt, J.C., et al.: Support vector method for novelty detection. In: NIPS, vol. 12, pp. 582–588. Citeseer (1999)
15. Smola, A.J., Song, L., Teo, C.H., et al.: Relative novelty detection. In: AISTATS, vol. 12, pp. 536–543 (2009)
16. Williams, C., Seeger, M.: The effect of the input density distribution on kernel-based classifiers. In: Proceedings of the 17th International Conference on Machine Learning, number EPFL-CONF-161323, pp. 1159–1166 (2000)
17. Williams, C., Seeger, M.: Using the nyström method to speed up kernel machines. In: Proceedings of the 14th Annual Conference on Neural Information Processing Systems, number EPFL-CONF-161322, pp. 682–688 (2001)
18. Zimek, A., Schubert, E., Kriegel, H.-P.: A survey on unsupervised outlier detection in high-dimensional numerical data. Stat. Anal. Data Mining ASA Data Sci. J. **5**(5), 363–387 (2012)

Optimization for Large-Scale Machine Learning with Distributed Features and Observations

Alexandros Nathan$^{(\boxtimes)}$ and Diego Klabjan

Department of Industrial Engineering and Management Sciences,
Northwestern University, Evanston, IL 60208, USA
anathan@u.northwestern.edu, d-klabjan@northwestern.edu

Abstract. As the size of modern data sets exceeds the disk and memory capacities of a single computer, machine learning practitioners have resorted to parallel and distributed computing. Given that optimization is one of the pillars of machine learning and predictive modeling, distributed optimization methods have recently garnered ample attention in the literature. Although previous research has mostly focused on settings where either the observations, or features of the problem at hand are stored in distributed fashion, the situation where both are partitioned across the nodes of a computer cluster (doubly distributed) has barely been studied. In this work we propose two doubly distributed optimization algorithms. The first one falls under the umbrella of distributed dual coordinate ascent methods, while the second one belongs to the class of stochastic gradient/coordinate descent hybrid methods. We conduct numerical experiments in Spark using real-world and simulated data sets and study the scaling properties of our methods. Our empirical evaluation of the proposed algorithms demonstrates the outperformance of a block distributed ADMM method, which, to the best of our knowledge is the only other existing doubly distributed optimization algorithm.

Keywords: Machine learning · Distributed optimization · Big data · Spark

1 Introduction

The collection and analysis of data is widespread nowadays across many industries. As the size of modern data sets exceeds the disk and memory capacities of a single computer, it is imperative to store them and analyze them distributively. Designing efficient and scalable distributed optimization algorithms is a challenging, yet increasingly important task. There exists a large body of literature studying algorithms where either the features or the observations associated with a machine learning task are stored in distributed fashion. Nevertheless, little attention has been given to settings where the data is doubly distributed, i.e., when both features and observations are distributed across the nodes of a computer cluster. This scenario may arise in practice as a result of distinct

P. Perner (Ed.): MLDM 2017, LNAI 10358, pp. 132–146, 2017.
DOI: 10.1007/978-3-319-62416-7_10

data collection efforts focusing on different features – we are assuming that the result of each data collection process is stored using the split across observations. The benefit of using doubly distributed algorithms stems from the fact that one can bypass the costly step (due to network bandwidth) of moving data between servers to avoid the two levels of parallelism.

In this work, we propose two algorithms that are amenable to the doubly distributed setting, namely D3CA (Doubly Distributed Dual Coordinate Ascent) and RADiSA (RAndom Distributed Stochastic Algorithm). These methods can solve a broad class of problems that can be posed as minimization of the sum of convex functions plus a convex regularization term (e.g. least squares, logistic regression, support vector machines).

D3CA builds on previous distributed dual coordinate ascent methods [7,11,26], allowing features to be distributed in addition to observations. The main idea behind distributed dual methods is to approximately solve many smaller sub-problems (also referred to herein as partitions) instead of solving a large one. Upon the completion of the local optimization procedure, the primal and dual variables are aggregated, and the process is repeated until convergence. Since each sub-problem contains only a subset of the original features, the same dual variables are present in multiple partitions of the data. This creates the need to aggregate the dual variables corresponding to the same observations. To ensure dual feasibility, we average them and retrieve the primal variables by leveraging the primal-dual relationship (3), which we discuss in Sect. 3.

In contrast with D3CA, RADiSA is a primal method and is related to a recent line of work [14,24,28] on combining Coordinate Descent (CD) methods with Stochastic Gradient Descent (SGD). Its name has the following interpretation: the randomness is due to the fact that at every iteration, each sub-problem is assigned a random sub-block of local features; the stochastic component owes its name to the parameter update scheme, which follows closely that of the SGD algorithm. The work most pertinent to RADiSA is RAPSA [14]. The main distinction between the two methods is that RAPSA follows a distributed gradient (mini-batch SGD) framework, in that in each global iteration there is a single (full or partial) parameter update. Such methods suffer from high communication cost in distributed environments. RADiSA, which follows a local update scheme similar to D3CA, is a communication-efficient generalization of RAPSA, coupled with the stochastic variance reduction gradient (SVRG) technique [8].

The contributions of our work are summarized as follows:

– We address the problem of training a model when the data is distributed across observations and features. We propose two doubly distributed optimization methods.
– We perform a computational study to empirically evaluate the two methods. Both methods outperform on all instances the block splitting variant of ADMM [17], which, to the best of our knowledge, is the only other existing doubly distributed optimization algorithm.

The remainder of the paper is organized as follows: Sect. 2 discusses related works in distributed optimization; Sect. 3 provides an overview of the problem under consideration, and presents the proposed algorithms; in Sect. 4 we present the results for our numerical experiments, where we compare D3CA and two versions of RADiSA against ADMM.

2   Related Work

Stochastic Gradient Descent Methods. SGD is one of the most widely-used optimization methods in machine learning. Its low per-iteration cost and small memory footprint make it a natural candidate for training models with a large number of observations. Due to its popularity, it has been extensively studied in parallel and distributed settings. One standard approach to parallelizing it is the so-called mini-batch SGD framework, where worker nodes compute stochastic gradients on local examples in parallel, and a master node performs the parameter updates. Different variants of this approach have been proposed, both in the synchronous setting [4], and the asynchronous setting with delayed updates [1]. Another notable work on asynchronous SGD is Hogwild! [18], where multiple processors carry out SGD independently and one can overwrite the progress of the other. A caveat of Hogwild! is that it places strong sparsity assumptions on the data. An alternative strategy that is more communication efficient compared to the mini-batch framework is the Parallelized SGD (P-SGD) method [29], which follows the research direction set by [12,13]. The main idea is to allow each processor to independently perform SGD on the subset of the data that corresponds to it, and then to average all solutions to obtain the final result. Note that in all aforementioned methods, the observations are stored distributively, but not the features.

Coordinate Descent Methods. Coordinate descent methods have proven very useful in various machine learning tasks. In its simplest form, CD selects a single coordinate of the variable vector, and minimizes along that direction while keeping the remaining coordinates fixed [16]. More recent CD versions operate on randomly selected blocks, and update multiple coordinates at the same time [20]. Primal CD methods have been studied in the parallel [21] and distributed settings [10,19]. Distributed CD as it appears in [19] can be conducted with the coordinates (features) being partitioned, but requires access to all observations. Recently, dual coordinate ascent methods have received ample attention from the research community, as they have been shown to outperform SGD in a number of settings [6,22]. In the dual problem, each dual variable is associated with an observation, so in the distributed setting one would partition the data across observations. Examples of such algorithms include [7,11,26]. CoCoA [7], which serves as the starting point for D3CA, follows the observation partitioning scheme and treats each block of data as an independent sub-problem. Due to the separability of the problem over the dual variables, the local objectives that are maximized are identical to the global one. Each sub-problem is approximately solved using a dual optimization method; the Stochastic Dual Coordinate Ascent (SDCA) method [22] is a popular algorithm for this task. Following the optimization step, the locally updated primal and dual variables are averaged, and the process is repeated until convergence. Similar to SGD-based algorithms, dual methods have not yet been explored when the feature space is distributed.

SGD-CD Hybrid Methods. There has recently been a surge of methods combining SGD and CD [9,14,24,25,28]. These methods conduct parameter updates based on stochastic partial gradients, which are computed by randomly sampling observations and blocks of variables. With the exception of RAPSA [14], which is a parallel algorithm, all other methods are serial, and typically assume that the sampling process has access to all observations and features. Although this is a valid assumption in a parallel (shared-memory) setting, it does not hold in distributed environments. RAPSA employs an update scheme similar to that of mini-batch SGD, but does not require all variables to be updated at the same time. More specifically, in every iteration each processor randomly picks a subset of observations and a block of variables, and computes a partial stochastic gradient based on them. Subsequently, it performs a single stochastic gradient update on the selected variables, and then re-samples feature blocks and observations. Despite the fact that RAPSA is not a doubly distributed optimization method, its parameter update is quite different from that of RADiSA. On one hand, RAPSA allows only one parameter update per iteration, whereas RADiSA permits multiple updates per iteration, thus leading to a great reduction in communication. Finally, RADiSA utilizes the SVRG technique, which is known to accelerate the rate of convergence of an algorithm.

ADMM-based Methods. A popular alternative for distributed optimization is the alternating direction method of multipliers (ADMM) [3]. The original ADMM algorithm, as well as many of its variants that followed (e.g. [15]), is very flexible in that it can be used to solve a wide variety of problems, and is easily parallelizable (either in terms of features or observations). A block splitting variant of ADMM was recently proposed that allows both features and observations to be stored in distributed fashion [17]. One caveat of ADMM-based methods is their slow convergence rate. In our numerical experiments we show empirically the benefits of using RADiSA or D3CA over block splitting ADMM.

3   Algorithms

In this section we present the D3CA and RADiSA algorithms. We first briefly discuss the problem of interest, and then introduce the notation used in the remainder of the paper.

Preliminaries

In a typical supervised learning task, there is a collection of input-output pairs $\{(x_i, y_i)\}_{i=1}^n$, where each $x_i \in \mathbb{R}^m$ represents an observation consisting of m features, and is associated with a corresponding label y_i. This collection is usually

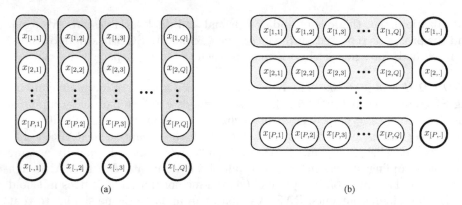

Fig. 1. An illustration of the partitioning scheme under consideration. (a) and (b) show the definitions of $x_{[.,q]}$ and $x_{[p,.]}$ respectively.

referred to as the training set. The general objective under consideration can be expressed as a minimization problem of a finite sum of convex functions, plus a smooth, convex regularization term (where $\lambda > 0$ is the regularization parameter, and f_i is parametrized by y_i):

$$\min_{w \in \mathbb{R}^m} F(w) := \frac{1}{n} \sum_{i=1}^{n} f_i(w^T x_i) + \lambda||w||^2. \tag{1}$$

We should remark that additional work would be needed to examine the adaptation of our methods for solving problems with non-smooth regularizers (e.g. L_1-norm). An alternative approach for finding a solution to (1) is to solve its corresponding dual problem. The dual problem of (1) has the following form:

$$\min_{\alpha \in \mathbb{R}^n} D(\alpha) := \frac{1}{n} \sum_{i=1}^{n} -\phi_i^*(-\alpha_i) - \frac{\lambda}{2} \left|\left| \frac{1}{\lambda n} \sum_{i=1}^{n} \alpha_i x_i \right|\right|^2, \tag{2}$$

where ϕ_i^* is the convex conjugate of f_i. Note that for certain non-smooth primal objectives used in models such as support vector machines and least absolute deviation, the convex conjugate imposes lower and upper bound constraints on the dual variables. One interesting aspect of the dual objective (2) is that there is one dual variable associated with each observation in the training set. Given a dual solution $\alpha \in \mathbb{R}^n$, it is possible to retrieve the corresponding primal vector by using

$$w(\alpha) = \frac{1}{\lambda n} \sum_{i=1}^{n} \alpha_i x_i. \tag{3}$$

For any primal-dual pair of solutions w and α, the duality gap is defined as $F(w) - D(\alpha)$, and it is known that $F(w) \geq D(\alpha)$. Duality theory guarantees that at an optimal solution α^* of (2), and w^* of (1), $F(w^*) = D(\alpha^*)$.

Notation: We assume that the data $\{(x_i, y_i)\}_{i=1}^n$ is distributed across observations and features over K computing nodes of a cluster. More specifically, we split the features into Q partitions, and the observations into P partitions (for simplicity we assume that $K = P \cdot Q$). We denote the labels of a partition by $y_{[p]}$, and the observations of the training set for its subset of features by $x_{[p,q]}$. For instance, if we let $Q = 2$ and $P = 2$, the resulting partitions are $(x_{[1,1]}, y_{[1]})$, $(x_{[1,2]}, y_{[1]})$, $(x_{[2,1]}, y_{[2]})$ and $(x_{[2,2]}, y_{[2]})$. Furthermore, $x_{[p,.]}$ represents all observations and features (across all q) associated with partition p ($x_{[.,q]}$ is defined similarly) – Fig. 1 illustrates this partitioning scheme. We let n_p denote the number of observations in each partition, such that $\sum_p n_p = n$, and we let m_q correspond to the number of features in a partition, such that $\sum_q m_q = m$. Note that partitions corresponding to the same observations all share the common dual variable $\alpha_{[p,.]}$. In a similar manner, partitions containing the same features share the common primal variable $w_{[.,q]}$. In other words, for some pre-specified values $\tilde{p}$ and $\tilde{q}$, the partial solutions $\alpha_{[\tilde{p},.]}$ and $w_{[.,\tilde{q}]}$ represent aggregations of the local solutions $\alpha_{[\tilde{p},q]}$ for $q = 1, ..., Q$ and $w_{[p,\tilde{q}]}$ for $p = 1, ..., P$. At any iteration of D3CA, the global dual variable vector can be written as $\alpha = [\alpha_{[1,.]}, \alpha_{[2,.]}, ..., \alpha_{[P,.]}]$, whereas for RADiSA the global primal vector has the form $w = [w_{[.,1]}, w_{[.,2]}, ..., w_{[.,Q]}]$, i.e. the global solutions are formed by concatenating the partial solutions.

Doubly Distributed Dual Coordinate Ascent

The D3CA framework presented in Algorithm 1 hinges on CoCoA [7], but it extends it to cater for the features being distributed as well. The main idea behind D3CA is to approximately solve the local sub-problems using a dual optimization method, and then aggregate the dual variables via averaging. The choice of averaging is reasonable from a dual feasibility standpoint when dealing with non-smooth primal losses – the LOCALDUALMETHOD guarantees that the dual variables are within the lower and upper bounds imposed by the convex conjugate, so their average will also be feasible. Although in CoCoA it is possible to recover the primal variables directly from the local solver, in D3CA, due to the averaging of the dual variables, we need to use the primal-dual relationship to obtain them. Note that in the case where $Q = 1$, D3CA reduces to CoCoA.

D3CA requires the input data to be doubly partitioned across K nodes of a cluster. In step 3, the algorithm calls the local dual solver, which is shown in Algorithm 2. The LOCALDUALMETHOD of choice is SDCA [22], with the only difference that the objective that is maximized in step 3 is divided by Q. The reason for this is that each partition now contains $\frac{m}{Q}$ variables, so the factor $\frac{1}{Q}$ ensures that the sum of the local objectives adds up to (2). Step 6 of Algorithm 1 shows the dual variable update, which is equivalent to averaging the dual iterates coming from SDCA. Finally, 9 retrieves the primal variables in parallel using the

---

Algorithm 1. Doubly Distributed Dual Coordinate Ascent (D3CA)

Data: $(x_{[p,q]}, y_{[p]})$ for $p = 1, ..., P$ and $q = 1, ..., Q$

Initialize: $\alpha^{(0)} \leftarrow 0$, $w^{(0)} \leftarrow 0$

1: **for** $t = 1, 2, ...$ **do**
2:    **for all partitions** $[p, q]$ **do in parallel**
3:       $\Delta\alpha_{[p,q]}^{(t)} = \text{LOCALDUALMETHOD}(\alpha_{[p,.]}^{(t-1)}, w_{[.,q]}^{(t-1)})$
4:    **end for**
5:    **for all** p **do in parallel**
6:       $\alpha_{[p,.]}^{(t)} = \alpha_{[p,.]}^{(t-1)} + \frac{1}{P \cdot Q} \sum_{q=1}^{Q} \Delta\alpha_{[p,q]}^{(t)}$
7:    **end for**
8:    **for all** q **do in parallel**
9:       $w_{[.,q]}^{(t)} = \frac{1}{\lambda n} \sum_{p=1}^{P} ((\alpha_{[p,q]}^{(t)})^T x_{[p,q]})$
10:    **end for**
11: **end for**

---

primal-dual relationship. The new primal and dual solutions are used to warm-start the next iteration. The performance of the algorithm turns out to be very sensitive to the regularization parameter λ. For small values of λ relative to the problem size, D3CA is not always able to reach the optimal solution. One modification we made to alleviate this issue was to add a step-size parameter when calculating the $\Delta\alpha$'s in the local dual method (Algorithm 2, step 3). In the case of linear Support Vector Machines (SVM) where the closed form solution for step 3 is given by $\Delta\alpha = y_i \max(0, \min(1, \frac{\lambda n(1 - x_i^T w^{(h-1)} y_i)}{||x_i||^2} + \alpha_i^{(h-1)} y_i)) - \alpha_i^{(h-1)}$, we replace $||x_i||^2$ with a step-size parameter β [23]. In our experiments we use $\beta = \frac{\lambda}{t}$, where t is the global iteration counter. Although, a step-size of this form does not resolve the problem entirely, the performance of the method does improve.

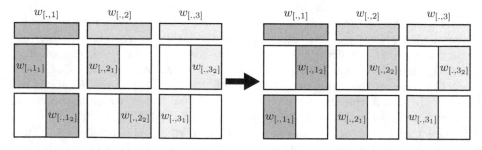

Fig. 2. An illustration of two iterations of RADiSA, with six overall partitions ($P = 2$ and $Q = 3$).

Algorithm 2. LOCALDUALMETHOD: Stochastic Dual Coordinate Ascent (SDCA)

Input: $\alpha_{[p,q]} \in \mathbb{R}^{n_p}$, $w_{[p,q]} \in \mathbb{R}^{m_q}$

Data: Local $(x_{[p,q]}, y_{[p]})$

Initialize: $\alpha^{(0)} \leftarrow \alpha_{[p,q]}, w^{(0)} \leftarrow w_{[p,q]}, \Delta\alpha_{[p,q]} \leftarrow 0$

1: **for** $h = 1, 2, \ldots$ **do**
2: choose $i \in \{1, 2, \ldots, n_p\}$ at random
3: find $\Delta\alpha$ maximizing $-\frac{1}{Q}\phi_i^*(-(\alpha_i^{(h-1)} + \Delta\alpha))-$
    $\frac{\lambda n}{2}\|w^{(h-1)} + (\lambda n)^{-1}\Delta\alpha(x_{[p,q]})_i\|^2)$
4: $\alpha_i^{(h)} = \alpha_i^{(h-1)} + \Delta\alpha$
5: $(\Delta\alpha_{[p,q]})_i = (\Delta\alpha_{[p,q]})_i + \Delta\alpha$
6: $w^{(h)} = w^{(h-1)} + \frac{1}{\lambda n}\Delta\alpha(x_{[p,q]})_i$
7: **end for**
8: **Output:** $\Delta\alpha_{[p,q]}$

In terms of parallelism, the $P \times Q$ sub-problems can be solved independently. These independent processes can either be carried out on separate computing nodes, or in distinct cores in the case of multi-core computing nodes. The only steps that require communication are step 6 and step 9. The communication steps can be implemented via *reduce* operations – in Spark we use *treeAggregate*, which is superior to the standard *reduce* operation.

Random Distributed Stochastic Algorithm

Similar to D3CA, RADiSA, outlined in Algorithm 3, assumes that the data is doubly distributed across K partitions. Before reaching step 1 of the algorithm, all partitions associated with the same block of variables (i.e. $[.,q]$ for $q = 1, \ldots, Q$) are further divided into P non-overlapping sub-blocks. The reason for doing this is to ensure that at no time more than one processor is updating the same variables. Although the blocks remain fixed throughout the runtime of the algorithm, the random exchange of sub-blocks between iterations is allowed (step 5). The process of randomly exchanging sub-blocks can be seen graphically in Fig. 2. For example, the two left-most partitions that have been assigned the coordinate block $w_{[.,1]}$, exchange sub-blocks $w_{[.,1_1]}$ and $w_{[.,1_2]}$ from one iteration to the next. The notation $\bar{q}_p^q$ in step 5 of the algorithm essentially implies that sub-blocks are partition-specific, and, therefore, depend on P and Q.

A possible variation of Algorithm 3 is one that allows for complete overlap between the sub-blocks of variables. In this setting, however, concatenating all local variables into a single global solution (step 12) is no longer an option. Other techniques, such as parameter averaging, need to be employed in order to aggregate the local solutions. In our numerical experiments, we explore a parameter averaging version of RADiSA (RADiSA-avg).

The optimization procedure of RADiSA makes use of the Stochastic Variance Reduce Gradient (SVRG) method [8], which helps accelerate the convergence of the algorithm. SVRG requires a full-gradient computation (step 3), typically after a full pass over the data. Note that for models that can be expressed as the sum functions, like in (1), it is possible to compute the gradient when the data is doubly distributed. Although RADiSA by default computes a full-gradient for each global iteration, delaying the gradient updates can be a viable alternative. Step 9 shows the standard SVRG step, which is applied to the sub-block of coordinates assigned to that partition. The total number of inner iterations is determined by the batch size L, which is a hyper-parameter. As is always the case with variants of the SGD algorithm, the learning rate η_t (also known as step-size) typically requires some tuning from the user in order to achieve the best possible results. In Sect. 4 we discuss our choice of step-size. The final stage of the algorithm simply concatenates all the local solutions to obtain the next global iterate. The new global iterate is used to warm-start the subsequent iteration.

Similar to D3CA, the $P \times Q$ sub-problems can be solved independently. As far as communication is concerned, only the gradient computation (step 3) and parameter update (step 9) stages require coordination among the different processes. In Spark, the communication operations are implemented via *treeAggregate*.

Algorithm 3. Random Distributed Stochastic Algorithm (RADiSA)

Input: batch size L, learning rate η_t

Data: $(x_{[p,q]}, y_{[p]})$ for $p = 1, ..., P$ and $q = 1, ..., Q$

Initialize: $\tilde{w}_0 \leftarrow 0$

Partition each $[., q]$ into P blocks, such that $w_{[.,q]} = [w_{[.,q_1]}, w_{[.,q_2]}, ..., w_{[.,q_P]}]$

1: **for** $t = 1, 2, ...$ **do**
2:     $\tilde{w} = \tilde{w}^{(t-1)}$
3:     $\tilde{\mu} = \frac{1}{n} \sum_{i=1}^{n} \nabla f_i(\tilde{w})$
4:     **for all partitions** $[p, q]$ **do in parallel**
5:         Randomly pick sub-block $\bar{q} = \bar{q}_p^q$ in non-overlapping manner
6:         $w^{(0)} = \tilde{w}_{[p,\bar{q}]}$
7:         **for** $i = 0, ..., L - 1$ **do**
8:             randomly pick $j \in \{1, ..., n_p\}$
9:                 $w^{(i+1)} = w^{(i)} - \eta_t(\hat{\nabla} f_j(w^{(i)}) - \hat{\nabla} f_j(\tilde{w}_{[p,\bar{q}]}) + \tilde{\mu}_{[p,\bar{q}]})$
10:         **end for**
11:     **end for**
12:     $\tilde{w}^{(t)} = [w_{[.,1]}, w_{[.,2]}, ..., w_{[.,Q]}]$, where $w_{[.,q]} = [w_{[.,\bar{q}_1^q]}^{(L)}, ..., w_{[.,\bar{q}_P^q]}^{(L)}]$
13: **end for**

---

4   Numerical Experiments

In this section we present two sets of experiments. The first set is adopted from [17], and we compare the block distributed version of ADMM with RADiSA and D3CA. In the second set of experiments we explore the scalability properties of the proposed methods. We implemented all algorithms in Spark and conducted the experiments in a Hadoop cluster with 4 nodes, each containing 8 Intel Xeon E5-2407 2.2 GHz cores. For the ADMM method, we follow the approach outlined in [17], whereby the Cholesky factorization of the data matrix is computed once, and is cached for re-use in subsequent iterations. Since the computational time of the Cholesky decomposition depends substantially on the underlying BLAS library, in all figures reporting the execution time of ADMM, we have excluded the factorization time. This makes the reported times for ADMM lower than in reality.

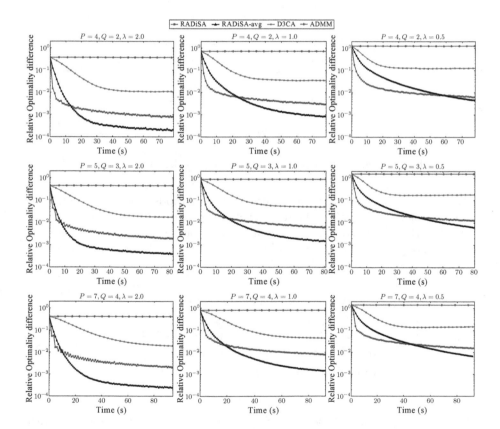

Fig. 3. Relative optimality difference against elapsed time for three data sets with the following configurations of P and Q: (4, 2), (5, 3) and (7, 4).

The problem solved in [17] was lasso regression, which is not a model of the form (1). Instead, we trained one of the most popular classification models: binary classification hinge loss support vector machines (SVM). The data for the first set of experiments was generated according to a standard procedure outlined in [27]: the x_i's and w were sampled from the $[-1, 1]$ uniform distribution; $y_i = \text{sgn}(w^T x_i)$, and the sign of each y_i was randomly flipped with probability 0.1. The features were standardized to have unit variance. We take the size of each partition to be dense $2,000 \times 3,000$,[1] and set P and Q accordingly to produce problems at different scales. For example, for $P = 4$ and $Q = 2$, the size of the entire instance is $8,000 \times 6,000$. The information about the three data sets is summarized in Table 1. As far as hyper-parameter tuning is concerned, for ADMM we set $\rho = \lambda$. For RADiSA we set the step-size to have the form $\eta_t = \frac{\gamma}{(1+\sqrt{t-1})}$, and select the constant γ that gives the best performance.

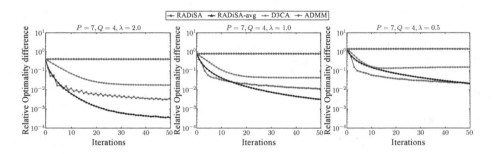

Fig. 4. Relative optimality difference against iteration count.

To measure the training performance of the methods under consideration, we use the relative optimality difference metric, defined as $(f^{(t)} - f^*)/f^*$, where $f^{(t)}$ is the primal objective function value at iteration t, and f^* corresponds to the optimal objective function value obtained by running an algorithm for a very long time.

Table 1. Datasets for numerical experiments (Part 1)

$P \times Q$	4×2	5×3	7×4
Nonzero entries	48 M	90 M	168 M
Number of cores used	8	15	28

In Fig. 3, we observe that RADiSA-avg performs best in all cases, with RADiSA coming in a close second, especially for smaller regularization values.

---

[1] In [17] the size of the partitions was $3,000 \times 5,000$, but due to the BLAS issue mentioned earlier, we resorted to smaller problems to obtain comparable run-times across all methods.

Both variants of RADiSA and D3CA clearly outperform ADMM, which needs a much larger number of iterations to produce a satisfactory solution. We provide an additional comparison in Fig. 4 that further demonstrates this point. We plot the relative optimality difference across 50 iterations. One note about RADiSA-avg is that its performance depends heavily on the number of observation partitions. The averaging step tends to dilute the updates, leading to a slower convergence rate. This is evident when training models on larger data sets than the ones shown in this round of experiments. Another important remark we should make is that when dealing with larger data sets, the behavior of D3CA is erratic for small regularization values. For large regularization values, however, it can produce good solutions.

In the second set of experiments we study the strong scaling properties of our algorithms. Note that the goal of these experiments is to gain insight into the properties of the two methods, rather than to find the best partitioning strategy. The reason for this is that the partitioning of the data is dictated by the application, and is, therefore, out of the practitioner's control. The model under consideration is again linear SVM. To conduct strong scaling experiments, the overall size of the data set does not change, but we increase the number of available computing resources. This means that as the overall number of partitions K increases, the workload of each processor decreases. For RADiSA, we keep the overall number of data points processed constant as we increase K, which implies that as the sub-problem/partition size decreases, so does the batch size L. One matter that requires attention is the step-size parameter. For all SGD-based methods, the magnitude of the step-size η_t is inversely proportional to the batch size L. We adjust the step-size as K increases by simply taking into account the number of observation partitions P. D3CA does not require any parameter tuning. We test our algorithms on two real-world data sets that are available through the LIBSVM website.[2] Table 2 summarizes the details on these data sets.

Table 2. Datasets for numerical experiments (Part 2 - strong scaling)

Dataset	Observations	Features	Sparsity
real-sim	72,309	20,958	0.240%
news20	19,996	1,355,191	0.030%

As we can see in Fig. 5, RADiSA exhibits strong scaling properties in a consistent manner. In both data sets the run-time decreases significantly when introducing additional computing resources. It is interesting that early configurations with $P < Q$ perform significantly worse compared to the alternate configurations where $P > Q$. Let us consider the configurations (4, 1) and (1, 4). In each case,

---

[2] http://www.csie.ntu.edu.tw/~cjlin/libsvmtools/datasets/binary.html.

the number of variable sub-blocks is equal to 4. This implies that the dimensionality of the sub-problems is identical for both partition arrangements. However, the second partition configuration has to process four times more observations compared to the first one, resulting in an increased run-time. It is noteworthy that the difference in performance tails away as the number of partitions becomes large enough. Overall, to achieve consistently good results, it is preferable that $P > Q$.

The strong scaling performance of D3CA is mixed. For the smaller data set (realsim), introducing additional computing resources deteriorates the run-time performance. On the larger data set (news20), increasing the number of partitions K pays dividends when $P > Q$. On the other hand, when $Q > P$, providing additional resources has little to no effect. The pattern observed in Fig. 5 is representative of the behavior of D3CA on small versus large data sets (we conducted additional experiments to further attest this). It is safe to conclude that when using D3CA, it is desirable that $Q > P$.

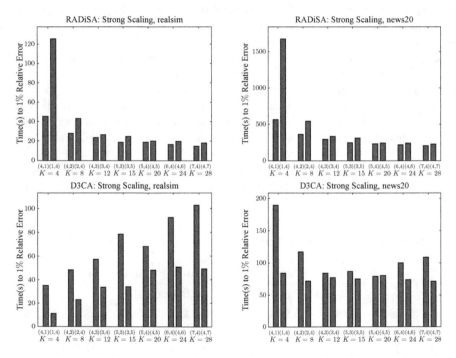

Fig. 5. Strong scaling of realsim and news20. The x-axis shows the various partition configurations for each level of K. The y-axis shows the total time in seconds that is needed to reach a 1% optimality difference. The run-time for the two methods is not comparable due to different regularization values being used. For RADiSA we used $\lambda = 10^{-3}$ and for D3CA we used $\lambda = 10^{-2}$.

5   Conclusion

In this work we presented two doubly distributed algorithms for large-scale machine learning. Such methods can be particularly flexible, as they do not require each node of a cluster to have access to neither all features nor all observations of the training set. It is noteworthy that when massive datasets are already stored in a doubly distributed manner, methods such as the ones introduced in this paper may be the only viable option. Our numerical experiments show that both methods outperform the block distributed version of ADMM. There is, nevertheless, room to improve both methods. The most important task would be to derive a step-size parameter for D3CA that will guarantee the convergence of the algorithm for all regularization parameters. Furthermore, removing the bottleneck of the primal vector computation would result into a significant speedup. As far as RADiSA is concerned, one potential extension would be to incorporate a streaming version of SVRG [5], or a variant that does not require computation of the full gradient at early stages [2]. Finally, studying the theoretical properties of both methods is certainly a topic of interest for future research.

References

1. Agarwal, A., Duchi, J.C.: Distributed delayed stochastic optimization. In: Advances in Neural Information Processing Systems, pp. 873–881 (2011)
2. Babanezhad, R., Ahmed, M.O., Virani, A., Schmidt, M., Konečnỳ, J., Sallinen, S.: Stop wasting my gradients: practical SVRG 2015. arXiv preprint arXiv:1511.01942
3. Boyd, S., Parikh, N., Chu, E., Peleato, B., Eckstein, J.: Distributed optimization and statistical learning via the alternating direction method of multipliers. Found. Trends® Mach. Learn. **3**(1), 1–122 (2011)
4. Dekel, O., Gilad-Bachrach, R., Shamir, O., Xiao, L.: Optimal distributed online prediction using mini-batches. J. Mach. Learn. Res. **13**(1), 165–202 (2012)
5. Frostig, R., Ge, R., Kakade, S.M., Sidford, A.: Competing with the empirical risk minimizer in a single pass (2014). arXiv preprint arXiv:1412.6606
6. Hsieh, C.-J., Chang, K.-W., Lin, C.-J., Keerthi, S.S., Sundararajan, S.: A dual coordinate descent method for large-scale linear SVM. In: Proceedings of the 25th International Conference on Machine Learning, pp. 408–415. ACM (2008)
7. Jaggi, M., Smith, V., Takác, M., Terhorst, J., Krishnan, S., Hofmann, T., Jordan, M.I.: Communication-efficient distributed dual coordinate ascent. In: Advances in Neural Information Processing Systems, pp. 3068–3076 (2014)
8. Johnson, R., Zhang, T.: Accelerating stochastic gradient descent using predictive variance reduction. In: Advances in Neural Information Processing Systems, pp. 315–323 (2013)
9. Konečnỳ, J., Qu, Z., Richtárik, P.: Semi-stochastic coordinate descent (2014). arXiv preprint arXiv:1412.6293
10. Liu, J., Wright, S.J., Ré, C., Bittorf, V., Sridhar, S.: An asynchronous parallel stochastic coordinate descent algorithm. J. Mach. Learn. Res. **16**(1), 285–322 (2015)
11. Ma, C., Smith, V., Jaggi, M., Jordan, M.I., Richtárik, P., Takáč, M.: Adding vs. averaging in distributed primal-dual optimization (2015). arXiv preprint arXiv:1502.03508

12. Mann, G., McDonald, R.T., Mohri, M., Silberman, N., Walker, D.: Efficient large-scale distributed training of conditional maximum entropy models. NIPS **22**, 1231–1239 (2009)
13. McDonald, R., Hall, K., Mann, G.: Distributed training strategies for the structured perceptron. In: Human Language Technologies: The 2010 Annual Conference of the North American Chapter of the Association for Computational Linguistics, pp. 456–464 (2010)
14. Mokhtari, A., Koppel, A., Ribeiro, A.: Doubly random parallel stochastic methods for large scale learning (2016). arXiv preprint arXiv:1603.06782
15. Mota, J.F., Xavier, J.M., Aguiar, P.M., Püschel, M.: D-ADMM: a communication-efficient distributed algorithm for separable optimization. IEEE Trans. Signal Process. **61**(10), 2718–2723 (2013)
16. Nesterov, Y.: Efficiency of coordinate descent methods on huge-scale optimization problems. SIAM J. Optim. **22**(2), 341–362 (2012)
17. Parikh, N., Boyd, S.: Block splitting for distributed optimization. Math. Program. Comput. **6**(1), 77–102 (2014)
18. Recht, B., Re, C., Wright, S., Niu, F.: Hogwild: a lock-free approach to parallelizing stochastic gradient descent. In: Advances in Neural Information Processing Systems, pp. 693–701 (2011)
19. Richtárik, P., Takáč, M.: Distributed coordinate descent method for learning with big data (2013). arXiv preprint arXiv:1310.2059
20. Richtárik, P., Takáč, M.: Iteration complexity of randomized block-coordinate descent methods for minimizing a composite function. Math. Program. **144**(1–2), 1–38 (2014)
21. Richtárik, P., Takáč, M.: Parallel coordinate descent methods for big data optimization. Math. Program. 1–52 (2015)
22. Shalev-Shwartz, S., Zhang, T.: Stochastic dual coordinate ascent methods for regularized loss. J. Mach. Learn. Res. **14**(1), 567–599 (2013)
23. Takáč, M., Bijral, A., Richtárik, P., Srebro, N.: Mini-batch primal and dual methods for SVMS (2013). arXiv preprint arXiv:1303.2314
24. Wang, H., Banerjee, A.: Randomized block coordinate descent for online and stochastic optimization (2014). arXiv preprint arXiv:1407.0107
25. Xu, Y., Yin, W.: Block stochastic gradient iteration for convex and nonconvex optimization. SIAM J. Optim. **25**(3), 1686–1716 (2015)
26. Yang, T.: Trading computation for communication: distributed stochastic dual coordinate ascent. In: Advances in Neural Information Processing Systems, pp. 629–637 (2013)
27. Zhang, C., Lee, H., Shin, K.G.: Efficient distributed linear classification algorithms via the alternating direction method of multipliers. In: International Conference on Artificial Intelligence and Statistics, pp. 1398–1406 (2012)
28. Zhao, T., Yu, M., Wang, Y., Arora, R., Liu, H.: Accelerated mini-batch randomized block coordinate descent method. In: Advances in Neural Information Processing Systems, pp. 3329–3337 (2014)
29. Zinkevich, M., Weimer, M., Smola, A.J., Li, L.: Parallelized stochastic gradient descent. In: NIPS, vol. 4, p. 4 (2010)

CCPM: A Scalable and Noise-Resistant Closed Contiguous Sequential Patterns Mining Algorithm

Yacine Abboud$^{(\boxtimes)}$, Anne Boyer, and Armelle Brun

Université de Lorraine, LORIA UMR 7503, Vandoeuvre-lès-Nancy, France
yacine.abboud@loria.fr

Abstract. Mining closed contiguous sequential patterns has been addressed in the literature only recently, through the CCSpan algorithm. CCSpan mines a set of patterns that contains the same information than traditional sets of closed sequential patterns, while being more compact due to the contiguity. Although CCSpan outperforms closed sequential pattern mining algorithms in the general case, it does not scale well on large datasets with long sequences. Moreover, in the context of noisy datasets, the contiguity constraint prevents from mining a relevant result set. Inspired by BIDE, that has proven to be one of the most efficient closed sequential pattern mining algorithm, we propose CCPM that mines closed contiguous sequential patterns, while being scalable. Furthermore, CCPM introduces usable wildcards that address the problem of mining noisy data. Experiments show that CCPM greatly outperforms CCSpan, especially on large datasets with long sequences. In addition, they show that the wildcards allows to efficiently tackle the problem of noisy data.

Keywords: Data mining · Sequential pattern mining · Closed contiguous sequential pattern · Scalable · Noise resistant

1 Introduction

Pattern mining [2] is one of the most studied topic in the data mining literature. An ordered pattern is usually called a sequential pattern. Many applications rely on sequential patterns: pattern discovery in protein sequences [15,23], analysis of customer behavior in web logs [6], sequence-based classification [4], etc. Consequently, sequential pattern mining [3] has become a significant part of pattern mining studies. Many sequential pattern mining algorithms have been proposed to solve various issues, such as general sequential pattern mining [12,18], string mining [9,16], closed sequential pattern mining [20,21], constraint-based sequential pattern mining [11,19], etc.

A pattern is said closed if there is no super-pattern with the same support. Closed sequential mining has become popular as the set of closed patterns

© Springer International Publishing AG 2017
P. Perner (Ed.): MLDM 2017, LNAI 10358, pp. 147–162, 2017.
DOI: 10.1007/978-3-319-62416-7_11

extracted contains the same information than the set of general sequential patterns, but with less redundancy [24].

Many real-life tasks greatly benefit from patterns with contiguous items. For example in text mining [10], contiguous patterns are used for statistical natural language processing or document classification. In Web log mining [6,7], they are used to predict navigation paths and thus to improve the design of web pages. The search of frequent contiguous sequential patterns in DNA and amino acid sequences is unavoidable to reveal common shared functions [13,14]. The contiguity constraint leads to the extraction of much fewer patterns with a shorter average length, while carrying the same information [24]. That is why closed contiguous sequential patterns (CCSPs) are highly relevant to mine data. However, the added value provided by the contiguity constraint has its cost. In case of noisy data, the support of the mined contiguous patterns is underestimated, making some of these patterns not frequent.

CCSpan is the only algorithm designed to mine CCSPs, its design stands on a snippet-growth scheme to generate candidate patterns. A snippet is a contiguous sub-pattern. CCSpan outperforms, in runtime and memory usage, all the closed sequential pattern mining algorithms in the literature [24]. Nevertheless, it suffers from the same issue than other contiguous pattern mining algorithms: the lack of adaptability to noisy data. In addition, despite being the most efficient algorithm to extract sequential patterns, CCSpan does not scale well on large databases having long sequences.

Two scientific questions thus arise: (1) How to improve contiguous sequential pattern mining algorithms to more accurately extract information from noisy data? (2) How to increase the efficiency of CCSP mining algorithms to scale with large databases and long sequences?

In this paper, we introduce a new algorithm CCPM that overcomes those two issues. CCPM is based on Prefixspan [18] and on BIDE [20], two of the most efficient sequential mining and closed sequential mining algorithms [8]. CCPM mines closed contiguous patterns with usable wildcards. A wildcard [8] is a joker of one item regarding the contiguity constraint. Let $\langle a, b, d, c \rangle$ be a pattern, it can be considered as the pattern $\langle a, b, c \rangle$ with one wildcard. Unlike other algorithms with wildcards of the literature, CCPM introduces a maximal number of wildcards that can be used per pattern. To the best of our knowledge, CCPM is the first noise-resistant CCSP mining algorithm due to the use of wildcards. CCPM also offers the possibility to mine patterns that start with a constrained item and thus gets a new set of patterns not included in the closed contiguous sequential patterns. Experiments show that CCPM greatly outperforms CCSpan, especially on large datasets with long sequences. Hence the main contributions of this paper are:

- we introduce the second closed contiguous sequential pattern mining algorithm that significantly outperforms the first one, CCSpan;
- we use expendabled wildcards to make CCPM noise-resistant;
- we present experiments on standard datasets and on jobs offers datasets to show how CCPM performs on large and noisy data;

This paper is organized as follows. Section 2 introduces concepts and notations required by the formulation of CCSP problem. In Sect. 3 we explore properties of CCSP and present CCPM. Section 4 is dedicated to the experiments conducted, while Sect. 5 concludes.

2   Preliminaries and Related Work

2.1   Preliminaries

In this section we introduce some definitions and notations used for CCSP mining.

Let $I = \{i_1, i_2, ..., i_n\}$ be a set of distinct items. A sequential pattern P is an ordered list denoted as $P = \langle e_1, e_2, ..., e_m \rangle$ where $e_k \in I$ is an item for $1 \leq k \leq m$. For brevity, a sequential pattern is also written $e_1 e_2 ... e_m$. A sequential pattern $P_1 = a_1 a_2 ... a_n$ is contiguously contained in another sequential pattern $P_2 = b_1 b_2 ... b_m$, if $n < m$ and there exist consecutive integers $1 \leq j_1 < j_2 < ... < j_n < m$ such that $a_1 = b_{j_1}, a_2 = b_{j_2}, ..., a_n = b_{j_n}$. P_1 is called a contiguous sub-pattern or a snippet of P_2 and P_2 a contiguous super-pattern of P_1, denoted by $P_1 \sqsubseteq_c P_2$. The concatenation of P_1 and P_2 is a new contiguous sequential pattern, denoted as $\langle P_1, P_2 \rangle$.

An input sequence database SDB is a set of tuples (sid, S), where sid is a sequence id, and S an input sequence. Table 1 contains an example of a sequence database, made up of 4 tuples and the set of items $I = \{A, B, C\}$. The number of tuples in SDB is called the size of SDB and is denoted by $|SDB|$. The absolute support of a contiguous sequential pattern P_α in a sequence database SDB is the number of tuples in SDB that contiguously contain P_α, denoted by $sup_A^{SDB}(S_\alpha)$. Similarly, the relative support of P_α in SDB is the proportion of sequences in SDB that contain P_α, denoted by $sup_R^{SDB}(S_\alpha)$. The universal contiguous support is the number of occurrences of P_α in SDB, denoted by $uni\_sup^{SDB}(P_\alpha)$. Given a threshold min_sup, a contiguous sequential pattern P_α is frequent in SDB if $sup_A^{SDB}(P_\alpha) \geq min\_sup$ or $sup_R^{SDB}(P_\alpha) \geq min\_sup$. P_α is a closed contiguous sequential pattern if there exists no contiguous sequential pattern P_β such that: $P_\alpha \sqsubseteq_c P_\beta$, and $sup_A^{SDB}(P_\alpha) = sup_A^{SDB}(P_\beta)$. An item e is a *starting item*, denoted _e, if it satisfies a chosen constraint. A pattern is a *starting item* pattern if its first item is a *starting item*. Aside from the above notations, we further present some definitions.

Example 1. We present different sets of patterns from the sequence database SDB (shown in Table 1) when $min\_sup = 2$ with one wildcard. The complete set of frequent contiguous sequential patterns consists of 7 sequential patterns $S_{fc} = \{A : 4, AB : 4, ABC : 3, B : 4, BC : 4, C : 4, CA : 3\}$. The set of frequent closed contiguous sequential patterns consists of only 4 sequential patterns $S_{fcc} = \{AB : 4, ABC : 3, BC : 4, CA : 3\}$. The complete set of frequent contiguous sequential patterns with one wildcard is $S_{fcw} = \{A : 4, AB : 4, ABB : 2, ABC : 4, B : 4, BB : 2, BC : 4, C : 4, CA : 3, CB : 2, CAB : 2, CABC : 2\}$. The use of one wildcard allows to discover the sequential patterns $ABB, BB, CB, CABC$

Table 1. An example of sequence database SDB

Sequence id	Sequence
1	CAABC
2	ABCB
3	CABC
4	ABBCA

and to increase the support of ABC by one, making it a frequent CCSP. The set of frequent CCSPs with one wildcard is thus $S_{fccw} = \{ABB : 2,\ ABC : 4,\ CA : 3,\ CB : 2,\ CABC : 2\}$. The pattern ABC has absorbed AB and BC.

2.2 Related Work

The sequential pattern mining problem was first introduced by Agrawal and Srikant in [3] with the Apriori algorithm. Apriori is based on the monotonicity property "all nonempty subsets of a frequent itemset must also be frequent" [2]. Since then, many sequential pattern mining algorithms have been proposed for performance improvements: Prefixspan [18], SPAM [5], etc. Prefixspan algorithm uses projected databases to mine frequent sequential patterns. Prefixspan recursively extends a prefix sequential pattern by adding a frequent item from the projected database of this prefix. This new representation of data, based on projected databases, allows a more efficient mining [5] and thus Prefixspan performs really well regarding execution time. As closed sequential patterns lead to a more compact set of patterns but also a better efficiency, several algorithms focus their mining on such patterns [17,22]. Clospan [21] is one of them, it mines a candidate set of closed patterns based on Prefixspan algorithm, and keeps the candidate patterns set for post-pruning. The candidate set is expensive in term of memory usage. To cope with this issue, the BIDE algorithm [20] does not memorize any candidate set of closed patterns. BIDE generates new patterns with the framework of Prefixspan and uses a BI-Directional closure checking scheme that does not use any set of candidates, hence improves greatly both runtime and memory usage. The philosophy of this scheme is to check if a frequent pattern can be extended to the left (backward extension) or to the right (forward extension) with at least one item. The pattern is closed if it can not be extended in each sequence. BIDE is known to be one of the most efficient closed sequential pattern mining algorithm [8]. One major issue in closed sequential pattern mining comes from large databases: the set of frequent closed sequential patterns becomes unmanageable. To solve this problem, CCSpan [24] algorithm introduced closed contiguous sequential pattern (CCSP) mining. Frequent CCSPs are far fewer and shorter than closed sequential patterns. CCSpan uses a snippet-growth scheme to generate candidate patterns. The original sequence is split into a set of snippets of the same length and their support is calculated to keep the frequent ones as candidate patterns. Then for each candidate pattern, two sub-patterns are considered, the one without the first item of the pattern and the one without the last item. The closure checking is performed recursively by comparing the support of the pattern and the support of both sub-patterns. Despite being more efficient than closed sequential pattern mining algorithms due to the contiguity constraint, CCSpan scales rather poorly on datasets with long sequences. Indeed, the framework that builds snippets becomes highly time consuming as sequences become longer. Furthermore, CCSpan suffers from a lack of adaptability of CCSPs to noisy data. In fact, the support of CCSPs on noisy data may not reflect the real support of those patterns in the data. Therefore, the CCSPs can not be used to accurately extract information from noisy data.

The issues raised by the state-of-the art make us consider the following questions: (1) Can the highly efficient framework from Prefixspan that mines frequent sequential patterns, used as well in Clospan and BIDE, be used to get frequent CCSPs? (2) How to take inspiration from BIDE, one of the most efficient algorithm for closed sequential patterns mining, to mine CCSPs? (3) How to bypass CCSpan algorithm issues with noisy data and keep the added value of contiguity?

3   CCPM: Mining Closed Contiguous Sequential Patterns

In this section, we introduce the CCPM algorithm (Closed Contiguous sequential Patterns Mining), which shares the philosophy of both Prefixspan [18] and BIDE algorithms [20]. The use of contiguous sequential patterns mining algorithm is only relevant in case of clean data but most of the time data still contains noise. In order to overcome this issue, we introduce a maximal number of wildcards per pattern. As the mined data is usually just a bit noisy, the number of wildcards allowed is often rather small. In addition, CCPM allows to mine patterns starting with an item that satisfies a chosen constraint: *starting item* patterns. Such patterns are interesting in many domains. In text mining, patterns starting with an action verb in order to identify competencies [1] or in marketing, purchase patterns starting by a given category of items to establish more targeted selling strategies, etc. As a contiguous sequential *starting item* pattern (CSSP) can only start with a *starting item*, a closed contiguous sequential *starting item* pattern (CCSSP) is necessarily equal or included in a closed contiguous sequential super-pattern. If it is included, the CCSSP will not appear in the set of mined CCSPs. Therefore, if one CCSSP is not equal to the closed contiguous sequential super-pattern then the set of CCSSPs is not included in the set of CCSPs.

CCPM uses a framework that enumerates frequent contiguous sequential *starting item* patterns with wildcards (CSSPWs), inspired by the framework of Prefixspan to enumerate frequent sequential patterns. CCPM uses a closure scheme and a pruning technique inspired by the BI-Directional closure scheme and the Backscan method of BIDE to mine closed patterns. CCPM relies on three steps listed here: First, a framework that mines frequent CSSPWs. Second, a BI-Directional checking scheme, adapted to CSSPWs, that forms closed contiguous sequential *starting item* patterns with wildcards (CCSSPWs). Third, a contiguous pruning techniques used to improve the efficiency of CCPM.

3.1   Framework to Enumerate Frequent CSSPWs

Here we introduce some definitions about projected contiguous items and database in order to explain the framework.

Definition 1. *Projected contiguous items of a prefix sequential pattern with wildcards: Given an input sequence S and a sequential pattern P with $P \sqsubseteq_c S$, the projected contiguous items of P are all the adjacent items after each occurrence of P in S and P is called a prefix sequential pattern. For example, in the sequence ABBCA of Table 1, the projected contiguous items of the prefix sequential pattern B are B and C. If a wildcard can be used, all the second adjacent items after P are projected items too. A wildcard is denoted by * on the item it corresponds to. On the previous example, the item A is a projected contiguous item of B with a wildcard, noted $A*$. A wildcard is not used if the first and the second adjacent items after P are the same. For example, with the prefix sequential pattern A, the projected contiguous items are B and $B*$, so we only keep B. To summarize, the projected contiguous items with a wildcard of the prefix sequential pattern B in ABBCA are $\{B, C, A*\}$.*

Definition 2. *Projected contiguous database of a prefix sequential pattern with wildcards: The complete set of projected contiguous items in SDB of a prefix sequential pattern P with wildcards is called the projected contiguous database of P with wildcards.*

Theorem 1 (Projected database pruning). *Given two items e_1 and e_2, if e_2^* is always a projected contiguous item of P together with e_1, and if e_2 is never a projected contiguous item, e_2^* can be safely removed from the projected contiguous database.*

Proof. Given two items e_1 and e_2, a prefix sequential pattern P and a sequence database SDB. If e_2^* is always a projected contiguous item of P together with e_1 and if e_2 is never a projected contiguous item of P in SDB, it means that there exists a contiguous sequential pattern $Q = \langle P, e_1, e_2 \rangle$ with $uni\_sup^{SDB}(Q) = uni\_sup^{SDB}(\langle P, e_2 \rangle)$. Thus, the sequential pattern $\langle P, e_2 \rangle$ is always included in Q and so is its expansions. Therefore, it is useless to expand it.

The complete search space of contiguous sequential patterns forms a sequential pattern tree [5]. The root node of the tree is at the top level and is labeled $\emptyset$. The framework of CCPM recursively extends a node (referred to as a prefix sequential pattern) at a certain level in the tree by adding a contiguous item, resulting in a child node at the next level. As CCPM mines *starting item* patterns, the framework only extends frequent items satisfying the constraint. Applying this constraint greatly reduces the search space, hence greatly improves its efficiency. It also improves the results by removing irrelevant patterns. When all frequent *starting items* are found, each one becomes a prefix sequential pattern to expand, they are at the first level of the tree. The framework then builds the projected contiguous database of each prefix. To know if a wildcard is avalaible, the count of wildcards used is memorized for each occurrence of the prefix sequential pattern. Then, only the frequent items in the projected contiguous database grow the prefix sequential pattern.

Now we describe how to enumerate frequent closed contiguous sequential patterns with one wildcard on this example. The prefix sequential pattern A is the first studied, its projected contiguous database is: $\{A, B, C^*; \ B, C^*; \ B, C^*; \ B\}$. In this projected database, B has a support of 4 and C a support of 3. As their support exceed min_sup, both of them can be extended. However, B always occurs together with C^* in the projected contiguous database and C is never a projected contiguous item. According to Theorem 1, A can be extended with B, C can be removed. Then, given the prefix sequential pattern AB, the projected database is: $\{C; \ C, B^*; \ C; \ B, C^*\}$. C has a support of 4 and B a support of 2, both of them can be extended. The projected database of ABC is: $\{\emptyset; \ B; \ \emptyset; \ A\}$ and the one of ABB is: $\{\emptyset; \ C\}$. As no item have a support greater or equal to $min\_sup = 2$, the prefix sequential pattern A is now completely extended. The same procedure can be used with B and C.

3.2   BI-Directional Contiguous Closure Checking with Wildcards

The previous framework mines frequent CSSPWs. To get the set of frequent CCSSPWs, a closure checking has to be used. CCPM closure checking scheme is based on the same BI-Directional design than BIDE. The definitions of backward and forward extensions of BIDE are adapted to CSSPWs with the same number of wilcards for all CSSPWs.

Given two CSSPWs $P_1 = \_b_1 b_2 ... b_n$ and $P_2 = \_a_1 ... a_m \_b_1 b_2 ... b_n$ in a sequence database SDB. If $sup_A^{SDB}(P_1) = sup_A^{SDB}(P_2)$, $\langle \_a_1 ... a_m \rangle$ is called a backward extension of P_1. The specificity of CSSPW forces the backward extension to be a *starting item* pattern. Given a sequence S of SDB and a CSSPW P with $P \sqsubseteq_c S$. For each occurrence of P in S, the backward space of P is the CSSPW Q such that $Q = \_c_1 ... c_j$ and $\langle Q, P \rangle \sqsubseteq_c S$ where $\_c_1$ is the first *starting item* on the left of this occurrence of P in S. If there is no *starting item* on the left of P, P can not have a backward extension. Let $P_3 = \_b_1 b_2 ... b_n d_1 ... d_k$ be a CSPW, if $sup_A^{SDB}(P_1) = sup_A^{SDB}(P_3)$, we say $d_1 ... d_k$ is a forward extension item of P_1.

With those definitions in mind, the following theorem and lemma are straightforward.

Theorem 2. *A CSSPW P is closed, if and only if P has no backward extension and no forward extension items.*

Lemma 1. *Given a CSSPW P in a sequence database SDB, if there is a backward space Q of P in SDB, with $sup_A^{SDB}(\langle Q, P \rangle) = sup_A^{SDB}(P)$, then Q is a backward extension of P.*

Lemma 2. *For a given CSSPW P, its complete set of forward extensions items is the set of items in its projected contiguous database that have a support equal to $sup_A^{SDB}(P)$.*

Proof. For each item e_k in the projected contiguous database of P, a new CSSPW $P_k = \langle P, e_k \rangle$ can be created, if $sup_A^{P\text{-}SDB}(e_k) = sup_A^{SDB}(P)$ thus $sup_A^{SDB}(P_k) = sup_A^{SDB}(P)$ then $\langle e_k \rangle$ is a forward extension of P.

We introduce two stop conditions during the backward spaces scanning to improve the efficiency of the backward extension checking. The backward checking extension of a CSSPW is stopped as soon as: (1) a *starting item* cannot be found in the backward spaces of a sequence; (2) one of the two backward contiguous items (if a wildcard is available) or the backward contiguous item (no wildcard) of the prefix sequential pattern cannot be found in the backward spaces of a sequence.

3.3 The Contiguous BackScan Pruning

To prune the useless parts of the search space, we propose to use the Backscan technique from BIDE, adapted to CCSSPW mining. The idea behind Backscan is to detect if the current prefix sequential pattern can be absorbed by a prefix sequential pattern that will be mined later. If it is the case, it is not expanded.

Theorem 3 (The Backscan Pruning). *Given a CSSPW P in SDB, if there exists a backward space Q with $uni\_sup(\langle Q, P \rangle) = uni\_sup^{SDB}(P)$, P can be safely pruned.*

Proof. Given a CSSPW P and a backward space Q with $uni\_sup(\langle Q, P \rangle) = uni\_sup^{SDB}(P)$. Assume we extend P with a contiguous sequential pattern K to get a longer CSSPW R $(R = \langle P, K \rangle)$, we can also construct another CSSPW $R' = \langle Q, P, K \rangle$. As $uni\_sup^{SDB}(\langle Q, P \rangle) = uni\_sup^{SDB}(P)$, we have $uni\_sup^{SDB}(R) = uni\_sup^{SDB}(R')$. Hence, we can not use P as a prefix sequential pattern to generate any CCSSPW.

3.4 The CCPM Algorithm

CCPM (Algorithm 1), integrates the three previous steps: the framework that searches frequent CSSPWs, the BI-Directional closure checking and the BackScan pruning.

Algorithm 1. CCPM(SDB, min_sup, W)

Input: a sequence database SDB, a minimum support threshold min_sup, a maximal number of wildcards W

Output: F, the complete set of frequent CCSSPWs

1: $F = \{\emptyset\}$
2: $F1 = frequent\_starting\_items(SDB, min\_sup)$;
3: **foreach** $f1$ in $F1$ **do**
4:     $f1\_SDB = projected\_database(SDB, f1, W)$;
5:     **if** (!**BackScan**($f1$))
6:         call **pGrowth**($f1\_SDB, f1, min\_sup, F$);
7: **return** F;

Algorithm 2. pGrowth(P_SDB, P, min_sup, F)

Input: projected sequence database with the used wildcards P_SDB, a prefix sequential pattern P, a minimum support threshold min_sup

Output:F, the current set of frequent CCSSPWs

1: $FI = frequent\_items(P\_SDB, min\_sup)$;
2: $BEI = $ **Backward_Check**(P, P_SDB, W)
3: **if** (!**BackScan**(P)
4: $\quad FEI = |\{z \ in \ FI \mid z.sup = sup^{SDB}(P)\}|$;
5: $\quad$ **if** ($\{BEI \cap FEI\} == \{\emptyset\}$)
6: $\quad\quad F = F \bigcup \{P\}$;
7: $\quad$ **foreach** i in FI **do**
8: $\quad\quad P^i = \langle P, i \rangle$;
9: $\quad\quad P^i\_SDB = projected\_database(P\_SDB, P^i, W)$;
10: $\quad\quad$ call **pGrowth**$(P^i\_SDB, P^i, min\_sup, F)$;

The input parameters of Algorithm 1 are a sequence database, a minimum support and a maximal number of wildcards per pattern. CCPM starts by scanning the database to find the set of frequent *starting items*: $F1$ (line 2). The *starting items* constraint can be removed without any impact on CCPM. The *projected_database* function (line 4) creates the projected database of each item in $F1$ and specifies if a wildcard has been used or not for each occurrence of each item. *BackScan* (line 5) is applied to check if a frequent item can immediately be pruned. The subroutine pGrowth (Algorithm 2) is called for each frequent *starting item* with its projected database. pGrowth checks if the input pattern P is closed and recursively extends P with each frequent items in its projected database. pGrowth continuously updates the set of CCSSPWs. The function *Backward_Check* (line 2) looks for backward extensions, with 2 stop conditions designed to improve the efficiency. *Backscan* (line 3) is used again to check if the prefix can be pruned, then if there are no forward and no backward extensions (line 4 and 5), the input pattern P is added to the set of frequent CCSSPW: F (line 6). Then the projected database of P is scanned to find the set of frequent contiguous items and specifies if a wildcard has been used or not for each item (line 9). Each frequent item extends P (line 8) and becomes the prefix sequential pattern to recall pGrowth until CCPM returns F, the complete set of frequent CCSSPW.

4   Experiments

We conduct a comprehensive performance study to evaluate CCPM. [24] has shown how CCSpan outscales and outperforms closed sequential pattern mining algorithms such as Clospan and BIDE. Therefore, we only compare CCPM to our implemantation of CCSpan to evaluate its efficiency. First, CCPM is compared to CCSpan [24] in terms of running time and memory usage. The parameters of CCPM are set so that CCSpan and CCPM correspond to the same configuration: CCSP mining. Second, CCPM with no constraints and CCPM with *starting item* patterns and wildcards are compared to evaluate the impact of these constraints.

4.1 Datasets and Environnement

The experiments are performed on a i7-4750HQ 2 GHz, 8 GB memory on Windows 10. In order to evaluate the performance of CCPM, we use five real datasets that cover a wide range of distribution characteristics. Two of them are reference datasets used to evaluate sequential pattern mining algorithms. *BMS*1 *Gazelle* [20, 21, 24] used in the KDD-CUP 2000 competition and *Mushroom*, available online on the SPMF website[1]. The three other datasets are datasets we formed from online job offers, so made up of noisy textual data.

The *BMS*1 *Gazelle* dataset contains clickstream and purchase data from Gazelle.com, a legwear and legcare web retailer. *BMS*1 *Gazelle* is sparse. The second dataset, *Mushroom*, is composed of characteristics of various species of mushrooms and was originally obtained from the UCI repository of machine learning databases. *Mushroom* is dense. The three job datasets (*jobs*900, *jobs*1000 and *jobs*60000) are composed of job offers collected on websites between January 2016 and June 2016. They reflect the kind of textual data that can be found on the web, they are noisy and very sparse. The characteristics of the five datasets are shown in Table 2. Some of these datasets share similarities on some attributes, so we can easily compare the influence of each attribute on CCPM. The datasets *BMS*1 *Gazelle* and *jobs*60000 as well as *jobs*900 and *jobs*1000 have approximately the same number of sequences with a very different average

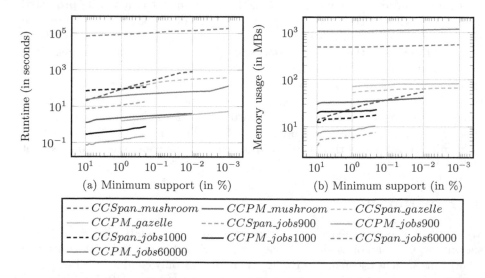

Fig. 1. Runtime (a) and Memory (b) on Mushroom, Gazelle, Jobs900, Jobs1000 and Jobs60000 datasets

---

[1] http://www.philippe-fournier-viger.com/spmf/index.php.

Table 2. Datasets characteristics

Dataset	#seq	#items	Avg.len
*BMS*1 *Gazelle*	59,601	497	2.4
Mushroom	8,416	119	23.0
*jobs*900	900	4,460	26.6
*jobs*1000	1,000	8,813	76.0
*jobs*60000	60,000	92,394	77.0

length, they will be compared to evaluate the impact of average length. The datasets *jobs*900 and *Mushroom* as well as *jobs*1000 and *jobs*60000 have the same average length, they will be compared to evaluate the impact of the number of sequences. The experiment will highlight the impact of both average length and number of sequences on the efficiency of CCSpan and CCPM).

4.2 CCSP Mining: Comparison of CCSpan and CCPM

As CCPM can be set to mine CCSPs like CCSpan, we can conduct a series of experiments on the five datasets in terms of execution time and memory usage, with various relative support thresholds (Fig. 1). Each line stops when the absolute support reaches the value of 1 for the corresponding dataset. First of all, Fig. 1 shows that neither CCSpan nor CCPM are significantly impacted by the variation of the support. Indeed, both runtime and memory usage are never increased by more than 10 whereas the number of patterns is increased from 110 times on *jobs*1000 to 6,200 on *jobs*60000. The only exception is the runtime of CCSpan on the *Mushroom* dataset, which is increased by almost 100. The contiguity constraint seems to make CCSpan and CCPM almost independent on the support and on the number of patterns mined. As the variation of the support has little impact on runtime and memory usage, we choose to set the support value for the rest of the experiments, this value is set to $min\_sup = 1\%$.

Average Length. We first evaluate the impact of the average length of sequences on the execution time of CCSpan and CCPM (Fig. 1a). *jobs*900 and *jobs*1000 share a similar number of sequences (900 vs 1,000) with a very different average length (23 vs 76). Same with *BMS*1 *Gazelle* and *jobs*60000 (59,601 vs 60,000) that also differ in the average sequence length (2.4 vs 77). CCSpan runs 8 times faster on *jobs*900 than on *jobs*1000 whereas the average length of *jobs*900 is 3 times smaller (ratio of 2.7). Same reasoning between *BMS*1 *Gazelle* and *jobs*60000, CCSpan runs 6,800 faster on *BMS*1 *Gazelle* than on *jobs*60000 whereas the average length is 32 times smaller (ratio of 212). We can conclude that the average length of the sequences exponentially impacts the runtime of CCSpan.

CCPM runs 3.5 times faster on *jobs*900 than on *jobs*1000 (ratio of 1.2) and 25 times faster on *Gazelle* than on *jobs*60000 (ratio of 0.8). We can conclude that, contrary to CCSpan, the runtime of CCPM is linearly impacted by the length of the sequences.

Number of Sequences. Now we focus on the impact of the number of sequences on the execution time of CCSpan and CCPM. *jobs*900 and *Mushroom* share a similar average length of sequences (26.6 vs 23), while having a different number of sequences (900 vs 8,416). The two datasets *jobs*1000 and *jobs*60000 also have a similar average length of sequences (76 vs 77), while having a very different number of sequences (1,000 vs 60,000). CCSpan runs 8 times faster on *jobs*900 than on *Mushroom* whereas the number of sequences of *jobs*900 is 9 times smaller (ratio of 0.9). CCSpan runs 915 times faster on *jobs*1000 than on *jobs*60000 whereas the number of sequences of *jobs*1000 is 60 times smaller (ratio 15). We can conclude that the number of sequences exponentially impact the runtime of CCSpan. This result has to be tempered due to specificity of the *Mushroom* dataset, indeed it is the only dataset that impacts the runtime of CCSpan, while having a support that varies.

CCPM runs 18 times faster on *jobs*900 than on *Mushroom* (ratio of 2) and 83 times faster on *jobs*1000 than on *jobs*60000 (ratio of 1.4). We observe that, contrary to CCSpan, the number of sequences linearly impacts the runtime of CCPM.

Finally, we compare the runtime between CCSpan and CCPM on all datasets. CCPM always greatly outperforms CCSpan in execution time on all the datasets and no matter the support. The ratio is going from 15 on *BMS*1 *Gazelle* to 2,175 on *jobs*60000.

We can conlude that CCPM is only linearly impacted by the number of sequences and their average length and is always significantly faster than CCSPan.

Memory Usage. We now focus on the evaluation of the memory usage of CCSpan and CCPM (Fig. 1b). CCSpan is always more efficient than CCPM in terms of memory usage, except on the *Mushroom* dataset. We observe that the ratio between CCSpan and CCPM, ranges from 1.25 on *BMS*1 *Gazelle* to 2.1 on *jobs*60000. The difference both algorithms remains rather small, CCSpan and CBIDE are in the same order of magnitude, regarding the memory usage. CCSpan has an average increase of memory usage of 1.8 with the support variation through all datasets. CCPM has an average increase of memory usage of 1.2 with support variation on all datasets. We observe that both algorithms are really stable; we suppose that it is another added value of the contiguity constraint.

This experiment has shown that when mining CCSPs the support variation has almost no impact on both runtime and memory usage, probably due to the contiguity constraint. In addition, the memory usage is extremely stable. CCPM is always significantly faster than CCSpan, regardless the dataset and the support, while having memory usage in the same order of magnitude. Moreover, the runtime of CCPM is linearly impacted by the number of sequences and their

Table 3. Runtime, patterns of CCPM with constraints

Conf	Pattern	Runtime	len>=3	Activities
No con	14,574	49	2,947	882
1WC	26,655	142	7,603	2,019
2WC	29,508	164	10,473	2,489
3WC	31,817	176	12,782	2,711
SI	3,828	27	1,365	1,365
SI& 1WC	7,938	42	3,597	3,597
SI& 2WC	9,757	49	5,417	5,417
SI& 3WC	11,649	55	7,308	7,308

average length, contrary to CCSpan. We have thus answered the second scientific question: CCPM scales well on large databases having long sequences.

4.3   CCSSPW Evaluation

In the previous experiment, CCPM was implemented with no constraint for the sake of comparison with CCSpan. In this section, we show how the wildcards and the *starting item* patterns impact CCPM.

This experiment is conducted on the *jobs*60000 dataset with a relative support of 0.1% (in order to mine a significant amount of patterns and perform a relevant analysis). In the employment sector, activities are at the core of job offers. An activity is a coherent set of completed tasks organized toward a predefined objective with a result that can be measured. An activity is formalized by action verbs [1]. In this context, activities are patterns starting with a verb with a length greater or equal to 3. The *starting item* constraint is thus defined as the *verb* category.

"Inform clients by explaining procedures", "Manage business by performing related duties" or "Start operations by entering commands" are examples of activities.

We study runtime and activities of CCPM with *starting item* (SI) and/or wildcards (WC) and compare them to those of CCPM with no constraint. Table 3 shows the results of this experiment. The memory usage (not shown in Table 3) remains stable (about 1000 MBs), whatever is the configuration, the memory used is thus independent of the configuration.

CCPM with no constraint mines 14,574 patterns in 49 s. 2,947 of them have a length greater or equal to 3 and 882 of them start with a verb, thus CCPM with no constraint mines 882 activities. We observe that CCPM with the *starting item* (SI) constraint mines 3,828 patterns in 27 s and 1,365 of them have a length greater or equal to 3. Therefore, CCPM with *starting item* mines 3.8 less patterns but the execution time is only divided by 2. In addition, the SI constraint allows to find 65% more activities. Those two results are the consequence of the backward checking with *starting item* patterns. Indeed, instead of looking for any item, the function is looking for the first *starting item* on the left of the pattern, thus taking more time to execute the backward extension checking. In addition, a *starting item* pattern can have a non *starting item* closed super-pattern. In this case, the list of patterns mined by a closed contiguous sequential mining algorithm does not contain the *starting item* pattern. This is the reason why the set of activities mined with the *starting item* constraint is larger than the set of activities mined without any constraint.

Now we evaluate the impact of wildcards on CCPM with and without the *starting item* constraint. We observe that the use of one, two and three wildcards increases the number of activities mined by respectively 230%, 280% and 307% for CCPM with no *starting item* and of 260%, 400% and 535% with *starting item*. Those numbers show that the use of wildcards greatly impacts the number of activities found and confirm that *jobs*60000 is actually a noisy dataset. Logically, introducing the wildcard constraint also increases the runtime. Recall that the wildcard constraint does not impact the memory usage. Going from no wildcard to one, two and three wildcards, the median and the mode of the length of the activities mined are respectively: 4 & 3; 6 & 3; 11 & 3; 16 & 3. The median length increases with the number of wildcards, but the mode remains stable. An expert of the domain stated that an activity is averagely made up of 3 to 6 words [1]. With 2 wildcards, the median reaches 11, patterns are becoming too long. Therefore, the optimal number of wildcards to mine activities in this dataset seems to be 1.

To conclude, this experiment shows that the *starting item* constraint enables to mine a new set of CCSSPs not included in the set of CCSPs. Additionally, this constraint greatly decreases the execution time of CCPM. This experiment also shows that *jobs*60000 is a noisy dataset and that the optimal configuration of CCPM exploits the *starting items* constraint and 1 wildcard. So, we answered the first scientific question raised in the introduction: CCPM is a way to more accurately mine patterns in noisy data.

5   Conclusion

In this paper, we proposed a novel algorithm for mining frequent closed contiguous sequential patterns: CCPM. CCPM can rely on usable wildcards to fit noisy data while preserving the added value of contiguity. The experiments showed that CCPM greatly outperforms CCSpan in execution time, while being competitive in memory usage. In addition, to the best of our knowledge, CCPM is the first algorithm able to efficiently mine closed contiguous sequential patterns in large datasets with long sequences. Hence, CCPM is a solution to the two scientific questions raised in the introduction: (1) How to improve the adaptability of contiguous sequential pattern mining algorithms to more accurately extract information from noisy data? (2) How to increase the efficiency of CCSP mining algorithms to scale with large databases and long sequences? The *starting items* constraint allows to mine a new set of patterns not included in the set of closed contiguous sequential patterns.

In a future work, CCPM will be evaluated on several others datasets with various characteristics. We will also focus on the improvement of the memory usage of CCPM. We will also work on enriching the types of constraint (on multiple items or on a item not necessarily at the start of the pattern) without increasing the execution time.

References

1. Abboud, Y., Boyer, A., Brun, A.: Predict the emergence - application to competencies in job offers. In: ICTAI (2015)
2. Agrawal, R., Imieliskiand, T., Swami, A.: Mining association rules between sets of items in large databases. In: ACM SIGMOD, pp. 207–216 (1993)
3. Agrawal, R., Srikant, R.: Mining sequential patterns. In: Proceedings of the Eleventh International Conference on Data Engineering, pp. 3–14 (1995)
4. C. Aggarwal, C., Ta, N., Wang, J., Feng, J., J. Zaki, M.: XProj: a framework for projected structural clustering of xml documents. In: Proceedings of the 13th ACM SIGKDD International Conference on Knowledge Discovery and Data Mining, pp. 46–55 (2007)
5. Ayres, J., Flannick, J., Gehrke, J., Yiu, T.: Sequential pattern mining using a bitmap representation. In: Proceedings of the Eighth ACM SIGKDD International Conference, pp. 429–435 (2002)
6. Chen, J., Cook, T.: Mining contiguous sequential patterns from web logs. In: Proceedings of the 16th International Conference on WWW (2007)
7. Chen, J.: Contiguous item sequential pattern mining using UpDown Tree. Intell. Data Anal. 12(1), 25–49 (2008)
8. Li, C., Wang, J.: Efficiently mining closed subsequences with gap constraints. In: Proceedings of SIAM International Conference on Data Mining (2008)
9. Fischer, J., Heun, V., Kramer, S.: Optimal string mining under frequency constraints. In: European Conference on Principles of Data Mining and Knowledge Discovery, pp. 139–150 (2006)
10. Fürnkranz, J.: A study using n-gram features for text categorization. In: Austrian Research Institute for Artificial Intelligence (1998)
11. Garofalakis, M., Rastogi, R., Shim, K.: MSPIRIT: sequential pattern mining with regular expression constraints. In: Proceedings of the 25th International Conference on Very Large Data Bases (1999)
12. Han, J., Pei, J., Mortazavi-Asl, B., Chen, Q., Dayal, U., Hsu, M.: FreeSpan: frequent pattern-projected sequential pattern mining. In: Proceedings of the Sixth ACM SIGKDD (2000)
13. Kang, T.H., Yoo, J.S., Kim, H.Y.: Mining frequent contiguous sequence patterns in biological sequences. In: 2007 IEEE 7th International Symposium on BioInformatics and BioEngineering (2007)
14. Karim, M., Rashid, M., Jeong, B.S., Choi, H.J.: An efficient approach to mining maximal contiguous frequent patterns from large DNA sequence databases. Genomics Inform. 10(1), 51–57 (2012)
15. Liao, V.C.C., Chen, M.S.: DFSP: a Depth-First SPelling algorithm for sequential pattern mining of biological sequences. Knowl. Inf. Syst. 38, 623–639 (2014)
16. Matsui, T., Uno, T., Umemori, J., Koide, T.: A New Approach to String Pattern Mining with Approximate Match, Discovery Science, pp. 110–125 (2013)

17. Pei, J., Han, J., Mao, R., Chen, Q., Dayal, U., Hsu, M.: CLOSET: an efficient algorithm for mining frequent closed itemsets. In: DMKD 2001 workshop (2001)
18. Pei, J., Han, J., Mortazavi-Asl, B., Chen, Q., Dayal, U., Hsu, M.: PrefixSpan: mining sequential patterns efficiently by prefix-projected pattern growth. In: Proceedings of the 17th International Conference on Data Engineering (2001)
19. Pei, J., Han, J., Wang, W.: Constraint-based sequential pattern mining: the pattern-growth methods. J. Intell. Inf. Syst. **28**(2), 133–160 (2007)
20. Wang, J., Han, J.: BIDE: efficient mining of frequent closed sequences. In: Proceedings of the 20th International Conference on Data Engineering (2004)
21. Yan, X., Han, J., Afshar, R.: CloSpan: mining closed sequential patterns in large datasets. In: Proceedings of SIAM Conference on Data Mining (2003)
22. Zaki, M., Hsiao, C.: CHARM: An efficient algorithm for closed itemset mining. In: Proceedings of SIAM Conference on Data Mining, vol. 2 (2002)
23. Zhang, M., Kao, B., Cheung, D., Yip, K.: Mining periodic patterns with gap requirement from sequences. ACM Trans. Knowl. Discov. Data **1**(2), Article No. 7 (2007)
24. Zhang, J., Wang, Y., Yang, D.: CCSpan: mining closed contiguous sequential patterns. Knowl.-Based Syst. **89**, 1–13 (2015)

Sparse Dynamic Time Warping

Youngha Hwang$^{(\boxtimes)}$ and Saul B. Gelfand

School of Electrical and Computer Engineering, Purdue University,
465 Northwestern Avenue, West Lafayette, IN 47907-2035, USA
{hwangyo,gelfand}@purdue.edu

Abstract. Dynamic time warping (DTW) has been applied to a wide range of machine learning problems involving the comparison of time series. An important feature of such time series is that they can sometimes be sparse in the sense that the data takes zero value at many epochs. This corresponds for example to quiet periods in speech or to a lack of physical or dietary activity. However, employing conventional DTW for such sparse time series runs a full search ignoring the zero data.

In this paper we focus on the development and analysis of a fast dynamic time warping algorithm that is exactly equivalent to DTW for the unconstrained case where there are no global constraints on the permissible warping path. We call this sparse dynamic time warping (SDTW). A careful formulation and analysis are performed to determine exactly how SDTW should treat the zero data. It is shown that SDTW reduces the computational complexity relative to DTW by about twice the sparsity ratio, which is defined as the arithmetic mean of the fraction of non-zero's in the two time series. Numerical experiments confirm the speed advantage of SDTW relative to DTW for sparse time series with sparsity ratio up to 0.4. This study provides a benchmark and also background to potentially understand how to exploit such sparsity in the more complex case of a global constraint, or when the underlying time series are approximated to reduce complexity.

Keywords: Dynamic time warping · Sparse time series · Sparsity ratio

1   Introduction

Dynamic time warping (DTW) was first introduced in speech recognition [12], and has been applied to a wide range of machine learning problems such as handwriting recognition, classification of motor activities, and others [2]. DTW is often viewed as one of the best methods for time series comparison in data mining, as discussed in several review articles [3,4].

Computational complexity has been a barrier to the use of DTW. The complexity is in general $O\left(N^2\right)$, where N is the common length of the time series. One approach to reducing this complexity is to add a global constraint on slope [6] or width [12] along the diagonal of the permissible warping path domain. Keogh and Pazzani [7], Zhou and Wong [15], and Assent et al. [1] investigated

© Springer International Publishing AG 2017
P. Perner (Ed.): MLDM 2017, LNAI 10358, pp. 163–175, 2017.
DOI: 10.1007/978-3-319-62416-7_12

a lower bounding technique which demonstrated significant empirical improvement in indexing speed based on constrained DTW. Another approach involves approximation of the time series. Keogh and Pazzani [9] proposed a piecewise constant approximation of the time series before applying DTW. Salvado and Chan [13] improved its speed by a multi-resolution approximation of DTW. The general strategy is to recursively project an optimal warping path computed at a coarse resolution level to the next higher resolution level and then to refine the projected path [10]. This method improves the time complexity to $O(N)$. However, none of the above speed enhancements of DTW provably achieve the exact unconstrained solution. In fact, these methods can yield solutions which are quite far from optimal, as discussed in Sect. 5.

Sparse time series frequently occur where the data takes zero value at many epochs. These zero values are not missing data, but are actually highly informative and characteristic of their source. While missing data happens from operational failure of sensor networks, or improper or avoided recordings in medical testing [5], the zero data arise from quiet periods in speech, or the lack of physical activity or food intake [11]. Here we focus on this zero data, where there has in fact been relatively little attention to exploiting its structure in machine learning and none with DTW in particular.

DTW for sparse time series intuitively wastes time because it runs a full search ignoring the zero data. In this paper, we focus on the development and analysis of a fast dynamic time warping algorithm that is exactly equivalent to DTW for the unconstrained case where there are no global constraints on the permissible warping path. We call this sparse dynamic time warping (SDTW). A careful treatment is needed to understand exactly how SDTW should treat the zero data. It turns out that SDTW updates the data along the boundary of certain rectangles determined by the nonsparse time indices. The complexity of SDTW is upper bounded by about $2\,s$ times the complexity of DTW, where s is the sparsity ratio, defined as the arithmetic mean of the fraction of non-zero's in the two time series. When $s = o\,(1)$ (with increasing data length) the complexity becomes $o\,(N^2)$. There are no comparable approaches and rigorous analysis for the exact solution to the unconstrained problem for this kind of sparse data. Furthermore, this study provides a benchmark and also background to potentially understand how such sparsity might be exploited in the presence of a global constraint, or by approximation of the time series, for further complexity reduction.

The next section reviews DTW from the dynamic programming perspective. Section 3 provides the principles underlying SDTW and its complexity reduction. Section 4 explicitly describes the SDTW algorithm. In addition, a simple example is given to illustrate how SDTW works. Finally, the last section presents some numerical analysis to demonstrate the efficiency of SDTW.

2   Review of DTW

DTW originated as a dynamic programming algorithm to compare time series in the presence of timing differences for spoken word recognition [12]. In order to describe the timing differences, consider two time series $X = (x_1, x_2, \cdots, x_M)$ and $Y = (y_1, y_2, \cdots, y_N)$. The timing differences between them can be associated with a sequence of points $F = (c(1), c(2), \cdots, c(k), \cdots, c(K))$, where $c(k) = (m(k), n(k))$. This sequence realizes a mapping from the time indices of X onto those of Y. It is called a warping function. The measure of difference between x_m and y_n is usually given as $d(c) = d(m, n) = |x_m - y_n|^p$, where $0 < p < \infty$ (Euclidean p-norm). For $p = 0$, $d(m, n) = 1$ if $x_m \neq y_n$ and zero otherwise. $d(m, n)$ is called the *local distance*.

The minimum of the weighted summation of local distances over all warping functions F is

$$D(X, Y) = \frac{1}{\sum_{k=1}^{K} w(k)} \min_F \left[\sum_{k=1}^{K} d(c(k)) w(k) \right], \tag{1}$$

where $w(k)$ is a nonnegative weighting coefficient. $D(X, Y)$ is the DTW distance and provides a measure of the difference between X and Y after alignment.

Restrictions on the warping function F (or points $c(k) = (m(k), n(k))$) are usually given as follows. (1) Monotonic conditions: $m(k-1) \leq m(k)$ and $n(k-1) \leq n(k)$. (2) Continuity conditions: $m(k) - m(k-1) \leq 1$ and $n(k) - n(k-1) \leq 1$. (3) Boundary conditions: $m(1) = 1, n(1) = 1$ and $m(K) = M, n(K) = N$.

Determining the distance $D(X, Y)$ given by Eq. 1 is a problem to which the well-known dynamic programming principle can be applied. The basic algorithm is as follows.

Initial condition:

$$g_1(c(1)) = d(c(1)) w(1). \tag{2}$$

Dynamic Programming equation:

$$g_k(c(k)) = \min_{c(k-1)} [g_{k-1}(c(k-1)) + d(c(k)) w(k)]. \tag{3}$$

Time-normalized distance:

$$D(X, Y) = \frac{1}{\sum_{k=1}^{K} w(k)} g_K(c(K)) \tag{4}$$

$g_k(c)$ is called the *accumulated distance*.

The algorithm Eqs. 2-4 can be simplified under the above restrictions on the warping function. Let w_d, w_v, and w_h represent the weights for each diagonal, vertical, and horizontal step. Typically, the weight of the diagonal steps is larger than those of the other steps, because the diagonal step can be regarded as the combination of the vertical step and the horizontal step. Then, the algorithm is given as follows.

Initial Condition:

$$g(1, 1) = d(1, 1) w_d \tag{5}$$

DP-equation:

$$g(m,n) = \min \begin{bmatrix} g(m,n-1) + d(m,n)w_v \\ g(m-1,n-1) + d(m,n)w_d \\ g(m-1,n) + d(m,n)w_h \end{bmatrix} \tag{6}$$

Time-normalized distance:

$$D(X,Y) = \frac{1}{\sum w(k)} g(M,N) \tag{7}$$

$\{g(m,n) : 1 \leq m \leq M,\ 1 \leq n \leq N\}$ is called the *accumulated distance matrix*.

3   Principles Underlying SDTW

This section starts with some notation and then derives properties of accumulated distances for sparse sequences. Once these properties are established, the SDTW algorithm is evident and its complexity can be derived. It is, however, somewhat involved to explicitly write down the algorithm, and this is done in Sect. 4.

As above let $X = (x_1, x_2, \cdots, x_M)$ and $Y = (y_1, y_2, \cdots, y_N)$ be the time series, and let $X_s = (x_{m_1}, x_{m_2}, \cdots, x_{m_{M_s}})$ and $Y_s = (y_{n_1}, y_{n_2}, \cdots, y_{n_{N_s}})$ be the subsequences made up only of non-zero values. We want to develop a fast DTW (i.e., SDTW) algorithm using X_s and Y_s that is equivalent to DTW using X and Y in terms of the distance and resulting alignment.

First we define notations to describe the geometric objects such as rectangles, vertices and edges in the $m - n$ plane. A rectangle is written as $R_{i,j} := \{(m,n) | m_{i-1} \leq m \leq m_i, n_{j-1} \leq n \leq n_j\}$, as shown in Fig. 1. It includes the indices between the two successive time indices of non-zero data (m_{i-1}, m_i) and (n_{j-1}, n_j). It has as its vertices $(m_{i-1}, n_{j-1}), (m_{i-1}, n_j), (m_i, n_{j-1})$ and (m_i, n_j). The vertices are labeled as $V_{l,b}, V_{l,t}, V_{r,b}$, and $V_{r,t}$, respectively. Top, bottom, left, and right edges of $R_{i,j}$ are represented as sequences E_t, E_b, E_l, and E_r, respectively, i.e.,

$$E_t = ((m_{i-1}+1, n_j), (m_{i-1}+2, n_j), \cdots, (m_i-1, n_j))$$
$$E_r = ((m_i, n_{j-1}+1), (m_i, n_{j-1}+2), \cdots, (m_i, n_j-1))$$
$$E_b = ((m_{i-1}+1, n_{j-1}), (m_{i-1}+2, n_{j-1}), \cdots, (m_i-1, n_{j-1}))$$
$$E_l = ((m_{i-1}, n_{j-1}+1), (m_{i-1}, n_{j-1}+2), \cdots, (m_{i-1}, n_j-1))$$

The interior of $R_{i,j}$ is defined as $R_{i,j}^\circ := \{(m,n) | m_{i-1}+1 \leq m \leq m_i-1, n_{j-1}+1 \leq n \leq n_j-1\}$, i.e., the dotted rectangle as shown in Fig. 1.

We will need the following lemmas.

Lemma 1. $R_{i,j}^\circ$ *has equal accumulated distances if they are nondecreasing along* E_l *and* E_b.

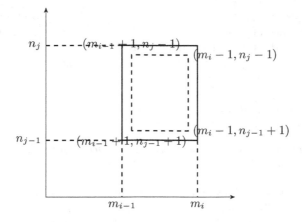

Fig. 1. $R_{i,j}$, and its vertices and edges. Dotted rectangle represents $R_{i,j}^\circ$.

Proof. It is enough to show that $g(m_{i-1} + 1, n_j - 1) = g(m_{i-1} + 1, n_j - 2)$ (the rest follows similarly). Now $g(m_{i-1} + 1, n_j - 1) \leq g(m_{i-1} + 1, n_j - 2)$ from the definition of $g(m_{i-1} + 1, n_j - 1)$ and the zero local distance. Suppose that $g(m_{i-1} + 1, n_j - 1) < g(m_{i-1} + 1, n_j - 2)$. Then $g(m_{i-1} + 1, n_j - 1) = \min(g(m_{i-1}, n_j - 2), g(m_{i-1}, n_j - 1)) = g(m_{i-1}, n_j - 2)$, because of the zero local distance and the nondecreasing assumption along E_l. After substituting $g(m_{i-1}, n_j - 2)$ for $g(m_{i-1} + 1, n_j - 1)$, $g(m_{i-1}, n_j - 2) < g(m_{i-1} + 1, n_j - 2)$ leads to a contradiction since $g(m_{i-1} + 1, n_j - 2) \leq g(m_{i-1}, n_j - 2)$ from the definition of $g(m_{i-1} + 1, n_j - 2)$.   □

Lemma 2. *The accumulated distances are nondecreasing along E_t and E_r if $R_{i,j}^\circ$ has equal accumulated distances.*

Proof. It is enough to show that $g(m_{i-1} + 1, n_j) \leq g(m_{i-1} + 2, n_j)$ (the rest follows similarly). Now $g(m_{i-1} + 2, n_j) = \min(g(m_{i-1} + 1, n_j) + w_h d(0, y_{n_j}), g(m_{i-1} + 1, n_j - 1) + w_d d(0, y_{n_j}), g(m_{i-1} + 2, n_j - 1) + w_v d(0, y_{n_j}))$. If $g(m_{i-1} + 2, n_j) = g(m_{i-1} + 1, n_j) + w_h d(0, y_{n_j})$ then $g(m_{i-1} + 2, n_j) \geq g(m_{i-1} + 1, n_j)$. Otherwise, $g(m_{i-1} + 2, n_j) = \min(g(m_{i-1} + 1, n_j - 1) + w_d d(0, y_{n_j}), g(m_{i-1} + 2, n_j - 1) + w_v d(0, y_{n_j})) = \min(w_d, w_v) d(0, y_{n_j}) + g(m_{i-1} + 1, n_j - 1) \geq g(m_{i-1} + 1, n_j)$ since $g(m_{i-1} + 1, n_j - 1) = g(m_{i-1} + 2, n_j - 1)$ from Lemma 1. The last inequality is from the definition of $g(m_{i-1} + 1, n_j)$.   □

The above two lemmas show that nondecreasing accumulated distances along the left and bottom edges of $R_{i,j}$ imply equal accumulated distances in the interior of $R_{i,j}$, which in turn implies nondecreasing accumulated distances along the right and top edges of $R_{i,j}$. This is the basis for an induction proof of Theorem 1 below. First, we need one additional lemma.

Lemma 3. *Suppose $m_{i-1} = m_i - 1$ or $n_{j-1} = n_j - 1$. Then the accumulated distances are nondecreasing along E_r and E_t of $R_{i,j}$ if the same holds for $R_{i-1,j}$ and $R_{i,j-1}$, respectively.*

Proof. Since either $m_{i-1} = m_i - 1$ or $n_{j-1} = n_j - 1$, $R_{i,j}^\circ$ is empty. If $n_{j-1} = n_j - 1$, there is no need to compute E_r. The distances along E_t are nondecreasing because they depend only on the nondecreasing distances along E_b of $R_{i,j}$, which is identical to E_t of $R_{i,j-1}$, and the local distances are constant as $d(0, n_j)$. Similarly for $m_{i-1} = m_i - 1$. □

The above interdependent lemmas actually hold for all $1 \le i \le M_s$ and $1 \le j \le N_s$ which can be stated as follow.

Theorem 1. *The accumulated distances are nondecreasing along the edges connecting the vertices in $R_{i,j}$, for all $1 \le i \le M_s$ and $1 \le j \le N_s$.*

Proof. The accumulated distances are called distances here. For proof by induction, we show that distances are nondecreasing along E_t and E_r of $R_{i,j}$, if the distances are nondecreasing along E_l and E_b, which are the same as E_r of $R_{i-1,j}$ iteration and E_t of $R_{i,j-1}$ iteration, respectively.

For $i = 1$ and $j = 1$, if $m_1 > 1$ and $n_1 > 1$, then all the distances are zeros in $R_{i,j}^\circ$ because the local distances are zeros there. Thus, the distances along edges E_r and E_t are nondecreasing from Lemma 2. If $m_1 = 1$ and $n_1 > 1$, then the distances are nondecreasing along the edge from $(1,1)$ to $(1, n_1 - 1)$ because of positive local distances $d(x_{m_1}, 0)$. Similarly for $m_1 > 1$ and $n_1 = 1$.

For $i = 1$, $j > 1$ and $m_1 > 1$, first suppose that $n_{j-1} < n_j - 1$. Since $m_1 > 1$, all distances are the same as $g(1, n_{j-1})$ in $R_{i,j}^\circ$ by Lemma 1 under the assumption the distances are nondecreasing along E_t of $R_{1,j-1}$ or E_b of $R_{i,j}$. Then by Lemma 2 the distances are nondecreasing along E_r and E_t for $R_{i,j}$. Next suppose that $n_{j-1} = n_j - 1$. Then the distances are still nondecreasing along E_t by Lemma 3. For $i = 1$, $j > 1$ and $m_1 = 1$, the distances are nondecreasing along the edge from $(1, n_{j-1} + 1)$ to $(1, n_j - 1)$ because of positive local distances $d(x_{m_1}, 0)$. We previously showed that the distances are nondecreasing along the edges connecting the vertices of $R_{1,1}$. By induction, all the distances are nondecreasing along the edges connecting the vertices for all $R_{i,j}$, where $i = 1$ and $j > 1$. Similarly for $i > 1$ and $j = 1$.

For $i > 1$ and $j > 1$, the distances are nondecreasing along E_t and E_r by combining Lemmas 1 and 2, if we assume the distances are nondecreasing along E_b (E_t of $R_{i,j-1}$) and E_l (E_r of $R_{i-1,j}$). By induction, the distances are nondecreasing along the edges connecting the vertices for all $R_{i,j}$, where $i > 1$ and $j > 1$. □

From the above theorem, the equal distance property in $R_{i,j}^\circ$ follows directly.

Corollary 1. *The accumulated distances are equal in $R_{i,j}^\circ$, for all $1 \le i \le M_s + 1$ and $1 \le j \le N_s + 1$.*

Hence, we do not have to compute the accumulated distances in $R_{i,j}^\circ$ except for the left lower point $(m_{i-1} + 1, n_{j-1} + 1)$, for all $1 \le i \le M_s + 1$ and $1 \le j \le N_s + 1$.

The following theorem considers the complexity of SDTW. For simplicity it is assumed that the time series have the same length, i.e., $N = M$, and it shows that the complexity of SDTW is proportional to $(N_s + M_s)N$ instead of N^2 for DTW. The time complexity of SDTW is measured by the number of computations of the accumulated distances.

Theorem 2. *The time complexity of the SDTW algorithm, C_{SDTW}, is bounded as follows: $(N_s + M_s)N - N_sM_s \leq C_{SDTW} \leq (N_s + M_s)(N + 1) + 1$, when the time series both have length N.*

Remark 1. The bound can be expressed as $(2s - s_1s_2)N^2 \leq C_{SDTW} \leq 2sN(N + 1) + 1$ where $s_1 = \frac{N_s}{N}$, $s_2 = \frac{M_s}{N}$ and $s = \frac{1}{2}(s_1 + s_2)$. s is called the *sparsity ratio*.

Proof. No accumulated distances need to be computed except $(m_{i-1}+1, n_{j-1}+1)$. Thus, there is max$((m_i - m_{i-1} - 1)(n_j - n_{j-1} - 1) - 1, 0)$ reduction of computations. Let $\delta m_i = m_i - m_{i-1} - 1$ and $\delta n_j = n_j - n_{j-1} - 1$ represent the number of zeros between (m_{i-1}, m_i) and (n_{j-1}, n_j), respectively. Thus, $\sum_{i=1}^{M_s+1} \delta m_i = N - M_s$ equals the number of zeros in the sequence x_{m_i} under the assumption that $m_0 = 0$ and $m_{M_s+1} = N + 1$. Likewise, $\sum_{j=1}^{N_s+1} \delta n_j = N - N_s$.

Since reduction in each (i, j) update is bounded below by $\delta m_i \delta n_j - 1$, summation over all data gives $r_l := (N - N_s)(N - M_s) - (N_s + 1)(M_s + 1)$. Hence $C_{SDTW} \leq N^2 - r_l = (N_s + M_s)(N + 1) + 1$.

Also, since reduction in each (i, j) update is bounded above by $\delta m_i \delta n_j$, summation over all data gives $r_u := \sum_{i=1}^{M_s+1} \sum_{j=1}^{N_s+1} \delta m_i \delta n_j = (N - N_s)(N - M_s)$. Hence $C_{SDTW} \geq N^2 - r_u = (N_s + M_s)N - N_sM_s$. □

4   SDTW Algorithm

This section explicitly describes the SDTW algorithm. A simple example is then given to demonstrate how SDTW works.

DTW computes the accumulated distances for all points (m, n), where $1 \leq m \leq M$, and $1 \leq n \leq N$ as we have seen in the previous section. On the other hand, SDTW computes accumulated distances only on edges connecting vertices made up of (m_i, n_j) where $1 \leq i \leq M_s$ and $1 \leq j \leq N_s$, because the accumulated distances are equal elsewhere due to zero local distances.

In our notation, SDTW computes g only on E_t, E_r, $V_{r,t}$, and $(m_{i-1}+1, n_{j-1}+1)$ of each $R_{i,j}$, where $1 \leq i \leq M_s + 1$ and $1 \leq j \leq N_s + 1$, by assuming that $g(\cdot)$ is zero on E_b and E_l for $R_{1,1}$. Set dummy non-zero data at $m = M + 1$ and $n = N + 1$, where $M_s + 1 = M + 1$ and $N_s + 1 = N + 1$. Then, the interior $R^\circ_{M_s+1, N_s+1}$ includes (m, n) satisfying $m_{M_s} + 1 \leq m \leq M$ and $n_{N_s} + 1 \leq n \leq N$. Formally, the SDTW algorithm can be written as follows.

1. Build X_s and Y_s of lengths M_s and N_s from non-zeros of the time series X and Y of lengths M and N

2. Add dummy non-zeros at $M + 1$ and $N + 1$, respectively.

3. Compute the local distances $d(m, n)$, $d(m, 0)$, and $d(0, n)$ for every $m \in \{m_1, m_2, \cdots, m_{M_s}\}$ and $n \in \{n_1, n_2, \cdots, n_{N_s}\}$

4. Compute the accumulated distances g on E_t, E_r, and $V_{r,t}$ for all $R_{i,j}$, starting with $R_{1,1}$, where i and j increase up to N_s and M_s, respectively.

5. Find the optimal path from $R_{i,j}$, where i and j decrease to one. That is, starting from E_t, E_r, or $V_{r,t}$ of $R_{i,j}$, travel towards E_l, E_b, $V_{l,t}$, $V_{l,b}$, or $V_{r,b}$ that also belong to $R_{i-1,j}$, $R_{i,j-1}$, or $R_{i-1,j-1}$.

The starting index of $R_{i,j}$ at the above step 5 requires some further discussion. When $m_{M_s} = M$, the i index starts from M_s. Otherwise, the i index starts from $M_s + 1$ to cover $R_{i,j}^\circ$. Recall $M_s + 1$ is set as $M + 1$ using the dummy non-zero. Similarly, when $n_{N_s} = N$, the j index starts from N_s. Otherwise, the j index starts from $N_s + 1$.

The $R_{i,j}$ case where $i > 1$ and $j > 1$ is no different than DTW. g is set as $g(m_{i-1} + 1, n_{j-1} + 1)$ in $R_{i,j}^\circ$. The accumulated distance at $(m_{i-1} + 1, n_{j-1} + 1)$ is computed as $g(m_{i-1} + 1, n_{j-1} + 1) = \min(g(m_{i-1} + 1, n_{j-1}), g(m_{i-1}, n_{j-1} + 1), g(m_{i-1}, n_{j-1}))$. g is computed on $V_{r,t}$ along with either E_t or E_r, depending on whether $m_{i-1} = m_i - 1$ or $n_{j-1} = n_j - 1$.

The $R_{i,j}$ cases where $i = 1$ or $j = 1$ are presented below.

$R_{1,1}$ case:

if $m_1 = 1$ and $n_1 > 1$
    E_r assignment: $g(m_1, k) \leftarrow k \times d(m_1, 0)w_v$, $1 \leq k \leq n_1 - 1$
    $V_{r,t} \leftarrow d(m_1, n_1) + g(m_1, n_1 - 1)$

elseif $m_1 > 1$ and $n_1 = 1$
    E_t assignment: $g(k, n_1) \leftarrow k \times d(0, n_1)w_h$, $1 \leq k \leq m_1 - 1$
    $V_{r,t} \leftarrow d(m_1, n_1) + g(m_1 - 1, n_1)$

elseif $m_1 = 1$ and $n_1 = 1$
    $g(m_1, n_1) \leftarrow d(m_1, n_1)w_d$

else
    $g()$ is set as zero in $R_{1,1}^\circ$
    $E_t \leftarrow d(0, n_1)w_v$
    $E_r \leftarrow d(m_1, 0))w_h$
    $V_{r,t} \leftarrow \min(d(m_1, n_1)w_d, d(m_1, n_1)w_v + d(m_1, 0)w_h, d(m_1, n_1)w_h + d(0, n_1)w_v)$
end

$R_{1,j}$ case, where $j > 1$:

---

if $m_1 > 1$

$\quad E_t \leftarrow d(0, n_j)w_v + g(1, n_{j-1})$, because $g()$ is set as $g(1, n_{j-1})$ in $R_{1,j}^\circ$

$\quad \underline{E_r \text{ computation:}}$

$\qquad g(m_1, n_{j-1}+1) \leftarrow \min(d(m_1, 0)w_v + g(m_1, n_{j-1}), d(m_1, 0)w_d + g(m_1 - 1, n_{j-1}), d(m_1, 0)w_h + g(1, n_{j-1}))$

$\qquad$ **for** $k = n_{j-1}+2 : n_j - 1$

$\qquad\qquad g(m_1, k) \leftarrow \min(d(m_1, 0)w_v + g(m_1, k-1), d(m_1, 0)w_h + g(1, n_{j-1}), d(m_1, 0)w_d + g(1, n_{j-1}))$

$\qquad$ **end**

$\qquad V_{r,t} \leftarrow \min(d(m_1, n_j)w_v + g(m_1, n_j - 1), d(m_1, n_j)w_h + g(m_1 - 1, n_j), d(m_1, n_j)w_d + g(1, n_{j-1}))$

else

$\quad \underline{E_r \text{ computation:}}$

$\qquad$ **for** $k = n_{j-1}+1 : n_j - 1$

$\qquad\qquad g(m_1, k) \leftarrow g(m_1, k-1) + d(m_1, 0)w_v$

$\qquad$ **end**

$\quad V_{r,t} \leftarrow d(m_1, n_j)w_v + g(m_1, n_j - 1)$

end

Finding the optimal path employs the same approach as DTW. However, there is no computation of the accumulated distances in the interior of $R_{i,j}$ due to their equality except for the bottom left vertex, $(m_{i-1} + 1, n_{j-1} + 1)$. Hence, the approach depends on DTW's policy on tie break, and we employ the diagonal path recovery here. So a walk toward $(1,1)$ is executed until meeting the bottom or left edges of $R_{i,j}^\circ$.

Table 1. Accumulated distances of the simple example

12	0.0317	0.0149	0.0149	0.0149	0.0149	0.0149	0.0949
11	0.0637	0.0049	0.0049	0.0049	0.0049	0.0049	**0.1649**
10	0.0637	0.0049	0.0049	0.0049	0.0049	0.0049	**0.1649**
9	0.0637	0.0049	0.0049	0.0049	0.0049	0.0049	**0.1649**
8	0.0490	0.0049	0.0049	0.0049	0.0049	0.0049	**0.1649**
7	0.0049	0.0245	0.0441	0.0637	0.0637	0.0637	0.1117
m/n	7	8	9	10	11	12	13

A simple example is shown in Table 1. The bold-faced accumulated distances on E_r, E_t, and $V_{r,t}$ are computed in $R_{i,j}$ for some i and j, given the underlined accumulated distances, $E_l, E_b, V_{l,b}, V_{l,t}$, and $V_{r,b}$, that are known from $R_{i-1,j}$ and $R_{i,j-1}$ (with l_2 norm assumption and all weights set to one). Here $m_{i-1} = 7$, $m_i = 13$, $n_{j-1} = 7$, and $n_j = 12$, along with $x_{m_{i-1}} = 0.21$, $x_{m_i} = 0.40$ and $y_{n_{j-1}} = 0.14$, $y_{n_j} = 0.10$. Note that the accumulated distances are equal in $R_{i,j}^\circ$, corresponding to the nondecreasing accumulated distances along E_b and E_l.

5   Numerical Analysis

This section presents some numerical results to compare the efficiency of SDTW versus DTW, and also to make some comparisons of SDTW with fast approximation versions of DTW. These experiments involve both synthetic and experimental data.

Publicly available sparse time series data are rare. In addition, it is impossible to control the sparsity of such experimental data for evaluation of the speed of SDTW. In such cases, sparsity can be controlled by removing random segments of contiguous time points [5]. We consider time series data sets from the UCR collection [8]. Similar to [5], our sparse time series are generated by selecting time indices whose intervals follow a geometric distribution and masking out the remainder as zeros. Hence, each time series independently has a probability of non-zero's that is the same as the mean sparsity ratio $\bar{s}$, where the bar denotes expectation over the geometric distributions. Note that in the Remark following Theorem 2 the bounds continue to hold with the mean quantities $\bar{s} = \bar{s}_1 = \bar{s}_2$ replacing the pointwise values (we omit the mean qualifier in the sequel, and simply refer to the sparsity ratio).

The solid line in Fig. 2 shows the time complexity of SDTW relative to DTW versus different sparsity ratios for the time series dataset of "50 words" using a computer equipped with Intel core i-5 1.9 GHz, 4 GB memory, MATLAB 2014Ra, and Ubuntu 14.04. With sparsity ratio around 0.45, SDTW finally loses its speed advantage. The dotted line represents a linear regression of this relative time complexity on the sparsity ratio $b_1 s + b_0$. Values of b_1 and b_0 are shown in Table 2 for several time series data sets. Values of b_1 are slightly higher than the corresponding upper bound $2 + \frac{2}{N}$ from Theorem 2, due perhaps to conditional branching. Values of b_0 decrease with increasing length of the time series, because the overhead of SDTW such as initialization decreases.

Table 2. Regression coefficients for relative time complexity of SDTW versus sparsity ratio

Data set	Length	b_1	b_0
Synthetic control	61	2.0874	0.1668
CBF	129	2.1130	0.0864
50 words	271	2.1121	0.0477
Beef	471	2.1214	0.0344
MALLAT	1025	2.1148	0.0220

We also considered some experimental data, which generally gave similar results to the synthetic data. NHANES (National Health and Nutrition Examination Survey) [11] is a program of studies designed to assess the health and nutritional status of adults and children in the United States. The survey is

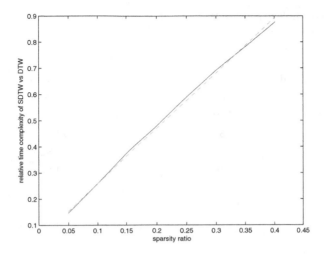

Fig. 2. Regression plot of relative time complexity of SDTW versus sparsity ratio for "50 words" dataset

unique in that it combines interviews and physical examinations [14]. The 24-hour dietary recall was used to assess dietary intake. Participants took approximately 30 to 45 min to complete a series of probes to recall as many foods and beverages as possible that were consumed in the past 24 h along with their respective times of consumption. Such 24-hour energy intakes (kcal) that are measured hourly can be represented as sparse time series with length 24. Their sparsity ratios range from 0.0833 to 0.333. Table 3 shows that SDTW performs better than DTW for sparsity ratios less than 0.3. The regression of relative time complexity has coefficients $b_1 = 2.0421$ and $b_0 = 0.3787$.

Table 3. Time complexity comparison between DTW, SDTW for NHANES data

Sparsity ratio	0.08333	0.125	0.16667	0.20833	0.250	0.29167	0.33333
Relative time complexity	0.5302	0.6386	0.7279	0.8198	0.8971	0.9644	1.0508

FastDTW [13] claims a speed of $O(N)$ at the expense of the accuracy of recovered warping paths. Table 4 shows that for FastDTW the relative error of DTW cost is significant for the CBF dataset under the similar speed assumption, which is controlled by the "radius". This radius sets the expansion of the search window after projection onto the higher resolution level of the coarse resolution warping path. Note that for SDTW the relative error over DTW cost is zero, as it finds the exact DTW solution. When the sparsity ratio is low, even the speed advantage of FastDTW disappears versus SDTW.

Table 4. Time complexity and relative error of Fast DTW over SDTW for CBF time series

Sparsity ratio	Radius	Fast DTW (sec)	SDTW (sec)	Relative error (%)
0.05	0	6.2625	2.9775	71.48
0.1	0	5.8803	4.7113	82.20
0.15	1	6.8004	6.7347	81.15
0.20	2	7.3689	7.7347	42.52
	3	8.0728		29.82
0.25	6	9.4680	9.5988	10.46
0.30	10	11.3151	11.3749	4.50
0.35	13	12.5310	12.3979	2.18
0.40	18	14.1626	14.2351	0.78

6   Conclusion

SDTW has been developed to exploit sparse time series which have repeated zeros, a situation which arises in some important applications including speech, physical activity, and dietary consumption. SDTW is provably equivalent to DTW and can find the same solution with less computation for these sparse time series. Its computational complexity is proportional to the sparsity ratio and can yield significant decreases for small enough sparsity ratio and large enough time series lengths. SDTW is different than other approaches which add global constraints to DTW or otherwise modify the time series and/or the DTW algorithm in a way which does not focus on the sparse structure. These methods can yield large complexity reduction but at the expense of increased error. Numerical experiments from synthetic data and real data support the analysis for the sparse time series. This work provides a benchmark and also the background to combine SDTW with these other complexity reduction approaches, in particular adding global constraints to SDTW seems promising. The geometry of the sparse constrained region appears to be complicated, but the approach seems feasible.

References

1. Assent, I., Wichterich, M., Krieger, R., Kremer, H., Seidl, T.: Anticipatory DTW for efficient similarity search in time series databases. Proc. VLDB Endowment **2**(1), 826–837 (2009)
2. Berndt, D.J., Clifford, J.: Using dynamic time warping to find patterns in time series. In: Knowledge Discovery in Databases Workshop, Seattle, WA, vol. 10, pp. 359–370 (1994)
3. Esling, P., Agon, C.: Time-series data mining. ACM Comput. Surv. (CSUR) **45**(1), 12 (2012)

4. Tak-chung, F.: A review on time series data mining. Eng. Appl. Artif. Intell. **24**(1), 164–181 (2011)
5. Grabocka, J., Nanopoulos, A., Schmidt-Thieme, L.: Classification of sparse time series via supervised matrix factorization. In: AAAI (2012)
6. Itakura, F.: Minimum prediction residual principle applied to speech recognition. IEEE Trans. Acoust. Speech Signal Process. **23**(1), 67–72 (1975)
7. Eamonn Keogh and Chotirat Ann Ratanamahatana: Exact indexing of dynamic time warping. Knowl. Inf. Syst. **7**(3), 358–386 (2005)
8. Keogh, E., Xi, X., Wei, L., Ratanamahatana, C.A.: The UCR time series classification/clustering homepage (2006). http://www.cs.ucr.edu/~eamonn/time_series_data
9. Keogh, E.J., Pazzani, M.J. Scaling up dynamic time warping for datamining applications. In: Proceedings of the Sixth ACM SIGKDD International Conference on Knowledge Discovery and Data Mining, pp. 285–289. ACM (2000)
10. Müller, M.: Information Retrieval for Music and Motion, vol. 2. Springer, Heidelberg (2007)
11. National Centers for Health Statistics, Centers for Disease Control, Prevention. National health, nutrition examination survey data 2011–2012. Centers for Disease Control and Prevention, Hyattsville, MD, US Department of Health and Human Services (2014)
12. Sakoe, H., Chiba, S.: Dynamic programming algorithm optimization for spoken word recognition. IEEE Trans. Acoust. Speech Signal Process. **26**(1), 43–49 (1978)
13. Salvador, S., Chan, P.: Toward accurate dynamic time warping in linear time and space. Intell. Data Anal. **11**(5), 561–580 (2007)
14. Zhao, Y., Hwang, Y., Gelfand, S., Delp, E., Eicher-Miller, H.: Temporal dietary patterns derived from spectral clustering and factor analysis are associated with diet quality using nhanes 1999–2004. FASEB J. **29**(1 Supplement), 587–19 (2015)
15. Zhou, M., Wong, M.H.: Boundary-based lower-bound functions for dynamic time warping and their indexing. Inf. Sci. **181**(19), 4175–4196 (2011)

Improving a Bayesian Decision Model for Supporting Diagnosis of Alzheimer's Disease and Related Disorders

Carolina Medeiros Carvalho[1]([✉]), Flávio Luiz Seixas[1], Aura Conci[1],
Débora Christina Muchaluat-Saade[1], and Jerson Laks[2]

[1] Institute of Computing, Fluminense Federal University,
Rua Passo da Pátria, 156, Niterói, RJ 24210-240, Brazil
{carolmc,fseixas,debora}@midiacom.uff.br,
aconci@ic.uff.br
[2] Center for Alzheimer's Disease and Related Disorders,
Institute of Psychiatry of the Federal University of Rio de Janeiro,
Av. Venceslau Brás, 71, Rio de Janeiro, RJ 22290-140, Brazil
jersonlaks@gmail.com

Abstract. Alzheimer's Disease (AD) is a degenerative disease with high prevalence in the elderly population. Its symptoms are often related to difficulty in remembering new information and include impaired judgment, disorientation, confusion, behavioral changes and difficulty in speaking and walking. Clinical Decision Support Systems can be designed to improve clinical decision-making by making the physician aware of the most probable diagnosis given the patient health records, and then reducing AD diagnostic error rates. This work extends a previous discrete Bayesian decision model for supporting diagnosis of AD and related disorders and proposes improvements in this model following two approaches: mixing continuous and discrete nodes by implementing a Hybrid Logistic Regression-Naïve Bayes model and relaxing independence assumptions by adopting the AnDE (Averaged n-Dependence Estimators) model. Our proposal presents better performance results. The 4-fold cross-validation results on CAD (Center for Alzheimer's Disease and Related Disorders) patient dataset showed that the A2DE classifier (AnDE with n = 2) outperforms the previous discrete Bayesian network for AD considering all proposed measures: Area Under Receiver Operating Curve (AUC), F1-score, Mean Square Error (MSE) and Mean Cross-Entropy (MXE). Also, the Hybrid Logistic Regression-Naïve Bayes model outperforms the previous discrete Bayesian network for dementia considering MSE and, for AD, considering AUC and MSE.

Keywords: Clinical decision support systems · Bayesian decision model · Dementia · Alzheimer's Disease · Mild cognitive impairment

© Springer International Publishing AG 2017
P. Perner (Ed.): MLDM 2017, LNAI 10358, pp. 176–191, 2017.
DOI: 10.1007/978-3-319-62416-7_13

1 Introduction

Population aging has been occurring as a global phenomenon in both developed and developing countries. Neurodegenerative diseases have high prevalence in the elderly population. Early diagnosis of this type of disease allows early treatment and improves patient quality of life.

Dementia (D) is a clinical state characterized by loss of function in multiple cognitive domains. There are various specific types of dementia and the most common is Alzheimer's Disease (AD), accounting for between 60% and 80% of dementia cases [1]. AD is a degenerative disease causing lesions in the brain. Early clinical symptoms of AD are often related to difficulty in remembering new information, and later symptoms include impaired judgment, disorientation, confusion, behavioral changes and difficulty in speaking and walking. Another disease related to AD is Mild Cognitive Impairment (MCI), which is usually associated to a preclinical stage of AD.

Regarding healthcare services, there is a growing global concern about patient safety and healthcare system effectiveness [2]. Clinical Decision Support Systems (CDSS) [3] are considered an important category of health information systems designed to improve clinical decision-making by reducing diagnostic error rates [4].

CDSSs can be grouped by their prevailing inference engine, including rule-based systems, clinical guideline-based systems and semantic network-based systems [5]. Rule-based systems use simple conditional expressions for making deductions and aiding clinical decisions. Guideline-based systems indicate the most likely clinical decision or pathway from a set of predetermined options, guided by a workflow that describes, for example, diagnosis rules or a treatment process. Semantic network-based systems use semantic relations between concepts to perform an inference algorithm. CDSSs based on Bayesian Networks (BNs) fall into the semantic network category, since BN nodes represent clinical concepts, and arcs represent dependence relations.

BNs represent a domain in terms of random variables and explicitly model the interdependence between them [6]. BNs are graphically represented by Directed Acyclic Graphs (DAGs), whose nodes represent random variables and arcs express direct dependencies. In a discrete BN model, only discrete random variables are considered and, if there are continuous variables in a domain, they are commonly discretized. In a discrete BN model, suppose X_i is the ith random variable containing r_i discrete states. Also suppose that x_i^k is the kth state of the random variable X_i. Considering a BN containing a set of nodes or random variables $X^G = \langle X_1, X_2, \ldots, X_N \rangle$, this BN consists in a pair $\langle G, \Theta \rangle$, where G represents a DAG and Θ the set of conditional probabilities $\Theta = \langle \Theta_{ijk} : \forall ijk \rangle$, where $\Theta_{ijk} = \Pr(X_i = x_i^k | pa(X_i) = p_i^j), p_i^j$ represents the parent (pa) nodes set of X_i from G and p_i^j represents the jth combination of parent nodes of X_i.

Pradhan et al. [7] proposed a generic three-level structure template for BNs, which includes three levels: level (1) representing background information and including nodes called predisposing factors; level (2) containing query nodes that represents diseases to be diagnosed; and level (3) representing finding information and containing nodes that represents symptoms, signs and test results.

Seixas et al. [8] described a decision model to be included in a CDSS to predict diagnosis of dementia, AD and MCI that was designed using BNs with structures built by a knowledge domain specialist using the generic template of [7]. The BN parameters were estimated from a clinical dataset using a supervised learning algorithm. The training database was composed by real patient's cases dataset and normal controls from CAD (Center for Alzheimer's Disease and Related Disorders), at the Institute of Psychiatry of the Federal University of Rio de Janeiro, Brazil, with the consent of the ethics committee for medical research, project identification number 284/2010. The training database attributes were composed by predisposal factors, patient demographic data, assessment scales, symptoms, signs and confirmed diagnosis, positive or negative for each patient. The continuous attributes were discretized before the BN parameters estimation.

It is well known that all discretization algorithms have intrinsic loss of information due to the discretization process itself. The work [9] stated that the problem with the discretization of continuous variables is to balance the desire for high accuracy in the approximations with a reasonable calculation burden to obtain reliable results. McGeachie et al. [10] said that researchers who wish to analyze continuous data in BNs generally discretize these data, leading to loss of information and concurrent loss of power. Our work extends the work of [8], proposing not to discretize BN nodes in order to improve its decision model performance.

The BN structures of [8] were built by expert knowledge but the data that were used to learn the BN parameters do not support independence assumptions established by the model as it will be seen later. The work [11] stated that some violations of the attribute independence assumptions in Bayesian models are important and, therefore, there is an increasing number of works developing techniques to retain simplicity and efficiency while alleviating the problems of the attribute independence assumptions. Alternatively, aiming at improving performance, this paper proposes relaxing independence assumptions in the decision model of [8]. Our proposal evaluation presents better results than the ones found in [8].

This paper is organized as follows. Section 2 presents related work that includes computer aided diagnosis of AD and related disorders based on patient's clinical datasets. Section 3 discusses references concerning BNs with mixed (continuous and discrete) nodes as well choices made in the implementation to improve the decision model of [8]. Also, Sect. 3 presents methods to model dependencies between variables in the training datasets without compromising the model readability and taking into consideration time constraints. Section 4 shows performance results obtained by applying the proposed approaches and discusses the obtained results. Section 5 presents our final considerations and future work.

2   Related Work

Pinheiro et al. [12] proposed a hybrid model combining MCDA (Multicriteria Decision Aiding) and BN showing a ranking model based on MCDA and BN for aiding the diagnosis of AD. Their model indicates which assessment patient items have the highest impact for determining the AD diagnosis. Their BN was manually built, where the Bayesian nodes were semantically related to each assessment item. Assessment items were composed by neuropsychological batteries provided by CERAD [13]. Instead of considering each neuropsychological question separately, our work used final neuropsychological test results and we propose improving a Bayesian decision model to predict AD diagnosis, and other related diseases, such as dementia and MCI.

Moreira and Namen [14] described an application that uses data mining techniques and aims at supporting specialists in AD diagnostic process. They used data from patients from the Alzheimer's and Parkinson's Center, located in the city of Campos dos Goytacazes, Brazil, for training and testing classifications models, including Naïve Bayes, BN and decision tree. The BN structure (G) and the BN parameters (Θ) were learned from data using WEKA[1] (Waikato Environment for Knowledge Analysis) implementation for BN. Our work differs from [14] because we combine expert knowledge and data from CAD aiming at improving a Bayesian decision model to support diagnosis of AD and related disorders.

Given a dataset composed by independent and identically distributed observations, for estimating a BN with random variables with discrete states, the parameters that best represents this data set for Maximum Likelihood Estimation (Θ_{ijk}^{MLE}) were used by [8] to compute parameters for BN following a Dirichlet distribution with hyper parameters for each attribute. The patient dataset used for learning the BN parameters have missing attribute values (partial observations) because of many reasons, either the patient might not have performed such neuropsychological test, or the physician might not have included such results in the dataset. A parametric learning method was applied in this case of partial observations: the EM algorithm [15], an iterative algorithm, which maximizes Θ_{ijk}^{MLE} at each iteration. This algorithm contains two steps: E-Step computes a posterior distribution over the observations using a BN inference engine and the M-step maximizes the log-likelihood. The inputs for the EM algorithm were: the training dataset and the Dirichlet initial hyper parameters set α. Seixas et al. [8] considered α = 1 for all attributes, assuming a prior uniform distribution, or non-informative priors [16]. The routine goes on until it reaches the convergence or stopping criterion. As a result of the iterative process, the parameters are learned. Seixas et al. [8] used the EM implementation available in the BN toolbox developed in Mathworks Matlab© [17] to perform the BNs parameters learning.

In the BN modeling process proposed by [8], diagnosis criteria are used together with the attributes from a preprocessed training dataset to build manually a BN structure. Then, the EM algorithm, using this dataset, learns the BN parameters. Three BNs were built, one for each disease to be diagnosed: dementia, AD and MCI. The training dataset was preprocessed including the following steps:

1. Selection of registers according to the disease to be diagnosed resulting in three training datasets, one for each disease:
   (a) all instances from CAD dataset were used to dementia diagnosis;
   (b) only instances with positive diagnosis for dementia were selected to AD diagnosis, because AD is a type of dementia; and

---

[1] http://www.cs.waikato.ac.nz/ml/weka/.

(c) only instances with negative diagnosis for dementia were selected to MCI diagnosis, because MCI can be viewed as a preclinical stage of AD where dementia is not present yet.

2. Exclusion of attributes with missing values in percentage greater than 70% and with information gains lower than 0.0001.

3. Application of a balancing technique called SMOTE (Synthetic Minority Oversampling Technique) [18] to oversample the negative cases without modifying the training dataset characteristics because there were much more cases with positive diagnosis than negative diagnosis for dementia and AD in CAD dataset.

4. Discretization of continuous attributes using an algorithm based on MDL (Minimum Description Length) described by [19].

Table 1 shows the number of cases with diagnosis negative (N) and positive (P) from the preprocessed training datasets before and after applying SMOTE as well the missing values ratio. The missing values ratio was calculated dividing the total of missing attribute values by the number of instances and by the number of the attributes of the corresponding dataset.

Table 1. Patients cases datasets

Diseases	D		AD		MCI	
Diagnosis	N	P	N	P	N	P
Number of cases	67	180	45	135	32	35
Number of cases after SMOTE	134	180	90	135	–	–
Missing values ratio	29%	24%	29%	22%	37%	21%

For more information about the selected attributes from the datasets and the obtained discretization levels for the attributes that were originally continuous, please refer to [8].

Figure 1 illustrates the obtained BN structure after applying the modeling process in which the attribute node values that were originally continuous are highlighted. This structure is essentially the same for each disease, only varying the diagnosis node (dementia, AD or MCI). The BN parameters and the marginal distributions in each node, which are different for each disease, are not shown. The BNs were drawn using the GeNIe (Graphical Network Interface) Modeler[2].

This paper extends [8] following two approaches: mixing continuous and discrete nodes and relaxing attribute independence assumptions.

---

[2] https://download.bayesfusion.com/files.html.

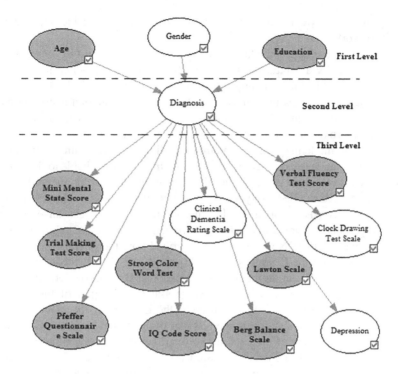

Fig. 1. BN structure built for diagnosis of dementia, AD and MCI.

3   Methodology

3.1   Mixing Continuous and Discrete Nodes

In the BNs from [8], the query nodes representing the diseases to be diagnosed have two states: positive and negative. These binary query nodes have parents that are continuous (see nodes "Age" and "Education" in Fig. 1) and discrete (see node "Gender" in Fig. 1). As far as we know, this type of BN currently lacks free available tools for modeling, learning and inference.

CGBayesNets [10] is a Matlab© software package that deals with Conditional Gaussian Bayesian Networks, which are a network formalism wherein discrete and continuous nodes are mixed, with the stipulation that continuous nodes have Gaussian distributions linearly dependent upon any continuous parents with parameters conditioned upon the values of any discrete parents. In this formalism, discrete nodes cannot be modeled as statistically dependent upon continuous nodes. Therefore, as in our model the discrete query nodes have continuous parents, CGBayesNet is not appropriate for the problem we want to address.

Lucas and Homerson [20] presented an algebraic framework that supports the modeling of discrete and continuous variables in causal BNs, either separately or mixed. The design of the framework was inspired by the convolution theorem of probability theory. However, as far as we know, there is no available implementation for this framework.

Shenoy [21] presented a method that consists of approximating general hybrid Bayesian networks by a Mixture of Gaussians (MoG) BNs, for which there are commercial available tools for modeling and inference, like Hugin[3]. In MoG BNs, all continuous chance variables must have conditional linear Gaussian distributions, and discrete chance nodes cannot have continuous parents. This method, through arc reversals and approximation of non-Gaussian distributions by sums of Gaussian distributions, allows modeling a BN containing a discrete node with continuous and discrete parents as an approximated BN that can be treated by available tools. However, [21] did not show sufficient examples to allow us to implement his method.

Tan et al. [22] described a Hybrid Logistic Regression-Naïve Bayes Model for classification that consists in a BN in that the binary class node can have continuous and discrete parents. Figure 2 illustrates this model. Tan et al. [22] showed an algorithm for discovering the BN structure and described how the inference can be made. The BN part including the C node (class node) and F_1, F_2, ..., F_n, features forms the Logistic Regression (LR) part of the model and the other BN part (including the C node and E_1, E_2, ..., E_m features) forms the Naïve Bayes (NB) part of the model. As illustrated by the BN of Fig. 2, the features in the LR part of the model are conditionally independent of the features in the NB part of the model given the class variable C. To learn the parameters of the conditional distribution of C given the F features in the LR part, the features E in the NB part are irrelevant for this task. Thus, one can use standard LR parameter estimation methods to learn those parameters. Similarly, to learn the parameters E of the NB part of the hybrid model, the features F in the LR part are irrelevant for this task, and thus, we can use standard NB parameter estimation methods for learning these parameters.

Suppose C is the binary class variable to predict, and F_1, F_2, ..., F_n, are real-valued features used to predict C. In practice, these features can be numeric or Boolean (with values 0 or 1). If we have a categorical feature with k distinct values, then we can represent such a feature with $k - 1$ Boolean features. For a LR model:

$$\text{odds}(C = 1 | f_1, \ldots, f_n) = \exp\left(\beta_0 + \sum_{j=1}^{n} \beta_j f_j\right) \tag{1}$$

where β_0, β_j are the regression coefficients, which are estimated by learning.

NB is a probabilistic classifier that is based on Bayes rule. It makes an assumption that features are mutually conditionally independent given the class variable. Suppose C is the binary class variable, whose value we wish to predict based on observation of a subset of m features $E = (E_1, E_2, \ldots, E_m)$. These features can be numeric or categorical. For an NB model:

$$\text{odds}(C = 1 | e_1, \ldots, e_m) = \text{odds}(C = 1) \prod_{i=1}^{m} \text{lr}(e_i, C = 1) \tag{2}$$

---

[3] http://www.hugin.com.

where $lr(e_i, C = 1) = \frac{P(e_i|C=1)}{P(e_i|C=0)}$. Therefore, the posterior odds of C = 1 is equal to prior odds of C = 1 times the likelihood ratios of observed features for C = 1. If a feature is not observed, we can regard its likelihood ratio as equal to 1.

After elimination of the features F in the LR part of the Hybrid LR-NB model, what is left is an NB model such that the posterior distribution of C (given F = f) is as given in (1). Thus, we can now compute the posterior distribution of C given F = f and E = e using the NB model (2) as follows:

$$\text{odds}(C = 1|e_1, \ldots, e_m, f_1 \ldots, f_n) = \text{odds}(C = 1|f_1, \ldots, f_n) \prod_{i=1}^{m} lr(e_i, C = 1) \quad (3)$$

Isolating the multiplicand in (2) and substituting it in (3), we have:

$$\text{odds}(C = 1|\mathbf{e}, \mathbf{f}) = \frac{1}{\text{odds}(C = 1)} \text{odds}(C = 1|\mathbf{e})\text{odds}(C = 1|\mathbf{f}) \quad (4)$$

where **e** is the vector containing the observed values for features of set E and **f** is the vector containing the observed values for features of set F.

In the Hybrid LR-NB model, the LR part, as well the NB part, can contain continuous (numeric) and discrete (categorical) nodes (features). The NB part can deal with missing values, while, for features of F set, the LR part replaces missing values by mean (for numeric features) or mode (for categorical features).

The similarity between the BNs structures of [8, 22] can easily be observed by comparing Figs. 1 and 2. However, the BNs structures of [8] were manually built using experts knowledge while the BNs structures of [22] have being discovered by an algorithm. We have implemented the mixing of discrete and continuous nodes using the Hybrid LR-NB model proposed by [22], but using CAD database and corresponding BNs structures from [8], including all the pre-processing steps previously described but without continuous nodes discretization. Our implementation follows (4) and is based on WEKA Java API for the Logistic Regression and Naïve Bayes classifiers. The attributes in the first level ("Age", "Education" and "Gender") form set F, the disease node in the second level is the binary class to predict (0 for "negative" and 1 for "positive" diagnosis) and the other attributes in the third level ("Mini Mental State Exam (MMSE) Score", "Clinical Dementia Rating (CDR) scale", etc.) form set E.

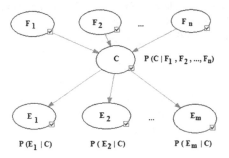

Fig. 2. A hybrid logistic regression-Naïve Bayes model as a Bayesian network [22].

3.2   Including Dependences

It is known that in a BN, each node is conditionally independent of its non-descendants given the values of its parents. We investigated this assumption for the datasets and BNs from [8] by carrying out chi-square independence tests. The null hypothesis H_0 is that the variables in question are independent. Table 6 (see Appendix) shows the p-values results for AD dataset indicating that, in many cases, the null hypothesis could not be sustained (p-value < 0.01).

Seixas et al. [8] also described BNs whose structures were automatically discovered from CAD dataset using the method described by Dash and Druzdzel [23]. Such method is based on independence tests between dataset attributes, and applies the Greedy Thick Thinning algorithm as a search method. However, these BNs with structures automatically discovered resulted in a decision model more complex and less readable by experts without obtaining expressive better performance results when compared with the BNs whose structures were manually built.

Therefore, we looked for other ways to relax independence assumptions in the decision model from Seixas et al. [8], aiming at improving its performance without compromising readability. We searched for a structure with two levels (Naïve Bayes Model) that would allow modeling dependences between the nodes in the second level. The second level, in this case, would include two node types: background and finding nodes, both depending on the disease query node (class variable to predict).

Zhang et al. [24] proposed a model, called Hidden Naïve Bayes (HNB), where a hidden parent is created for each attribute that combines the influences from all other attributes upon the attribute in question. However, this model was not chosen because it cannot deal with missing values.

The work [11] proposed AODE (Averaged One-Dependence Estimators), an ensemble of one-dependence classifiers that are learned from data and whose prediction is produced by aggregating the predictions of all qualified classifiers. An x-dependence estimator means that the probability of an attribute is conditioned by the class variable and at most x other attributes. In AODE, a one-dependence classifier is built for each attribute.

The work [25], in turn, described AnDE (Averaged n-Dependence Estimators), which are a family of algorithms that relaxes the Naïve Bayes independence assumption by generalizing AODE to higher levels of dependence.

Suppose we wish to estimate from a training sample T of t classified objects the probability $P(y|x)$ that an example $x = \langle x_1, \ldots, x_a \rangle$ belongs to class y, where x_i is the value of the ith attribute and $y \in \{c_1, \ldots, c_k\}$.

From the definition of conditional probability, we have:

$$P(y|x) = P(y,x)/P(x) \tag{5}$$

As the denominator in (5) is constant for all classes, it is enough to determine $P(y,x)$ because the class will be chosen according to the higher $P(y|x)$ value. But, applying the Bayes rule in (5), we have:

$$P(y, \boldsymbol{x}) = P(y)P(\boldsymbol{x}|y) \tag{6}$$

In an NB model, which assumes that all attributes are independent given the class, we have:

$$P(\boldsymbol{x}|y) = \prod_{i=1}^{a} P(x_i|y) \tag{7}$$

Hence, using (6) and (7), we classify in an NB model using:

$$\hat{P}_{NB}(y, \boldsymbol{x}) = \hat{P}(y) \prod_{i=1}^{a} \hat{P}(x_i|y) \tag{8}$$

where $\hat{P}(e)$ denotes an estimate of a probability of an event e.

AODE uses the Super-Parent One-Dependence Estimator (SPODE) concept, that consists in an estimator that relaxes the assumption of conditional independence by making all other attributes independent given the class and one privileged attribute, the super-parent, x_α, which is a weaker conditional independence assumption than NB's. It uses:

$$P(y, \boldsymbol{x}) = P(y, x_\alpha).P(\boldsymbol{x}|y, x_\alpha) \tag{9}$$

Together with the conditional independence assumption:

$$P(\boldsymbol{x}|y, x_\alpha) = \prod_{i=1}^{a} P(x_i|y, x_\alpha) \tag{10}$$

AODE averages over all estimates of $P(y|\boldsymbol{x})$ produced by using different super-parents. AODE seeks to use:

$$\hat{P}(y, \boldsymbol{x}) = \sum_{\alpha=1}^{a} \hat{P}(y, x_\alpha)\hat{P}(\boldsymbol{x}|y, x_\alpha)/a \tag{11}$$

However, in fact, AODE only uses estimates of probabilities for which relevant examples occur in the data:

$$\hat{P}_{AODE}(y, \boldsymbol{x}) = \begin{cases} \frac{\sum_{\alpha=1}^{a} \delta(x_\alpha)\hat{P}(y, x_\alpha)\hat{P}(\boldsymbol{x}|y, x_\alpha)}{\sum_{\alpha=1}^{a} \delta(x_\alpha)}, & \sum_{\alpha=1}^{a} \delta(x_\alpha) > 0 \\ \hat{P}_{NB}(y, \boldsymbol{x}), & otherwise \end{cases} \tag{12}$$

In (12), if attribute-value x_α is present in the data, $\delta(x_\alpha)$ is 1, otherwise is 0. Hence, AODE averages over all super parents whose values occur in the data, and defaults to NB if there are no such super parents.

AnDE, in turn, seeks to use:

$$\hat{P}(y, \boldsymbol{x}) = \sum_{s \in \binom{A}{n}} \hat{P}(y, x_S)\hat{P}(\boldsymbol{x}|y, x_S)/\binom{a}{n} \tag{13}$$

where $\begin{pmatrix} A \\ n \end{pmatrix}$ is the set of all size-n subsets of $\{1, \ldots, a\}$ and x_S is the subset of attributes values from x in the S subset. However, AnDE avoids using sets of super parents whose values do not occur in the data, and hence:

$$\hat{P}_{AnDE}(y, x) = \begin{cases} \dfrac{\displaystyle\sum_{S \in \binom{A}{n}} \delta(x_S)\hat{P}(y,x_S)\hat{P}(x|y,x_S)}{\displaystyle\sum_{S \in \binom{A}{n}} \delta(x_S)}, & \displaystyle\sum_{S \in \binom{A}{n}} \delta(x_S) > 0 \\ \hat{P}_{A(n-1)DE}(y, x), & otherwise \end{cases} \qquad (14)$$

Where attributes are assumed being independent given the super parents and the class. Hence, $P(x|y, x_S)$ is estimated by:

$$\hat{P}(x|y, x_S) = \prod_{i=1}^{a} \hat{P}(x_i|y, x_S) \qquad (15)$$

Notice from (8), (12) and (14) that A0DE is NB and A1DE is AODE. In practice, AnDE is used for $n \leq 2$ because higher orders produce more complex models with larger execution times for training and classification mainly in high-dimensional datasets [25]. Experimental results showed that A1DE and A2DE perform surprisingly well compared to other classification algorithms [25].

Aiming at modeling dependences between attributes from the pre-processed CAD dataset, we chose the implementations for A1DE and A2DE algorithms from an available WEKA package.

4   Results and Discussion

We carried out tests on the pre-processed datasets for dementia, AD and MCI based on 4-folds cross-validation [26] for the Hybrid LR-NB model we implemented and for A1DE and A2DE classifiers. For the Hybrid LR-NB model, we did not perform the discretization step in the pre-processing. As mentioned before, the training dataset (which was split in three according to the disease and pre-processed as described in Sect. 2) is composed by real patient's cases datasets and normal controls from CAD and includes predisposal factors, demographic data, assessment scales, symptoms, signs and confirmed diagnosis, positive or negative for each patient (see Table 1 and Fig. 1).

Concerning the evaluation of BNs, [27] proposed aligning BN classifiers with medical contexts by incorporating AUC (Area under the ROC curve), sensitivity (recall) and specificity measures in order to better evaluate these classifiers.

In our tests, the proposed Hybrid LR-NB, A1DE e A2DE classifiers were evaluated using the performance measures summarized in Table 2: AUC (Area Under ROC Curve), F1-score (obtained by the harmonic mean of precision and recall), Mean Square Error (MSE) and Mean Cross-Entropy (MXE) (see [8, 26] for more detailed

Table 2. Performance measures

Measure	Acronym	Domain	Best score
Area under ROC curve	AUC	[0.5, 1]	1
Harmonic mean of precision and recall	F1	[0, 1]	1
Mean square error	MSE	[0, 1]	0
Mean cross-entropy	MXE	[0, ∞)	0

information about these measures). To comparison purpose, Tables 3, 4 and 5 show the obtained results as well the results previously showed by [8] for the Naïve Bayes (NB) classifier and for the BNs obtained by [8].

Regarding the results for CAD dataset, evaluating for AUC, all classifiers presented similar results for dementia excepting LR-NB that performed worst. The proposed A1DE and A2DE showed the best results for AD (AUC of 0.92 and 0.93 respectively). BN and NB showed the best results for MCI (0.97).

Regarding F1-score, the best results were obtained by BN (0.94) for dementia, A1DE (0.83) and A2DE (0.83) for AD and A2DE (0.95) for MCI.

Regarding MSE, A1DE and A2DE performed best for all diseases. For MXE, NB performed best for dementia (0.07) and MCI (0.09), while A2DE was the best for AD (0.13).

Table 3. Classifiers performance for dementia

	NB[a]	BN[b]	Hybrid LR-NB	A1DE	A2DE
AUC	**0.97**	0.96	0.92	**0.97**	**0.97**
F1	0.93	**0.94**	0.81	0.92	0.91
MSE	0.13	0.18	0.15	**0.06**	**0.06**
MXE	**0.07**	0.16	0.26	0.1	0.09

[a]Naïve Bayes performance as showed by [8].
[b]Performance of the Bayesian Network for the disease in question obtained by [8].

Table 4. Classifiers performance for Alzheimer's disease

	NB	BN	Hybrid LR-NB	A1DE	A2DE
AUC	0.85	0.86	0.89	0.92	**0.93**
F1	0.80	0.82	0.79	**0.83**	**0.83**
MSE	0.16	0.23	0.15	0.11	**0.10**
MXE	0.21	0.15	0.20	0.14	**0.13**

Our Hybrid LR-NB implementation only outperforms the BN model of [8] in terms of MSE (0.15 against 0.18) for dementia and in terms of AUC (0.89 against 0.86) and MSE (0.15 against 0.23) for AD.

Table 5. Classifiers performance for mild cognitive impairment

	NB	BN	Hybrid LR-NB	A1DE	A2DE
AUC	**0.97**	**0.97**	0.88	0.95	0.95
F1	0.90	0.92	0.85	0.93	**0.95**
MSE	0.15	0.09	0.14	**0.08**	**0.08**
MXE	**0.09**	0.33	0.38	0.14	0.14

In general, A1DE and A2DE results were similar and the Hybrid LR-NB did not perform well when compared to the others. The A2DE classifier outperforms all other classifiers for AD when all measures are considered.

5 Conclusion and Future Work

Dementia, AD and MCI are relevant diseases due to the global aging phenomenon and their high prevalence among elderly population bringing high impact on patients' families, community and health care system.

This paper investigated ways to improve the Bayesian decision model of [8], used to support diagnosis of Dementia, AD and MCI. Two approaches were proposed: (2) a hybrid model mixing discrete and continuous variables and (2) a model relaxing attribute independence assumptions. The results presented by [8] were already good but we searched for other Bayesian decision models that could achieve even better results without losing the advantages of these models for clinical domains: they can deal with uncertainty and causality, perform inferences based on partial observations (missing values) and favor their readability by domain experts.

Regarding the first approach, we implemented a Hybrid LR-NB model with a manually built structure using WEKA Java API. The results of this model showed that the performance was not improved for a particular disease when considering all adopted measures. Hence, we conclude that trying to reduce discretization losses, mixing discrete and continuous nodes, through the adoption of a Hybrid LR-NB model with a manually built structure did not achieve reasonable performance results that could justify its adoption.

Regarding the second approach, we used a WEKA package for A1DE and A2DE algorithms. The results of these algorithms (which were similar with slight superiority of the A2DE algorithm) were promising for AD when compared to the results presented by [8]. Therefore, our work found that modeling dependencies between attributes in AD training dataset from CAD through A2DE adoption is a good option to improve the model of [8], making it even more reliable. Although not described in this paper, a web-based CDSS was developed aiming at being used by physicians through mobile devices. As future work, we intend to incorporate the A2DE classifier in the implemented CDSS to diagnose AD due to its best results for this disease.

Another goal is to deploy the implemented CDSS in a real clinical environment and evaluate acceptance and feedback reported by physicians.

Appendix

Table 6. [a]Chi-square independence tests in Alzheimer's Disease dataset: values of p.

Background	Age	Edu-cation	Gen-der							
Age	-									
Education	und[b]	-								
Gender	**0.01**	und	-							
Finding| AD =negative	MMSE	TMT	De-press.	CDR	Pfef-fer	IQ Code	VFT	CDT	Stro-op	Lawton
MMSE	-									
TMT	$<10^{-3}$	-								
Depression	0.55	0.51	-							
CDR	0.76	0.59	0.95	-						
Pfeffer	und	und	und	und	-					
IQCode	und	und	und	0.11	und	-				
VFT	0.08	0.14	0.09	**0.01**	und	und	-			
CDT	0.21	0.35	0.56	1	und	und	und	-		
Stroop	$<10^{-3}$	$<10^{-3}$	**0.002**	$<10^{-3}$	$<10^{-3}$	$<10^{-3}$	$<10^{-3}$	$<10^{-3}$	-	
Lawton	$<10^{-3}$	$<10^{-3}$	**0.002**	$<10^{-3}$	$<10^{-3}$	$<10^{-3}$	$<10^{-3}$	$<10^{-3}$	und	-
Berg	$<10^{-3}$	$<10^{-3}$	**0.002**	$<10^{-3}$	$<10^{-3}$	$<10^{-3}$	$<10^{-3}$	$<10^{-3}$	und	und
Finding| AD=positive	MMSE	TMT	De-press.	CDR	Pfef-fer	IQ Code	VFT	CDT	Stro-op	Lawton
MMSE	-									
TMT	10^{-3}	-								
Depression	0.19	0.97	-							
CDR	0.36	0.19	0.43	-						
Pfeffer	und	und	und	und	-					
IQCode	und	und	und	und	und	-				
VFT	0.67	0.06	0.38	$<10^{-3}$	0.06	und	-			
CDT	0.19	0.26	0.38	$<10^{-3}$	und	und	$<10^{-3}$	-		
Stroop	0.58	und	und	und	und	und	und	und	-	
Lawton	und	und	und	und	und	und	und	und	und	-
Berg	0.87	und	0.22	$<10^{-3}$	und	und	1	0.02	und	und

[a] Half-filled due to be symmetric.

[b] und: undefined, i.e., the number of joint occurrences was insufficient to establish a reliable statistical test.

References

1. Sosa-Ortiz, A., Acosta-Castillo, I., Prince, M.: Epidemiology of dementias and Alzheimer's disease. Arch. Med. Res. **43**(8), 600–608 (2012)
2. Newman-Toker, D., Pronovost, P.: Diagnostic errors: the next frontier for patient safety. J. Am. Med. Assoc. (JAMA) **301**(10), 1060–1062 (2009)
3. Berner, E.S.: Clinical Decision Support Systems: Theory and Practice. Springer, New York (2007)
4. Haynes, R.B., Wilczynski, N.L.: Effects of computerized clinical decision support systems on practitioner performance and patient outcomes: methods of a decision-maker-researcher partnership systematic review. Implement Sci. **5**(1), 12 (2010)
5. Kong, G., Xu, D., Yang, J.: Clinical decision support systems: a review on knowledge representation and inference under uncertainties. Int. J. Comput. Intell. Syst. **1**(2), 159–167 (2008)
6. Jensen, F.V., Nielsen, T.D.: Bayesian Networks and Decision Graphs. Springer, New York (2007)
7. Pradhan, M., Provan, G., Middleton, B., Henrion, M.: Knowledge engineering for large belief networks. In: Tenth Conference of Uncertainty in Artificial Intelligence, pp. 484–490. Morgan Kaufmann Publishers Inc., San Francisco (1994)
8. Seixas, F.L., Zadrozny, B., Laks, J., Conci, A., Muchaluat-Saade, D.C.: A Bayesian network decision model for supporting the diagnosis of dementia, Alzheimer's disease and mild cognitive impairment. Comput. Biol. Med. **51**, 140–158 (2014)
9. Langseth, H., Nielsen, T.D., Rumí, R., Salmerón, A.: Inference in hybrid Bayesian networks. Reliab. Eng. Syst. Saf. **94**(10), 1499–1509 (2009)
10. McGeachie, M.J., Chang, H., Weiss, S.T.: CGBayesNets: conditional Gaussian Bayesian network learning and inference with mixed discrete and continuous data. PLoS Comput. Biol. **10**, 6 (2014)
11. Webb, G.I., Boughton, J.R., Wang, Z.: Not so Naïve Bayes: aggregating one-dependence estimators. Mach. Learn. **58**(1), 5–24 (2005)
12. Pinheiro, P.R., Castro, A., Pinheiro, M.: A multicriteria model applied in the diagnosis of Alzheimer's disease: a Bayesian network. In: Proceedings of the 11th IEEE International Conference on Computational Science and Engineering, CSE 2008, São Paulo (2008)
13. Fillenbaum, G., van Belle, G., Morris, J., Mohs, R., Mirra, S., Davis, P., Tariot, P., Silverman, J., Clark, C., Welsh-Bohmer, K.: Consortium to establish a registry for Alzheimer's disease (CERAD): the first twenty years. Alzheimer's Dement. J. Alzheimer's Assoc. **4**(2), 96–109 (2008)
14. Moreira, L.B., Namen, A.A.: System predictive for Alzheimer's disease in clinical trial. J. Health Inform. **8**, 3 (2016)
15. Dempster, A., Laird, N., Rubin, D.: Maximum likelihood from incomplete data via the EM algorithm. J. R. Stat. Soc. Ser. B Methodol. **39**(1), 1–38 (1977)
16. Cooper, G.F., Herskovits, E.: A Bayesian method for the induction of probabilistic networks from data. Mach. Learn. **9**(4), 309–347 (1992)
17. Murphy, K.: The Bayes net toolbox for Matlab. Comput. Sci. Stat. **33**(2), 1024–1034 (2001)
18. Chawla, N., Bowyer, K., Hall, L., Kegelmeyer, W.: Smote: synthetic minority over-sampling technique. J. Artif. Intell. Res. **16**, 321–357 (2002)
19. Kononenko I.: On biases in estimating multi-valued attributes. In: International Joint Conference on Artificial Intelligence, vol. 14, pp. 1034–1040. Lawrence Erlbaum Associates, Montreal (1995)

20. Lucas, P.J.F., Hommersom, A.: Modeling the interactions between discrete and continuous causal factors in Bayesian networks. Int. J. Intell. Syst. **30**(3), 209–235 (2015)
21. Shenoy, P.P.: Inference in hybrid Bayesian networks using mixtures of Gaussians. arXiv preprint arXiv:1206.6877 (2012)
22. Tan, Y., Moses, P.P., Chan, W., Romberg, P.M.: On construction of hybrid logistic regression-Naïve Bayes model for classification. In: Proceedings of the Eighth International Conference on Probabilistic Graphical Models, Lugano, 6–9 September 2016
23. Dash, D., Druzdzel, M.J.: Robust independence testing for constraint-based learning of causal structure. In: Proceedings of the Nineteenth Conference on Uncertainty in Artificial Intelligence, pp. 167–174. Morgan Kaufmann Publishers Inc., San Francisco (2002)
24. Zhang, H., Jiang, L., Su J.: Hidden Naïve Bayes. In: Proceedings of the Twentieth National Conference on Artificial Intelligence, Pennsylvania, 9–13 July 2005
25. Webb, G.I., Boughton, J.R., Zheng, F., Ting, K.M., Salem, H.: Learning by extrapolation from marginal to full-multivariate probability distributions: decreasingly Naïve Bayesian classification. Mach. Learn. **86**(2), 233–272 (2012)
26. Han, J., Kamber, M., Pei, J.: Data Mining: Concepts and Techniques, 3rd edn. Elsevier, Amsterdam (2011)
27. Gaag, L.C., Renooij, S., Feelders, A., Groote, A., Eijkemans, M.J.C., Broekmans, F.J., Fauser, B.C.J .M.: Aligning Bayesian network classifiers with medical contexts. In: Perner, P. (ed.) MLDM 2009. LNCS (LNAI), vol. 5632, pp. 787–801. Springer, Heidelberg (2009). doi:10.1007/978-3-642-03070-3_59

Over-Fitting in Model Selection
with Gaussian Process Regression

Rekar O. Mohammed$^{(\boxtimes)}$ and Gavin C. Cawley

University of East Anglia, Norwich, UK
rekarmajidi@gmail.com, g.cawley@uea.ac.uk
http://theoval.cmp.uea.ac.uk/

Abstract. Model selection in Gaussian Process Regression (GPR) seeks
to determine the optimal values of the hyper-parameters governing the
covariance function, which allows flexible customization of the GP to
the problem at hand. An oft-overlooked issue that is often encountered
in the model process is over-fitting the model selection criterion, typically the marginal likelihood. The over-fitting in machine learning refers
to the fitting of random noise present in the model selection criterion
in addition to features improving the generalisation performance of the
statistical model. In this paper, we construct several Gaussian process
regression models for a range of high-dimensional datasets from the UCI
machine learning repository. Afterwards, we compare both MSE on the
test dataset and the negative log marginal likelihood (nlZ), used as the
model selection criteria, to find whether the problem of overfitting in
model selection also affects GPR. We found that the squared exponential
covariance function with Automatic Relevance Determination (SEard) is
better than other kernels including squared exponential covariance function with isotropic distance measure (SEiso) according to the nLZ, but
it is clearly not the best according to MSE on the test data, and this is
an indication of over-fitting problem in model selection.

Keywords: Gaussian process · Regression · Covariance function ·
Model selection · Over-fitting

1 Introduction

Supervised learning tasks can be divided into two main types, namely classification and regression problems. Classification is usually used when the outputs
are categorical (discrete class labels), whereas, regression is concerned with the
prediction of continuous quantities. Gaussian process is defined as a distribution
over functions, and inference takes place directly in the space of functions, i.e.
the function-space view. Gaussian process regression is not a new area of study,
it has been extensively used in research areas such as machine learning, statistics
and engineering. In the literature, Gaussian process regression has been widely
used for many real-world problems, including time series analysis. For instance,
Duvenaud et al. (2013) applied GPR to the total solar irradiance dataset and

© Springer International Publishing AG 2017
P. Perner (Ed.): MLDM 2017, LNAI 10358, pp. 192–205, 2017.
DOI: 10.1007/978-3-319-62416-7_14

obtained good results, and Williams and Rasmussen (2006) also used GPR for modelling atmospheric CO2 concentrations.

Model selection approaches for GPR seek to determine good values for the hyper-parameters of the model, typically via maximising the marginal likelihood or via cross validation (Williams and Rasmussen 2006). Cawley and Talbot (2007) discusses an over-fitting issue that arises in model selection with Gaussian processes classification. They claim that for GP classification, covariance functions with large parameters clearly demonstrate the over-fitting issue, where reducing the value of the model selection criterion results in a model with worse generalisation performance. This is because the model selection criterion is evaluated over a finite set of data, and hence is a performance estimate with a non-negligible variance.

In this paper, we first describe the background methodology for applications of Gaussian progress regression, and then give some examples of covariance functions commonly used in GPR. The reminder of the paper the describes model selection practices for GPR, and the causes of over-fitting in model selection, how one can detect it, and how this issue can be avoided. Finally we present empirical results using UCI benchmark datasets (2013), showing that over-fitting the model selection criterion is a potential pit-fall in practical applications and GPR, and present our conclusions.

2   Background

Regression analysis is a vital tool in applied statistics as well as in machine learning. It aims to investigate the influence of certain variables X on a certain outcome y (Walter and Augustin 2010).

The linear regression model is one of the most common models used to study the linear relationship between a dependent variable y and one or more independent variables X. The reason for its popularity is due to both the conceptual and computational simplicity of fitting a linear model. However, linear regression is dependent on some assumptions (Briegel and Tresp 2000), for example, the true relationship in the data must be approximately linear for good prediction using a linear model, but unfortunately this often is not the case for real-life data. Therefore, standard linear regression is generalized in many ways and here we use Bayesian linear regression as a treatment to the linear model (the following exposition is based on that given by Williams and Rasmussen 2006).

In Bayesian linear regression, we need to have a prior belief regarding the values of the model parameters that is combined with the likelihood function, describing the distribution of the data, to find the posterior distribution over the parameters. We can write down a generative model for our data.

$$f(\boldsymbol{x}) = \boldsymbol{x}^T \boldsymbol{w}, \quad y = f(\boldsymbol{x}) + \varepsilon,$$

where $f(\boldsymbol{x})$ is our modelling function, ε is some form of additive noise, and y is the observed target values. The input vector is defined as $\boldsymbol{x}$ and parameter vector of the linear model as $\boldsymbol{w}$. We also assume that ε are an independent and identically distributed (i.i.d.) sample from a zero-mean normal distribution, i.e. $N(0, \sigma_n^2)$. It follows that $y = \boldsymbol{x}^T \boldsymbol{w} + \varepsilon : \varepsilon \sim N(0, \sigma_n^2)$. Both noise and model assumptions enable us to identify the probability density of the observations given the parameters which is known as the Likelihood function, which is given by

$$p(\boldsymbol{y} \mid \boldsymbol{X}, \boldsymbol{w}) = \prod_{i=1}^{n} p(y_i \mid x_i, \boldsymbol{w}) = \prod_{i=1}^{n} \frac{1}{\sqrt{2\pi}\sigma_n} \exp\left(-\frac{(y_i - \boldsymbol{x}_i^T \boldsymbol{w})^2}{2\sigma_n^2}\right),$$

$$= \frac{1}{(2\pi\sigma_n^2)^{\frac{n}{2}}} \exp\left(-\frac{1}{2\sigma_n^2}|\boldsymbol{y} - \boldsymbol{X}^T \boldsymbol{w}|^2\right) = N(\boldsymbol{X}^T \boldsymbol{w}, \sigma_n^2 \boldsymbol{I}),$$

$$\propto \exp\left(-\frac{1}{2\sigma_n^2}(\boldsymbol{y} - \boldsymbol{X}^T \boldsymbol{w})^T (\boldsymbol{y} - \boldsymbol{X}^T \boldsymbol{w})\right).$$

In Bayesian linear regression, we assume that a prior distribution over the parameters is also given. For example, a typical choice is $\boldsymbol{w} : N(0, \Sigma_p)$

$$p(\boldsymbol{w}) = \frac{1}{(2\pi)^{\frac{p}{2}}|\Sigma_p|} \exp\left(-\frac{1}{2}\boldsymbol{w}^T \Sigma_p^{-1}\boldsymbol{w}\right),$$

$$\propto \exp\left(-\frac{1}{2}\boldsymbol{w}^T \Sigma_p^{-1}\boldsymbol{w}\right).$$

Now, by using Bayes' rule, we can obtain the posterior distribution for the parameters, which is given by

$$p(\boldsymbol{w} \mid \boldsymbol{y}, \boldsymbol{X}) = \frac{p(\boldsymbol{y} \mid \boldsymbol{X}, \boldsymbol{w})p(\boldsymbol{w})}{\int p(\boldsymbol{y} \mid \boldsymbol{X}, \boldsymbol{w})p(\boldsymbol{w})d\boldsymbol{w}}.$$

The denominator is known as the marginal likelihood $p(\boldsymbol{y} \mid \boldsymbol{X})$ and does not involve the parameters (weights), hence it can often be neglected. In the following steps, we get closer to the computation of the posterior distribution for the parameters.

$$p(\boldsymbol{w} \mid \boldsymbol{y}, \boldsymbol{X}) \propto \exp\left(-\frac{1}{2\sigma_n^2}(\boldsymbol{y} - \boldsymbol{X}^T \boldsymbol{w})^T (\boldsymbol{y} - \boldsymbol{X}^T \boldsymbol{w})\right) \exp\left(-\frac{1}{2}\boldsymbol{w}^T \Sigma_p^{-1}\boldsymbol{w}\right),$$

$$\propto \exp\left[-\frac{1}{2}(\boldsymbol{w} - \bar{\boldsymbol{w}})^T \left(\frac{1}{\sigma_n^2}\boldsymbol{X}\boldsymbol{X}^T + \Sigma_p^{-1}\right)(\boldsymbol{w} - \bar{\boldsymbol{w}})\right].$$

Therefore, the posterior is recognised as a Gaussian distribution with $\bar{\boldsymbol{w}} = \sigma_n^{-2} A^{-1}\boldsymbol{X}\boldsymbol{y}$ as a mean and as a covariance matrix $A^{-1} = (\frac{1}{\sigma_n^2}\boldsymbol{X}\boldsymbol{X}^T + \Sigma_p^{-1})^{-1}$, i.e.

$$p(\boldsymbol{w} \mid \boldsymbol{y}, \boldsymbol{X}) : N(\bar{\boldsymbol{w}}, A^{-1}).$$

Having specified $\boldsymbol{w}$, making predictions about unobserved values, $f(\boldsymbol{x}_*)$, at coordinates, $\boldsymbol{x}_*$, is then only a matter of drawing samples from the predictive distribution $p(f_* \mid \boldsymbol{x}_*, \boldsymbol{X}, \boldsymbol{y})$ which is defined as:

$$p(f_* \mid \boldsymbol{x}_*, \boldsymbol{X}, \boldsymbol{y}) = \int p(f_* \mid \boldsymbol{x}_*, \boldsymbol{w})p(\boldsymbol{w} \mid \boldsymbol{y}, \boldsymbol{X})d\boldsymbol{w}.$$

The predictive posterior is once again Gaussian:

$$p(f_* \mid \boldsymbol{x}_*, \boldsymbol{X}, \boldsymbol{y}) \sim N(\sigma_n^{-2} \boldsymbol{x}_*^T \boldsymbol{A}^{-1} \boldsymbol{X} \boldsymbol{y}, \boldsymbol{x}_*^T \boldsymbol{A}^{-1} \boldsymbol{x}_*).$$

In fact, both the parameter posterior and posterior predictive distribution provide a useful way to quantify our uncertainty in model estimates, and to exploit our knowledge of this uncertainty in order to make more robust predictions on new test points (Do 2007).

2.1 Gaussian Processes in Regression

Over the past few years, there has been a tremendous interest in applying non-parametric approaches to real-world problems. Numerous studies have been devoted to Gaussian processes (GPs) because of their flexibility when compared with parametric models. These techniques use Bayesian learning, which usually leads to analytically intractable posteriors (Csató 2002), however that is not the case for GPR.

A Gaussian distribution is a distribution over random variables, $\boldsymbol{x} \in \mathbb{R}^n$, which is completely specified by a mean vector $\boldsymbol{\mu}$ and a covariance matrix $\boldsymbol{\Sigma}$,

$$p(\boldsymbol{x}; \boldsymbol{\mu}, \boldsymbol{\Sigma}) = \frac{1}{(2\pi)^{\frac{n}{2}} |\Sigma|} \exp \left[-\frac{1}{2} (\boldsymbol{x} - \boldsymbol{\mu})^T \boldsymbol{\Sigma}^{-1} (\boldsymbol{x} - \boldsymbol{\mu}) \right].$$

We can write this as $\boldsymbol{x} \sim G(\boldsymbol{\mu}, \boldsymbol{\Sigma})$. Gaussian random variables are very useful in statistics and machine learning because they are very commonly used for modelling noise in statistical algorithms (Do 2007).

According to Rasmussen (2004), a Gaussian process (GP) is defined as "a collection of random variables, any finite number of which have (consistent) joint Gaussian distributions". A Gaussian process is a distribution over functions which is fully specified by the mean function, $m(x)$, and a covariance function, $k(x, x')$, of a process $f(x)$, where

$$m(x) = E[f(x)], \tag{1}$$
$$k(x, x') = E[(f(x) - m(x))(f(x') - m(x'))]. \tag{2}$$

We can now obtain a GP from the Bayesian linear regression model in which, $f(x) = \boldsymbol{\phi}(x)^T \boldsymbol{w}$, with $\boldsymbol{w} : (0, \Sigma_p)$. Both mean function and covariance function are obtained as

$$E[f(x)] = \boldsymbol{\phi}(\boldsymbol{x})^T E(\boldsymbol{w}) = 0, \tag{3}$$
$$E[f(x) f(x')] = \boldsymbol{\varphi}(x)^T E[\boldsymbol{w} \boldsymbol{w}^T] \boldsymbol{\varphi}(x)^T \boldsymbol{\Sigma_p} \boldsymbol{\varphi}(x'). \tag{4}$$

Hence, $f(x)$ and $f(x')$ are jointly Gaussian with zero mean and covariance function $\boldsymbol{\varphi}(x)^T \boldsymbol{\Sigma_p} \boldsymbol{\varphi}(x')$.

The mean function is commonly defined to be zero, "which is not a strong limitation if the data is centred in preprocessing" (Blum and Riedmiller 2013). The covariance function defines the similarity between values of the function as a function of the data points and plays an important role in controlling the properties of Gaussian Processes (Williams and Rasmussen 2006). Gaussian processes are a technique for expressing prior distributions over functions for one or more input variables. Given a set of inputs, $x^{(1)}, \ldots, x^{(n)}$, we can draw samples $f(x^{(1)}), \ldots, f(x^{(n)})$ from the GP prior:

$$f(x^{(1)}), \ldots, f(x^{(n)}) : (0, K).$$

Although drawing random functions from the prior is important, we want to extract the information that the training data delivers about the function.

Given a noise-free training data,

$$D = \{(x^{(i)}, y^{(i)}) \mid i = 1, \ldots, n\} = \{X, f\}.$$

according to GP prior, the joint distribution of the training outputs, f, and the test outputs f_* is given by

$$\begin{bmatrix} f \\ f_* \end{bmatrix} : \left(0, \begin{bmatrix} K(X, X) & K(X, X_*) \\ K(X_*, X) & K(X_*, X_*) \end{bmatrix} \right).$$

In order to make predictions, we need to obtain the posterior distribution over functions. It is also necessary to restrict the prior to contain only functions which agree with D. The posterior distribution is obtained from the condition $\{X_*, f_*\}$ on $D = X, f$, and it is Gaussian.

$$f_* \mid X_*, X, f : N(K(X_*, X)K(X, X)^{-1}f, K(X_*, X_*) - K(X_*, X)K(X, X)^{-1}K(X, X_*))$$

However, the data of real world problems are typically noisy. Thus we need to define a GP for noisy observations.

$$D = \{X, y\}, \text{ where } y = f + \epsilon.$$

We assume additive noise, $\epsilon \sim N(0, \sigma^2 I)$, and can derive the predictive distribution by conditioning on $D = \{X, y\}$ that gives a Gaussian with

$$\mu = K(X_*, X)[K(X, X) + \sigma^2 I]^{-1}y, \tag{5}$$
$$\Sigma = K(X_*, X_*) - K(X_*, X)[K(X, X) + \sigma^2 I]^{-1}K(X, X_*). \tag{6}$$

Now if we give a new 'test' input x_*, the predictive distribution of the corresponding $f(x)$ is readily obtained. In practice, the predictive mean, denoted μ, of the GP is used as a point estimate for the function output, while the variance can be interpreted as uncertainty bounds ($\pm 2\sigma$ error-bars) on this estimate (Girard and Murray-Smith 2005).

The main aim of using Gaussian processes regression is for prediction. In the case of having D-dimensional input vector x mapped onto an N-dimensional feature space, m is an $n \times 1$ vector and Σ is an $n \times n$ matrix. More computational power is needed for implementing Gaussian processes regression when we have multivariate inputs.

The covariance function of the Gaussian process, that allows the model to find the high-level description of the data properties, can be specified as a hierarchical prior. For example, covariance function is used to identify the inputs that are useful in predicting the response. Inference for these covariance hyper-parameters can be performed using Markov chain sampling (Bernardo et al. 1998).

2.2 The Covariance Functions

There are three main concerns in Gaussian processes regression, namely the choice of the covariance function, the selection of variables, and the choice of good values of hyper-parameters which effectively control the complexity of the model (Shi and Choi 2011). Choosing a suitable covariance kernel is crucial because it determines almost all generalization properties of a Gaussian processes model (MacKay 1999).

There are a variety of different covariance functions that can be used in a Gaussian processes regression model, including stationary and non-stationary covariance functions. Stationary covariance functions, which are invariant under translation, are the most often used in GPR. One can simply assume that the mean is constant (zero), which means the process is stationary (Shi and Choi 2011). Stationary covariance functions depend only on the distance between the inputs, $\boldsymbol{x}$, such that the covariance function expresses the covariance between y_p and y_q (Williams and Rasmussen 2006). The formula is written as,

$$\mathrm{cov}(f(\boldsymbol{x}_p), f(\boldsymbol{x}_q)) = k(\boldsymbol{x}_p, \boldsymbol{k}_q) = \exp\left(-\frac{1}{2}|\boldsymbol{x}_p - \boldsymbol{x}_q|^2\right).$$

1. Squared Exponential Covariance Function (SE):
   This function is a smooth function of the inputs and is a common choice of covariance function because it has some nice properties, namely it can be integrated against most functions that we need in Gaussian processes.
   The form is given by

$$k_{\mathrm{SE}}(\boldsymbol{x}_p, \boldsymbol{x}_q) = \sigma_f^2 \exp\left(-\frac{(\boldsymbol{x}_p - \boldsymbol{x}_q)^2}{2r^2}\right) + \sigma_\epsilon^2 \delta_{pq},$$

   where σ_f^2 is the magnitude, r is the length scale that characterize variation, and σ_ϵ^2 represents noise.
2. Automatic Relevance Determination Covariance Function (SE-ARD):
   The SE-ARD covariance function for multi-dimensional inputs is considered as a more general form of the squared exponential kernel:

$$k_{\mathrm{SE-ARD}}(\boldsymbol{x}_p, \boldsymbol{x}_q) = \sigma_f^2 \exp\left(-\frac{1}{2}\sum_{d=1}^{D}\frac{(\boldsymbol{x}_p^{(d)} - \boldsymbol{x}_q^{(d)})^2}{r_d^2}\right),$$

   The parameter r_d is the characteristic length scale of dimension d. The relevancy of input feature can be determined by r_d, for instance, If r_d is very large, then the feature is irrelevant (Snelson 2006)

3. The Matérn Covariance Function:
   The formula of this type of covariance function is given by

$$k_{\text{Matérn}}(x, x') = \frac{2^{1-v}}{\Gamma(v)} \left(\frac{\sqrt{2v}|\boldsymbol{x} - \boldsymbol{x}'|}{r} \right)^v K_v \left(\frac{\sqrt{2v}|\boldsymbol{x} - \boldsymbol{x}'|}{r} \right),$$

where both v and r are positive parameters, v determines the smoothness and K_v is an amended Bessel function (Abramowitz 1966). When $v \to \infty$, then $k_{\text{Matérn}}(\boldsymbol{x}, \boldsymbol{x}')$ becomes squared exponential covariance function.

4. The Rational Quadratic Covariance Function (RQ):
   This kernel is equivalent to adding many SE kernels together with different length-scales. The form of the rational quadratic (RQ) covariance function is;

$$K_{\text{RQ}}(\boldsymbol{x}, \boldsymbol{x}') = \left(1 + \frac{|\boldsymbol{x} - \boldsymbol{x}'|^2}{2\alpha r^2} \right)^{-\alpha},$$

where α determines the smoothness and r is the characteristic length, when $\alpha \to \infty$ then RQ is identical to the SE.

5. Polynomial Covariance Function:
   The Polynomial kernel is a non-stationary kernel that takes the following form

$$k_{\text{Poly}}(\boldsymbol{x}, \boldsymbol{x}') = (\boldsymbol{x} \cdot \boldsymbol{x}' + \sigma_0^2)^p,$$

where $\sigma_0^2 > 0$ is a constant, trading off the effect of higher-order against lower-order terms in the polynomial, and the kernel is known as a homogeneous polynomial when $\sigma_0^2 = 0$, $p > 0$ is the polynomial degree, which is a natural number.

2.3   Model Selection for GP Regression

As mentioned previously, Gaussian processes are specified by their mean and covariance functions. The purpose of covariance function is to determine the similarity between data points that involved some free parameters known as hyper-parameters. Indeed, the hyper-parameters are useful since they allow for flexible customization of the GP to the problem. Therefore, it is necessary to select the covariance functions and its hyper-parameters appropriately by the so-called model selection process (Blum and Riedmiller 2013).

In literature, two techniques are most often discussed for model selection in Gaussian process regression, namely marginal likelihood maximisation and cross validation (Williams and Rasmussen 2006). We only describe the Marginal Likelihood method of selecting the model for GP regression, as that is the approach we adopt in our experiments.

A reliable framework for inference over the hyper-parameters is obtained via the Bayesian approach but good approximations are not easily derived, due to the required complex integrals over the hyper-parameters being analytically intractable. In fact, it is not easy to know what the parameters of the model are because Gaussian process model is a non-parametric model.

One can obtain the probability of the data given the hyper-parameters $p(\boldsymbol{y} \mid \boldsymbol{X}, \theta)$ for GPs regression with Gaussian noise by marginalization over the function values f. The log marginal likelihood is given by

$$\log p(\boldsymbol{y} \mid \boldsymbol{X}, \theta) = -\frac{1}{2}\boldsymbol{y}^T \boldsymbol{K}_y^{-1}\boldsymbol{y} - \frac{1}{2}\log |\boldsymbol{K}_y| - \frac{n}{2}\log 2\pi.$$

where $\boldsymbol{K}_y = \boldsymbol{K}_f + \sigma_n^2 \boldsymbol{I}$ is the covariance function for the noisy output, $\boldsymbol{y}$, and $\boldsymbol{K}_f$ is the covariance function for the noise-free latent function, f. The first term from the above equation is known as a data-fit term, the second term is a complexity penalty, and the last term is a normalizing constant (Blum and Riedmiller 2013).

In order to tune hyper-parameters by maximizing the marginal likelihood, the derivatives of the log marginal likelihood with respect to the hyper-parameters are required:

$$\frac{\partial}{\partial \theta_j}\log p(\boldsymbol{y} \mid \boldsymbol{X}, \theta) = \frac{1}{2}\operatorname{tr}\left[(\boldsymbol{\alpha}\boldsymbol{\alpha}^T - \boldsymbol{K}_y^{-1})\frac{\partial K_y}{\partial \theta_j}\right], \text{ where } \boldsymbol{\alpha} = \boldsymbol{K}^{-1}\boldsymbol{y}.$$

From the above equation, "any gradient based optimization algorithm can be used to obtain the hyper-parameters that maximize the marginal likelihood of a GP. We will call this optimization procedure *training* the GP" (Blum and Riedmiller 2013).

3   Over-Fitting in Model Selection with Gaussian Processes in Regression

In this section, we first define an over-fitting issue that rises in the context of model selection in machine learning. Afterwards, the reasons for the occurrence of this problem will be discussed; we will also explain how one can detect this over-fitting issue in model selection with Gaussian processes algorithms. The methods of preventing this problem will also be described. Finally, results obtained on a suite of eleven real-world benchmark data sets will be demonstrated.

3.1   Over-Fitting in Model Selection

Over-fitting in machine learning refers to the fitting of a random noise in the data in addition to it's underlying structure by a statistical model. Over-fitting usually occurs when a model is too complicated, for example, when the parameters are excessively more than the number of observations. The potential consequence of an over-fitted model is poor predictive performance, as it can amplify very small fluctuations in the data (Joshi 2013). While the dangers of over-fitting in determining the parameters of a model (training) are well documented, the risk of over-fitting in tuning the hyper-parameters (model selection) is less well appreciated.

3.2   The Causes of Over-Fitting in Model Selection

When selecting a model, over-fitting often occurs due to the variance of the model selection criteria. Models are typically trained via performance maximization based on a finite set of training data, the efficiency of the model on the other hand is not dictated based on the performance of the model using the training data. It is instead established using the success and effectiveness of the model of handling unseen data. The problem of over-fitting is encountered when a model begins to memorize training data as opposed to learning to generalize from the observed trend in the training data. For instance, if the number of parameters is the equal to or greater than the number of data points available, a basic linear model or learning process will be able to perfectly estimate the training data merely through memorization of the entire training data set. However, such elemental models and processes will frequently fail significantly when estimating new data. As the basic model has not learned to generalize to any degree, we experience the over-fitting problem (Joshi 2013).

According to Dieterich (1995) the major complication of over-fitting usually emerges from the structure of the machine-learning tasks. A learning algorithm is trained on a training dataset, but then applied to provide estimations using new unseen data points. We are not necessarily concerned with the algorithm's accuracy on the training data, but instead achieving optimal predictive accuracy on these unseen data points. The scenario of "over-fitting" arises when we try too hard to find the very best fit to the training data (or to the model selection criteria) and thus risk that noise will be consumed in the data due to the model memorizing particular characteristics of the training data instead of discovering a general predictive rule.

3.3   Detecting Over-Fitting in Model Selection

According to Cawley and Talbot (2010), fitting a Gaussian process with the non-ARD (Auto Relevance determination) equivalent covariance function (the Radial Basis Function (RBF) covariance function) and comparing the test error rates, would seem like the most straightforward progression to do. For several reasons, the ARD covariance function fails to perform as well as the non-ARD covariance function due to the over-fitting in tuning the hyper-parameters. The RBF is a special case of ARD where parameters constrained to be equal. Having fewer parameters gives less scope for over-fitting.

3.4   Avoiding Over-Fitting in Model Selection

Over-fitting mainly occurs when a small dataset is used. Therefore, it is always better to have a large data set. Thus, by using a lot of patterns the problem can potentially be avoided. However, having an excessively high number of data points, the algorithm is obliged to generalize and come up with a good model to fits all the points, without having sufficient capacity to model the noise. The convenience of choosing a large database does not always exist. There are times where a small database is the only available option, limiting our choice of model development. In such cases, a technique called cross validation can be used. This technique divides the dataset into training and testing datasets. The model is developed using the training dataset and the validity of the model is tested using the testing database. This process is then repeated using various partitions of training and testing datasets. As a result of this technique, a fairly good approximation of the underlying model is given, due to the fact that it is tested on several partitions to achieve generalization at the maximum possible degree (Joshi 2013).

According to Cawley and Talbot (2010) over-fitting in model selection may seem logical, if a model selection criterion estimated over a specific number of data observations is directly optimized. For example, over-fitting in model selection, similarly to over-fitting in training, can be significantly harmful when the data sample is small and the population of hyper-parameters to be tuned is large. Similarly, under the assumption that further data are unavailable, possible solutions to the over-fitting the model selection criterion may be analogous to the solutions for the over-fitting the training criterion which has been tried and tested.

4   UCI Benchmark Datasets Used in Empirical Demonstrations

In this section, we use eleven benchmark data sets from the UCI machine learning repository (Bache and Lichman 2013) to examine the problem of over-fitting in model selection for Gaussian processes regression. Table 1 shows the details of the datasets, including the number of features, and test patterns for each dataset.

Table 1. Details of data sets used in empirical comparison.

Data set	Training patterns	Testing patterns	Number of replications	Input features
Airfoil self noise	1353	150	100	5
Community crime	1792	199	100	99
Concrete	927	103	100	8
dat	203	22	100	2
Energy Efficiency	692	76	100	8
Fertility	90	10	100	8
Housing	456	50	100	13
Istanbul Stock Exchange	483	53	100	8
Mpg	359	39	100	7
Servo	151	16	100	4
Yacht Hydrodynamics	278	30	100	6

4.1   Results and Discussion

In order to examine whether the problem of over-fitting during model selection is encountered with Gaussian processes regression, we find both mean squared error (MSE) and negative log marginal likelihood (nLZ) of seven kernel functions over a suite of eleven benchmark datasets. MSE is found based on the test set as a performance evaluation criteria, while nLZ is evaluated over the training set and used as a model selection criteria. Afterwards, the Friedman test is used to determine whether there are statistically significant differences in either MSE or nLZ for different covariance functions. This test is illustrated by critical difference diagrams (Friedman test with Post-Hoc test) (Demšar 2006), which shows the average ranks of seven kernels, as shown in Fig. 1.

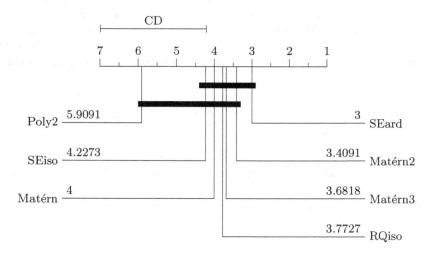

Fig. 1. Critical difference diagram showing the average ranks of seven kernels with using mean squared error (MSE)

This diagram shows the bold bars that joins the lines, such that if two or more lines (representing models with different covariance functions) are joined by a bar, it means these models are not statistically significantly different from each other. It clearly shows that only poly2 is statistically worse than SEard, in terms of generalisation performance, and the remaining differences are non-statically significant.

Figure 2 shows the average ranks of seven kernels with using negative log marginal likelihood. For the majority of the benchmarks, the lowest negative log-likelihood is obtained using SEard which is not surprising because it has more hyper-parameters. However, this is not a good result since SEard does not always give the minimum MSE compared to SEiso. This is called "over-fitting in model selection". In other words, when we have such a problem the negative log-likelihood is no longer a good indication of performance of the model. Indeed,

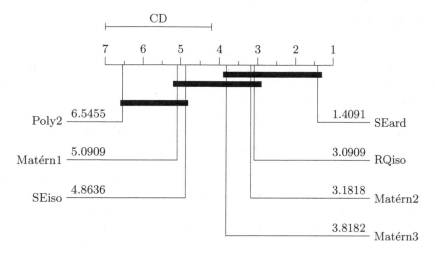

Fig. 2. Critical difference diagram showing the average ranks of seven kernels with using negative log marginal likelihood (nLZ)

the SEiso kernel is a special case of SEard kernel because both are squared exponential function. Thus, we should always obtain better negative log likelihood for SEard than SEiso simply because of having a lot of different parameters to be changed. On the other hand, sometimes the choice of hyper-parameters will result in a model over-fitting the model selection criteria or it may result in under-fitting the data rather than over-fitting it. The significantly lower negative log marginal likelihood of the SEard covariance over the SEiso is not reflected in the statistically insignificant difference in generalisation performance.

Figure 2 shows that SEiso is not significantly worse than SEard, while having fewer hyper-parameters. This is interesting result because it suggests that unlike classification datasets investigated by Cawley and Talbot (2010), the regression data sets are less susceptible to be over-fitting in model selection. Although, there is a great difference between SEard and the rest of the kernels used, SEard still performs well in terms of MSE. This suggests that over-fitting is still a problem but not as much as a problem in classification. In brief, we found that SEard kernel is better than most other kernels including SEiso according to the marginal likelihood, but it is not clearly the best according to MSE on the test datasets, and this is an indication of over-fitting problem. It is worth mentioning that the datasets used in this study were all rather small, however there are algorithms for large scale GP as it is described in the GPML web page by Williams and Rasmussen (2006), but the problem with over-fitting the model selection is most apparent with small datasets, hence there is unlikely to be a significant problem for larger datasets.

5   Conclusion

The contribution of this paper is to find whether the problem of over-fitting in model selection takes place with Gaussian processes regression, both mean squared error (cross validated MSE) and negative log marginal likelihood (nLZ) were found for seven kernel functions over a suit of eleven benchmark datasets. The negative log marginal likelihood is the model selection criteria that can be optimized, whereas the MSE is the test criteria. Afterwards, Friedman test was used to determine whether there is a statistically significant difference in either MSE or nLZ for different covariance functions. For the majority of the benchmarks, the lowest negative log marginal likelihood was obtained using SEard kernel which is not surprising because it has more hyper-parameters. We found that SEard kernel was clearly better than other kernels including SEiso according to the marginal likelihood, but it was clearly not the best according to MSE on the test datasets, and this is an indication of over-fitting problem. This is because the negative log marginal likelihood is the model selection criteria thus it is always decreasing and MSE is getting worse or not improving. We conclude that over-fitting is still a problem in GPs regression but not as much as a problem in GPs classification.

References

Abramowitz, M.: Handbook of mathematical functions. Am. J. Phys. **34**(2), 177 (1966)

Bernardo, J., Berger, J., Dawid, A., Smith, A., et al.: Regression and classification using Gaussian process priors. Bayesian Stat. **6**, 475 (1998)

Blum, M., Riedmiller, M.A.: Optimization of Gaussian process hyperparameters using Rprop. In: ESANN (2013)

Briegel, T., Tresp, V.: Dynamic neural regression models, Collaborative Research Center 386, Discussion Paper 181(2000)

Cawley, G.C., Talbot, N.L.C.: Preventing over-fitting during model selection via Bayesian regularisation of the hyper-parameters. J. Mach. Learn. Res. **8**, 841–861 (2007)

Cawley, G.C., Talbot, N.L.C.: On over-fitting in model selection and subsequent selection bias in performance evaluation. J. Mach. Learn. Res. **11**, 2079–2107 (2010)

Csató, L.: Gaussian processes: iterative sparse approximations. Ph.D. thesis, Aston University (2002)

Demšar, J.: Statistical comparisons of classifiers over multiple data sets. J. Mach. Learn. Res. **7**(January), 1–30 (2006)

Dietterich, T.: Overfitting and undercomputing in machine learning. ACM Comput. Surv. (CSUR) **27**(3), 326–327 (1995)

Do, C.B.: Gaussian processes (2007). http://see.stanford.edu/materials/aimlcs229/cs229-gp.pdf. Accessed 28 June 2014

Duvenaud, D.K., Lloyd, J.R., Grosse, R.B., Tenenbaum, J.B., Ghahramani, Z.: Structure discovery in nonparametric regression through compositional kernel search. In: ICML, vol. 3, pp. 1166–1174 (2013)

Girard, A., Murray-Smith, R.: Gaussian processes: prediction at a noisy input and application to iterative multiple-step ahead forecasting of time-series. In: Murray-Smith, R., Shorten, R. (eds.) Switching and Learning in Feedback Systems. LNCS, vol. 3355, pp. 158–184. Springer, Heidelberg (2005). doi:10.1007/978-3-540-30560-6_7

Joshi, P.: Overfitting in machine learning (2013). https://prateekvjoshi.com/2013/06/09/overfitting-in-machine-learning/

Lichman, M.: UCI machine learning repository (2013). http://archive.ics.uci.edu/ml

MacKay, D.J.C.: Comparison of approximate methods for handling hyperparameters. Neural Comput. **11**(5), 1035–1068 (1999)

Rasmussen, C.E.: Gaussian processes in machine learning. In: Bousquet, O., Luxburg, U., Rätsch, G. (eds.) ML -2003. LNCS, vol. 3176, pp. 63–71. Springer, Heidelberg (2004). doi:10.1007/978-3-540-28650-9_4

Shi, J.Q., Choi, T.: Gaussian Process Regression Analysis for Functional Data. CRC Press, Boca Raton (2011)

Snelson, E.: Tutorial: Gaussian process models for machine learning. Gatsby Computational Neuroscience Unit, UCL (2006)

Walter, G., Augustin, T.: Bayesian linear regression — different conjugate models and their (in)sensitivity to prior-data conflict. In: Kneib, T., Tutz, G. (eds.) Statistical Modelling and Regression Structures, pp. 59–78. Springer, Heidelberg (2010)

Williams, C.K., Rasmussen, C.E.: Gaussian processes for machine learning. **2**(3), 4 (2006). The MIT Press. ISBN 0-262-18253-X

Machine Learning-as-a-Service and Its Application to Medical Informatics

Ahmad P. Tafti[1](✉), Eric LaRose[1], Jonathan C. Badger[1], Ross Kleiman[2], and Peggy Peissig[1,2](✉)

[1] Biomedical Informatics Research Center,
Marshfield Clinic Research Institute, Marshfield, WI 54449, USA
{pahlavantafti.ahmad,peissig.peggy}@mcrf.mfldclin.edu
[2] Computation and Informatics in Biology and Medicine,
University of Wisconsin-Madison, Madison, WI 53706, USA

Abstract. Machine learning as an advanced computational technology has been around for several years in discovering patterns from diverse biomedical data sources and providing excellent capabilities ranging from gene annotation to predictive phenotyping. However, machine learning strategies remain underused in small and medium-scale biomedical research labs where they have been collaboratively providing a reasonable amount of scientific knowledge. While most machine learning algorithms are complicated in code, theses labs and individual researchers could accomplish iterative data analysis using different machine learning techniques if they had access to highly available machine learning components and powerful computational infrastructures. In this contribution, we provide a comparison of several state-of-the-art Machine Learning-as-a-Service platforms along with their capabilities in medical informatics. In addition, we performed several analyses to examine the qualitative and quantitative attributes of two Machine Learning-as-a-Service environments namely "BigML" and "Algorithmia".

1 Introduction

Every single day, an enormous amount of data (almost 2.5 quintillion bytes) is produced by a variety of data sources, such as social medias, blogs, physical sensors, digital cameras, financial transaction records, electronic health records, and GPS signals [1]. Processing such large amount of raw data in an efficient manner is a difficult problem without the help of scalable strategies that can turn data into high quality information in a timely fashion. Advanced computational technologies, including machine learning and cloud computing could be combined to accurately and efficiently process the huge amount of data generated on a daily basis.

Machine learning (ML), embedded in statistical-computational theory, assists computers to mimic the human brain in developing prediction models by using sample data and/or past experiences. Machine learning is applied to four common problems: (1) classification, (2) clustering, (3) regression, and (4) rule

© Springer International Publishing AG 2017
P. Perner (Ed.): MLDM 2017, LNAI 10358, pp. 206–219, 2017.
DOI: 10.1007/978-3-319-62416-7_15

extraction [2]. Classification problems focus on assigning labels to new unlabeled data, while clustering divides data into different groups based on similarities or other structural measures [3]. Regression statistically estimates the relationships between variables and is used in modeling, while rule extraction algorithms are used to discover propositional rules or relationships between attributes in the data [2,3]. Big data services have evolved to address these ML problems and now offer on-demand access to a shared pool of configurable and computational "Software-as-a-Service" (SaaS) technologies. Several contributions, such as Pop [4] Kumar et al. [5], and Nguyen [6] discuss the hype and hope of distributed and SaaS-based ML solutions. Inspired by these studies, software developers have entered the development of distributed ML libraries (e.g., the works established by Cano et al. [7], Kraska et al. [8], Gillick et al. [9], and Xing et al. [10]). Many different ML on the cloud services are now available which, aside from lending theoretical contributions to other fields, have demonstrated successful application in health informatics [11,12], computer vision [13], genomic studies [14], face recognition [15], medical diagnosis [16], and speech recognition [17–19]

There have been several ML-as-a-Service platforms, such as "Amazon Machine Learning", "IBM's Watson Analytics", "BigML", "FICO Analytic Cloud", "Google Cloud Prediction API", "Microsoft Azure Machine Learning Studio", "Algorithmia", and "DataRobot". While some of them have been commercialized, some others could be freely and/or inexpensively available to the data science community.

We limit this study to a set of two ML-as-a-Service environments which is providential to small and medium-scale laboratories with small funding opportunities. The goal of the current work is to assess the state-of-the-art ML-as-a-Service technologies to provide a better understanding of its application for use in the medical informatics community. We briefly summarize our **main contributions** as follows:

- We explain the state-of-the-art of ML-as-a-Service technologies which have been developed over the years to tackle the problem of big data distributed ML. To the best of the authors' knowledge, this contribution is the first to review different ML-as-a-Service environments and discuss their capabilities and shortcomings to medical informatics.
- We identify 10 ML-as-a-Service environments, and address several challenges which could be resolved with recent advances in SaaS-based architectural design.
- We perform several experimental analyses to examine a set of qualitative and quantitative attributes of two ML-as-a-Service environments, namely "BigML" and "Algorithmia", and compare their accuracy and efficiency with "Weka"; a very well-known ML tool with APIs developed in Java.
- As an important contribution, the current work provides insights and tendencies for the next generation of ML components, and it opens the doors for several interesting directions to track ML-as-a-Service environments in general, and highlight their benefits and applications in big data health informatics.

2   Machine Learning-as-a-Service Environments

ML-as-a-Service is a modern computational delivery architecture that makes ML components available through both human-oriented (e.g., web browsers) and application-oriented (e.g., APIs) interfaces over the Internet. A list of 10 ML-as-a-Service environments are presented below. These services provide reliable APIs for classification, regression, clusters analysis, anomaly detection, and association discovery via an Internet connection.

Amazon Machine Learning is a scalable and high performance ML-as-a-Service environment which provides a comprehensive set of ML functionalities to developers in an easy-to-use fashion. Utilizing several wizards and visualization tools, it facilitates the process of creating prediction models without having to learn complex underlying strategies in ML algorithms. Cloud and big data infrastructures are managed automatically by the environment, and prediction of large amount of data records can be accomplished all at once using the batch processing APIs. *Internet Address*: https://aws.amazon.com/machine-learning/.

Google Cloud Prediction API offers researchers fast, reliable, and highly available ML services in which a large amount of data records are systematically replicated across multiple nodes using Google Cloud Storage. The service can integrate with the Google App Engine, and support using many programing languages, such as Python. Google Prediction API frees us from the overhead of maintaining big data infrastructures and configuring networks, and it assists creating intelligent software application in a timely manner. *Internet Address*: https://cloud.google.com/prediction/.

IBM Watson Analytics as a smart data discovery platform provides advanced analytics and ML components over the Internet, offering automatic data visualization and flexible predictive interfaces on big data and cloud infrastructures in a reliable fashion. *Internet Address*: https://www.ibm.com/analytics/watson-analytics/us-en/.

Microsoft Azure Machine Learning Studio can assist data scientists to develop, share, and publish powerful cloud based ML services using hundreds of built-in packages. It utilizes best-in-class ML algorithms with easy-to-use drag-and-drop interfaces, providing a prediction model as a web service that can be called from any device, anywhere, anytime. *Internet Address*: https://azure.microsoft.com/en-us/services/machine-learning/.

BigML is a big data ML platform which provides an easy-to-use graphical interface and simple strategies to embed predictive models into software applications using RESTful APIs. The modest pricing makes it attractive to small and medium-scale groups who need the benefits associated with scalable ML services without high-priced costs and implementation delays. BigML algorithms can be combined together very flexibly to implement higher-level algorithms that provide better predictive models. *Internet Address*: https://bigml.com/.

Yottamine Analytics allows for creating predictive models with reliable and scalable cloud computing infrastructures. ML services provided by Yottamine Analytics are very straightforward for the data scientist. Users can connect to Yottamine Predictive web services employing the R programming language through YottamineR package. *Internet Address*: https://yottamine.com/yottamine-predictive-web-service.

Algorithmia offers a cloud hosted platform for data scientists to share their contributions, incorporating different types of prediction algorithms into their applications. Hundreds of algorithms are already available addressing the most conceivable tasks including text mining and ML. It allows algorithm developers to join to the community to improve the algorithms development, and the RESTful APIs provided by Algorithmia assists developers to easily build intelligent applications. *Internet Address*: https://algorithmia.com/.

Ersatz is a general purpose web-based platform which offers ML services with support of GPU-based deep learning strategies. It has many components developed to build modern ML pipelines efficiently. It also provides different use cases to tackle a variety of scenarios in biology, medical imaging, and, text analytics. *Internet Address*: http://www.ersatzlabs.com/.

FICO Analytic Cloud provides a wide range of ML, statistical analysis, and optimization services. The cloud platform scales automatically for different workloads so that data sources of virtually any size can be accommodated, serving an increasing number of analytics users with high performance, availability, and security. *Internet Address*: http://www.ficoanalyticcloud.com/.

DataRobot is a cloud hosted ML service which provides most of the predictive types of ML algorithms. It automatically enables feature engineering to select the best features, offering an API for model deployment. It takes best-in-class algorithms from R, Python, H20, Spark, and other sources. The infrastructure is powerful enough for rapid processing, and big data is well supported with certification on Cloudera Enterprise 5. *Internet Address*: https://www.datarobot.com/.

3   Methods

To validate the accuracy, performance, and time efficiency of ML-as-a-Service environments and asses the analytical capabilities, several experiments using real medical data were conducted. We limited this analysis to the two services: "BigML" and "Algorithmia", because they included well-known service-based ML algorithms, and compared to the other ML-as-a-Service environments discussed in this paper, a longer term free account and quota has been offered by them, so small research laboratories and individual data scientists could establish free evaluation studies for a longer period of time. We also compared ML-as-a-Service environments results with a very well-known non service-based machine learning library, namely "Weka" (http://www.cs.waikato.ac.nz/ml/weka/).

3.1   Datasets

Six different datasets were used to analyze and compare "Algorithmia" and "BigML". Four datasets from the SEER Research Data [20] were obtained at http://seer.cancer.gov/data/, and two datasets from KEEL-dataset repository [21] available at http://sci2s.ugr.es/keel/datasets.php. The SEER (Surveillance, Epidemiology, and End Results) includes incidence and population data associated by age, sex, race, and year of cancer diagnosis. The KEEL (Knowledge Extraction based on Evolutionary Learning) dataset repository provides a set of datasets used for different data mining tasks, such as classification and clustering. Each dataset was publicly available. Table 1 shows the detailed attributes of each dataset.

3.2 Machine Learning Algorithms

To verify the generalizability of the ML-as-a-Service platforms, and based on the available algorithms provided by the platforms at the time of writing this paper, we focused on a set of supervised methods. This includes Logistic Regression, Cost Sensitive Classifier, Naïve Bayes, and Ensemble Learning (Bagging employing the Decision Tree). ML algorithms in each ML-as-a-Service environment were compared to the same algorithm in "Weka".

3.3 Testbed

The proposed experimental study on "BigML" and "Algorithmia" were performed using the Mozilla Firefox web browser, Version 48.0.2, with the Internet bandwidth characteristics of: Ping (3 ms), Upload (25.49 Mbps), and Download (46.96 Mbps). All experiments along with a comparative study using the "Weka" library have been accomplished by employing a 64-bit Windows 8.1 operating system on a PC with Intel Core i5-4200M CPU, and 6 GB of RAM.

3.4 Quantitative Analyses

To assess the prediction model performance of "BigML" and "Algorithmia" environments, we compared like ML algorithms and datasets using the following criteria: (1) Accuracy, (2) Precision (3) Recall, and (4) Area Under the Curve (AUC). Accuracy describes the percent of prediction that are correct, and precision refers to the percent of positive prediction that are correct. Recall refers to the percent of positive cases that are detected, while AUC is equal to the probability that a prediction model will rank a randomly chosen positive instance higher than a randomly chosen negative example. In this experiment and to evaluate a prediction model, we utilized 75% of each dataset to train and make a prediction model, and 25% to test the model, without any overlapping between the train and test data. Five fold cross validation was performed to calculate accuracy, precision, recall, and AUC.

3.5 Results

Snapshots of the "BigML" and "Algorithmia" environments are shown in Fig. 1. We applied a set of supervised ML algorithms to each dataset illustrated in

Table 1. Data sets and attributes used in testing "BigML" and "Algorithmia". The first four datasets listed in the table belong to SEER Research Data [20], and include survivability information for four different cancer types (corpus uteri, breast, colon, and lung cancer). In these datasets, survival refers to those patients who have survived 5 years or more. The last two datasets are publicly available at KEEL-dataset repository [21]. The SMML4 dataset includes severity (benign or malignant) of mammographic mass lesion information, and the THYD5 consists of data records associated with patients who are normal or suffer from hyperthyroidism or hypothyroidism. There are no missing values across all datasets. The number of instances indicates the number of data records, and the number of attributes refers to the number of individual features in each dataset.

Dataset ID	Type	Attributes	Number of instances	Number of attributes	Number of classes
CUCSTG0	Multivariate	Nominal, Integer	701,219	19	2
BCSTG1	Multivariate	Nominal, Integer	176,973	19	2
CCSTG2	Multivariate	Nominal, Integer	62,667	19	2
LCSTG3	Multivariate	Nominal, Integer	25,807	19	2
SMML4	Univariate	Nominal, Integer	961	5	2
THYD5	Univariate	Nominal, Real, Integer	215	5	3

Table 1. Several quantitative analyses and comparisons have been performed to analyze the systems. Table 2 presents comparative quantitative results obtained by "BigML" and "Weka" using two supervised machine learning algorithms, such as Logistics Regression, and Ensemble learning employing a bagging methodology with the Decision Tree classifier. Since the Logistic Regression is available in the all three ML systems, the first section of Table 2 presents a comparative result using three ML systems together. For each set of the classifiers, we employed same configurations with equal parameters values (e.g., using L1 regularization for Logistics Regression classifiers offered by "Weka", "BigML", and "Algorithmia"). The small differences between the accuracy, precision, recall, and AUC values obtained by "BigML", "Weka", and "Algorithmia" (Table 2) might be associated to the detailed internal parameters of each algorithm (e.g., the ridge value for L1 regularization in Logistic Regression), or they could come from the way the various computer's processors handle information processing (e.g., CPU architecture). Performing a t-test on AUCs matched by classifier showed no statistical differences. Therefore, the accuracy of "BigML" and "Algorithmia" has been promising, as "Weka" components have been in use for several years in research communities to tackle different ML problems ranging from classification and dimension reduction to clustering and rule extraction.

Table 3 illustrates the results obtained by "Algorithmia" in comparison with "Weka" using two supervised ML algorithms. The same configuration setup has been used for classifiers in "Algorithmia" and "Weka". As seen in Table 3, the same prediction analysis results have been achieved by "Algorithmia" and "Weka".

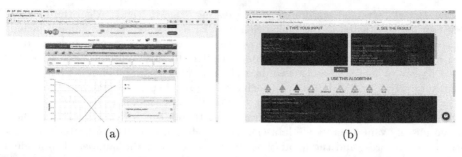

(a)   (b)

Fig. 1. (a) The snapshot of "BigML" environment when it was implementing Logistics Regression. (b) The snapshot of "Algorithmia" environment while it was performing a Naïve Baye classification task.

Table 2. A comparative study of "BigML", "Algorithmia", and "Weka" using two classification algorithms, including Logistic Regression and Ensemble Learning. Since Logistic Regression is available in three discussed ML systems, the first part of the table shows the results obtained by all of them together. The second part of the table presents results achieved by "BigML" and "Weka" using the Ensemble Learning.

Dataset ID	ML system	ML algorithms	Results			
			Accuracy	Precision	Recall	AUC
CUCSTG0	BigML	Logistic Regression	94.16%	90.01%	89.36%	0.849
	Weka & Algorithmia	Logistic Regression	93.71%	91.63%	88.72%	0.844
BCSTG1	BigML	Logistic Regression	93.03%	81.22%	70.17%	0.817
	Weka & Algorithmia	Logistic Regression	93.02%	81.59%	70.03%	0.822
CCSTG2	BigML	Logistic Regression	82.19%	81.81%	78.10%	0.793
	Weka & Algorithmia	Logistic Regression	81.98%	81.16%	78.21%	0.791
LCSTG3	BigML	Logistic Regression	72.59%	72.60%	72.19%	0.722
	Weka & Algorithmia	Logistic Regression	72.11%	72.09%	72.24%	0.718
SMML4	BigML	Logistic Regression	81.33%	81.62%	81.50%	0.817
	Weka & Algorithmia	Logistic Regression	81.39%	81.27%	81.54%	0.819
THYD5	BigML	Logistic Regression	94.35%	96.85%	87.78%	0.839
	Weka & Algorithmia	Logistic Regression	94.21%	96.66%	88.19%	0.841
CUCSTG0	BigML	Ensemble Using Bagging	93.16%	84.59%	79.11%	0.831
	Weka	Ensemble Using Bagging	93.73%	85.06%	79.74%	0.837
BCSTG1	BigML	Ensemble Using Bagging	92.83%	83.40%	77.81%	0.834
	Weka	Ensemble Using Bagging	92.98%	84.73%	78.09%	0.839
CCSTG2	BigML	Ensemble Using Bagging	81.13%	81.19%	81.17%	0.795
	Weka	Ensemble Using Bagging	81.54%	81.30%	81.52%	0.798
LCSTG3	BigML	Ensemble Using Bagging	71.25%	70.89%	70.88%	0.703
	Weka	Ensemble Using Bagging	70.44%	70.41%	70.43%	0.697
SMML4	BigML	Ensemble Using Bagging	81.02%	79.32%	79.16%	0.797
	Weka	Ensemble Using Bagging	81.10%	79.64%	79.71%	0.808
THYD5	BigML	Ensemble Using Bagging	93.21%	92.82%	84.60%	0.822
	Weka	Ensemble Using Bagging	93.58%	93.04%	84.91%	0.836

Table 3. A comparative study of "Algorithmia" and "Weka" using Naïve Bayes and Cost Sensitive Classifier. For the Cost Sensitive Classifiers, we utilized Naïve Bayes as the underlying classification algorithm, employing same parameters as well as the same cost matrices in each experiment.

Dataset ID	ML systems	ML algorithm	Results			
			Accuracy	Precision	Recall	AUC
CUCSTG0	Algorithmia & Weka	Naïve Bayes	91.87%	92.31%	92.11%	0.848
BCSTG1	Algorithmia & Weka	Naïve Bayes	91.65%	95.70%	91.82%	0.839
CCSTG2	Algorithmia & Weka	Naïve Bayes	80.49%	83.19%	82.04%	0.835
LCSTG3	Algorithmia & Weka	Naïve Bayes	70.88%	74.29%	74.71%	0.730
SMML4	Algorithmia & Weka	Naïve Bayes	80.14%	80.89%	82.47%	0.818
THYD5	Algorithmia & Weka	Naïve Bayes	90.66%	91.82%	89.73%	0.840
CUCSTG0	Algorithmia & Weka	CostSensitiveClassifier	91.51%	92.69%	91.02%	0.825
BCSTG1	Algorithmia & Weka	CostSensitiveClassifier	92.71%	96.00%	90.14%	0.840
CCSTG2	Algorithmia & Weka	CostSensitiveClassifier	83.24%	85.37%	81.48%	0.856
LCSTG3	Algorithmia & Weka	CostSensitiveClassifier	72.13%	76.40%	76.09%	0.770
SMML4	Algorithmia & Weka	CostSensitiveClassifier	80.19%	83.55%	82.64%	0.836
THYD5	Algorithmia & Weka	CostSensitiveClassifier	82.58%	91.77%	94.53%	0.861

Figure 2(a) shows the running time analysis of "BigML" environment and "Weka" library using different classifiers on different datasets. The running time analysis of "Algorithmia" and "Weka" is also presented in Fig. 2(b). The running time here includes dataset loading and transformation plus training and testing the prediction models. By employing the proposed dataset, the experiments shows that the average running time of "BigML" and "Algorithmia" were almost six times faster than "Weka" in a local machine.

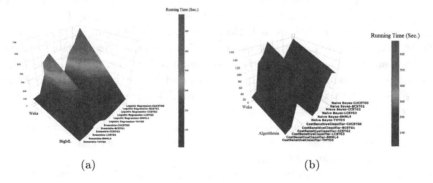

(a)   (b)

Fig. 2. (a) A comparative study of running time by utilizing "BigML" and "Weka". (b) Running time comparative study by using "Algorithmia" and "Weka".

4   Discussion

The application of machine learning in medical informatics has the potential to impact nearly every conceivable subdomain. Machine learning has been used for text mining (gene annotation [22,23], protein annotation [24], name entity recognition [25], literature mining [26]), genomics and proteomics [27,28], EHR-based phenotyping [29,30], image analysis (medical image segmentation [31–33], 3D microscopy vision [34]), system biology (metabolic pathways [35]), microarray data, and image analysis [36] purposes to name a few. With all of these possible applications, particularly in the Big Data era, scalability, performance, and ease of use quickly become important considerations when determining what methods to utilize for analysis. ML-as-a-Service can provide a secure environment with high performance, batch processing capabilities, low cost, and interactive tools designed for ease of use. To this end, we propose that ML-as-a-Service must be considered when deciding what environment to use for ML tasks.

Since there are a wide variety of choices when it comes to implementing ML algorithms, the computational needs of a given task must be considered in the selection process. Table 4 shows a comparison of both price and performance based on the environments discussed in this paper. For very small tasks a local PC environment may suit requirements. For medium size tasks that require more CPU and Memory, a Virtual Machine (VM) or ML-as-a-Service environment becomes the best option. As the task grows even larger, private deployments and other scalable options become necessary. For many tasks options like "BigML" or "Algorithmia" become very useful as they provide a solution that requires minimal setup with above minimal resources and low cost.

Consider performance in regard to ML done in a local environment compared to ML done in a service-based environment. Performance is important because as the ML task increases in size or complexity the requirements of the system needed to run it also increases. In a local platform, performance will be limited to the physical resource available, and if the physical limitations of the hardware are reached, increasing them can be both costly and time consuming. In a local VM environment, the resources are a little more scalable, but pricing increases. In contrast, many of these limitations can be mitigated in a service platform because of the increased availability to resource. As discussed earlier, many of the service platforms provide scalable environments with cost associated with the resources that are used. Data used for medical informatics research continues to grow both in quantity as well as points of observation making a scalable environment even more ideal. This can also become important when you are trying to select the appropriate ML algorithm, as a scalable environment can allow you to run concurrent ML tasks against your data.

ML-as-a-Service environments also offer options for distributed processing through parallel processing and support of big data tools that utilize distributed architectures. Both of these factors will decrease the processing time required and increase the time available to interpret results. This becomes prudent when the machine learning task has many predictions to calculate. The distributed

Table 4. Comparison of price and performance based on the different environments discussed in this paper.

Solution	Setup costs and cost model	CPU/memory	Disk space	Scalability	Security
Local	Personal time	Low	Low	No	High
Local VM	Personal time Server administrator time Administrative maintenance: ($\sim 1,200 - 10,000$/server)/year[a]	Mid	Mid	Yes	High
BigML	Free	Mid	Low	No	Mid
BigML Private deployment Single (multiple)	Configuration: 10,000(20,000) License-30,000(70,000)/year Maintenance - 6,000(14,000)/year Support fee - 9,000(21,000)/year	Mid (High)	Mid (High)	Yes	High
BigML Private deployment Site wide (world wide) [Deployed via Amazon web services]	Configuration: 50,000(100,000) License - 500,000(1,500,000)/year Maintenance - 100,000(300,000)/year Support fee - 150,000(450,000)/year + AWS fees	Unlimited	Unlimited	Yes	Mid
Algorithmia	Credit(s) per API call +, 1 cred- it/sec. Compute time, 10,000 credits = \$1.00 Free forever - Free,(5,000 credits monthly) Pay as you go - starting at \$20, + Purchase credits on demand Enterprise - Per case basis	Mid (High)	Low	No	Mid

[a]Many considerations must go into this price estimate including OS, Backups, Cores, Memory, Storage, etc.

architecture can also take advantages of some of the big data packages available to help break apart the ML task to improve overall performance time.

Cost is another factor that must be considered when choosing a machine learning environment. Pricing models vary significantly when you consider the many options that you have. Table 4 identifies that the price models vary both in cost as well as how they are implemented. In general, depending on things like available resources, number of users, and size of your requests cost will fluctuate. Cost starts low with local and some ML-as-a-Service environments coming in at no extra cost, but increase to be much higher with ML-as-a-Service enterprise environments.

Security is another important factor to consider when working with sensitive information, and especially with medical informatics data. Much of this data will include protected health information and must remain confidential. The cloud based options provide less security than a private network would. To this end, de-identification of the data should be done before transferring and storing data onto one of the service based environments. The data is protected in the transfer process through SSL connections as well as at rest via encryption on these platforms. However, security is still general tighter on a closed network system.

Another consideration necessary for medical informatics research would be flexibility. While all environments provide a level of flexibility, further evaluation can be broken down into three main areas: data source support, feature engineering, and model selection. We suggest that the ML-as-a-Service environment provides optimized flexibility in all three of these areas.

Medical data can come from many different data sources. Data can often be found in a variety of formats including Weka's ARFF file types or other flatter representations like CSV or plaintext. In both the local and VM environments the task would fall on the researcher to normalize the data into a format that could be used by whatever analysis tools they have selected. Many of the service based environments (including "BigML" and "Algorithmia") are set up to handle these different types of data sources without further transformation. Furthermore, the service based environments allow for multiple data files to be uploaded and saved to your account for re-use and management. For these reasons, ML-as-a-Service provides the best environment for diverse data sources.

Feature engineering can often become an overwhelming task due to the diversity of medical data. In both the local and VM environments, this task falls on the researcher entirely to determine the appropriate filtering through whatever tools they have set up to do so. In our analysis, the service based environments provide different tools such as visualization, feature counts for filled, missing and unknown values, and also suggestions as to what features may not be good candidates based on frequency and homogeneity of the data to help address this task. You can then use this analysis to create a dataset for ML as many times as necessary to develop the best set of features. We suggest that ML-as-a-Service provides the necessary tools to evaluate your features without having to install more software to do manual frequency analysis. Finally selecting the correct ML model is very important to your final results, especially with complex medical data. As discussed earlier, a scalable environment provides mechanisms to run multiple ML tasks concurrently which can help to facilitate selection. We see this capability with the service based environments. However, the number of algorithms available to a single service environment will likely be more limited than what can be set up in a local or VM environment. Utilizing multiple service based environments the risk can be mitigated as different services will provide different algorithms without extra software installation. Furthermore, the service based environments provide many good visualizations and metrics to assist in the evaluation of the resulting models. For these reasons, we suggest that ML as a service provides the best environment for ML model selection.

ML-as-a-Service as a type of cloud computing architecture is a distributed model in which ML components and the underlying computational resources are hosted by a service provider. The services are typically accessed using both human-oriented (e.g., web browsers) and application-oriented (e.g., APIs) interfaces. The advantages of ML-as-a-Service to the medical informatics research community could be listed as follows: (1) Less physical computational resources required, (2) Scalability as diverse data records and number of individual users grow, (3) High availability as an Internet connection is all that required,

(4) Resiliency since the data scientists' data resides with the ML-as-a-Service provider, and (5) Batch processing capabilities. Shortcomings would be: (1) Latency issue as data and also ML components are stored in service provider side, so it may take more transactional time as compared to traditional approaches, (2) Data transformation issue in which transferring very large data files over the Internet might be a difficult and slow task, and (3) Security which is the most concern for healthcare informatics where sensitive data and processes are to be entrusted to a third-party service provider, then issues such as identity and access management required to be addressed significantly.

5   Conclusion and Future Work

The amount of medical data being digitally collected on a daily basis is expanding, and as a result, the science of data analysis and processing must be more advanced to enable biomedical researchers to turn this large body of structured and unstructured data into high quality information and knowledge. Computer scientists have entered the development of ML-as-a-Service environments to tackle the problem of pattern and information discovery form large-scale data sources in an efficient and scalable fashion with ease of use. ML-as-a-Service greatly expands the potential capacity to generate new knowledge from big biomedical data, assisting individual researchers to accomplish iterative data analysis using different ML algorithms with powerful computational resources, so they can advance biomedical knowledge dissemination. However, using such services may require considerable attention to privacy concerns to protect confidential medical and clinical data. Based on the unique architecture of ML-as-a-Service environments and their core functionalities, they have a remarkable impact on healthcare informatics. Our future work will be focused on adding more practical experiments with most of the ML-as-a-Service environments by utilizing a variety of big datasets and ML algorithms.

References

1. IBM (2017). https://www-01.ibm.com/software/data/bigdata/what-is-big-data.html
2. Alpaydin, E.: Introduction to Machine Learning. MIT Press, Cambridge (2014)
3. Michalski, R.S., Carbonell, J.G., Mitchell, T.M.: Machine Learning: An Artificial Intelligence Approach. Springer, Heidelberg (2013)
4. Pop, D.: Machine learning and cloud computing: survey of distributed and SaaS solutions. arXiv preprint arXiv:1603.08767 (2016)
5. Kumar, A., Kiran, M., Prathap, B.R.: Verification and validation of mapreduce program model for parallel k-means algorithm on hadoop cluster. In: 2013 Fourth International Conference on Computing, Communications and Networking Technologies (ICCCNT), pp. 1–8. IEEE (2013)
6. Nguyen, T.: Machine learning on the cloud for pattern recognition (2016)
7. Cano, I., Weimer, M., Mahajan, D., Curino, C., Fumarola, G.M.: Towards geo-distributed machine learning. arXiv preprint arXiv:1603.09035 (2016)

8. Kraska, T., Talwalkar, A., Duchi, J.C., Griffith, R., Franklin, M.J., Jordan, M.I.: MLbase: a distributed machine-learning system. In: CIDR, vol. 1, pp. 1–2 (2013)
9. Gillick, D., Faria, A., DeNero, J.: Mapreduce: distributed computing for machine learning, Berkley, 18 December 2006
10. Xing, E.P., Ho, Q., Dai, W., Kim, J.K., Wei, J., Lee, S., Zheng, X., Xie, P., Kumar, A., Yu, Y.: Petuum: a new platform for distributed machine learning on big data. IEEE Trans. Big Data **1**(2), 49–67 (2015)
11. Herath, D.H., Wilson-Ing, D., Ramos, E., Morstyn, G.: Assessing the natural language processing capabilities of IBM Watson for oncology using real Australian lung cancer cases. In: ASCO Annual Meeting Proceedings, vol. 34, p. e18229 (2016)
12. Guidi, G., Miniati, R., Mazzola, M., Iadanza, E.: Case study: IBM Watson analytics cloud platform as analytics-as-a-service system for heart failure early detection. Future Internet **8**(3), 32 (2016)
13. Agrawal, H.: CloudCV: deep learning and computer vision on the cloud. Ph.D. thesis, Virginia Tech (2016)
14. Evani, U.S., Challis, D., Jin, Y., Jackson, A.R., Paithankar, S., Bainbridge, M.N., Jakkamsetti, A., Pham, P., Coarfa, C., Milosavljevic, A., et al.: Atlas2 cloud: a framework for personal genome analysis in the cloud. BMC Genomics **13**(6), 1 (2012)
15. Li, C., Tan, Y., Wang, D., Ma, P.: Research on 3D face recognition method in cloud environment based on semi supervised clustering algorithm. Multimed. Tools Appl. 1–19 (2016)
16. Jiang, G., Fan, M., Li, L.: A cloud platform for remote diagnosis of breast cancer in mammography by fusion of machine and human intelligence. In: SPIE Medical Imaging, p. 97890S. International Society for Optics and Photonics (2016)
17. Chang, Y.-S., Hung, S.-H., Wang, N.J.C., Lin, B.-S.: CSR: a cloud-assisted speech recognition service for personal mobile device. In: 2011 International Conference on Parallel Processing, pp. 305–314. IEEE (2011)
18. Assefi, M., Wittie, M., Knight, A.: Impact of network performance on cloud speech recognition. In: 2015 24th International Conference on Computer Communication and Networks (ICCCN), pp. 1–6. IEEE (2015)
19. Assefi, M., Liu, G., Wittie, M.P., Izurieta, C.: An experimental evaluation of Apple Siri and Google speech recognition. In: Proceedings of the 2015 ISCA SEDE (2015)
20. Surveillance, Epidemiology, and End Results (SEER) Program (www.seer.cancer.gov) Research Data (1973–2012): National Cancer Institute. DCCPS, Surveillance Research Program, Surveillance Systems Branch, April 2015. Based on the November 2014 submission
21. Alcalá, J., Fernández, A., Luengo, J., Derrac, J., García, S., Sánchez, L., Herrera, F.: Keel data-mining software tool: data set repository, integration of algorithms and experimental analysis framework. J. Mult. Valued Log. Soft Comput. **17**(2–3), 255–287 (2010)
22. Hayes, W.S., Borodovsky, M.: How to interpret an anonymous bacterial genome: machine learning approach to gene identification. Genome Res. **8**(11), 1154–1171 (1998)
23. Rätsch, G., Sonnenburg, S., Srinivasan, J., Witte, H., Müller, K.-R., Sommer, R.-J., Schölkopf, B.: Improving the caenorhabditis elegans genome annotation using machine learning. PLoS Comput. Biol. **3**(2), e20 (2007)
24. Baldi, P., Brunak, S.: Bioinformatics: The Machine Learning Approach. MIT Press, Cambridge (2001)
25. Zhou, G., Zhang, J., Jian, S., Shen, D., Tan, C.: Recognizing names in biomedical texts: a machine learning approach. Bioinformatics **20**(7), 1178–1190 (2004)

26. Singhal, A., Simmons, M., Lu, Z.: Text mining for precision medicine: automating disease-mutation relationship extraction from biomedical literature. J. Am. Med. Inform. Assoc. ocw041 (2016)
27. Ashley, D.M., Gupta, S., Tran, T., Wei, L., Lorgelly, P.K., Thomas, D.M., Fox, S.B., Venkatesh, S.: Machine-learning prediction of cancer survival: a prospective study examining the impact of combining clinical and genomic data. In: ASCO Annual Meeting Proceedings, vol. 33, p. 6521 (2015)
28. Käll, L., Canterbury, J.D., Weston, J., Noble, W.S., MacCoss, M.J.: Semi-supervised learning for peptide identification from shotgun proteomics datasets. Nature Methods 4(11), 923–925 (2007)
29. Wu, J., Roy, J., Stewart, W.F.: Prediction modeling using ehr data: challenges, strategies, and a comparison of machine learning approaches. Med. Care 48(6), S106–S113 (2010)
30. Peissig, P.L., Costa, V.S., Caldwell, M.D., Rottscheit, C., Berg, R.L., Mendonca, E.A., Page, D.: Relational machine learning for electronic health record-driven phenotyping. J. Biomed. Inform. 52, 260–270 (2014)
31. Bardosi, Z., Granata, D., Lugos, G., Tafti, A.P., Saxena, S: Metacarpal bones localization in x-ray imagery using particle filter segmentation. arXiv preprint arXiv:1412.8197 (2014)
32. de Bruijne, M.: Machine learning approaches in medical image analysis: from detection to diagnosis (2016)
33. Malakooti, M.V., Tafti, A.P., Naji, H.R.: An efficient algorithm for human cell detection in electron microscope images based on cluster analysis and vector quantization techniques. In: 2012 Second International Conference on Digital Information and Communication Technology and it's Applications (DICTAP), pp. 125–129. IEEE (2012)
34. Tafti, A.P., Holz, J.D., Baghaie, A., Owen, H.A., He, M.M., Yu, Z.: 3DSEM++: adaptive and intelligent 3D SEM surface reconstruction. Micron 87, 33–45 (2016)
35. Patel, K.G., Welch, M., Gustafsson, C.: Leveraging gene synthesis, advanced cloning techniques, and machine learning for metabolic pathway engineering. In: Van Dien, S. (ed.) Metabolic Engineering for Bioprocess Commercialization, pp. 53–71. Springer, Cham (2016)
36. Brown, M.P.S., Grundy, W.N., Lin, D., Cristianini, N., Sugnet, C.W., Furey, T.S., Ares, M., Haussler, D.: Knowledge-based analysis of microarray gene expression data by using support vector machines. Proc. Natl. Acad. Sci. 97(1), 262–267 (2000)

Anomaly Detection from Kepler Satellite Time-Series Data

Nathaniel Grabaskas[✉] and Dong Si

Computing and Software Systems Department, University of Washington Bothell, Bothell, USA
{ngrab,dongsi}@uw.edu

Abstract. Kepler satellite data is analyzed to detect anomalies within the short cadence light curve using traditional statistical algorithms and neural networks. Windowed mean division normalization is presented as a method to transform non-linear data to linear data. Modified Z-score, general extreme studentized deviate, and percentile rank algorithms were applied to initially detect anomalies. A refined windowed modified Z-score algorithm was used to determine "true anomalies" that were then used to train both a Pattern Neural Network and Recurrent Neural Network to detect anomalies. For speed in detection, trained neural networks have the clear advantage. However, the additional tuning and complexity of training means that unless speed is the primary concern traditional statistical methods are easier to use and equally effective at detection.

Keywords: Recurrent neural network · Pattern neural network · Modified Z-Score · General extreme studentized deviate · Percentile rank · Time-Series data · Anomaly detection

1 Introduction

The Kepler satellite has been in operation since 2009 continuously looking at a star field in the Cygnus-Lyra region [9]. In 2013, two of its reaction wheels failed preventing it from looking at one continuous section of space. It was re-tasked in 2014 to look at a pre-defined section for 80 day periods and labeled as the K2 mission. During this time, it has been measuring the brightness of stars using Simple Aperture Photometry (SAP).

The mission of this project is to detect exoplanets that are in orbit around distant stars using the transit method. This method can be simply stated: by watching a star continuously a dimming in the brightness of the star can be observed as the exoplanet passes in front of it [6]. When a possible transit is detected, it is labeled as a Kepler Object of Interest (KOI). Typically, multiple instances of the transit need to be observed before it will be labeled as a confirmed exoplanet. A transit can be considered an anomaly or an outlier in the dataset.

A series of observances on a single star are placed in a file and labeled as containing a KOI or no known KOI, but individual observances lack ground truth labels. Related work [2] used the entire light curve file as one data instance, therefore using the KOI or non-KOI of the file as a ground truth label and trained neural networks to predict whether a file contains a KOI with roughly 85% accuracy. Our work differs in that we attempt to identify the individual observations that indicate a KOI.

© Springer International Publishing AG 2017
P. Perner (Ed.): MLDM 2017, LNAI 10358, pp. 220–232, 2017.
DOI: 10.1007/978-3-319-62416-7_16

We did not find any related work that used the same approach to detect anomalies on the Kepler data. Improving exoplanet detection and analysis is addressed annually at NASA's Sagan workshop and is not considered to be the main contribution of this paper. The contribution of our work is a detailed analysis of the challenges and techniques to overcome non-linear data, unbalanced classes, and lack of classification labels on time-series data.

This paper highlights two methods operating on the data from the Kepler satellite and used both supervised and unsupervised machine learning to detect these transits. Method one, used the statistical anomaly detection methods of modified Z-score, General Extreme Studentized Deviate (GESD), and percentile rank as a starting point and are modified to fit time-series data. Method two, used the identification from the traditional algorithms and trained neural networks as a different method of detecting potential transits.

In the next section of this paper we discuss the methods of pre-processing, traditional algorithms, and neural networks used. In Sect. 3, we will show the results obtained from the various methods. Section 4 offers a discussion on these results and finally what future work is still needed.

2   Methods

2.1   Pre-processing

Data Acquisition. NASA releases Kepler light curve data in the form of Flexible Image Transport System (.fits) files every 90 days. These files are hosted on the Mikulski Archive for Space Telescopes (MAST). All data used in this project is from KOI_Q16_short (contains KOI files) [7].

'Long' vs 'Short' cadence refers to the frequency of readings. In long cadence light curves the reading is taken every 15 min, and in short cadence light curves the reading is every one minute [2]. Long cadence files contain a limited number of samplings over 90 days; short cadence files contain many more samplings but vary in the period they cover. Short cadence files were used in these experiments due to the greater number of samplings and the higher likelihood of distinct anomalies being present.

Data Forms. Short cadence light curve data will be label as two general forms: "horizontal" or "curved". Horizontal or linear data is easier to detect anomalies on as they will be above or below the mean. In Fig. 1, the anomaly A1 is easily distinct from the dataset mean. Any data that is not horizontal is considered curved or non-linear and presents more challenges because anomalies can be between the trough and crest of a curve which would put them around the mean. In Fig. 2, the anomaly A2 is nearly identical to the mean of the dataset and thus more difficult to detect.

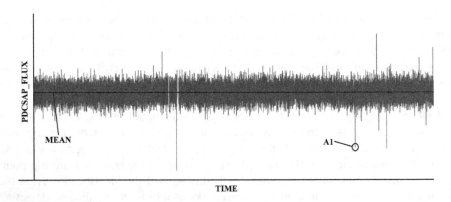

Fig. 1. Horizontal Raw PDCSAP_FLUX data (normalized using mean division) from a short cadence light curve containing a KOI. Anomaly A1 is easily distinct from the mean.

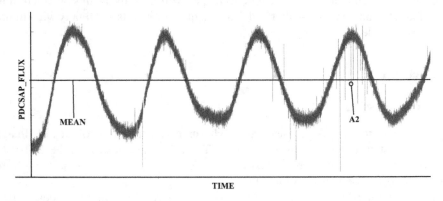

Fig. 2. Curved Raw PDCSAP_FLUX data (normalized using mean division) from a short cadence light curve containing a KOI. Anomaly A2 is difficult to distinguish from the mean.

Data Reduction. Raw light curve data contains 20 unique attributes and each file consists of 5,000–50,000 entries. When attempting to detect transits the SAP_FLUX (Simple Aperture Photometry Fluctuation) and PDCSAP_FLUX (Pre-search Data Conditioning SAP Fluctuation) attributes are the main indicators. SAP_FLUX is the flux in electrons-per-second contained in the optimal aperture pixels collected by the Kepler Satellite [2].

All attribute fields except for SAP_FLUX and PDCSAP_FLUX are removed. Additionally, any instances of zero for either of these fields are removed as this will corrupt the detection methods.

Finally, a time indicator field is needed. The time attribute that is contained within the .fits file may contain gaps. This again causes issues when analyzing data, to mitigate this a new time field is added for the entire dataset that increments by 1 for each instance.

Data Normalization. The data must be normalized as the brightness of each star varies and this causes drastic differences in the SAP_FLUX and PDCSAP_FLUX fields. This variance will cause difficulty when training on star set and then testing on another star set. Data was normalized to 1 using the mean division method, Eq. 1, where t is time and μ is the mean.

$$PDC_{normalized} = \frac{PDCSAP(t)}{\mu} \tag{1}$$

Additionally, a Windowed Mean Division Normalization method (WMDN) is introduced. The entire dataset is broken down into subsets of the specified range and normalized using the mean for only the subset instead of the entire dataset. This allows the curved data to be flattened without loss of the anomalies, Fig. 3.

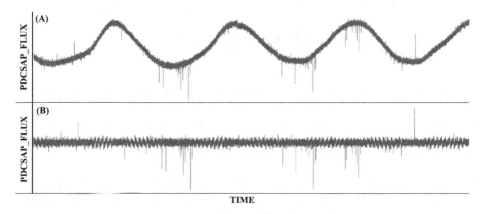

Fig. 3. (A) Shows curved light curve data before WMDN. (B) Shows curved light curve data flattened after WMDN with anomalies still easily visible.

2.2   Traditional Methods

Windowed Modified Z-Score (WMZ). The Z-score is a way of determining how many standard deviations a given value is from the mean of the dataset. Equation 2 is used where $PDCSAP(t)$ is the given data instance, μ is the mean, and σ is the standard deviation.

$$Z_{score} = \frac{(PDCSAP(t) - \mu)}{\sigma} \tag{2}$$

Using the concept of the Z-score, the modified Z-score algorithm [4] reduces the effect of anomalies on μ and σ by replacing μ with a sample median and σ with the Median of Absolute Deviations (MAD).

$$Z_{modified\_score} = 0.6745 \times \frac{(PDCSAP(t) - \mu_{sample})}{MAD} \tag{3}$$

This method was modified using a windowed technique where the function is run on sub-ranges of the data. Once the entire set of data has been run through the function, anomalies from each sub-range or window were compiled into a complete list. The two variables used to alter this algorithm were the window size and threshold (Z-score) used to determine if an instance was an anomaly.

Windowed General Extreme Studentized Deviate (WGESD). GESD is a variation on the Grubb's test that allows for more than one anomaly to be detected. There can be up to N number of anomalies in the data. Equation 4 is performed N times and the dataset shrinks by a size of one on each iteration as the value that maximized the function is removed. Once N number of values are found the critical value is used to determine which of the X values are anomalies [10].

$$X_{max} = \frac{MAX_i \left| (X_i - \mu_{sample}) \right|}{\sigma_{sample}} \tag{4}$$

In Eq. 4, μ is the mean, and σ is the standard deviation. This method was also modified using a windowed technique where the function is run on a sub-range of the data until the entire set has been run through the function. Once complete, anomalies from each window are compiled into a complete list. The two variables used to alter this algorithm were the number of anomalies being checked for (N) and the window size.

Percentile Rank (PERC). PERC is determined by finding the frequency distribution of the dataset and selecting the specified furthest extremes or least frequent values [5]. The percentile rank is found using Eq. 5.

$$PERC_{rank} = \frac{PDC_{pos}}{N} \times 100\% \tag{5}$$

PDC_pos is the position of the score within the distribution and N is the total number of instances in the dataset. Two factors determine percentile rank anomalies: the number of instances in the dataset and the extreme values. The two variables used to alter this algorithm were the low and high extreme boundaries.

2.3   Neural Networks

Artificial Neural Network (ANN). ANNs, as the name suggests, are inspired by the network of neurons in the human brain. This paradigm seeks to mimic the way the neurons in the human brain not only process information but also learn as new information is given. This is accomplished with the creation of a heavily interconnected network of neurons that together can solve problems through learning. The training of an ANN is perhaps the most important aspect. Information with known answers (ground truth) is fed through the ANN and the input weight and activation of each neuron is adjusted to move the network state towards being able to correctly identify all the training data answers [1].

Figure 4(a) shows a diagram of a more standard neural network. In this example the value from each of the three inputs is fed to each neuron of the hidden layer. Through training, the weights of each input and the threshold for the activation function will be adjusted. It will continue to learn and adjust if better results can be achieved. When no variation can achieve better results the network is considered trained.

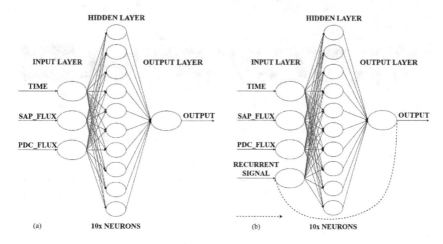

Fig. 4. (a) Shows a 10-neuron pattern neural network. (b) Shows a 10-neuron recurrent neural network where the output from the activation function is fed back in as a fourth input.

Many aspects of a network can be modified such as activation function, training function, number of neurons in a hidden layer, how many hidden layers are used, how each layer connects, etc. This creates endless variations of ANNs, the key is to determine which configuration is likely to have the best results for any given problem.

In this project two types of networks were used: Pattern Neural Networks (PNN) and Recurrent Neural Networks (RNN). Our PNNs used scaled conjugate gradient backpropagation and the RNNs used Levenberg-Marguardt as the training functions.

Recurrent Neural Networks. RNNs differ from standard neural networks by allowing the output of the hidden layer neurons to feedback and serve as the input to themselves or other neurons. This allows the network to use history as a way of understanding the sequential nature of the data [8]. We used a RNN where the output from the activation function was an additional input to the hidden layer, Fig. 4(b).

Data Replication. Unbalanced classes heavily influence the training of neural networks. In the short cadence files, there is an average of 30 anomalies per 50,000 instances. Resampling is one method to counter unbalanced classes, but due to the limited number of anomalies in the dataset we chose to use replication as a method to counter this data imbalance problem [3]. Anomalies were first detected using a traditional algorithm and then replicated.

On horizontal data, the anomalies were multiplied a given number of times, randomly redistributed throughout the dataset, and then run through the detection algorithm a second time. Results are shown in Fig. 5.

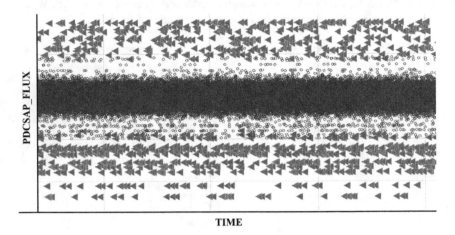

Fig. 5. Approximately 1,500 anomalies were created in the horizontal dataset to train neural networks. Anomalies are displayed as arrows.

On curved data, random redistribution would not work as a random redistribution of the points could place them within the normal range. A method of detecting the anomalies using a traditional algorithm, moving both forward and backward within the time sequence a set number of steps, and then replacing the existing data with the anomaly data a given number of times was used.

Additional Factors. Having no ground truth labels for the anomalies in our data makes determining which network is most accurate more difficult. Two additional factors in our detection are which algorithm and parameters are used for the training data and the test data.

Measurement Terminology. Precision - Positive Predictive Value (PPV) uses the anomalies detected by traditional algorithms as a base and measures how many of the instances labeled as an anomaly are true anomalies.

$$Precision = PPV = \frac{TP}{TP + FP} \tag{6}$$

For Eq. 6, TP (true positive) is the number of anomalies correctly detected, FP (false positive) is the number incorrectly labeled as anomalies.

Sensitivity – True Positive Rate (TPR) again uses the traditional algorithm's labels as a base and measures how many of the true anomalies were correctly identified.

$$Sensitivity = TPR = \frac{TP}{TP + FN} \tag{7}$$

For Eq. 7, *TP* (true positive) is the number of anomalies correctly detected, *FN* (false negative) is the number of true anomalies incorrectly labeled as normal.

3   Results

3.1   Traditional Algorithm Anomaly Detection

Traditional algorithms were run on short cadence light curves to detect anomalies within the dataset. The horizontal dataset contained 106,839 instances and the curved dataset contained 60,376 instances.

Performance.   The window function allows the modified Z-score algorithm to execute on only a smaller portion of the data at a time and thus it is unaffected by the horizontal or curved shape of the datasets. Figure 6 columns 2–5, demonstrate using WMZ with a constant threshold of 4.5 and varying the window size. The other columns show where the threshold was varied. Window size has a small effect on the number of anomalies that are detected. However, threshold has a more profound effect; the lower the value the more sensitive the algorithm.

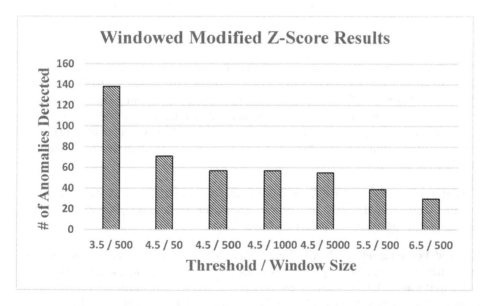

Fig. 6. Number of anomalies detected when both the threshold and window size are alterred using WMZ.

Because of the same window function as seen with WMZ, the WGESD algorithm can operate on both horizontal and curved datasets. In Tables 1 and 2 WGESD is compared against WMZ. When using WGESD the number of anomalies being checked for has a small effect. Inversely from WMZ, the window size has a larger effect on the number of anomalies detected.

Table 1. WGESD and PERC results are compared with WMZ (4.5, 500) detected anomalies on horizontal data.

Method	Time	# Detected	Precision	Sensitivity
WMZ - 4.5/500	1.55 s	57	–	–
WGESD - 2/50	3.38 s	162	34.57%	98.25%
WGESD - 2/500	0.34 s	64	78.13%	87.72%
WGESD - 5/500	0.50 s	72	79.17%	100.00%
WGESD – 10/500	0.73 s	73	78.08%	100.00%
WGESD - 2/5000	0.14 s	26	100.00%	45.61%
PERC - 0.03/99.97	0.008 s	64	79.69%	89.47%

Table 2. WGESD and PERC results are compared with WMZ (4.0, 750) detected anomalies on curved data.

Method	Time	# Detected	Precision	Sensitivity
WMZ - 4.0/750	0.94 s	29	–	–
WGESD - 2/500	0.20 s	35	65.71%	79.31%
WGESD - 2/750	0.13 s	27	85.19%	79.31%
WGESD - 2/1000	0.10 s	17	94.12%	55.17%
PERC - 0.03/99.97	0.006 s	36	11.11%	13.79%

With the PERC method, the greater the number of instances in the dataset the more anomalies that will be detected. And the closer the extreme values are to the center the more anomalies that will be detected. PERC method requires tuning and knowledge of where anomalies are in the dataset to ensure their detection. This algorithm also performs poorly on curved data, Table 2 last row. It is only able to detect anomalies that are outside the mean range of the entire dataset, but anomalies that are within the mean range are unable to be detected.

Parameters. In Tables 1 and 2 these parameter formats are used. WMZ – threshold/ window size, WGESD - # of anomalies to check for/window size, and PERC – lower percentage boundary/higher percentage boundary.

Ground Truth Label. Both WMZ and WGESD have seemingly accurate detection of anomalies within the data. WMZ was chosen as the algorithm to use for ground truth labels as it was slightly less sensitive to data points that were on the edge of being normal points. From Table 1, this oversensitivity was why WGESD consistently had a higher number of anomalies detected and showed lower precision. From Table 2 both sensitivity and precision were affected using WGESD.

3.2   Neural Network Anomaly Detection

Parameters. In Tables 3 and 4 these parameter formats are used. RNN X:Y where X indicates the lowest time delay and Y indicates the highest time delay. For example, 1:5 means that the input is delayed at 1, 2, 3, 4, and 5 time steps backwards.

Table 3. Neural network results on the horizontal dataset.

#	Type	Time	Neurons	Comparison	Training data	Precision	Sensitivity
1	PNN	0.031 s	10	3.5, 400	4.5, 400	89.86%	43.97%
2	PNN	0.043 s	32	4.1, 750	4.1, 750	91.94%	100.00%
3	PNN	0.096 s	100	4.1, 750	4.1, 750	79.41%	94.74%
4	PNN	0.785 s	1000	4.1, 750	4.1, 750	71.05%	94.74%
5	RNN 1:1	0.054 s	10	3.5, 400	4.5, 400	97.06%	46.81%
6	RNN 1:3	0.081 s	10	4.1, 750	4.1, 750	91.80%	98.25%
7	RNN 1:3	0.267 s	20	4.1, 750	4.1, 750	93.33%	98.25%
8	RNN 1:3	0.345 s	32	4.1, 750	4.1, 750	91.67%	96.49%
9	RNN 1:3	0.142 s	[10,10]	4.1, 750	4.1, 750	90.16%	96.49%

Table 4. Neural network result on curved dataset. All comparison and training data used a window size of 500.

#	Type	Neurons	Comp	Train data	Replication	Precision	Sensitivity
1	PNN	10	4.5	4.5	30, 2	100.00%	9.30%
2	PNN	1000	4.5	4.5	30, 10	100.00%	25.58%
3	PNN	5000	4.5	4.5	30, 10	100.00%	20.93%
4	PNN	[32,32]	4.5	4.5	30, 10	100.00%	23.26%
5	RNN 1:2	10	4.5	4.5	30, 10	100.00%	16.28%
6	RNN 1:3	10	4.5	4.5	30, 10	100.00%	20.93%
7	RNN 1:3	10	4.5	4.5	40,10	100.00%	23.26%
8	RNN 1:5	32	4.5	4.5	30, 10	100.00%	6.98%

All comparison and training data was processed using the WMZ algorithm. The first number indicates the threshold and the second number indicates the window size.

In horizontal replication used for Table 3, the anomalies were replicated 50 times, redistributed randomly in the dataset, and then labeled again using WMZ with a window size of 400 and threshold of 6.0. In Curved replication, the first number is how many times the data was replicated and the second number is how many time steps outward each iteration was replicated.

Number of neurons [X, Y] indicates that multiple hidden layers were used and how many neurons were in each layer. Otherwise only one hidden layer was used.

Performance. With the properly adjusted parameters, both the PNN and RNN achieved high precision and sensitivity when detecting anomalies in horizontal data. The PNN achieved both higher sensitivity and precision, but both operated well.

Table 3 method 2 and method 7 demonstrate the highest performances achieved. Figure 7(A) compares WMZ and RNN anomalies detected and shows that both detections are nearly identical.

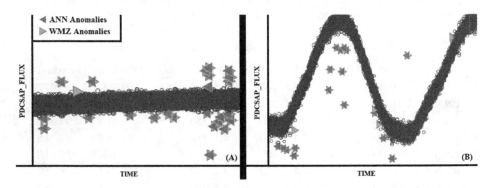

Fig. 7. (A) Comparison of anomalies detected on horizontal data using Table 3 method #7. (B) Comparison of anomalies detected on curved data using Table 5 method #4.

Both PNNs and RNNs had difficulty operating on the curved dataset. No variation of parameters that we tested allowed the network to detect the anomalies that were within the curve of the data. Only anomalies underneath the overall range of the data were detected. Our best performance, Table 4 method 7, was only able to detect 23.26% of the anomalies within the dataset.

However, we found using the WMDN to flattened the curved data into horizontal data was indeed able to preserve the anomalies and allow ANNs to be effectively trained for detection. The data was flattened and used to train the ANN, the test data was also flattened and run through the ANN for detection, finally the anomalies detected from the flattened test data were plotted back into the original curved data. Table 5 shows that with some tuning of the WMDN training and test data window size and replication parameters we achieved high precision and sensitivity. Method 4 is compared against WMZ in Fig. 7(B).

Table 5. Neural network result on flattened curved data. All comparison and training data used WMZ with a window size of 500 and threshold of 4.5.

#	Type	Neurons	Train win	Test win	Replication	Precision	Sensitivity
1	**PNN**	32	150	500	50, 5.0	89.47%	79.07%
2	**PNN**	32	200	400	50, 4.5	95.00%	90.48%
3	**PNN**	32	400	400	50, 4.0	94.74%	85.71%
4	**PNN**	32	400	400	50, 3.65	90.70%	92.86%
5	**PNN**	64	400	400	50, 3.65	90.70%	92.86%
6	**PNN**	100	400	400	50, 3.65	79.59%	92.86%
7	**RNN 1:3**	10	400	400	50, 3.65	90.91%	95.24%
8	**RNN 1:5**	10	400	400	50, 3.65	90.91%	95.24%
9	**RNN 1:3**	20	400	400	50, 3.65	90.91%	95.24%

4   Discussion

Initially, all traditional algorithms were only able to operate on horizontal data. Through the introduction of the windowed method both modified Z-score and general extreme studentized deviate achieved successful results on curved data. The PERC method is weak on curved data due to anomalies often being hidden within the mean of the entire dataset. The PERC algorithm is much faster than any other method and this is due to the small number of calculations necessary to identify anomalies. This advantage is also what makes this algorithm less versatile.

Ground truth labels would eliminate both the comparison and training data parameters and leave only the replication parameters to be adjusted. Which would allow for more confident conclusions regarding the effectiveness of ANNs.

ANNs achieved similar performance on horizontal data and once trained they can execute up to 36 times faster than WMZ and up to 11 times faster than WGESD. Unfortunately, neither of these networks were effective in detection on curved data. The variance between brightness of stars and locations of the anomalies is too complex for the simpler ANNs that we used in these experiments.

We found using windowed mean division normalization allowed the curved data to be transformed to horizontal without loss of the anomaly data. This enabled ANNs to detect anomalies for the curved datasets without adding additional complexity to the neural networks. Manipulating the data and tuning the combination of variables (window size, threshold, replication parameters, etc.) adds complexity and time to the training of ANNs. But once trained, ANNs offer a speedup in detection when compared to the traditional algorithms

5   Conclusion and Future Work

In this paper, we show that robust anomaly detection from Kepler satellite time-series data was accomplished using various algorithms. The type of method and parameters can be adjusted to fit any type of horizontal or curved short cadence light curve file. For detection speed, trained ANNs have the clear advantage. However, the additional tuning and complexity of training ANNs means that unless speed is the primary concern traditional windowed statistical methods are easier to use and equally effective at detection.

Given the fact that neural networks are highly non-linear end-to-end "black boxes", it is possible that other types of ANN models may be able to detect the anomalies within the curve without data modification. Further research on additional methods such as Long-Short Term Memory (LSTM) integrated neural networks is needed.

The prediction of KOIs from the anomalies detected fell outside of the scope of this paper. Initial tests showed that anomalies on short cadence light curves may not be an effective indicator to whether a star is harboring an exoplanet. More research and experiments are needed to find methods of using these anomalies as a predictor to whether a file contains a Kepler Object of Interest.

References

1. Aleksander, I., Morton, H.: An introduction to Neural Computing, 2nd edn. Intl Thomson Computer Pr (T), London (1995)
2. Botros, A.: Artificial intelligence on the final frontier: using machine learning to find new earths. Technical report, Stanford University (2014)
3. Guo, H., Murphey, Y., Feldkamp, L.: Neural learning from unbalanced data. In: Applied Intelligence, vol. 21, pp. 117–128. Kluwer Academic Publishers (2004)
4. Iglewicz, B., Hoaglin, D.: The ASQC basic references in quality control: statistical techniques. In: How to Detect and Handle Outliers, vol. 16 (1993)
5. Lane, D., Scott, D., Hebl, M., Guerra, R., Osherson, D., Zimmer, H.: Introduction to statistics, pp. 29–33. Rice University, University of Houston, Tufts University (2007)
6. Leiner, E.: Other Worlds: Analyzing the Light Curves of Transiting Extrasolar Planets. Wesleyan University, Thesis (2010)
7. Mikulski Archive for Space Telescopes: Space Telescope Science Institute. http://archive.stsci.edu/pub/kepler/lightcurves/tarfiles/
8. Nanduri, A., Sherry, L.: Anomaly detection in aircraft data using recurrent neural networks. In: Integrated Communications Navigation and Surveillance (ICNS) Conference, pp. 5C2 1–8. IEEE Press (2016)
9. NASA: Kepler: About the Mission. https://kepler.nasa.gov/Mission
10. Rosner, B.: Percentage points for a generalized ESD many-outlier procedure. Technometrics **25**(2), 165–172 (1983)

Prediction of Insurance Claim Severity Loss Using Regression Models

Ruth M. Ogunnaike$^{(\boxtimes)}$ and Dong Si

Computing and Software Systems, University of Washington Bothell,
Bothell, WA, USA
{tunrayo,dongsi}@uw.edu

Abstract. The objective of this work is to predict the severity loss value of an insurance claim using machine learning regression techniques. The high dimensional data used for this research work is obtained from All-state insurance company which consists of 116 categorical and 14 continuous predictor variables. We implemented Linear regression, Random forest regression (RFR), Support vector regression (SVR) and Feed forward neural network (FFNN) for this problem. The performance and accuracy of the models are compared using mean squared error (MSE) value and coefficient of determination (Rsquare) value. We predicted the claim severity loss value with a MSE value of 0.390 and a Rsquare value 0.562 using bagged RFR model. In addition where applicable, the final loss value was also predicted with an error of 0.440 using FFNN regression model. We also demonstrate the use of lasso regularization to avoid over-fitting for some of the regression models.

Keywords: Regression · Insurance claim severity · Machine learning · Regularization

1 Introduction

A claim severity can be defined as the amount of loss associated with an insurance claim. The average severity is calculated by dividing the total amount of losses that an insurance company experiences by the number of claims that were made against policies that it underwrites. Loss is the amount paid or to be paid to the claimants under their insurance policy contracts. Currently, the details of computing a forecast of the paid claim loss is complicated [2].

Insurance companies rely on actuaries and the models that actuaries create to predict future claims, as well as the losses that those claims may result in. The models are dependent on a number of factors, including the type of risk being insured against, the demographic and geographic information of the individual or business that bought a policy, and the number of claims that are made. Actuaries look at past experience data to determine if any patterns exist, and then compare this data to the industry at large.

Claim severity loss forecasting has played a major role in determining auto insurance rate and premiums [7]. It is important to obtain accurate estimate of

© Springer International Publishing AG 2017
P. Perner (Ed.): MLDM 2017, LNAI 10358, pp. 233–247, 2017.
DOI: 10.1007/978-3-319-62416-7_17

the losses that could arise from an insurance contract. Also in an event a car accident occurs, an insurance policy holder will prefer a fast and quality service from the insurance company when it comes to processing claims and it can also take a considerable amount of time to settle claims in some cases [5].

To provide quality claim service to millions of policy holders protected by insurance companies, and also to create an accurate forecast that predicts rates and premiums, it is necessary for insurance companies to have automated systems that can accurately predict claim severity loss given a set of input. This research work aim to predict the severity loss value of an insurance claim using continuous and categorical features from previously processed insurance claims.

In the domain of loss prediction model, the work in [10] attempts to use convolution approach to estimate loss severity distribution; using convolution of normal and exponential distribution for modelling a loss distribution of property insurance claims. The work in [12] demonstrates actuarial applications that uses hierarchical models to fit micro-level insurance data consisting of claims policy and payment files to predict loss type and accident frequency in automobile.

Researchers have also tried to use insurance claims data to build loss prediction models used for financial risk assessment in construction projects [9]. The research work uses regression analysis to explore the relationships among independent risk factors such as natural disasters, geographic information, and model construction and the dependent variable (percentage of loss) to build a loss prediction model based on the insurance payout records. A similar work [11] uses regression analysis to build loss estimation models for insurance companies that shows the correlation between post-earthquake damage of structural components to direct financial loss of residential buildings based on post-earthquake damage evidence and obtained a Rsquare value of 0.41 using quadratic regression function.

As far as we are aware, we are the first to publish results from a regression model that directly predicts severity loss value of an insurance claim. However, the work in [4] uses regression models (K-nearest neighbour, support vector regression and feed forward neural networks) to predict the overall cost of energy consumed during various categories of entertainment events. The work in [3] also used regression models to predict real estate property prices. Similarly, the work in [6] uses regression models (boosting, linear regression, support vector machine) to predict stock prices.

In addition, the work in [1] uses regression models to predict the quality of signal transmitted by optical fiber across communication channels. Other works attempted to illustrates how to quantify the inherent uncertainty in fitting claims severity distributions and estimates the cost of high layered factors that contributes to the severity loss of a claim [5].

This study explores the use of four regression machine learning approaches to predict claim severity loss: Support Vector Regression (SVR), Linear Regression, Random Forest Regression (RFR), and Feed Forward Neural Networks (FFNN). The four approaches generated varying prediction error rates and coefficient of determination values (R-square). Of the four approaches used for this study, random forest regression model generated the lowest mean square error.

The rest of this paper is organized as follows; The research methodology is presented in Sect. 2. Section 3 explains algorithm selection & optimization and an evaluation of the achieved results in Sect. 4. Section 6 concludes the paper.

2 Methodology

This section introduces the data set and describes the analyses and preparation (data pre-processing) that occurred before the machine learning approaches were applied. It also outlines features selection, regularization and model building.

Figure 1 describes the overview of our approach. It starts by designing data which involves data transformation, normalization and splitting the data into training, test and validation set. This is followed by feature extraction steps that figures out which subset of selected features are really informative for assessing the predicted value (severity loss). Model selection decides on which function will be used to fit the input data as well as the extraction of model coefficients. Model quality is evaluated using Rsquare, and MSE between the actual and the predicted values. The proposed framework uses a software package called MATLAB which is equipped with tools for training models and enables graphical representation of some of the experimental results.

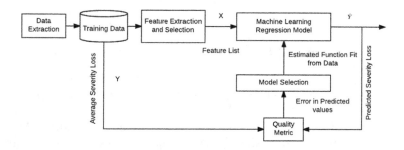

Fig. 1. Work flow of the proposed regression machine learning framework

2.1 Dataset

The dataset includes continuous and categorical features. The target value for this study is the severity loss value of an insurance claim. The dataset contains a total of 188318 records and a total of 130 features (116 categorical features and 14 continuous features). The data was provided by Allstate insurance company on Kaggle website for the purpose of implementing models that predicts a claim severity loss value in order to improve services for their customers.

The original dataset contains normalized continuous predictors while the categorical feature values are alphabet characters of length 1 and 2. The categorical feature values were converted to numeric values ranging from 0 to 205. Figure 2 shows the graph of the original data points in our dataset.

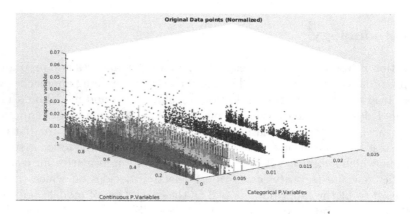

Fig. 2. Original data points showing the predictor and response variable. The x-axis is the average of the normalized value of categorical predictors, y-axis is the average of continuous predictor and z-axis shows the response variable (the severity loss value to be predicted).

2.2   Feature Design and Selection

To identify the important features i.e. the parameter with large variance with the response variable, we used ensemble function to compute the importance of each predictor variable.

Figure 3 shows the importance estimates of the predictor variables. As shown in Fig. 3, approximately 10 predictors have importance estimate value greater than 0.02. Using only the predictor with high importance value does not guarantee a high performance model. Only the predictors of high importance can be used if those predictors explains the variance of the data with a cumulative value of 0.9 ± 0.05. It is easy to identify the number of predictors to be used in training the regression models (the number of predictors that cumulatively explains the data variance).

In addition, least absolute shrinkage and selection operator (LASSO) was used to select important predictor variables. Lasso is a regression analysis method that performs both variable selection and regularization in order to enhance the prediction accuracy and interpretability of the model it produces. Lasso was also introduced to minimize the magnitude of the coefficients of each of the predictor variables in order to avoid over-fitting and solves the problem by;

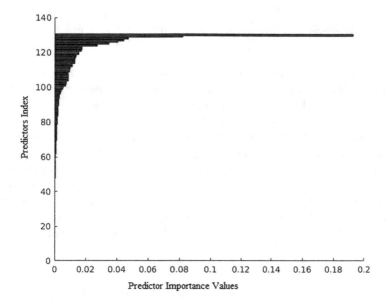

Fig. 3. Importance of predictor variables

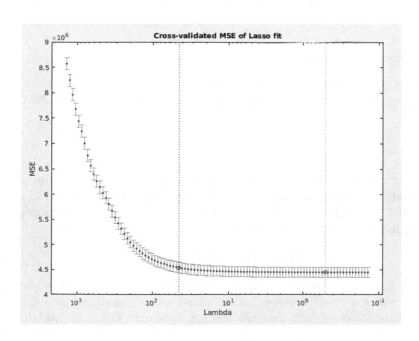

Fig. 4. Cross-validated MSE of Lasso fit (Color figure online)

$$LASSO : min \sum_{i=1}^{m} (y_i - w.x_i - b)^2 + C \sum_{j=1}^{n} \|w_j\|, \tag{1}$$

where x_i is the predictor variable value, y_i is response variable, w is the coefficient of a specific predictor variable and b is the fitted least-squares regression coefficients for a set of regularization coefficients, Lambda.

Figure 4 shows the cross-validated MSE of our lasso fit plot. The plot identifies the minimum deviance point with a green circle and dashed line as a function of the regularization parameter Lambda (λ). The green circle points are the lambda values that minimizes the cross validated MSE and the blue circled points are the lambda values with the greatest amount of shrinkage whose cross-validated MSE is within one standard error of the minimum. We selected the predictors between the blue and the green line, however the performance of the models were really poor. Hence, the primary training uses the entire predictors for training.

3   Algorithm Selection and Implementation

We selected linear regression, support vector regression, random forest regression and feed forward neural networks regression models to predict insurance claim severity loss value.

3.1   Linear Regression

To establish baseline performance, we used linear regression to model the severity loss values, Y, as a linear function of the data, X, the predictor variables.

$$f_w(X) = w_0 + w_1 x_1 + ... + w_m x_m = w_0 + \sum_{i=1}^{m} w_j x_j \tag{2}$$

where w_j are the weights of the features, m is the number of features and w_0 is the weight to the bias term, $x_0 = 1$. The weights of the linear model can be found with the least square solution method, we find the w that minimizes error. Writing in matrix notation, we have:

$$Err(w) = (Y - Xw)^T (Y - Xw) \tag{3}$$

where X is the $n \times m$ matrix of input data, Y is the $n \times 1$ vector of output data, and w is the $m \times 1$ vector of weights. To speed up computation, the weights can be fitted iteratively with a gradient descent approach. Given an initial weight vector w_0, for $k = 1, 2, ..., m$, $w_{k+1} = w_k - \alpha_k \delta \ Err(w_k)/\delta \ w_k$, and end when $|w_{k+1} - w_k| < \epsilon$

Here, the parameter $\alpha_k > 0$ is the learning rate for iteration K. The performance of linear regression is shown in Table 1.

3.2   Support Vector Regression (SVR)

We used the linear and gaussian kernel SVR to predict severity loss. The linear SVR estimates a function by maximizing the number of deviations from the actual obtained targets y_n within the normalized margin stripe. The SVR algorithm is a convex minimization problem that finds the normal vector of the linear function as follows [8].

$$min_{w,\gamma} \ (\frac{1}{2}|w|^2 + C \sum_{n=1}^{N} \gamma_n + \gamma_n^*) \tag{4}$$

where γ_n, γ_n^* are slack variables allowing for errors to cross the margin. The constant $C \geq 0$ determines the trade off between the flatness (minimized of the function and the amount up to which deviations larger than margins stripes are tolerated. The results of the SVR can be found in Fig. 5 and Table 1.

3.3   Random Forest Regression (RFR)

The Random forest regression is an ensemble algorithm that combines multiple Regression Trees (RTs). Each RT is trained using a random subset of the features, and the output is the average of the individual RTs. The sum X of squared X errors for a tree T is:

$$S = \sum_{c \epsilon leaves(T)} \sum_{i \epsilon C} \sum_{i \epsilon C} (y_i - m_c)^2 \tag{5}$$

where $m_c = \frac{1}{n_c} \sum_{i \epsilon C} y_i$, the prediction for leaf c, y_i is the actual loss, T is a regression tree, c are leaves of trees T and S is the sum of squared errors. Each split in the RTs is performed in order to minimize S. The basic RT growing algorithm is as follows:

- **Step 1:** Begin with a single node containing all points. Calculate m_c and S.
- **Step 2:** If all the points in the node have the same value for all the independent variables, then stop. Otherwise, search over all binary splits of all variables for the one which will reduce S the most. If the largest decrease in S would be less than some threshold δ, or one of the resulting nodes would contain less than q points, then stop. Otherwise, take that split, creating two new nodes.
- **Step 3:** In each new node, go back to step 1. One problem with the basic tree-growing algorithm is early termination. An approach that works better in practice is to fully grow the tree (i.e., set $q = 1$ and $\delta = 0$), then prune the tree using a holdout test set.

For our implementation we used the LSBoost and bagged decision trees.

1. **Bagged:** Bootstrap-aggregated (bagged) decision trees combine the results of many decision trees, which reduces the effects of overfitting and improves

generalization. Individual decisions trees tend to overfit. Tree bagger grows the decision trees in the ensemble using a bootstrap samples of the data. It selects a random subset of predictors to use at each decision split in the random forest algorithm.

2. **LSBoost:** The ensemble methods use multiple learning algorithms to obtain better predictive performance that could be obtained from any of the constituent learning algorithms alone.

3.4   Feed-Forward Neural Network

Neural networks (NN) consists of interconnected neurons, or nodes, and have the ability to approximate nonlinear relationships between the input variables and output of a complicated system. Feed Forward Neural Networks (FFNN) are one of the most frequently used NNs for value forecasting and was chosen for this study. Feed forward neural network is composed of an input layer, one or more hidden layers of neurons, and an output layer. Each layer contains a chosen number of neurons, which are then individually interconnected with adaptable weighted connections to neurons in the succeeding layer (with the exception of the output layer). The output of each neuron in the hidden layer is determined using;

$$f_j(x) = \varphi(\sum_{i=1}^{n} w_{ij}x_i + \theta_i) \tag{6}$$

where $f_j(x)$ is the output of the jth neuron, φ is transfer function (such as a Gaussian or sigmoid function), x_i is the ith input to the neuron, w_{ij} is the connection weight between the ith neuron in the input layer and the jth neuron in the hidden layer, and θ_i is the bias or threshold.

The neurons in the output layer also have weighted connections, exclusively with the last hidden layer in the network. Training the network involves adjusting the weights between neurons so that the neural network can produce desirable results when given a set of inputs. A variety of training algorithms can then be used to minimize the network error function. This study uses a feed forward network with a single hidden layer and Levenberg-Marquardt learning algorithm.

4   Results

The regression models used both 5-fold and 10-fold cross validation unless otherwise stated. We analyzed the MSE and Rsquare values of each models with the values ranging from 0 to 1. Smaller values of MSE indicates a better model fit while a larger rsquare value indicates that the model well explains the variability of the response data around its mean.

1. **Mean Squared Error (MSE)** is the most important criterion for this study since the main purpose of the regression models used is prediction. MSE is an

absolute measure of fit and it indicates how closely the predicted values from the model match the response variable (severity loss) the model is intended to predict. Lower values of MSE indicate better accuracy and it is calculated as follows;

$$\frac{1}{n} \sum_{i=1}^{n} \frac{(y_i - \hat{y}_i)^2}{y_i} \tag{7}$$

where n is the number of observations, y_i is the actual loss and $\hat{y}_i$ is the predicted loss.

2. **Coefficient of determination (Rsquare)** indicates the proportionate amount of variation in the response variable y explained by the independent/predictor variables X in the regression model. Rsquare is the square of the correlation between the response values and the predicted response values. A higher rsquare value indicates a better model fit and is calculated as follows;

$$R^2 = 1 - \frac{\sum_{i=1}^{n}(y_i - \hat{y}_i)^2}{\sum_{i=1}^{n}(y_i - \bar{y}_i)^2} \tag{8}$$

where y_i is the actual loss, $\hat{y}_i$ is the predicted loss and $\bar{y}_i$ is the mean of the target value (severity loss).

Table 1 shows the summary of the performance of each model using only categorical predictors, only continuous predictors and using the entire predictors. Tests was also done to identify the effect of using only the predictor that minimizes the mean square error. However, there was drop in the performance of the regression models.

Table 1. Performance (MSE and R-square value) of regression models

Model	All		Categorical		Continuous	
	MSE	R-Square	MSE	R-Square	MSE	R-Square
Random forest (Bagged)	0.391	0.562	0.420	0.527	0.419	0.527
Neural network	0.441	0.519	0.411	0.484	0.441	0.529
Linear regression (Simple)	0.477	0.463	0.491	0.449	0.491	0.4495
SVR (Linear)	0.498	0.439	0.510	0.452	0.546	0.467
Random forest (LSBoost)	0.790	0.108	0.832	0.066	0.832	0.066
Linear regression (CV)	0.899	0.038	0.773	0.131	0.772	0.131
SVR (Gaussian)	0.970	−0.033	0.970	−0.033	0.990	−0.031

4.1   Support Vector Regression

Both linear and Gaussian kernels were used for training the SVR algorithm. 5-fold cross validation was used to train both models and the linear kernel achieved a better performance with a MSE value of 0.498 while the Gaussian kernel MSE value of 0.970. As shown in Table 1, the linear kernel SVR model also explains the variability of the response data around its mean better than the Gaussian kernel model with rsquare values 0.439 and -0.033 respectively.

The overall performance of support vector machine is average. Based on our observations the linear kernel functions fits better with the dataset used. Figure 5 shows the mean squared error of the linear partitioned SVR model across the 5 partitions. The minimum MSE was achieved on the fifth partitioned model. The MSE reported for this model is the average of all the partitioned data models.

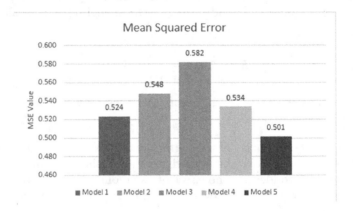

Fig. 5. MSE values of linearly partitioned SVR model

4.2   Linear Regression

Using 5-fold cross validation and lasso regularization that returns a Lambda value that helps to minimize MSE, we obtained MSE value of 0.899. Training the model using predictors with higher coefficients did not improve the performance of the model.

When using a partitioning algorithm to train a model, it randomly divides the data into K which is 5 in this case. It trains the model using $K-1$ partitioned data and test the model using the outstanding partitioned data set. This is done iteratively for each partitions and computes the MSE for each partition. The blue line in Fig. 6 shows the MSE values across the partitioned data model. We also ran a test using a simple linear regression in which all the predictors variable are fit linearly and prediction is done using the coefficient value of each predictor variable and a lower MSE value 0.477 was achieved.

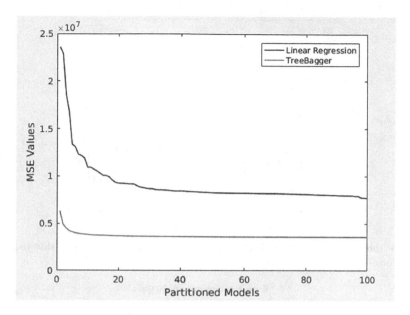

Fig. 6. Graph showing the mean squared error of linear regression model and the bagged regression model

4.3   Random Forest Regression

Bagged random forest and boosted (LSBoost) decision tree were the algorithm used. We used 10-fold cross validation to decide the optimal number of regression trees for both models. For our implementation, we used 100 compact regression trees because we obtained the minimum MSE with that value (100). Bagged RFR model achieved a better performance with MSE value 0.391 and 0.790 for LSBoost. Bagged RFR achieved the best result in terms of its MSE and it also best explains the variability between the response variables around its mean as shown in Table 1. On the other hand, LSBoost model did not perform well compared to other models.

Figure 7 shows a plot of the actual and predicted severity loss of 100 data points from the test data. It can be observed that the residuals (actual - predicted) of some of the data point are small while some have a large residual margin. The residuals of the prediction determines the accuracy of the model. As shown in Fig. 6, the MSE value falls as we move across the bagged regression tress. Each tree predicts the loss value and the average of all the tree's prediction is returned as the final predicted value. Figure 7 shows a major margin between the MSE of the linear regression and the bagged regression model.

4.4   Feed Forward Neural Network

Levenberg-Marquardt back propagation algorithm was used to train the neural network. This algorithm is an efficient technique for calculating the output

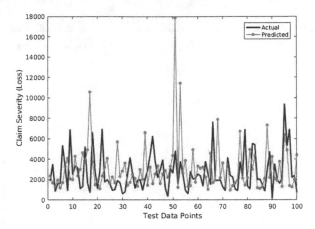

Fig. 7. Graph showing the actual loss value versus the predicted loss value using the random forest (tree bagger) model

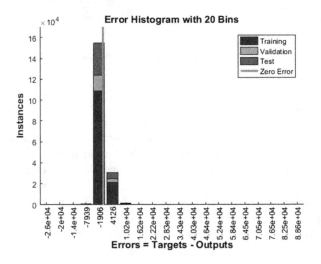

Fig. 8. Error histogram for 20 hidden neurons

variable. The initial weights are configured and random weights are associated with each transition. Using the feed forward propagation, the network progresses further and predicts an output based on these weights. This model achieved the second best performance with MSE value 0.416 and R-square value 0.519 (see Table 1).

The model was trained using all predictor variables and the selected predictor variables (49 predictor variables identified by the lasso regularization) which resulted in a MSE value 0.442. The difference in the performance is insignificant. The model fits up to 75% of the test data and shows the feasibility of neural network regression model (See Fig. 10). This shows further computations can

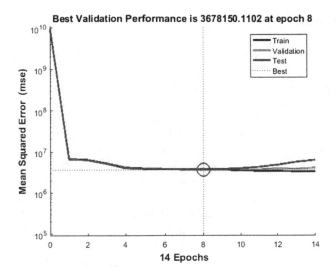

Fig. 9. Mean squared error for 20 hidden layers

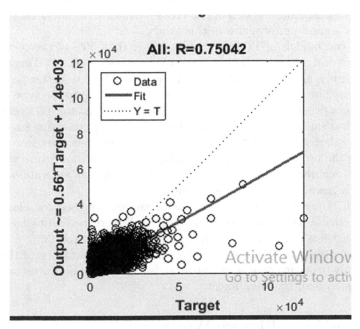

Fig. 10. Regression plot for 20 hidden neurons

improve the performance of this model. Figure 8 shows the Error Histogram for 20 layer neuron. The result is satisfactory as the maximum frequency of instances in the minimum error region denoted by the yellow vertical line in Fig. 8. As shown in Fig. 9, the minimum MSE achieved is 0.441 when normalized.

5   Discussion

Random forest regression and feed forward neural network performed significantly better than the baseline linear algorithms (linear regression and linear SVR). This is possible due to their ability to account for nonlinear interactions between the predictor variables and the severity loss. As shown in Table 1, RFR and neural network has the best overall performance in terms of a minimum mean squared error and coefficient of determination for all test scenarios.

It was observed that, the higher the rsquare value of a model, the smaller the MSE of that model, since the rsquare value of a model indicates how well it explains the variability between the response variable (severity loss) and it surrounding mean. A model is considered a good fit if its rsquare value is high and its mean squared error is low. Models that well explains the variability between its independent variable (predictors) and its dependent variable (severity loss), predicts the response variable more accurately.

Gaussian kernel SVR model has the worst performance for our study and almost performs as a naive model. Having a negative r-square values shows it is totally unfit for our data prediction. Although cross validated models improves performance for random forest models, cross validation on linear regression did not result in a good performance in this study.

The top two models in Table 1, explains more than 50% of the covariance of the predictors and response variable. None of the models performed exceptionally well in this study, although a more refined data could lead to better performance of the models. The original dataset does not give insight into the type of feature being used as a predictor; The features provided where just tagged as continuous and categorical variable. We implemented some feature engineering such as using the lasso regularization and principal component analysis, however these did not improve the prediction errors. Ensembling two models for future work could improve the accuracy of the proposed framework such as ensembling random forest with K-nearest neighbour regression models.

Although K-nearest neighbour are most commonly used for classification tasks, theoretically they can perform equally well in regression tasks since any arbitrary functions can be fitted with a multilayer perceptron [3].

Future efforts could be spent into improving the performance of the models and also apply other machine learning models such as the K-nearest neighbour, fast forest quantilem Bayesian linear.

6   Conclusion and Fiture Work

Our prediction of insurance claim severity loss using categorical and continuous set of features with regression techniques provides a baseline for further studies. Current study results in a mean square error 0.391 and rsquare value 0.562. Although the implemented models needs further improvement, this application can be useful for insurance companies to create severity loss forecast that enables the insurance companies set premium rates and policies.

While this study focused on using different regression machine learning models to predict insurance claim severity loss, only four regression models was used in this study. Future studies can be done to improve the performance of the models and also apply other machine learning models such as the K-nearest neighbour, fast forest quantile and Bayesian linear. A redefined dataset that gives insight to the types of features can also be used. This will aid in feature engineering in addition to the regularization techniques used in this study.

Adopting well-used datasets in machine learning and improving the error by 0.01 with a particular algorithm can be considered a significant breakthrough. Since insurance companies usually involve large monetary claim transactions, improving the prediction error by 0.01 can lead into the development of interesting future applications for insurance companies for computing severity loss that helps to determine premium and insurance rate.

A   Appendix

The dataset and implementation codes used for this project can be found on the link provided https://github.com/ruthogunnaike/lossPredictor.

References

1. Rudra, R., Biswas, A., Dutta, P., Aarthi, G.: Applying regression models to calculate the Q factor of multiplexed video signal based on optisystem. In: 2015 SAI Intelligent Systems Conference (IntelliSys), pp. 201–209 (2015)
2. Cummins, J.D., Griepentrog, G.: Forecasting automobile insurance paid claims using econometric and ARIMA models. Res. Gate J. **1**(3), 203–215 (1985)
3. Nissan, P., Emil, J., Liu, D.: Applied machine learning project 4 prediction of real estate property prices in Montreal. Accessed 6 Dec 2016
4. Andrea, Z., Katarina, G., Miriam, C., Luke, S.: Energy cost forecasting for event venues. In: EPEC 2015 (2015)
5. Glenn, M.: On predictive modeling for claim severity. Sci. Direct J
6. Vatsal, H.S.: Machine learning techniques for stock prediction (2007)
7. Philipe, H., Glenn, G.M.: The calculation of aggregate loss distributions from claim severity and claim count distributions (1993)
8. Smola, A.J., Scholkopf, B.: A tutorial on support vector regression. Statist. Comput. **14**(3), 199–222 (2004)
9. Ryu, H., Son, K., Kim, J.-M.: Loss prediction model for building construction projects using insurance claim payout. J. Asian Archit. Build. Eng. **15**(3), 441–446 (2016)
10. Pak, R.J.: Estimating loss severity distribution convolution approach. J. Math. Statist. **10**(3), 247–254 (2014)
11. Xue, Q., Chen, C.-C., Chen, K.-C.: Damage and loss assessment for the basic earthquake insurance claim of residential RC buildings in Taiwan. J. Build. Appraisal **6**, 213–226 (2011)
12. Frees, E.W., Shi, P., Valdez, E.A.: Actuarial applications of hierarchical insurance claims model. Astin Bull. **39**, 165–197 (2009)

A Spectral Clustering Method for Large-Scale Geostatistical Datasets

Francky Fouedjio$^{(\boxtimes)}$

CSIRO Mineral Resources, Perth, WA, Australia
francky.fouedjiokameni@csiro.au

Abstract. Spectral clustering is one of the most popular modern clustering techniques for conventional data. However, the application of the general spectral clustering method in the geostatistical data framework poses a double challenge. Firstly, applied to geostatistical data, the general spectral clustering method produces clusters that are spatially non-contiguous which is undesirable for many geoscience applications. Secondly, it is limited in its applicability to large-scale problems due to its high computational complexity. This paper presents a spectral clustering method dedicated to large-scale geostatistical datasets in which spatial dependence plays an important role. It extends a previous work to large-scale geostatistical datasets by computing the similarity matrix only at a reduced set of locations over the study domain referred to as anchor locations. It has the advantage of using all data during the computation of the similarity matrix at anchor locations; so there is no sacrifice of data. The spectral clustering algorithm can then be efficiently performed on this similarity matrix at anchor locations rather than all data locations. Given the resulting cluster labels of anchor locations, a weighted k-nearest-neighbour classifier is trained using their geographical coordinates as covariates and their cluster labels as the response. The assignment of clustering membership to the entire data locations is obtained by applying the trained classifier. The effectiveness of the proposed method to discover spatially contiguous and meaningful clusters in large-scale geostatistical datasets is illustrated using the US National Geochemical Survey database.

Keywords: Geostatistics · Large-scale datasets · Spectral clustering · Spatial dependency · Spatial contiguity

1 Introduction

Over the past few years, spectral clustering has become one of the most popular modern clustering approaches for classical data [8, 19, 22, 23, 25]. Spectral clustering is a class of partitional clustering algorithms that uses information contained in the spectrum of a feature similarity matrix to find the underlying clusters of a given dataset. Advantages of using spectral clustering include its flexibility regarding incorporating diverse types of similarity measures, the superiority of its clustering solution compared to traditional clustering algorithms, its

© Springer International Publishing AG 2017
P. Perner (Ed.): MLDM 2017, LNAI 10358, pp. 248–261, 2017.
DOI: 10.1007/978-3-319-62416-7_18

simplicity in implementation with standard linear algebra methods, and its well-established theoretical properties [2,16,20,21,35]. However, in the geostatistical data framework, the application of the general spectral clustering method poses a double challenge.

Firstly, when applied to geostatistical data, the general spectral clustering method produces clusters that are not spatially contiguous because of the assumption that the observations are independent. This is certainly undesirable for many geosciences applications [9–11,24]. Furthermore, the independence assumption does not hold for geostatistical data which often shows properties of spatial dependence. This means that observations close to one another in the geographical space might have similar characteristics. In addition, the mean, the variance, and the spatial dependence structure of the data can be different from one sub-domain to another. Secondly, the widespread development of remote sensing platforms and sensors networks has helped make large-scale geostatistical datasets common place. However, such datasets prevent the application of the general spectral clustering due to its high computational complexity [8,19,22,23,25].

Although a lot of research on spectral clustering for classical data has been done, very few references to spectral clustering in the realm of geostatistical data can be found in the literature [10,24]. A recent adaptation of the general spectral clustering approach has been proposed to deal with geostatistical data, in which spatial dependence plays an important role [10]. It relies on a similarity measure built from a non-parametric kernel estimator of the multivariate spatial dependence structure of the data, emphasizing the spatial correlation among data locations. Despite the ability of this geostatistical spectral clustering approach to provide spatially contiguous and meaningful clusters, it is limited in its applicability to large-scale problems due to its high computational complexity. It requires the construction of a similarity matrix and the computation of the eigen-decomposition of the corresponding graph Laplacian matrix. Both of these two steps are computationally expensive, especially when applying it to large-scale datasets. For applications with a number of data locations on the order of tens of thousands, this geostatistical spectral clustering method begins to become infeasible.

Several methods have been proposed for accelerating the general spectral clustering method when using to large-scale data [1,4,5,7,12,17,18,26–29,32–34,36]. However, they are not suitable in the realm of geostatistical data. Methods consisting in reducing the similarity matrix size [1,5,27,34] in order to efficiently perform the eigen-decomposition step operate directly in the attribute space. They are not suitable in the geostatistical data framework because even if geographical coordinates are considered as attributes, these methods cannot provide spatially contiguous clusters at the end. The spatial dependency of data has to be taken into account in order to ensure the spatially contiguity of the resulting clusters. Methods focusing only on reducing the computational cost of the eigen-decomposition step (e.g., via the Nyström method [7,12,29]) require the original similarity matrix whose the construction is also computationally expensive in the geostatistical data framework [10]. Moreover, the majority of these methods involves somehow the sampling technique which sacrifices a part of the original data; hence the loss of information from the original data.

This paper presents a spectral clustering method dedicated to large-scale geostatistical datasets, in which spatial dependence plays an important role. It extends the geostatistical spectral clustering approach proposed in reference [10] to large-scale problems. The basic idea is to compute the similarity matrix only at a reduced set of locations over the study domain referred to as anchor locations. The proposed method has the advantage of using all data during the computation of the similarity matrix at anchor locations; so there is no sacrifice of data. Then, the spectral clustering algorithm is efficiently performed on this similarity matrix at anchor locations rather than all data locations. Given the resulting cluster labels, a weighted k-nearest-neighbour classifier is trained using the geographical coordinates of anchor locations as predictors and their cluster labels as the response. Finally, the clustering of all data locations is obtained by applying the trained classifier. The effectiveness of the proposed method to discover spatially contiguous and meaningful clusters in large-scale geostatistical datasets is illustrated using the US National Geochemical Survey database.

The remainder of the paper is organized as follows. Section 2 introduces the proposed large-scale geostatistical spectral clustering method. Section 3 illustrates the effectiveness of the proposed method using a case study of the US National Geochemical Survey database consisting of 43,885 observations of eight critical heavy metals. Section 4 provides concluding remarks.

2   Methodology

Consider a set of p standardized variables of interest $\{Z_1, \ldots, Z_p\}$ defined over a continuous study domain $G \subset \mathbb{R}^d (d \geq 1)$, and all measured at a set of large number of distinct locations $\{\mathbf{x}_1, \ldots, \mathbf{x}_n\}$. The objective is to partition these large data locations into spatially contiguous and meaningful clusters so that data locations belonging to the same cluster are more similar than those in different clusters. This section describes the different ingredients required to implement the proposed large-scale geostatistical spectral clustering method.

2.1   Affinity Matrix Construction

Contrary to the traditional spectral clustering, here the construction of the affinity matrix takes into account the spatial dependency of the data. A non-parametric kernel estimator of the multivariate spatial dependence structure of the data described by the direct and cross variograms [30], at two locations $\mathbf{u} \in G$ and $\mathbf{v} \in G$ is given by [9–11]:

$$\widehat{\gamma}_{ij}(\mathbf{u}, \mathbf{v}) = \frac{\sum_{l,l'=1}^{n} K_\epsilon^\star\left((\mathbf{u}, \mathbf{v}), (\mathbf{x}_l, \mathbf{x}_{l'})\right)(Z_i(\mathbf{x}_l) - Z_i(\mathbf{x}_{l'}))(Z_j(\mathbf{x}_l) - Z_j(\mathbf{x}_{l'}))}{2\sum_{l,l'=1}^{n} K_\epsilon^\star\left((\mathbf{u}, \mathbf{v}), (\mathbf{x}_l, \mathbf{x}_{l'})\right)} \mathbb{1}_{\{\mathbf{u} \neq \mathbf{v}\}},$$

$$(1)$$

where $(i, j) \in \{1, \ldots, p\}^2$; $K_\epsilon^\star((\mathbf{u}, \mathbf{v}), (\mathbf{x}_l, \mathbf{x}_{l'})) = K_\epsilon(\|\mathbf{u} - \mathbf{x}_l\|)K_\epsilon(\|\mathbf{v} - \mathbf{x}_{l'}\|)$, with $K_\epsilon(\cdot)$ a non-negative kernel function with constant bandwidth parameter $\epsilon > 0$; $\mathbb{1}$ denotes the indicator function.

According to references [9–11], the kernel function $K_\epsilon(\cdot)$ is chosen as the Epanechnikov kernel whose support is compact, showing optimality properties in density estimation [31]. The use of a compactly supported kernel function reduces the computational complexity. The bandwidth parameter ϵ is chosen so that the support of the kernel function $K_\epsilon(\cdot)$ centered at each data location contains at least 35 observations. Thus, for each data location its distance to the 35th neighbour is computed; then, the maximum of resulting distances is taken as the value of the bandwidth parameter ϵ.

As highlighted in references [9–11], the computation of the set of estimated direct and cross variograms $\{\widehat{\gamma}_{ij}(\cdot, \cdot)\}_{i,j=1}^p$ at all pairs of data locations is very expensive for large-scale datasets. However, one can notice that the nonparametric kernel estimator of the multivariate spatial dependence structure of the data given in Eq. (1) is defined at any pair of locations throughout the study domain G and not only at a pair of data locations. Thus, the idea to reduce the computational burden associated with spectral clustering consists in computing the set of estimated direct and cross variograms $\{\widehat{\gamma}_{ij}(\cdot, \cdot)\}_{i,j=1}^p$ at only a reduced set of $m \ll n$ locations $\{\mathbf{o}_1, \ldots, \mathbf{o}_m\}$ named anchor locations, defined over the study domain. It is important to point out that anchor locations are not necessarily data locations and they are not "representatives" of data locations.

Given the set of estimated direct and cross variograms $\{\widehat{\gamma}_{ij}(\cdot, \cdot)\}_{i,j=1}^p$ defined in Eq. (1), the similarity between two anchor locations $\mathbf{o}_t$ and $\mathbf{o}_{t'}$ $(t, t' = 1, \ldots, m)$ is defined by:

$$s(\mathbf{o}_t, \mathbf{o}_{t'}) = 1 - \frac{1}{\Gamma} \sum_{i,j=1}^p |\widehat{\gamma}_{ij}(\mathbf{o}_t, \mathbf{o}_{t'})|, \tag{2}$$

with $\Gamma = \max_{(t,t') \in \{1, \ldots, m\}^2} \sum_{i,j=1}^p |\widehat{\gamma}_{ij}(\mathbf{o}_t, \mathbf{o}_{t'})|$. The resulting similarity matrix at anchor locations is denoted $\mathbf{S} = [s(\mathbf{o}_t, \mathbf{o}_{t'})]_{t,t'=1,\ldots,m}$.

The affinity matrix at anchor locations is obtained as follows: anchor locations with pairwise similarities are transformed into an undirected weighted graph $\mathcal{G} = (\mathcal{V}, \mathcal{E})$. $\mathcal{V}$ represents the set of vertices representing anchor locations. $\mathcal{E}$ is the set of edges between pairs of vertices, and each edge between two vertices v_t and $v_{t'}$ carries a non-negative weight representing strength of association between vertices. Here, all vertices having non-null similarities are connected each other (full connected graph) and the edge-weights are assigned according to the similarity matrix $\mathbf{S}$. Thus, the affinity matrix at anchor locations simply corresponds to the similarity matrix at anchor locations. This construction is suited according to reference [19] since the similarity measure defined in Eq. (2) itself already encodes local neighbourhoods (through the kernel function $K_\epsilon(\cdot)$ in Eq. (1)). Also, by connecting all anchor locations between them, this construction method has the advantage to be able to detect a same cluster in different parts of the study domain.

Regarding the choice of the set of anchor locations, we want to choose enough number of anchor locations to capture the main features of the spatial variability of data while at the same time reducing it as much as possible to avoid redundant computation. The set of anchor locations is chosen as an irregular grid over the study domain. A non-uniform distribution of the anchor locations allows taking into account a spatially varying sampling density. Thus, regions where there are no data locations will not contain anchor locations. Choosing a very dense grid of anchor locations is useless. Indeed, similarities calculated for two pairs of anchor locations that are very close can be unnecessary and redundant because of their high spatial correlation. On the other hand, using a too sparse grid of anchor locations will give a bad approximation of the spatial variability of data and will deteriorate the predicting performance of the classifier. The practitioner must find a balance between computational efficiency and the clustering performance accuracy.

2.2   Large-Scale Geostatistical Spectral Clustering Algorithm

The construction of the affinity matrix only at anchor locations automatically reduces the computational burden associated with other main steps of the spectral clustering what are the computation of the eigen-decomposition of the corresponding graph Laplacian matrix and the K-means algorithm. Then, the spectral clustering algorithm can be efficiently performed on this affinity matrix at anchor locations rather than all data locations. Given the resulting cluster labels, the idea is to train a classification rule using the geographical coordinates of anchor locations as covariates and their cluster labels as the response. Then, the clustering of all data locations is obtained by applying the trained classifier. To do this end, the weighted k-nearest-neighbour classifier [14] is used. This latter is an essentially model-free method that is appropriate in this spatial context, in addition to being relatively simple. However, other classifiers can be used as well (e.g., support vector machines, radial basis functions).

The weighted k-nearest-neighbour classifier is an extension of the k-nearest-neighbour classifier [14]. In this latter, a new observation is assigned to the class label from the training set that is closest to the new observation, according to the predictors used. The extension is based on that the contribution of each of k neighbours are weighted with respect to the distance to the target point, giving more weight to closer neighbours. More formally, let

$$\{(c_t, \mathbf{o}_t), t = 1, \ldots, m\} \tag{3}$$

be the training set of anchor locations, where $c_t \in \{1, \ldots, q\}$ denotes class (or cluster label) membership and the vector $\mathbf{o}_t^T = (o_{t1}, \ldots, o_{td})$ represents the predictors which are the geographical coordinates. The nearest neighbours are determined through the Euclidean distance. Let $\mathbf{x}$ be a new location, whose cluster label c has to be predicted.

The prediction of the class membership c of location $\mathbf{x}$ is given by the class which shows a weighted majority of the k nearest neighbours $\{\mathbf{o}_{(t)}\}_{t=1,\ldots,k}$:

$$\hat{c} = \arg\max_{r\in\{1,\ldots,q\}} \left(\sum_{t=1}^{k} K_\lambda(\|\mathbf{x} - \mathbf{o}_{(t)}\|) \mathbb{1}_{\{c_{(t)}=r\}} \right). \tag{4}$$

where $K_\lambda(.)$ a non-negative kernel function with bandwidth $\lambda > 0$ (e.g., Gaussian, Epanechnikov). The bandwidth λ is chosen according to the distance of the $(k+1)$th neighbour. The number of nearest neighbours k is determined by a cross-validation procedure consisting in the minimization of the misclassification error [15].

The proposed large-scale geostatistical spectral clustering performs the following steps:

1. define the anchor locations grid $\{\mathbf{o}_1, \ldots, \mathbf{o}_m\}$;
2. compute the affinity matrix $\mathbf{S}$ at anchor locations, according to Sect. 2.1;
3. compute the degree matrix $\mathbf{D} = [\sum_{t'=1}^{m} s(\mathbf{o}_t, \mathbf{o}_{t'})]_{t=1,\ldots,m}$;
4. compute the graph Laplacian matrix $\mathbf{D}^{-1/2}\mathbf{S}\mathbf{D}^{-1/2}$;
5. compute the q largest eigen-values of $\mathbf{D}^{-1/2}\mathbf{S}\mathbf{D}^{-1/2}$ and form the matrix $\mathbf{F} \in \mathbb{R}^{n\times q}$ whose columns are the associated q first eigenvectors of $\mathbf{D}^{-1/2}\mathbf{S}\mathbf{D}^{-1/2}$;
6. normalize the rows of $\mathbf{F}$ to unit norm;
7. cluster the rows of $\mathbf{F}$ with the K-means algorithm into clusters $c_1, \ldots, c_q$;
8. assign anchor location $\mathbf{o}_t$ to the same cluster the row t of $\mathbf{F}$ has been assigned;
9. train the weighted k-nearest-neighbour classifier using the geographical coordinates of anchor locations as covariates and their cluster labels as the response;
10. assign clustering membership of all data locations using the trained classifier.

3   Case Study

In this section, the proposed large-scale geostatistical spectral clustering method is illustrated with a real-world, large-scale geostatistical dataset.

3.1   Dataset Description

The dataset of interest corresponds to eight critical heavy metals in topsoils from the US National Geochemical Survey database (48-states area) [13]. Variables which are concentration elements include arsenic (As), cadmium (Cd), chromium (Cr), copper (Cu), mercury (Hg), nickel (Ni), lead (Pb), and zinc (Zn). This large dataset consists of $43,885$ observations irregularly spaced in the study domain. Before clustering, all variables have been transformed to normal scores [6] because distributions of the variables are skewed. This preliminary processing also allows to have comparable scales and identify easily a spatial pattern in the variables. Moreover, the relative order is maintained so high normal scores corresponds to high raw values and vice versa. Spatial plots of transformed variables are shown in Fig. 1.

Fig. 1. Spatial plots of transformed variables (normal scores) for clustering purpose. (Color figure online)

3.2   Clustering Results

Figure 2 shows the clustering results provided by the traditional spectral clustering using the Nyström method [12], for different number of clusters (2, 3, 4, and 5). In this case the geographical coordinates have been considered as attributes. As one can see, the classical spectral clustering is not able to produce spatially contiguous clusters, even if geographical coordinates are considered as attributes. Clusters on Fig. 2 are spatially scattered over the study domain. This is simply explained by the fact that the traditional spectral clustering including the Nyström method used for accelerating it ignore the spatial dependency of the data.

The large-scale geostatistical spectral clustering method is performed for three different anchor locations grid sizes: $m = 5,000$, $m = 7,500$, and $m = 10,000$. Grids of anchor locations corresponding to these sizes are represented in Fig. 3. The clustering result for different number of clusters (2, 3, 4, and 5) and corresponding to these three different sizes are shown respectively in Figs. 4, 5, and 6. It appears that spatially contiguous clusters corresponding to $m = 5,000$, $m = 7,500$, and $m = 10,000$ are quite similar.

To compare quantitatively the clustering results for $m = 5,000$, $m = 7,500$, and $m = 10,000$, three external cluster validity indices are computed, namely Rand statistic, Jaccard coefficient, and Fowlkes and Mallows index [3]. They are used to determine the similarity between two clusterings by the agreements and/or disagreements of the pairs of points in two partitions. A higher value

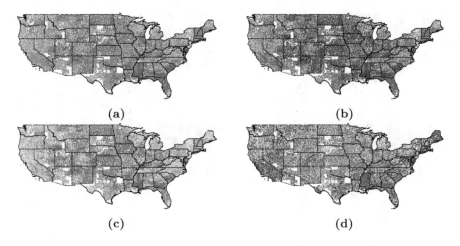

Fig. 2. (a, b, c, d) Traditional spectral clustering via the Nyström method for 2, 3, 4 and 5 clusters. The color of dots identifies the cluster membership. (Color figure online)

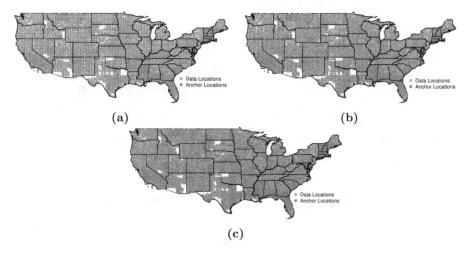

Fig. 3. Large-scale geostatistical spectral clustering method: anchor locations grid of size (a) 5,000, (b) 7,500 and (c) 10,000. (Color figure online)

for these indices indicates a greater similarity between two clusterings, and the highest value is one for perfect matching. The results are summarized in Tables 1 and 2. They indicate that the clustering results for $m = 5,000$, $m = 7,500$, and $m = 10,000$ are very similar. Concerning the optimal number of clusters, plots of the Caliński-Harabasz index versus the number of clusters for $m = 5,000$, $m = 7,500$, and $m = 10,000$ are also similar as shown in Fig. 7. In all cases, the optimal number of clusters corresponds to two. All these results show that choosing a very dense grid of anchor locations is not necessary. As indicated before, by using a very dense grid of anchor locations, similarities computed

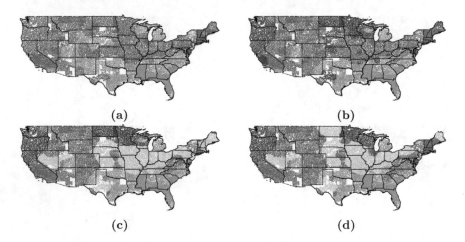

Fig. 4. (a, b, c, d) Large-scale geostatistical spectral clustering method for 2, 3, 4, and 5 clusters based on the anchor locations grid of size $m = 5,000$. The color of dots identifies the cluster membership. (Color figure online)

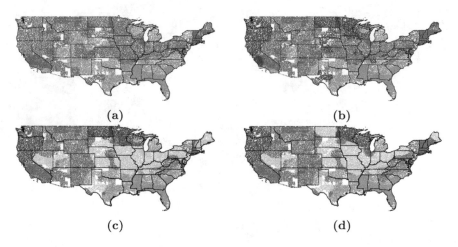

Fig. 5. (a, b, c, d) Large-scale geostatistical spectral clustering method for 2, 3, 4, and 5 clusters based on the anchor locations grid of size $m = 7,500$. The color of dots identifies the cluster membership. (Color figure online)

for two pairs of anchor locations that are very close can be unnecessary and redundant because of their high spatial correlation.

The time-consuming reduction is an important aspect, especially when several clustering with different sets of variables have to be performed on the same dataset. The application of the geostatistical spectral clustering method proposed in reference [10] is out of reach on this dataset. In other words, the application of the proposed large-scale geostatistical spectral clustering using all data locations as anchor locations is practically infeasible. On this dataset, the

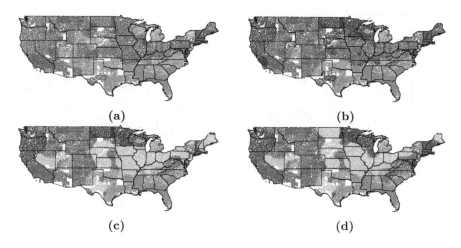

Fig. 6. (a, b, c, d) Large-scale geostatistical spectral clustering method for 2, 3, 4, and 5 clusters based on the anchor locations grid of size $m = 10,000$. The color of dots identifies the cluster membership. (Color figure online)

Table 1. Large-scale geostatistical spectral clustering method: Rand statistic, Jaccard coefficient, and Fowlkes and Mallows index between the clustering based on the anchor locations grid of size $m = 5,000$ and the clustering based on the anchor locations grid of size $m = 7,500$.

Number of clusters	2	3	4	5	6	7	8
Rand statistic	0.981	0.980	0.984	0.984	0.983	0.982	0.982
Jaccard coefficient	0.965	0.945	0.941	0.928	0.909	0.902	0.900
Fowlkes and Mallows index	0.982	0.971	0.969	0.962	0.952	0.940	0.937

Table 2. Large-scale geostatistical spectral clustering method: Rand statistic, Jaccard coefficient, and Fowlkes and Mallows index between the clustering based on the anchor locations grid of size $m = 5,000$ and the clustering based on the anchor locations grid of size $m = 10,000$.

Number of clusters	2	3	4	5	6	7	8
Rand statistic	0.981	0.981	0.984	0.983	0.979	0.981	0.984
Jaccard coefficient	0.965	0.950	0.939	0.925	0.904	0.902	0.900
Fowlkes and Mallows index	0.982	0.974	0.968	0.960	0.939	0.937	0.937

large-scale geostatistical spectral clustering for $m = 7,500$ is approximately 2 times time-consuming than for $m = 5,000$. Whereas, the large-scale geostatistical spectral clustering for $m = 10,000$ is approximately 2^2 times time-consuming than for $m = 5,000$.

Table 3 reports the means and standard deviations of the variables (normal scores) corresponding to the two optimal spatial clusters using the anchor

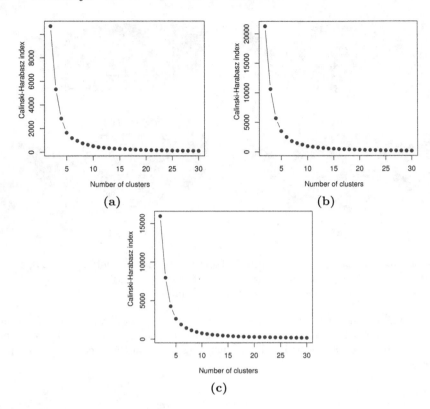

Fig. 7. Large-scale geostatistical spectral clustering method: selection of the optimal number of clusters through the Caliński-Harabasz index using (a) the anchor locations grid of size $m = 5,000$, (b) the anchor locations grid of size $m = 7,500$ and (c) the anchor locations grid of size $m = 10,000$.

Table 3. Large-scale geostatistical spectral clustering method based on the anchor locations grid of size $m = 5,000$: means and standard deviations of the variables (normal scores) corresponding to the two optimal spatial clusters.

	Spatial cluster 1 Mean	$(n_1 = 27,169)$ Std.	Spatial cluster 2 Mean	$(n_2 = 16,716)$ Std.
As	0.39	0.79	−0.63	0.99
Cd	0.45	0.77	−0.73	0.89
Cr	0.36	0.85	−0.58	0.96
Cu	0.43	0.82	−0.70	0.87
Ni	0.46	0.79	−0.74	0.84
Zn	0.44	0.77	−0.72	0.89
Hg	0.34	0.86	−0.55	0.96
Pb	0.18	0.92	−0.29	1.05

locations grid of size $m = 5,000$. There is a marked difference between the properties of samples in each spatial cluster. It appears that spatial cluster 1 (red points in Fig. 4a) is characterized by the highest concentrations; whereas spatial cluster 2 shows lowest concentrations (green points in Fig. 4a). The group of high values contains $27,169$ observations located for example in Washington, Oregon, Nevada, Idaho, Montana, North and South Dakota, Minnesota, Iowa, Illinois, Indiana, Ohio and Pennsylvania. The group of low values contains $16,716$ observations located for example in Florida, Georgia, Alabama, Mississippi, and Louisiana.

4 Conclusion

In this paper, a spectral clustering method aimed to discover spatially contiguous and meaningful clusters in large-scale geostatistical datasets has been developed. The proposed method reduces the similarity matrix size by computing it only at a reduced set of locations over the study domain referred to as anchor locations. It has the advantage of using all data during the computation of the similarity matrix at anchor locations; so there is no sacrifice of data. The clustering of all data locations is obtained by training a classifier on the dataset formed by the cluster labels (response) and the geographical coordinates (predictors) of the anchor locations. Although the weighted k-nearest-neighbour classifier has been used in this method, other classifiers can be utilized as well. A case study has been carried out to evaluate the feasibility of the proposed method. Results show that the proposed method can effectively process large-scale geostatistical datasets and discover spatially contiguous and meaningful clusters. It is shown that choosing a very dense grid of anchor locations is unnecessary. However, future work is needed to balance better and automatically determine the appropriate number of anchor locations. The effectiveness of the proposed method needs further evaluation with other large-scale geostatistical databases.

References

1. Cai, D., Chen, X.: Large scale spectral clustering via landmark-based sparse representation. IEEE Trans. Cybern. **45**(8), 1669–1680 (2015)
2. Cao, Y., Chen, D.R.: Consistency of regularized spectral clustering. Appl. Comput. Harmonic Anal. **30**(3), 319–336 (2011)
3. Charu, C., Chandan, K.: Data Clustering: Algorithms and Applications. Chapman and Hall/CRC (2013)
4. Chen, B., Gao, B., Liu, T.-Y., Chen, Y.-F., Ma, W.-Y.: Fast spectral clustering of data using sequential matrix compression. In: Fürnkranz, J., Scheffer, T., Spiliopoulou, M. (eds.) ECML 2006. LNCS, vol. 4212, pp. 590–597. Springer, Heidelberg (2006). doi:10.1007/11871842_56
5. Chen, X., Cai, D.: Large scale spectral clustering with landmark-based representation. In: Proceedings of the Twenty-Fifth AAAI Conference on Artificial Intelligence, pp. 313–318. AAAI Press (2011)

6. Chilès, J.P., Delfiner, P.: Geostatistics: Modeling Spatial Uncertainty. Wiley, NJ (2012)
7. Choromanska, A., Jebara, T., Kim, H., Mohan, M., Monteleoni, C.: Fast spectral clustering via the Nyström method. In: Jain, S., Munos, R., Stephan, F., Zeugmann, T. (eds.) ALT 2013. LNCS, vol. 8139, pp. 367–381. Springer, Heidelberg (2013). doi:10.1007/978-3-642-40935-6_26
8. Filippone, M., Camastra, F., Masulli, F., Rovetta, S.: A survey of kernel and spectral methods for clustering. Pattern Recogn. **41**(1), 176–190 (2008)
9. Fouedjio, F.: A clustering approach for discovering intrinsic clusters in multivariate geostatistical data. In: Perner, P. (ed.) MLDM 2016. LNCS, vol. 9729, pp. 491–500. Springer, Cham (2016)
10. Fouedjio, F.: Discovering spatially contiguous clusters in multivariate geostatistical data through spectral clustering. In: Li, J., Li, X., Wang, S., Li, J., Sheng, Q.Z. (eds.) ADMA 2016. LNCS (LNAI), vol. 10086, pp. 547–557. Springer, Cham (2016). doi:10.1007/978-3-319-49586-6_38
11. Fouedjio, F.: A hierarchical clustering method for multivariate geostatistical data. Spat. Stat. **18**, 334–351 (2016)
12. Fowlkes, C., Belongie, S., Chung, F., Malik, J.: Spectral grouping using the Nyström method. IEEE Trans. Pattern Anal. Mach. Intell. **26**(2), 214–225 (2004)
13. Grossman, J.N., Grosz, A., Schweitzer, P.N., Schruben, P.G.: The national geochemical survey - database and documentation. Version 5. U.S. geological Survey, Reston, VA (2008)
14. Hastie, T., Tibshirani, R., Friedman, J.: The Elements of Statistical Learning: Data Mining, Inference, and Prediction. Springer Series in Statistics. Springer, New York (2009)
15. Hechenbichler, K., Schliep, K.: Weighted k-nearest-neighbor techniques and ordinal classification. Discussion Paper 399, SFB 386, Ludwig-Maximilians University Munich (2004)
16. Kannan, R., Vempala, S., Vetta, A.: On clusterings: Good, bad and spectral. J. ACM **51**(3), 497–515 (2004)
17. Khoa, N.L.D., Chawla, S.: Large scale spectral clustering using resistance distance and spielman-teng solvers. In: Ganascia, J.-G., Lenca, P., Petit, J.-M. (eds.) DS 2012. LNCS, vol. 7569, pp. 7–21. Springer, Heidelberg (2012). doi:10.1007/978-3-642-33492-4_4
18. Kong, T., Tian, Y., Shen, H.: A fast incremental spectral clustering for large data sets. In: 12th International Conference on Parallel and Distributed Computing, Applications and Technologies, pp. 1–5. IEEE (2011)
19. Luxburg, U.V.: A tutorial on spectral clustering. Stat. Comput. **17**(4), 395–416 (2007)
20. Luxburg, U.V., Belkin, M., Bousquet, O.: Consistency of spectral clustering. Ann. Statist. **36**(2), 555–586 (2008)
21. Luxburg, U.V., Bousquet, O., Belkin, M.: Limits of spectral clustering. In: Advances in Neural Information Processsing Systems, pp. 857–864 (2004)
22. Nascimento, M.C., de Carvalho, A.C.: Spectral methods for graph clustering a survey. Eur. J. Oper. Res. **211**(2), 221–231 (2011)
23. Ng, A.Y., Jordan, M.I., Weiss, Y.: On spectral clustering: analysis and an algorithm. In: Advances in Neural Information Processsing Systems, pp. 849–856. MIT Press (2001)
24. Romary, T., Ors, F., Rivoirard, J., Deraisme, J.: Unsupervised classification of multivariate geostatistical data: two algorithms. Comput. Geosci. **85**(Pt. B), 96–103 (2015)

25. Schaeffer, S.E.: Graph clustering. Comput. Sci. Rev. **1**(1), 27–64 (2007)
26. Semertzidis, T., Rafailidis, D., Strintzis, M., Daras, P.: Large-scale spectral clustering based on pairwise constraints. Inform. Process. Manage. **51**(5), 616–624 (2015)
27. Shinnou, H., Sasaki, M.: Spectral clustering for a large data set by reducing the similarity matrix size. In: Proceedings of the Sixth International Conference on Language Resources and Evaluation (LREC 2008) (2008)
28. Tremblay, N., Puy, G., Gribonval, R., Vandergheynst, P.: Compressive spectral clustering. In: Proceedings of the 33rd International Conference on Machine Learning (ICML 2016) (2016)
29. Vladymyrov, M., Carreira-Perpiñán, M.: The variational Nyström method for large-scale spectral problems. In: Proceedings of the 33rd International Conference on Machine Learning (ICML 2016) (2016)
30. Wackernagel, H.: Multivariate Geostatistics: An Introduction with Applications. Springer, Heidelberg (2003)
31. Wand, M., Jones, C.: Kernel Smoothing. Monographs on Statistics and Applied Probability. Chapman and Hall, Sanford (1995)
32. Wang, C.: Large-scale spectral clustering on graphs. In: IJCAI. Elsevier (2013)
33. Wang, L., Leckie, C., Ramamohanarao, K., Bezdek, J.: Approximate Spectral Clustering, pp. 134–146. Springer, Heidelberg (2009)
34. Yan, D., Huang, L., Jordan, M.I.: Fast approximate spectral clustering. In: Proceedings of the 15th ACM SIGKDD International Conference on Knowledge Discovery and Data Mining, pp. 907–916. ACM (2009)
35. Zha, H., He, X., Ding, C., Gu, M., Simon, H.D.: Spectral relaxation for k-means clustering. In: Advances in Neural Information Processsing Systems, pp. 1057–1064 (2001)
36. Zhang, X., Zong, L., You, Q., Yong, X.: Sampling for Nyström extension-based spectral clustering: incremental perspective and novel analysis. ACM Trans. Knowl. Discov. Data **11**(1), 7:1–7:25 (2016)

Vulnerability of Deep Reinforcement Learning to Policy Induction Attacks

Vahid Behzadan[1,2]([✉]) and Arslan Munir[1,2]

[1] Department of Computer Science and Engineering, University of Nevada, Reno,
1664 N Virginia St, Reno, NV 89557, USA
[2] Department of Computer Science, Kansas State University,
Manhattan, KS 66506, USA
behzadan@k-state.edu, amunir@ksu.edu

Abstract. Deep learning classifiers are known to be inherently vulnerable to manipulation by intentionally perturbed inputs, named adversarial examples. In this work, we establish that reinforcement learning techniques based on Deep Q-Networks (DQNs) are also vulnerable to adversarial input perturbations, and verify the transferability of adversarial examples across different DQN models. Furthermore, we present a novel class of attacks based on this vulnerability that enable policy manipulation and induction in the learning process of DQNs. We propose an attack mechanism that exploits the transferability of adversarial examples to implement policy induction attacks on DQNs, and demonstrate its efficacy and impact through experimental study of a game-learning scenario.

Keywords: Reinforcement Learning · Deep Q-Learning · Adversarial examples · Policy induction · Manipulation · Vulnerability

1 Introduction

Inspired by the psychological and neuroscientific models of natural learning, Reinforcement Learning (RL) techniques aim to optimize the actions of intelligent agents in complex environments by learning effective controls and reactions that maximize the long-term reward of agents [1]. The applications of RL range from combinatorial search problems such as learning to play games [2] to autonomous navigation [3], multi-agent systems [4], and optimal control [5]. However, classic RL techniques generally rely on hand-crafted representations of sensory input, thus limiting their performance in the complex and high-dimensional real world environments. To overcome this limitation, recent developments combine RL techniques with the significant feature extraction and processing capabilities of deep learning models in a framework known as Deep Q-Network (DQN) [6]. This approach exploits deep neural networks for both feature selection and action-value function approximation, hence enabling unprecedented performance in complex settings, such as learning efficient playing strategies from unlabeled

© Springer International Publishing AG 2017
P. Perner (Ed.): MLDM 2017, LNAI 10358, pp. 262–275, 2017.
DOI: 10.1007/978-3-319-62416-7_19

video frames of Atari games [7], robotic manipulation [8], and autonomous navigation of aerial [9] and ground vehicles [10].

The growing interest in the application of DQNs in critical systems necessitate the investigation of this framework with regards to its resilience and robustness to adversarial attacks on the integrity of RL processes. The reliance of RL on interactions with the environment gives rise to an inherent vulnerability which makes the process of learning susceptible to perturbation as a result of changes in the observable environment. Exploiting this vulnerability provides adversaries with the means to disrupt or change control policies, leading to unintended and potentially harmful actions. For instance, manipulation of the obstacle avoidance and navigation policies learned by autonomous Unmanned Aerial Vehicles (UAV) enables the adversary to use such systems as kinetic weapons by inducing actions that lead to intentional collisions.

In this paper, we study the efficacy and impact of policy induction attacks on the Deep Q-Learning RL framework. To this end, we propose a novel attack methodology based on adversarial example attacks against deep learning models [13]. Through experimental results, we verify that similar to classifiers, Q networks are also vulnerable to adversarial examples, and confirm the transferability of such examples between different models. We then evaluate the proposed attack methodology on the original DQN architecture of Mnih et al. [7], the results of which verify the feasibility of policy induction attacks by incurring minimal perturbations in the environment or sensory inputs of an RL system. We also discuss the insufficiency of defensive distillation [14] and adversarial training [15] techniques as state of the art countermeasures proposed against adversarial example attacks on deep learning classifiers, and present potential techniques to mitigate the effect of policy induction attacks against DQNs.

The remainder of this paper is organized as follows: Sect. 2 presents an overview of Q-Learning, Deep Q-Networks, and adversarial examples. Section 3 formalizes the problem and defines the target and attacker models. In Sect. 4, we outline the attack methodology and algorithm, followed by the experimental evaluation of the proposed methodology in Sect. 5. A high-level discussion on effectiveness of the current countermeasures is presented in Sect. 6, and the paper is concluded in Sect. 7 with remarks on future research directions.

2   Background

2.1   Q-Learning

The generic RL problem can be formally modeled as a Markov Decision Process (MDP), described by the tuple $MDP = (S, A, P, R)$, where S is the set of reachable states in the process, A is the set of available actions, R is the mapping of transitions to the immediate reward, and P represents the transition probabilities. At any given time-step t, the MDP is at a state $s_t \in S$. The RL agent's choice of action at time t, $a_t \in A$ causes a transition from s_t to a state s_{t+1} according to the transition probability $P_{s_t,s_{t+a}}^{a_t}$. The agent receives a reward

$r_t = R(s_t, a_t) \in \mathbb{R}$, where $\mathbb{R}$ denotes the set of real numbers, for choosing the action a_t at state s_t.

Interactions of the agent with MDP are captured in a policy π. When such interactions are deterministic, the policy $\pi : S \to A$ is a mapping between the states and their corresponding actions. A stochastic policy $\pi(s, a)$ represents the probability of optimality for action a at state s.

The objective of RL is to find the optimal policy π^* that maximizes the cumulative reward over time at time t, denoted by the return function $\hat{R}_t = \sum_{k=0}^{\infty} \gamma^k r_{t+k}$, where $\gamma \in [0, 1]$ is the discount factor representing the diminishing worth of rewards obtained further in time, hence ensuring that $\hat{R}$ is bounded.

One approach to this problem is to estimate the optimal value of each action, defined as the expected sum of future rewards when taking that action and following the optimal policy thereafter. The value of an action a in a state s is given by the action-value function Q defined as:

$$Q(s, a) = R(s, a) + \gamma max_{a'}(Q(s', a')) \tag{1}$$

Where s' is the state that emerges as a result of action a, and a' is a possible action in state s'. The optimal Q value given a policy π is hence defined as: $Q^*(s, a) = max_\pi Q^\pi(s, a)$, and the optimal policy is given by $\pi^*(s) = \arg\max_a Q(s, a)$

The Q-learning method estimates the optimal action policies by using the Bellman equation $Q_{i+1}(s, a) = \mathbf{E}[R + \gamma \max_a Q_i]$ as the iterative update of a value iteration technique. Practical implementation of Q-learning is commonly based on function approximation of the parametrized Q-function $Q(s, a; \theta) \approx Q^*(s, a)$. A common technique for approximating the parametrized non-linear Q-function is to train a neural network whose weights correspond to the parameter vector θ. Such neural networks, commonly referred to as Q-networks, are trained such that at every iteration i, it minimizes the loss function

$$L_i(\theta_i) = \mathbf{E}_{s, a \sim \rho(.)}[(y_i - Q(s, a; \theta_i))^2] \tag{2}$$

where $y_i = \mathbf{E}[R + \gamma \max_{a'} Q(s', a'; \theta_{i-1})|s, a]$, and $\rho(s, a)$ is a probability distribution over states s and actions a. This optimization problem is typically solved using computationally efficient techniques such as Stochastic Gradient Descent (SGD) [11].

2.2 Deep Q Networks

Classical Q-networks present a number of major disadvantages in the Q-learning process. First, the sequential processing of consecutive observations breaks the *iid* (Independent and Identically Distributed) requirement of training data as successive samples are correlated. Furthermore, slight changes to Q-values leads to rapid changes in the policy estimated by Q-network, thus enabling policy oscillations. Also, since the scale of rewards and Q-values are unknown, the gradients of Q-networks can be sufficiently large to render the backpropagation process unstable.

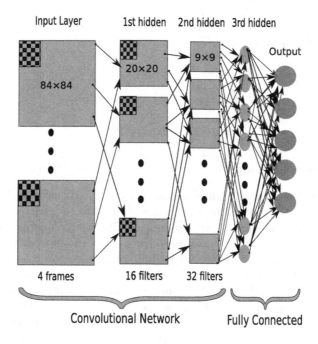

Fig. 1. DQN architecture for end-to-end learning of Atari 2600 game plays

A deep Q network (DQN) [6] is a multi-layered Q-network designed to mitigate such disadvantages. To overcome the issue of correlation between consecutive observations, DQN employs a technique named *experience replay*: Instead of training on successive observations, experience replay samples a random batch of previous observations stored in the replay memory to train on. As a result, the correlation between successive training samples is broken and the iid setting is re-established. In order to avoid oscillations, DQN fixes the parameters of the optimization target y_i. These parameters are then updated at regulat intervals by adopting the current weights of the Q-network. The issue of unstability in backpropagation is also solved in DQN by clipping the reward values to the range $[-1, +1]$, thus preventing Q-values from becoming too large.

Mnih et al. [7] demonstrate the application of this new Q-network technique to end-to-end learning of Q values in playing Atari games based on observations of pixel values in the game environment. The neural network architecture of this work is depicted in Fig. 1. To capture the movements in the game environment, Mnih et al. use stacks of 4 consecutive image frames as the input to the network. To train the network, a random batch is sampled from the previous observation tuples (s_t, a_t, r_t, s_{t+1}), where r_t denotes the reward at time t. Each observation is then processed by 2 layers of convolutional neural networks to learn the features of input images, which are then employed by feed-forward layers to approximate the Q-function. The target network $\hat{Q}$, with parameters θ^-, is synchronized with the parameters of the original Q network at fixed periods intervals. i.e., at every

ith iteration, $\theta_t^- = \theta_t$, and is kept fixed until the next synchronization. The target value for optimization of DQN learning thus becomes:

$$y_t' \equiv r_{t+1} + \gamma max_{a'} \hat{Q}(S_{t+1}, a'; \theta^-)$$ (3)

Accordingly, the training process can be stated as:

$$min_{a_t}(y_t' - Q(s_t, a_t, \theta))^2$$ (4)

2.3 Adversarial Examples

In [16], Szegedy et al. report an intriguing discovery: several machine learning models, including deep neural networks, are vulnerable to adversarial examples. That is, these machine learning models misclassify inputs that are only slightly different from correctly classified samples drawn from the data distribution. Furthermore, a wide variety of models with different architectures trained on different subsets of the training data misclassify the same adversarial example.

This suggests that adversarial examples expose fundamental blind spots in machine learning algorithms. The issue can be stated as follows: Consider a machine learning system M and a benign input sample C which is correctly classified by the machine learning system, i.e. $M(C) = y_{true}$. According to the report of Szegedy [16] and many proceeding studies [13], it is possible to construct an adversarial example $A = C + \delta$, which is perceptually indistinguishable from C, but is classified incorrectly, i.e. $M(A) \neq y_{true}$.

Adversarial examples are misclassified far more often than examples that have been perturbed by random noise, even if the magnitude of the noise is much larger than the magnitude of the adversarial perturbation [17]. According to the objective of adversaries, adversarial example attacks are generally classified into the following two categories:

1. Misclassification attacks, which aim for generating examples that are classified incorrectly by the target network.
2. Targeted attacks, whose goal is to generate samples that the target misclassifies into an arbitrary class designated by the attacker.

To generate such adversarial examples, several algorithms have been proposed, such as the Fast Gradient Sign Method (FGSM) by Goodfellow et al. [17], and the Jacobian Saliency Map Algorithm (JSMA) approach by Papernot et al. [13]. A grounding assumption in many of the crafting algorithms is that the attacker has complete knowledge of the target neural networks such as its architecture, weights, and other hyperparameters. Recently, Papernot et al. [18] have proposed the first black-box approach to generating adversarial examples. This method exploits the generalized nature of adversarial examples: an adversarial example generated for a neural network classifier applies to most other neural network classifiers that perform the same classification task, regardless of their architecture, parameters, and even the distribution of training data. Accordingly, the approach of [18] is based on generating a replica of the target network. To train this replica, the attacker creates and trains over a dataset from a mixture of samples obtained by observing target's interaction with the environment, and synthetically generated inputs and label pairs. Once trained, any of the adversarial example crafting algorithms that require knowledge of the target network can be applied to the replica. Due to the transferability of adversarial examples, the perturbed samples generated from the replica network will induce misclassifications in many of the other networks that perform the same task. In the following sections, we describe how a similar approach can be adopted in policy induction attacks against DQNs.

3   Threat Model

We consider an attacker whose goal is to perturb the optimality of actions taken by a DQN learner via inducing an arbitrary policy π_{adv} on the target DQN. The attacker is assumed to have minimal *a priori* information of the target, such as the type and format of inputs to the DQN, as well as its reward function R and an estimate for the frequency of updating the $\hat{Q}$ network. It is noteworthy that even if the target's reward function is not known, it can be estimated via Inverse Reinforcement Learning techniques [19]. No knowledge of the target's exact architecture is considered in this work, but the attacker can estimate this architecture based on the conventions applied to the input type (e.g. image and video input may indicate a convolutional neural network, speech and voice data point towards a recurrent neural network, etc.).

In this model, the attacker has no direct influence on the target's architecture and parameters, including its reward function and the optimization mechanism. The only parameter that the attacker can directly manipulate is the configuration of the environment observed by the target. For instance, in the case of video game learning [6], the attacker is capable of changing the pixel values of the game's frames, but not the score. In cyber-physical scenarios, such perturbations can be implemented by strategic rearrangement of objects or precise illumination of certain areas via tools such as laser pointers. To this end, we assume that the attacker is capable of changing the state before it is observed by the target, either by predicting future states, or after such states are generated by the environment's dynamics. The latter can be achieved if the attacker has a faster action speed than the target's sampling rate, or by inducing a delay between generation of the new environment and its observation by the target.

To avoid detection and minimize influence on the environment's dynamics, we impose an extra constraint on the attack such that the magnitude of perturbations applied in each configuration must be smaller than a set value denoted by ϵ. Also, we do not limit the attacker's domain of perturbations (e.g. in the case of video games, the attacker may change the value of any pixel at any position on the screen).

4 Attack Mechanism

As discussed in Sect. 2, the DQN framework of Mnih et al. [7] can be seen as consisting of two neural networks, one is the native network which performs the image classification and function approximation, and the other is the auxiliary $\hat{Q}$ network whose architecture and parameters are copies of the native network sampled once every c iterations. Training of DQN is performed optimizing the loss function of Eq. 4 by Stochastic Gradient Descent (SGD). Due to the similarity of this process and the training mechanism of neural network classifiers, we hypothesize that the function approximators of DQN are also vulnerable to adversarial example attacks. In other words, the set of all possible inputs to the approximated function $\hat{Q}$ contains elements which cause the approximated functions to generate outputs that are different from the output of the original Q function. Furthermore, we hypothesize that similar to the case of classifiers, the elements that cause one DQN to generate incorrect Q values will incur the same effect on other DQNs that approximate the same Q-function.

Consequently, the attacker can manipulate a DQN's learning process by crafting states s_t such that $\hat{Q}(s_{t+1}, a; \theta_t^-)$ identifies an incorrect choice of optimal action at s_{t+1}. If the attacker is capable of crafting adversarial inputs s_t' and s_{t+1}' such that the value of Eq. 4 is minimized for a specific action a', then the policy learned by DQN at this time-step is optimized towards suggesting a' as the optimal action given the state s_t.

Considering that the attacker is not aware of the target's network architecture and its parameters at every time step, crafting adversarial states must rely on black-box techniques such as those introduced in [18]. The attacker can exploit the transferability of adversarial examples by obtaining the state perturbations from the replica Q' and $\hat{Q}'$ networks that correspond to the target's Q and $\hat{Q}$ networks, respectively. Algorithm 1 details the procedural flow of this phase.

At every time step of training this replica, the attacker observes the interaction of its target with the environment (s_t, a_t, r_t, s_{t+1}) (step 2). If the resulting state is not terminal, the attacker then calculates the perturbation vectors $\hat{\delta}_{t+1}$ for the next state s_{t+1} such that $max_{a'}\hat{Q}(s_{t+1} + \hat{\delta}_{t+1}, a'; \theta_t^-)$ causes $\hat{Q}$ to generate its maximum when $a' = \pi_{adv}^*(s_{t+1})$, i.e., the maximum reward at the next state is obtained when the optimal action taken at that state is determined by the attacker's policy (steps 4–6). The attacker then reveals the perturbed state s_{t+1} to the target (step 7), and re-trains the replica based on the new state and action (steps 8–10).

This is procedurally similar to targeted misclassification attacks described in Sect. 2 that aim to find minimal perturbations to an input sample such that the classifier assigns the maximum value of likelihood to an incorrect target class. Therefore, the adversarial example crafting techniques developed for classifiers, such as the Fast Gradient Sign Method (FGSM) and the Jacobian Saliency Map Algorithm (JSMA), can be applied to obtain the perturbation vector $\hat{\delta}_{t+1}$.

Algorithm 1. Exploitation Procedure

    Input   : adversarial policy π^*_{adv}, initialized replica DQNs Q', $\hat{Q}'$, synchronization
               frequency c, number of iterations N

    Output: perturbed states s'_{t+1}

1 **for** *observation = 1, N* **do**

2 | Observe current state s_t, action a_t, reward r_t, and resulting state s_{t+1}

3 | **if** s_{t+1} *is not terminal* **then**

4 | | set $a'_{adv} = \pi^*_{adv}(s_{t+1})$

5 | | Calculate perturbation vector $\hat{\delta}_{t+1} = Craft(\hat{Q}', a'_{adv}, s_{t+1})$

6 | | $s'_{t+1} \leftarrow s_{t+1} + \hat{\delta}_{t+1}$

7 | | Reveal s'_{t+1} to target

8 | | **if** *observation* mod $c = 0$ **then** $\theta'^- \leftarrow \theta'$;

9 | | Set $y_t = (r_t + max_{a'}\hat{Q}'(s_{t+1} + \hat{\delta}_{t+1}, a'; \theta'^-)$

10 | | Perform SGD on $(y_t - Q'(s_t, a_t, \theta'))^2$ w.r.t θ'

11 | **end**

12 **end**

The procedure of this attack can be divided into the two phases of initialization and exploitation. The initialization phase implements processes that must be performed before the target begins interacting with the environment, which are:

1. Train a DQN based on attacker's reward function r' to obtain the adversarial policy π^*_{adv}.
2. Create a replica of the target's DQN and initialize with random parameters.

The exploitation phase implements the attack processes such as crafting adversarial inputs. This phase constitutes an attack cycle depicted in Fig. 2. The cycle initiates with the attacker's first observation of the environment, and runs in tandem with the target's operation.

5 Experimental Verification

To study the performance and efficacy of the proposed mechanism, we examine the targeting of Mnih et al.'s DQN designed to learn Atari 2600 games [7]. In our setup, we train the network on a game of Pong implemented in Python using the PyGame library [12]. The game is played against an opponent with a modest level of heuristic artificial intelligence, and is customized to handle the delays in DQN's reaction due to the training process. The game's backened provides the DQN agent with the game screen sampled at 8 Hz, as well as the game score ($+1$ for win, -1 for lose, 0 for ongoing game) throughout each episode of the game. The set of available actions $A = \{UP, DOWN, Stand\}$ enables the DQN agent to control the movements of its paddle. Figure 3 illustrates the game screen of Pong used in our experiments.

The training process of DQN is implemented in TensorFlow [20] and executed on an Amazon EC2 g2.2xlarge instance [21] with 8 Intel Xeon E5-2670 CPU cores

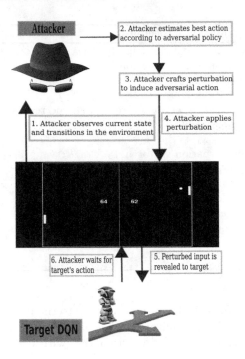

Fig. 2. Exploitation cycle of policy induction attack

and a NVIDIA GPU with 1536 CUDA cores and 4 GB of video memory. Each state observed by the DQN is a stack of 4 consecutive 80 × 80 gray-scale game frames. Similar to the original architecture of Mnih et al. [7], this input is first passed through two convolutional layers to extract a compressed feature space for the following two feed-forward layers for Q function estimation. The discount factor γ is set to 0.99, and the initial probability of taking a random action is set to 1, which is annealed after every 500000 actions. The agent is also set to train its DQN after every 50000 observations. Regular training of this DQN takes approximately 1.5 million iterations ($\sim$16 h on the g2.2xlarge instance) to reach a winning average of 51% against the heuristic AI of its opponent. As expected, longer training of this DQN leads to better results. After a 2-week period of training we verified the convergent trait of our implementation by witnessing winning averages of more than 80%.

Following the threat model presented in Sect. 3, this experiment considers an attacker capable of observing the states interactions between his target DQN and the game, but his domain of influence is limited to implementation of minor changes on the environment. Considering the visual representation of the environment in this setup, the minor changes incurred by attacker take the form of perturbing pixel values in the 4 consecutive frames of a given state.

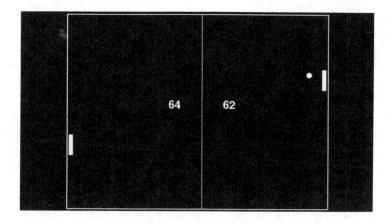

Fig. 3. Game of Pong

5.1 Evaluation of DQN's Vulnerability to Adversarial Examples

Successful implementations of the proposed policy induction attack mechanisms rely on the vulnerability of DQNs to targeted adversarial perturbations. To verify the existence of this vulnerability, the $\hat{Q}$ networks of target were sampled at regular intervals during training in the game environment. In the next step, 100 observations comprised of a pair of consecutive states (s_t, s_{t+1}) were randomly selected from the experience memory (i.e., the pool of previous state-action observations) of DQN, to ensure the possibility of their occurrence in the game. Considering s_{t+1} to be the variable that can be manipulated by the attacker, it is passed along with the model $\hat{Q}$ to the adversarial example crafting algorithms. To study the extent of vulnerability, we evaluated the success rate of both FGSM and JSMA algorithms for each of the 100 random observations in inducing a random game action other than the current optimal a_t^*. The results, presented in Fig. 4, verify that DQNs are indeed vulnerable to adversarial example attacks. It is noteworthy that the success rate of FGSM with a fixed perturbation limit decreases by one percent per 100000 observations as

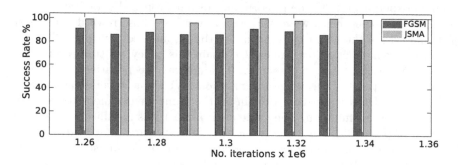

Fig. 4. Success rate of crafting adversarial examples for DQN

the number of observations increases, which in practice, does not constrain the viability of this attack. Yet, JSMA seems to be more robust to this effect as it maintains a success rate of 100% throughout the experiment.

5.2   Verification of Transferability of Adversarial Examples

To measure the transferability of adversarial examples between models, we trained another Q-network with a similar architecture on the same experience memory of the game at the sampled instances of the previous experiment. It is noteworthy that due to random initializations, the exploration mechanism, and the stochastic nature of SGD, even similar Q-networks trained on the same set of observations will obtain different sets of weights. The second Q-network was tested to measure its vulnerability to the adversarial examples obtained from the last experiment. Figure 5 shows that more than 70% of the perturbations obtained from both FGSM and JSMA methods also affect the second network, hence verifying the transferability of adversarial examples between DQNs.

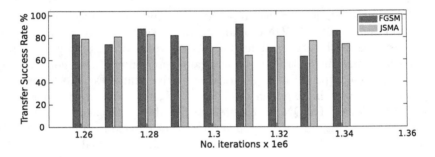

Fig. 5. Transferability of adversarial examples in DQN

5.3   Performance of Proposed Policy Induction Attack

Our final experiment tests the performance of our proposed exploitation mechanism. In this experiment, we consider an adversary whose reward value is the exact opposite of the game score, meaning that it aims to devise a policy that maximizes the number of lost games. To obtain this policy, we trained an adversarial DQN on the game, whose reward value was the negative of the value obtained from target DQN's reward function. With the adversarial policy at hand, a target DQN was setup to train on the game environment to maximize the original reward function. The game environment was modified to allow perturbation of pixel values in game frames by the adversary. A second DQN was also setup to train on the target's observations to provide an estimation of the target DQN to enable blackbox crafting of adversarial example. At every observation, the adversarial policy obtained in the initialization phase was consulted

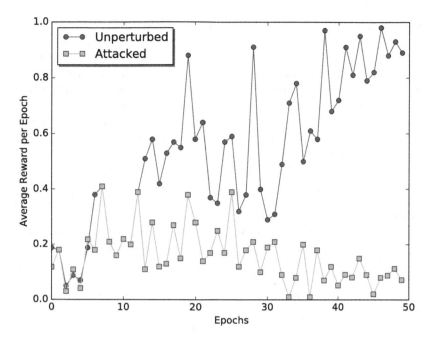

Fig. 6. Comparison of rewards between unperturbed and attacked DQNs

to calculate the action that would satisfy the adversary's goal. Then, the JSMA algorithm was utilized to generate the adversarial example that would cause the output of the replica DQN network to be the action selected by the adversarial policy. This example was then passed to the target DQN as its observation. Figure 6 compares the performance of unperturbed and attacked DQNs in terms of their reward values, measured as the difference of current game score with the average score. It can be seen that the reward value for the targeted DQN agent rapidly falls below the unperturbed case and maintains the trend of losing the game throughout the experiment. This result confirms the efficacy of our proposed attack mechanism, and verifies the vulnerability of Deep Q-Networks to policy induction attacks.

6   Discussion on Current Counter-Measures

Since the introduction of adversarial examples by Szgedey et al. [16], various counter-measures have been proposed to mitigate the exploitation of this vulnerability in deep neural networks. Goodfellow et al. [17] proposed to retrain deep networks on a set of minimally perturbed adversarial examples to prevent their misclassification. This approach suffers from two inherent short-comings: Firstly, it aims to increase the amount of perturbations required to craft an adversarial example. Second, this approach does not provide a comprehensive counter-measure as it is computationally inefficient to find all possible adversarial examples. Furthermore, Papernot et al. [18] argue that by training the network on adversarial examples, the emerging network will have new adversarial examples and hence this technique does not solve the problem of exploiting this vulnerability for critical systems. Consequently, Papernot et al. [14] proposed a technique named Defensive Distillation, which is also based on retraining the network on a dimensionally-reduced set of training data. This approach, too, was recently shown to be insufficient in mitigating adversarial examples [22]. It is hence concluded that the current state of the art in countering adversarial examples and their exploitation is incapable of providing a concrete defense against such exploitations.

In the context of policy induction attacks, we conjecture that the temporal features of the training process may be utilized to provide protection mechanisms. The proposed attack mechanism relies on the assumption that due to the decreasing chance of random actions, the target DQN is most likely to perform the action induced by adversarial inputs as the number of iterations progress. This may be mitigated by implementing adaptive exploration-exploitation mechanisms that both increase and decrease the chance of random actions according to the performance of the trained model. Also, it may be possible to exploit spatio-temporal pattern recognition techniques to detect and omit regular perturbations during the pre-processing phase of the learning process. Investigating such techniques is the priority of our future work.

7 Conclusions and Future Work

In this work, we established the vulnerability of reinforcement learning based on Deep Q-Networks to policy induction attacks. Furthermore, we proposed an attack mechanism which exploits the vulnerability of deep neural networks to adversarial examples, and demonstrated its efficacy and impact through experiments on a game-learning DQN.

This preliminary work solicitates a wide-range of studies on the security of Deep Reinforcement Learning. As discussed in Sect. 6, novel countermeasures need to be investigated to mitigate the effect of such attacks on DQNs deployed in cyber-physical and critical systems. Also, an analytical treatment of the problem to establish the bounds and relationships of model parameters, such as network architecture and exploration mechanisms, with DQN's vulnerability to policy induction will provide deeper insight and guidelines into designing safe and secure deep reinforcement learning architectures.

References

1. Sutton, R.S., Barto, A.G.: Introduction to Reinforcement Learning, vol. 135. MIT Press, Cambridge (1998)
2. Ghory, I.: Reinforcement learning in board games, Department of Computer Science, University of Bristol. Technical report (2004)
3. Dai, X., Li, C.-K., Rad, A.B.: An approach to tune fuzzy controllers based on reinforcement learning for autonomous vehicle control. IEEE Trans. Intell. Transp. Syst. 6(3), 285–293 (2005)

4. Busoniu, L., Babuska, R., De Schutter, B.: A comprehensive survey of multiagent reinforcement learning. IEEE Trans. Syst. Man Cybern. Part C Appl. Rev. **38**(2), 156 (2008)
5. Sutton, R.S., Barto, A.G., Williams, R.J.: Reinforcement learning is direct adaptive optimal control. IEEE Control Syst. **12**(2), 19–22 (1992)
6. Mnih, V., Kavukcuoglu, K., Silver, D., Graves, A., Antonoglou, I., Wierstra, D., Riedmiller, M.: Playing atari with deep reinforcement learning. arXiv preprint arXiv:1312.5602 (2013)
7. Mnih, V., Kavukcuoglu, K., Silver, D., Rusu, A.A., Veness, J., Bellemare, M.G., Graves, A., Riedmiller, M., Fidjeland, A.K., Ostrovski, G., et al.: Human-level control through deep reinforcement learning. Nature **518**(7540), 529–533 (2015)
8. Gu, S., Holly, E., Lillicrap, T., Levine, S.: Deep reinforcement learning for robotic manipulation. arXiv preprint arXiv:1610.00633 (2016)
9. Zhang, T., Kahn, G., Levine, S., Abbeel, P.: Learning deep control policies for autonomous aerial vehicles with MPC-guided policy search. arXiv preprint arXiv:1509.06791 (2015)
10. Hussein, A., Gaber, M.M., Elyan, E.: Deep active learning for autonomous navigation. In: Jayne, C., Iliadis, L. (eds.) EANN 2016. CCIS, vol. 629, pp. 3–17. Springer, Cham (2016). doi:10.1007/978-3-319-44188-7_1
11. Baird, L., Moore, A.W.: Gradient descent for general reinforcement learning. In: Advances in Neural Information Processing Systems, pp. 968–974 (1999)
12. McGugan, W.: Beginning game development with Python and Pygame: from novice to professional. Apress (2007)
13. Papernot, N., McDaniel, P., Jha, S., Fredrikson, M., Celik, Z.B., Swami, A.: The limitations of deep learning in adversarial settings. In: 2016 IEEE European Symposium on Security and Privacy (EuroS&P), pp. 372–387. IEEE (2016)
14. Papernot, N., McDaniel, P., Wu, X., Jha, S., Swami, A.: Distillation as a defense to adversarial perturbations against deep neural networks. arXiv preprint arXiv:1511.04508 (2015)
15. Carlini, N., Wagner, D.: Towards evaluating the robustness of neural networks. arXiv preprint arXiv:1608.04644 (2016)
16. Szegedy, C., Zaremba, W., Sutskever, I., Bruna, J., Erhan, D., Goodfellow, I., Fergus, R.: Intriguing properties of neural networks. arXiv preprint arXiv:1312.6199 (2013)
17. Goodfellow, I.J., Shlens, J., Szegedy, C.: Explaining and harnessing adversarial examples. arXiv preprint arXiv:1412.6572 (2014)
18. Papernot, N., McDaniel, P., Goodfellow, I., Jha, S., Celik, Z.B., Swami, A.: Practical black-box attacks against deep learning systems using adversarial examples. arXiv preprint arXiv:1602.02697 (2016)
19. Gao, Y., Peters, J., Tsourdos, A., Zhifei, S., Joo, E.M.: A survey of inverse reinforcement learning techniques. Int. J. Intell. Comput. Cybern. **5**(3), 293–311 (2012)
20. Abadi, M., Agarwal, A., Barham, P., Brevdo, E., Chen, Z., Citro, C., Corrado, G.S., Davis, A., Dean, J., Devin, M., et al.: Tensorflow: large-scale machine learning on heterogeneous distributed systems. arXiv preprint arXiv:1603.04467 (2016)
21. Gilani, M., Inibhunu, C., Mahmoud, Q.H.: Application and network performance of Amazon elastic compute cloud instances. In: 2015 IEEE 4th International Conference on Cloud Networking (CloudNet), pp. 315–318. IEEE (2015)
22. Carlini, N., Wagner, D.: Defensive distillation is not robust to adversarial examples. arXiv preprint (2016)

Mobile Robot Localization via Machine Learning

Alexander Kuleshov[1], Alexander Bernstein[1,2], and Evgeny Burnaev[1,2(✉)]

[1] Skolkovo Institute of Science and Technology, Moscow, Russia
{A.Kuleshov,A.Bernstein,E.Burnaev}@skoltech.ru
[2] Kharkevich Institute for Information Transmission Problems RAS, Moscow, Russia

Abstract. We consider an appearance-based robot self-localization problem in the machine learning framework. Using recent manifold learning techniques, we propose a new geometrically motivated solution. The solution includes estimation of the robot localization mapping from the appearance manifold to the robot localization space, as well as estimation of the inverse mapping for image modeling. The latter allows solving the robot localization problem as a Kalman filtering problem.

Keywords: Machine learning · Robotics · Mobile robot self-localization · Appearance-based learning · Manifold learning · Regression on manifolds

1 Introduction

Machine learning is an essential and ubiquitous framework for solving a wide range of tasks in different application areas. One of these tasks is the problem of estimating position (localization) of a mobile robot moving in an uncertain environment, which is necessary for understanding the environment to make navigational decisions.

1.1 Robot Localization Problem

Localization is a fundamental component of mobile robotics to perform certain tasks in a given workspace. If a robot does not know where it is, it cannot navigate effectively and achieve goals. In order for a mobile robot to travel from one location to another it has to know its position and orientation (pose) at any given moment of time. Estimation of a mobile robot pose (localization) with respect to known locations in an environment is called the localization problem.

This task is especially important for navigation of autonomous robots when the localization problem is solved with the use of sensor systems installed on the robot. In this case, the task is called the self-localization problem.

The most common and basic method for performing localization is through dead-reckoning using data received from odometer sensors. This technique integrates a history of sensor readings and executed actions (e.g., velocity history) of the robot over time to determine changes in positions from its starting location [1,2]. Unfortunately, pure dead-reckoning methods are prone to errors growing

© Springer International Publishing AG 2017
P. Perner (Ed.): MLDM 2017, LNAI 10358, pp. 276–290, 2017.
DOI: 10.1007/978-3-319-62416-7_20

over time, so we need to periodically adjust the robot position with the help of other sensor systems, e.g. a visual sensing system (cameras). Usually in order to probabilistically update the robot position [4] an extended Kalman filter [3] is used.

Passive vision-based localization methods, including methods based only on visual data, have received extensive attention over the past few decades from the robotic and machine vision communities. Surveys [5,6] describe approaches to visual navigation of autonomous mobile robots, whose performance significantly depends on accurate robot localization in an environment. Thus vision-based localization solutions, developed for autonomous robots, are reviewed in these surveys.

We consider the self-localization problem of a mobile robot as a specific machine vision problem in which the robot position is estimated from images captured by its visual system. Omnidirectional imaging system, consisting of a vertically oriented standard color camera and a hyperbolic mirror, mounted in front of the lens [7–10] or camera with steerable orientation [11], is a typical example of such visual system.

Continuous set of images, which can be captured by a camera under all possible image formation parameters, is called the Appearance space. Parameters, defining the appearance space, involve relative position and orientation of the robot moving in a certain workspace, as well the camera intrinsic parameters and its illumination function (including color, intensity and angle). In this paper, we consider only the most typical case when these parameters contain robot localization (pose) consisting of robot position and orientation only.

Assuming that captured images allow distinguishing and recognizing poses from which they have been taken, the solution to the considered appearance-based (passive vision-based) robot localization problem is a mapping from the Appearance space to the Localization space consisting of all possible robot localizations.

1.2 Robot Localization: Machine Learning Framework

We consider the appearance-based localization problem in the machine learning framework: the appearance-based model describes the relation between observed images and their locations; the model is constructed using training data, which consists of captured images taken in known positions; the model allows estimating an unknown robot position from a newly acquired image. Such appearance-based learning framework has become very popular in the field of robot learning [12].

Because images are represented as very high-dimensional vectors, various appearance-based models (aka maps), describing the underlying low-dimensional structure in the appearance space, are usually constructed from training positions-images data using supervised learning techniques [13–16]. Appearance-based models provide an internal representation of the appearance space consisting of certain visual features, extracted from images. A number of methods for building such models using sonar, odometer and optical sensors have been proposed in the last few years. Given such model, the robot localization task can be considered as a prediction problem: to predict position of a robot from a new sensor observation as accurately as possibly.

1.3 Related Works

A geometrical approach to Machine Vision is to extract and estimate relative positions of detected environmental features or landmarks (e.g., doors, corners, columns, etc. in office workspace) [17,18] and then track over time their three-dimensional positions in order to estimate robot ego-motion [19,20].

In some cases, visual landmarks (as well as objects) are recognized by projecting rectangular sub-regions from newly acquired images into a space of descriptors and associating them with nearby ones. Various methods for recognizing landmark objects and scenes in images by a robot visual system are described in [21–27].

Appearance-based methods consider input images holistically, in relation to other images. Usually in case of appearance-based methods, as well as in geometrical approaches [23–27], natural visual features are computed by projecting images or range data onto low-dimensional subspaces [11,28–31]. For that Principal Component Analysis (PCA) [32] is often used. A review of PCA and other related subspace methods for the appearance-based localization problem is provided in [10]. There are a number of methods to determine significant variables in PCA. According to [33], the merit of these methods highly depends on the extent to which variables are correlated.

A mapping from visual features to the robot pose is highly nonlinear and sensitive to the type of selected features. Since PCA-features might not be able to maximize localization performance, other criteria are used to select linear subspaces [7–9]. For example, the subspace can be selected via Canonical Correlation Analysis (CCA) [34] to maximize correlations between poses of the robot and captured images [35,36]. In contrast to PCA, CCA performs a sort of a regression task in a sense that it provides a one-to-one mapping between an image and a camera pose [9]. A closely related problem is investigated in [29,37], in which ego-motion technique is used for estimating the robot pose based on Scale-Invariant Features Transform (SIFT features) [38] and Gaussian process regression [63–65]. In [39] linear projections of supervised high-dimensional robot observations are constructed to minimize conditional entropy of robot positions given the projected observations. Iterative technique for the minimization problem is proposed, which starts from the CCA-solution.

In [10] a probabilistic localization algorithm is proposed, which directly maps high-dimensional appearance images to robot positions via a nonlinear dimensionality reduction. Starting from Locally Linear Embedding manifold learning algorithm [40], Locally Linear Projection (LLP) method with subsequent usage of Bayesian filtering technique is proposed for robot pose estimation.

Regression dependency between PCA-based features of images and robot coordinates is constructed in [41] on the basis of the proposed General regression neural network, trained on panoramic snapshots.

Simultaneous Localization And Mapping (SLAM) algorithm is presented in [43], allowing to get a real-time localization. The algorithm combines dimension reduction methods (kernel principal component analysis [42]), manifold regularization techniques and parameter selection.

In [44] a multivariate Gaussian process (GP) with unknown hyper-parameters, estimated using maximum likelihood technique, models sought-for visual features of the data from an omnidirectional camera. Then GP regression is used for estimating the robot localization.

Papers [45,46] present appearance-based localization for an omnidirectional camera that builds on a combination of the group Least Absolute Shrinkage and Selection Operator (LASSO) [47] and the Extended Kalman filter (EKF) [3,48].

Speeded-Up Robust Features (SURF points) [49], being scale- and rotation-invariants utilizing Haar wavelet responses to produce a 64-dimensional descriptor vector for points of interest in an image, are computed for each image via the group LASSO regression [50]. The EKF uses an output of the LASSO regression-based initial localization as observations for a final localization.

1.4 Paper Contribution

The paper proposes new geometrically motivated manifold learning approach [51] to the solution of the appearance-based robot localization problem using specific Grassmann & Stiefel Eigenmaps algorithm [52] for nonlinear dimensionality reduction and regression on manifolds [53].

The paper is organized as follows. Section 2 contains rigorous statement of the appearance-based robot localization problem. Proposed approach is described in Sect. 3. Section 4 provides details of this solution. Section 5 describes how to use the solution in Kalman filtering procedures for robot localization.

2 Robot Localization: Rigorous Problem Statement

Let a mobile robot, equipped with a visual system (for example, an omnidirectional imaging system), moves on a 2D-workspace $\mathbf{W} \subset \mathbb{R}^2$. Its localization $\boldsymbol{\theta} = (\theta_{RC}, \theta_{RO}) \in \mathbb{R}^3$ is a three-dimensional vector consisting of **R**obot **P**osition $\theta_{RP} \in \mathbf{W} \subset \mathbb{R}^2$ (robot coordinates in the workspace $\mathbf{W}$) and **R**obot **O**rientation (an angle) $\theta_{RO} \in \mathbb{R}^1$ relative to the coordinate system in the workspace. Note that in case of a camera with steerable orientation, **C**amera **O**rientation (an angle) $\theta_{CO} \in \mathbb{R}^1$ relative to the robot should be included in the robot localization parameter, but for simplicity we consider only three-dimensional robot localization parameter $\boldsymbol{\theta}$. Let us denote by $\boldsymbol{\Theta} \subset \mathbb{R}^3$ a subset consisting of all possible robot localization parameters and called Localization space.

Let an image, captured by the robot imaging system, consists of p pixels. Therefore we can represent the image by a p-dimensional image-vector X. We denote by $X = \varphi(\boldsymbol{\theta}) \in \mathbb{R}^p$ an image, captured by the robot with the localization parameter $\boldsymbol{\theta}$, which is described by the Image modeling function φ with the domain of definition $\boldsymbol{\Theta}$. Let

$$\mathbf{X} = \varphi(\boldsymbol{\Theta}) = \{X = \varphi(\boldsymbol{\theta}), \, \boldsymbol{\theta} \in \boldsymbol{\Theta} \subset \mathbb{R}^3\} \subset \mathbb{R}^p, \tag{1}$$

be an Appearance space consisting of all possible images, which can be captured by the mobile robot and parameterized by the robot localization parameter $\boldsymbol{\theta} \in \mathbb{R}^3$.

We assume that images, captured by the robot in different localizations, are different; hence, the Image modeling function $\varphi : \boldsymbol{\Theta} \to \mathbf{X}$ is a one-to-one mapping from the Localization space to the Appearance space. Thus the Appearance space $\mathbf{X}$ is a manifold (Appearance manifold, AM) without self-intersections and with intrinsic dimension $q = 3$. This manifold is parameterized by the chart φ and is embedded in the ambient p-dimensional Euclidean space. Therefore, there exists an inverse mapping

$$\psi = \varphi^{-1} : X \in \mathbf{X} \to \boldsymbol{\Theta} = \psi(X) \in \boldsymbol{\Theta}, \tag{2}$$

called Localization function from the AM $\mathbf{X}$ to the Localization space $\boldsymbol{\Theta}$.

The functions φ and ψ, as well the AM $\mathbf{X}$, are unknown, and the Robot localization problem consists in constructing the robot localization $\boldsymbol{\theta} = \psi(X)$ from the image $X = \varphi(\boldsymbol{\theta}) \in \mathbf{X}$. We consider this problem in the machine learning framework [12]. We have a training set

$$\mathbf{S}_n = \{(X_i, \boldsymbol{\theta}_i), i = 1, 2, \dots, n\}, \tag{3}$$

consisting of images $\{X_i = \varphi(\boldsymbol{\theta}_i)\}$, captured by the robot imaging system in known conditions $\{\boldsymbol{\theta}_i \in \boldsymbol{\Theta}\}$ when robot moves in the workspace $\mathbf{W}$ randomly or on a regular grid. For example, the mobile robot, described in [7,8], captures omnidirectional images every 25 centimeters along robot random paths; reference positions are located on a regular grid with cells of size either 25 cm in a 2.7 m × 5.4 m workspace [9] or 1 m in a 20 m × 20 m workspace [10], respectively.

We consider two interconnected statements related to the robot localization problem and based on two representations of the training dataset $\mathbf{S}_n$ (3):

- Estimating the Localization function problem: to recover an unknown Localization function $\boldsymbol{\theta} = \psi(X)$ at an arbitrary out-of-sample point $X \notin \mathbf{X}$, from its known values $\{\boldsymbol{\theta}_i = \psi(X_i)\}$ at known points $\{X_i\}$,
- Estimating the Image modeling function problem: to predict a captured image $X = \varphi(\boldsymbol{\theta})$ at an arbitrary out-of-sample point $\boldsymbol{\theta} \in \boldsymbol{\Theta}$ from the known images $\{X_i = \varphi(\boldsymbol{\theta}_i)\}$ captured at known points $\boldsymbol{\theta}_i$.

The solution to the inverse problem can be used in the incremental statement of the robot localization problem considered below.

Both considered problems are regression problems with high-dimensional manifold valued inputs and high-dimensional manifold valued outputs, respectively.

Most of known appearance-based learning methods are related only to the first problem "Estimating the Localization function". In this regression problem the dimensionality p of input vectors $X \in \mathbf{X}$ is large, for example, $p = 16384, 10240$, and 3925 in case of panoramic images considered in [7–10], respectively; $p = 6912$ and $p = 4096$ in two examples considered in [11]. Thus standard regression methods perform poorly due the statistical and computational "curse of dimensionality" phenomenon: collinearity or "near-collinearity" of high-dimensional inputs cause difficulties when constructing regression; regression error cannot achieve a convergence rate faster than $n^{-s/(2s+p)}$ when estimating at least s times differentiable function $f(X)$ [54,55].

In order to avoid the curse of dimensionality phenomena, we use the fact that the Appearance space $\mathbf{X}$ (1) is a nonlinear manifold with intrinsic dimensionality $q = 3$, and q-dimensional visual features are constructed and used in the proposed method. Note that most of conventional methods use various dimensionality reduction techniques (usually, PCA) for constructing low-dimensional visual features by projecting images onto constructed low-dimensional subspaces, and the robot localization function is estimated from these features. However constructed low-dimensional subspaces may have a larger dimension than the real intrinsic dimension $q = 3$, i.e. they are not optimal.

3   Robot Localization: Proposed Approach

3.1   Regression Manifold

Consider an unknown smooth manifold called Regression manifold (RM)

$$\mathbf{M} = \{Z = F(\boldsymbol{\theta}), \, \boldsymbol{\theta} \in \boldsymbol{\Theta} \subset \mathbb{R}^3\} \subset \mathbb{R}^{p+3} \qquad (4)$$

with the intrinsic dimension $q = 3$, which is embedded in an ambient $(p + 3)$-dimensional Euclidean space and parameterized by an unknown chart

$$F : \boldsymbol{\theta} \in \boldsymbol{\Theta} \subset \mathbb{R}^3 \to Z = F(\boldsymbol{\theta}) = \begin{pmatrix} \varphi(\boldsymbol{\theta}) \\ \boldsymbol{\theta} \end{pmatrix} \in \mathbf{M}, \qquad (5)$$

defined on the Localization space $\boldsymbol{\Theta}$. Let

$$J_F(\boldsymbol{\theta}) = \nabla_{\boldsymbol{\theta}} F(\boldsymbol{\theta}) = \begin{pmatrix} J_\varphi(\boldsymbol{\theta}) \\ \mathbf{I}_3 \end{pmatrix}, \qquad (6)$$

be $(p + 3) \times q$ Jacobian matrix of the mapping F (5) which is split into $p \times 3$ Jacobian matrix $J_\varphi(\boldsymbol{\theta})$ of the Image modeling function φ (1) and $\mathbf{I}_3$ being a 3×3 unit matrix. The Jacobian $J_F(\boldsymbol{\theta})$ (6) determines a three-dimensional linear space $L(Z) = \mathrm{Span}(J_F(\boldsymbol{\theta}))$ in $\mathbb{R}^{p+3}$ which is a tangent space to the RM $\mathbf{M}$ at the point $Z = F(\boldsymbol{\theta}) \in \mathbf{M}$; hereinafter, $\mathrm{Span}(H)$ is a linear space spanned by columns of an arbitrary matrix H.

The set $\mathrm{TB}(\mathbf{M}) = \{(Z, L(Z)) : Z \in \mathbf{M}\}$ composed of points Z of the RM $\mathbf{M}$ equipped by tangent spaces $L_F(X)$ at these points is known in the Manifold theory [56,57] as the Tangent Bundle of the RM $\mathbf{M}$.

3.2 Tangent Bundle Manifold Learning Problem for Regression Manifold

The dataset $\mathbf{S}_n$ (3), written in the form

$$\mathbf{S}_n = \left\{ Z_i = \begin{pmatrix} X_i = \varphi(\boldsymbol{\theta}_i) \\ \boldsymbol{\theta}_i \end{pmatrix}, i = 1, 2, \ldots, n \right\} \tag{7}$$

$$= \left\{ Z_i = \begin{pmatrix} X_i \\ \boldsymbol{\theta}_i = \psi(X_i) \end{pmatrix}, i = 1, 2, \ldots, n \right\},$$

can be considered as a sample from the unknown RM $\mathbf{M}$ (4).

Let us consider certain dimensionality reduction problem called Tangent bundle manifold learning problem [58, 59] for the RM $\mathbf{M}$: estimate the Tangent Bundle TB($\mathbf{M}$) given the sample $\mathbf{S}_n$ (7) from the unknown RM $\mathbf{M}$.

3.3 Grassmann and Stiefel Eigenmaps Solution

Using Grassmann & Stiefel Eigenmaps (GSE) method [58, 59] and the sample $\mathbf{S}_n$ we construct the solution to the Tangent bundle manifold learning problem, resulting in the following quantities:

- sample-based area $\mathbf{M}^* \subset \mathbb{R}^{p+3}$ which is close to the unknown RM $\mathbf{M}$,
- embedding mapping $h_{GSE}(Z)$ from the area $\mathbf{M}^*$ to the 3-dimensional Feature space (FS) $\mathbf{Y}_{GSE} = h_{GSE}(\mathbf{M}^*) \subset \mathbb{R}^3$,
- recovery mapping $g_{GSE}(y)$ from the FS $\mathbf{Y}_{GSE}$ to $\mathbb{R}^{p+3}$,
- $(p+3) \times q$ matrix $G_{GSE,g}(y)$ defined on the FS $\mathbf{Y}_{GSE}$,

which together provides both

- proximity

$$Z_{GSE}(Z) \equiv g_{GSE}(h_{GSE}(Z)) \approx Z \text{ for all points } Z \in \mathbf{M}^*, \tag{8}$$

between initial and recovered points Z and $Z_{GSE}(Z)$. Thanks to (8) we get small Hausdorff distance $d_H(\mathbf{M}, \mathbf{M}_{GSE})$ between the RM $\mathbf{M}$ and the three-dimensional recovered regression manifold (RRM)

$$\mathbf{M}_{GSE} = \{g_{GSE,g}(y) \in \mathbb{R}^{p+3} : y \in \mathbf{Y}_{GSE} \subset \mathbb{R}^3\}, \tag{9}$$

embedded in the ambient $(p+3)$-dimensional Euclidean space;
- proximity

$$G_{GSE,g}(y) \approx J_{GSE,g}(y) \text{ for all points } y \in \mathbf{Y}_{GSE}, \tag{10}$$

in which $J_{GSE,g}(y)$ is a Jacobian matrix of the mapping $g_{GSE}(y)$. Thanks to (10) we get proximity between the tangent space $L(Z)$ to the RM $\mathbf{M}$ at the point Z and the tangent space $L_{GSE}(Z) = \text{Span}(G_{GSE,g}(h_{GSE}(Z)))$ to the RRM $\mathbf{M}_{GSE}$ (9) at the nearby recovered point $Z_{GSE}(Z)$. The proximity between these tangent spaces, considered as elements of the Grassmann manifold, is defined using chosen metric on the Grassmann manifold.

Therefore, the tangent bundle $\text{TB}(\mathbf{M}_{GSE}) = \{(Z_{GSE}(Z), L_{GSE}(Z)) : Z \in \mathbf{M}\}$ of the RRM $\mathbf{M}_{GSE}$ accurately approximates the tangent bundle $\text{TB}(\mathbf{M})$.

Note also that the original GSE algorithm [58, 59] has computational complexity $O(n^3)$ for a sample size n; the incremental version of the GSE [60] has significantly smaller running time $O(n^{(q+4)/(q+2)})$.

3.4   Robot Localization: GSE-Based Approach

A splitting of an arbitrary vector $Z = \begin{pmatrix} Z_u \\ Z_v \end{pmatrix} \in \mathbb{R}^{p+3}$ into two vectors $Z_u \in \mathbb{R}^p$ and $Z_v \in \mathbb{R}^3$ implies the corresponding partitions

$$g_{GSE}(y) = \begin{pmatrix} g_{GSE,u}(y) \\ g_{GSE,v}(y) \end{pmatrix}, \; G_{GSE,g}(y) = \begin{pmatrix} g_{GSE,g,u}(y) \\ g_{GSE,g,v}(y) \end{pmatrix} \tag{11}$$

of the mapping $g_{GSE}(y)$ and the matrix $G_{GSE,g}(y)$.

Using the representation $Z = F(\boldsymbol{\theta})$ (5), the embedding mapping $y = h_{GSE}(Z)$, defined on the RM $\mathbf{M}$, can be written as a function

$$y = R_{GSE}(\boldsymbol{\theta}) = h_{GSE}(F(\boldsymbol{\theta})), \tag{12}$$

defined on the Localization space $\boldsymbol{\Theta}$.

Using the Localization function ψ (2), the RM $\mathbf{M}$ and the embedding mapping $y = h_{GSE}(Z)$ can be written as $\mathbf{M} = \{Z = f(X), X \in \mathbf{X}\} \subset \mathbb{R}^{p+3}$ and

$$y = r_{GSE}(X) = h_{GSE}(f(X)) \tag{13}$$

respectively, where the functions

$$f(X) = \begin{pmatrix} X \\ \psi(X) \end{pmatrix} \tag{14}$$

and $r_{GSE}(X)$ (13) are defined on the Appearance space $\mathbf{X}$.

In the paper [53], the $3 \times p$ and 3×3 Jacobian matrices $J_{GSE,r}(X)$ (the covariant differentiation is used here) and $J_{GSE,R}(\boldsymbol{\theta})$ of the mappings $r_{GSE}(X)$ and $R_{GSE}(\boldsymbol{\theta})$ are estimated by the matrices

$$G_{GSE,r}(X) = G^-_{GSE,g,u}(r_{GSE}(X)) \times \pi_{GSE,\mathbf{x}}(X), \tag{15}$$

$$G_{GSE,R}(\boldsymbol{\theta}) = G^{-1}_{GSE,g,v}(R_{GSE}(\boldsymbol{\theta})), \tag{16}$$

respectively. Here $H^- = (H^T \times H)^{-1} \times H^T$ denotes a pseudo-inverse Moore-Penrose matrix [61] of an arbitrary matrix H and $\pi_{GSE,\mathbf{x}}(X)$ is a certain estimator [62] of a $p \times p$ projection matrix onto the tangent space of the AM $\mathbf{X}$ (1) at the point $X \in \mathbf{X}$.

Using representations (12) and (13), the proximity (8), implies approximate equalities

$$\psi_{GSE}(X) \equiv g_{GSE,v}(r_{GSE}(X)) \approx \psi(X), \tag{17}$$

$$\varphi_{GSE}(\boldsymbol{\theta}) \equiv g_{GSE,u}(R_{GSE}(\boldsymbol{\theta})) \approx \varphi(\boldsymbol{\theta}). \tag{18}$$

Although the GSE-based functions $\psi_{GSE}(X)$ (17) and $\varphi_{GSE}(\boldsymbol{\theta})$ (18) accurately approximate the sought-for functions $\psi(X)$ and $\varphi(\boldsymbol{\theta})$, respectively, they cannot be considered as the solution to the Robot Localization problem because the mappings $g_{GSE,u}(y)$ and $g_{GSE,v}(y)$ (11) depend on the argument

$$y = r_{GSE}(X) = R_{GSE}(\boldsymbol{\theta}), \tag{19}$$

whose values are known only at sample points:

$$y_i = r_{GSE}(X_i) = R_{GSE}(\boldsymbol{\theta}_i) = h_{GSE}(Z_i), i = 1, 2, \ldots, n. \tag{20}$$

Based on known values (20) of the functions $r_{GSE}(X)$ and $R_{GSE}(\boldsymbol{\theta})$ (19) at sample points, as well on the known values of their Jacobian matrices (15), (16) at these points, the estimators $r^*(X)$ and $R^*(\boldsymbol{\theta})$ of these functions at arbitrary points $X \in \mathbf{X}$ and $\boldsymbol{\theta} \in \boldsymbol{\Theta}$, respectively, can be constructed using the Jacobian Regression method, proposed in [53]. Substitution of estimators $r^*(X)$ and $R^*(\boldsymbol{\theta})$ in formulas (17) and (18) instead of $r_{GSE}(X)$ and $R_{GSE}(\boldsymbol{\theta})$, provides the final estimators $\psi^*(\mathbf{X})$ and $\varphi^*(\boldsymbol{\theta})$ for the Localization function $\psi(X)$ and the Image modeling function $\varphi(\boldsymbol{\theta})$.

4   Robot Localization: Solution

4.1   Estimating the Localization Function

GSE solution applied to the RM $\mathbf{M}$ includes construction of the kernels $K_X(X, X')$ and $K_\theta(\boldsymbol{\theta}, \boldsymbol{\theta}')$ on the AM $\mathbf{X}$ and the Localization space $\boldsymbol{\Theta}$, respectively. These kernels reflect not only geometrical closeness between points $Z = F(\boldsymbol{\theta}) = f(X)$ and $Z' = F(\boldsymbol{\theta}') = f(X')$ but also closeness between the tangent spaces $L(Z)$ and $L(Z')$ to the RM $\mathbf{M}$.

Estimating the GSE-based embedding mapping $r_{GSE}(X)$. For close points $X, X' \in \mathbf{X}$, write the Taylor series expansion

$$r_{GSE}(X) - r_{GSE}(X') \approx J_{GSE,r}(X') \times (X - X')$$

for the embedding function $y = r_{GSE}(X)$.

Taking as points X' the sample images $\{X_i\}$, using known values y_i (20) of $r_{GSE}(X_i)$, and replacing the Jacobians $J_{GSE,r}(X_i)$ by estimators $\{G_{GSE,r}(X_i)\}$ (15) known at sample points, construct the estimator $y^* = r^*(X)$ for the embedding mapping $y = r_{GSE}(X)$ as a minimizer of the cost function

$$\sum_{i=1}^{n} K_X(X, X_i) \times |y - y_i - G_{GSE,r}(X_i) \times (X - X_i)|^2$$

over y. The solution $r^*(X)$ to this problem is computed in an explicit form

$$r^*(X) = \frac{1}{K_X(X)} \times \sum_{i=1}^{n} K_X(X, X_i) \times \{y_i + G_{GSE,r}(X_i) \times (X - X_i)\}, \tag{21}$$

where $K_X(X) = \sum_{i=1}^{n} K_X(X, X_i)$.

Final estimator for the Localization function at arbitrary point $X \in \mathbf{X}$ is given by the formula

$$\psi^*(X) \equiv g_{GSE,v}(r^*(X)). \tag{22}$$

4.2   Estimating the Image Modeling Function

Estimating the GSE-based embedding mapping $R_{GSE}(\boldsymbol{\theta})$. For close points $\boldsymbol{\theta}, \boldsymbol{\theta}' \in \boldsymbol{\Theta}$, write the Taylor series expansion

$$R_{GSE}(\boldsymbol{\theta}) - R_{GSE}(\boldsymbol{\theta}') \approx J_{GSE,R}(\boldsymbol{\theta}') \times (\boldsymbol{\theta} - \boldsymbol{\theta}')$$

for the embedding function $y = R_{GSE}(\boldsymbol{\theta})$.

Taking as points $\boldsymbol{\theta}'$ the sample localizations $\{\boldsymbol{\theta}_i\}$, using known values $\{y_i\}$ (20) of $\{R_{GSE}(\boldsymbol{\theta}_i)\}$, and replacing the Jacobians $\{J_{GSE,R}(\boldsymbol{\theta}_i)\}$ by their estimators $\{G_{GSE,R}(\boldsymbol{\theta}_i)\}$ (16) known at sample points, choose the estimator $y^* = R^*(\boldsymbol{\theta})$ for the embedding mapping $y = R_{GSE}(\boldsymbol{\theta})$ as a quantity that minimizes the cost function

$$\sum_{i=1}^{n} K_{\boldsymbol{\theta}}(\boldsymbol{\theta}, \boldsymbol{\theta}_i) \times |y - y_i - G_{GSE,R}(\boldsymbol{\theta}_i) \times (\boldsymbol{\theta} - \boldsymbol{\theta}_i)|^2$$

over y. The solution $R^*(\boldsymbol{\theta})$ to this problem is computed in an explicit form

$$R^*(\boldsymbol{\theta}) = \frac{1}{K_{\boldsymbol{\theta}}(\boldsymbol{\theta})} \times \sum_{i=1}^{n} K_{\boldsymbol{\theta}}(\boldsymbol{\theta}, \boldsymbol{\theta}_i) \times \{y_i + G_{GSE,R}(\boldsymbol{\theta}_i) \times (\boldsymbol{\theta} - \boldsymbol{\theta}_i)\}, \tag{23}$$

where $K_{\boldsymbol{\theta}}(\boldsymbol{\theta}) = \sum_{i=1}^{n} K_{\boldsymbol{\theta}}(\boldsymbol{\theta}, \boldsymbol{\theta}_i)$.

Final estimator for the Image modeling function at arbitrary point $\boldsymbol{\theta} \in \boldsymbol{\Theta}$ is given by the formula

$$\varphi^*(\boldsymbol{\theta}) \equiv g_{GSE,u}(R^*(\boldsymbol{\theta})). \tag{24}$$

5   Usage of Predicted Images for Robot Localization

The constructed estimator $\varphi^*(\boldsymbol{\theta})$ (24), which predicts an image $X = \varphi(\boldsymbol{\theta})$ captured at the localization $\boldsymbol{\theta} \in \boldsymbol{\Theta}$, can be used in Kalman filtering procedures (3) for robot localization.

Robot navigation consists in choosing control $u(t)$ at given time moments $t = 0, 1, 2, \ldots$. Let $\boldsymbol{\theta}(t)$ be a current robot position at time t, then, under some control $u(t)$, the robot must move in the expected position

$$\boldsymbol{\theta}(t+1) = F(\boldsymbol{\theta}(t), u(t)) \equiv F_t(\boldsymbol{\theta}(t)), \tag{25}$$

where $F(\boldsymbol{\theta}, u)$ is a known function defined by a solution of a navigation motion control problem.

In practice, the estimated position $\boldsymbol{\theta}_t(t)$ of the robot at time t is only known, which differs from the exact position $\boldsymbol{\theta}(t)$; the exact position $\boldsymbol{\theta}(t+1)$ at time $(t+1)$ also differs from the expected position $\boldsymbol{\theta}_t(t+1) = F_t(\boldsymbol{\theta}_t(t))$.

Let a robot visual sensing system provides a captured image $X(t+1) = \varphi(\boldsymbol{\theta}(t+1))$ at the moment $(t+1)$. We want to solve a filtering problem to improve the predicted localization $\boldsymbol{\theta}_t(t+1)$ from the captured image $X(t+1)$.

The constructed estimator $\varphi^*(\boldsymbol{\theta})$ (24) allows predicting $X^*(t+1) = \varphi^*(\boldsymbol{\theta}_t(t+1))$ for the captured image $X(t+1)$, and a standard Kalman filter [3] constructs the improved localization $\boldsymbol{\theta}_{t+1}(t+1)$ as

$$\boldsymbol{\theta}_{t+1}(t+1) = \boldsymbol{\theta}_t(t+1) + B(t+1) \times (X(t+1) - X^*(t+1)). \qquad (26)$$

Here $B(t+1)$ is a Kalman gain.

Using the estimator $\psi^*(X)$ (22) for the Localization function $\psi(X)$ (2), we can use a quantity $\psi^*(X(t+1))$, representing visual features, as an estimator of the robot pose in which the image $X(t+1)$ has been taken, and construct the estimator $\boldsymbol{\theta}_{t+1}(t+1)$ as

$$\boldsymbol{\theta}_{t+1}(t+1) = \boldsymbol{\theta}_t(t+1) + b(t+1) \times (\psi^*(X(t+1)) - \psi^*(X^*(t+1))). \qquad (27)$$

Here $b(t+1)$ is another gain function. Usage of the estimator (27) allows avoiding of handling high-dimensional images X, by replacing them with visual features $\psi^*(X)$.

As measurements, used in filtering procedures, it is possible to use low-dimensional representations $r^*(X)$ (21) of vectors $f(X)$ (14) from the RM $\mathbf{M}$, and to construct the estimator $\boldsymbol{\theta}_{t+1}(t+1)$ according to

$$\boldsymbol{\theta}_{t+1}(t+1) = \boldsymbol{\theta}_t(t+1) + d(t+1) \times (\psi^*(X(t+1)) - \psi^*(X^*(t+1))), \qquad (28)$$

where $d(t+1)$ is some gain function.

For choosing the optimal gain functions in (26)–(28), it is necessary to know covariance matrices for deviations between observations and their expected values, as well as between the expected robot position $\boldsymbol{\theta}_t(t+1) = F_t(\boldsymbol{\theta}_t(t))$ (25) and its real pose $\boldsymbol{\theta}(t+1)$. Corresponding covariance matrices can be estimated from the sample $\mathbf{S}_n$ (3), (7) in which robot poses are accurately known.

Acknowledgments. The research was supported solely by the Russian Science Foundation grant (project 14-50-00150).

References

1. Talluri, R., Aggarwal, J.K.: Position estimation techniques for an autonomous mobile robot – A review. In: Chen, C.H., Pau, L.F., Wang, P.S.P. (eds.) Handbook of Pattern Recognition and Computer Vision, chap. 4.4, pp. 769–801. World Scientific, Singapore (1993)
2. Borenstein, J.H., Everett, R., Feng, L., Wehe, D.: Mobile robot positioning: sensors and techniques. J. Robot. Syst. **14**, 231–249 (1997)

3. Candy, J.V.: Model-Based Signal Processing. John Wiley & Sons, Inc., New York (2006)
4. Olson, C.F.: Probabilistic self-localization for mobile robots. IEEE Trans. Robot. Autom. **16**(1), 55–66 (2000)
5. DeSouza, G.N., Kak, A.C.: Vision for mobile robot navigation: a survey. IEEE Trans. Pattern Anal. Mach. Intell. **24**(2), 237–267 (2002)
6. Bonin-Font, F., Ortiz, A., Oliver, G.: Visual navigation for mobile robots: a survey. J. Intell. Rob. Syst. **53**(3), 263–296 (2008)
7. Kröse, B.J.A., Vlassis, N., Bunschoten, R.: Omnidirectional vision for appearance-based robot localization. In: Hager, G.D., Christensen, H.I., Bunke, H., Klein, R. (eds.) Sensor Based Intelligent Robots. LNCS, vol. 2238, pp. 39–50. Springer, Heidelberg (2002). doi:10.1007/3-540-45993-6_3
8. Krose, B.J.A., Vlassis, N., Bunschoten, R., Motomura, Y.: A probabilistic model for appearance-based robot localization. Image Vis. Comput. **19**, 381–391 (2001)
9. Saito, M., Kitaguchi, K.: Appearance based robot localization using regression models. In: Proceedings of 4th IFAC-Symposium on Mechatronic Systems, vol. 2, pp. 584–589 (2006)
10. Hamm, J., Lin, Y., Lee, D.D.: Learning nonlinear appearance manifolds for robot localization. In: Proceedings of the IEEE/RSJ International Conference on Intelligent Robots and Systems (IROS 2005), pp. 1239–1244 (2005)
11. Crowley, J.L., Pourraz, F.: Continuity properties of the appearance manifold for mobile robot position estimation. Image Vis. Comput. **19**(11), 741–752 (2001)
12. Pauli, J.: Learning-Based Robot Vision. LNCS, vol. 2048, 292 p. Springer, Heidelberg (2001)
13. Oore, S., Hinton, G.E., Dudek, G.: A mobile robot that learns its place. Neural Comput. **9**, 683–699 (1997)
14. Thrun, S.: Bayesian landmark learning for mobile robot localization. Mach. Learn. **33**(1), 41–76 (1998)
15. Krose, B.J.A., Bunschoten, R.: Probabilistic localization by appearance models and active vision. In: Proceedings of the IEEE International Conference on Robotics and Automation (ICRA 1999), Detroit, Michigan, pp. 2255–2260 (1999)
16. Vlassis, N., Krose, B.J.A.: Robot environment modeling via principal component regression. In: Proceedings of the IEEE/RSJ International Conference on Intelligent Robots and Systems (IROS 1999), pp. 677–682 (1999)
17. Se, S., Lowe, D., Little, J.: Local and global localization for mobile robots using visual landmarks. In: Proceedings of the IEEE/RSJ International Conference on Intelligent Robots and Systems (IROS 2001), pp. 414–420 (2001)
18. Hayet, J., Lerasle, F., Devy, M.: Visual landmarks detection and recognition for mobile robot navigation. In: Proceedings of the IEEE Computer Society Conference on Computer Vision and Pattern Recognition (CVPR 2003), vol. 2, pp. 313–318 (2003)
19. Bunschoten, R., Krose, B.J.A.: 3-D scene reconstruction from cylindrical panoramic images. In Proceedings of the 9th International Symposium on Intelligent Robotic Systems (SIRS-2001), pp. 199–205 (2001)
20. Gluckman, J., Nayar, S.K.: Ego-motion and omnidirectional cameras. In: Proceedings of the Sixth International Conference on Computer Vision (ICCV 1998), pp. 999–1005 (1998)
21. Colin de Verdiere, V., Crowley, J.L.: Local appearance space for recognition of navigation landmarks. J. Robot. Auton. Syst. **32**(1–2), 61–89 (2000)

22. Dudek, G., Jugessur, D.: Robust place recognition using local appearance based methods. In: Proceedings of the International Conference on Robotics and Automation (ICRA 2000), pp. 1030–1035 (2000)
23. Betke, M., Gurvits, L.: Mobile robot localization using landmarks. IEEE Trans. Robot. Autom. **13**, 251–263 (1997)
24. Sugihara, K.: Some location problems for robot navigation using a single camera. Comput. Vis. Graph. Image Process. **42**, 112–129 (1988)
25. Sim, R., Dudek, G.: Robot positioning using learned landmarks. In: Proceedings of the IEEE/RSJ International Conference on Intelligent Robots and Systems (IROS 1998), vol. 2, pp. 1060–1065 (1998)
26. Friedman, A.: Robot localization using landmarks. In: Mathematics in Industrial Problems. The IMA Volumes in Mathematics and its Applications, vol. 67(7), pp. 86–94. Springer, New York (1995)
27. Jogan, M., Leonardis, A.: Robust localization using panoramic view-based recognition. In: Proceedings of the 15th International Conference on Pattern Recognition (ICPR 2000), pp. 136–139. IEEE Computer Society (2000)
28. Vlassis, N., Motomura, Y., Krse, B.J.A.: Supervised dimension reduction of intrinsically low-dimensional data. Neural Comput. **14**(1), 191–215 (2002)
29. Se, S., Lowe, D., Little, J.: Vision-based global localization and mapping for mobile robots. IEEE Trans. Rob. **21**(3), 364–375 (2005)
30. Cobzas, D., Zhang, H.: Cylindrical panoramic image-based model for robot localization. In: Proceedings of the IEEE/RSJ International Conference on Intelligent Robots and Systems (IROS), Maui, HI, pp. 1924–1930 (2001)
31. Crowley, J.L., Wallner, F., Schiele, B.: Position estimation using principal components of range data. In: Proceedings of the 1998 IEEE International Conference on Robotics and Automation, vol. 4, pp. 3121–3128 (1998)
32. Jollie, T.: Principal Component Analysis. Springer, New-York (2002)
33. Peres-Neto, P.R., Jackson, D.A., Somers, K.M.: How many principal components? stopping rules for determining the number of non-trivial axes revisited. Comput. Stat. Data Anal. **49**(4), 974–997 (2005)
34. Härdle, W.K., Simar, L.: Canonical correlation analysis. In: Applied Multivariate Statistical Analysis, pp. 443–454. Springer, Heidelberg (2015). doi:10.1007/978-3-662-45171-7_16
35. Melzer, T., Reiter, M., Bischof, H.: Appearance models based on kernel canonical correlation analysis. Pattern Recogn. **36**(9), 1961–1973 (2003)
36. Skocaj, D., Leonardis, A.: Appearance-based localization using CCA. In: Proceedings of the of the 9th Computer Vision Winter Workshop (CVWW 2004), pp. 205–214 (2004)
37. Se, S., Lowe, D., Little, J.: Mobile robot localization and mapping with uncertainty using scale-invariant visual landmarks. Int. J. Robot. Res. **21**(8), 735–758 (2002)
38. Lowe, D.: Distinctive image features from scale-invariant keypoints. Int. J. Comput. Vis. **60**(2), 91–110 (2004)
39. Vlassis, N., Motomura, Y., Krose, B.J.A.: Supervised linear feature extraction for mobile robot localization. In: Proceedings of the IEEE International Conference on Robotics and Automation (ICRA 2000), vol. 4, pp. 2979–2984 (2000)
40. Saul, L.K., Roweis, S.T.: Nonlinear dimensionality reduction by locally linear embedding. Science **290**, 2323–2326 (2000)
41. Wang, K., Wang, W., Zhuang, Y.: Appearance-based map learning for mobile robot by using generalized regression neural network. In: Liu, D., Fei, S., Hou, Z.-G., Zhang, H., Sun, C. (eds.) ISNN 2007. LNCS, vol. 4491, pp. 834–842. Springer, Heidelberg (2007). doi:10.1007/978-3-540-72383-7_97

42. Scholkopf, B., Smola, A., Muller, K.: Nonlinear component analysis as a kernel eigenvalue problem. Neural Comput. **10**(5), 1299–1319 (1998)
43. Wu, H., Wu, Y.-X., Liu, C.-A., Yang, G.-T., Qin, S.-Y.: Fast robot localization approach based on manifold regularization with sparse area features. Cognitive Comput. **8**(5), 856–876 (2016)
44. Do, H.N., Jadaliha, M., Choi, J., Lim, C.Y.: Feature selection for position estimation using an omnidirectional camera. Image Vis. Comput. **39**, 1–9 (2015)
45. Do, H.N., Choi, J., Lim, C.Y., Maiti, T.: Appearance-based localization using Group LASSO regression with an indoor experiment. In: Proceedings of the 2015 IEEE International Conference on Advanced Intelligent Mechatronics (AIM 2015), pp. 984–989 (2015)
46. Do, H.N., Choi, J.: Appearance-based outdoor localization using group lasso regression. In: Proceedings of the ASME Dynamic Systems and Control Conference (DSCC 2015), vol. 3, 8 p. (2015)
47. Tibshirani, R.: Regression shrinkage and selection via the lasso: a retrospective. J. Roy. Stat. Soc.: Ser. B (Methodol.) **73**(3), 273–282 (2011)
48. Ribeiro, M.I.: Kalman and extended Kalman filters: Concept, derivation and properties. Institute for Systems and Robotics, Technical report, 44 p. (2004)
49. Herbert, B., Andreas, E., Tuytelaars, T., Gool, L.V.: Speeded-Up Robust Features (SURF). Comput. Vis. Image Underst. **110**(3), 346–359 (2008)
50. Obozinski, G., Wainwright, M.J., Jordan, M.I., et al.: Support union recovery in high-dimensional multivariate regression. Annal. Stat. **39**(1), 1–47 (2011)
51. Ma, Y., Fu, Y. (eds.): Manifold Learning Theory and Applications. CRC Press, London (2011)
52. Kuleshov, A.P., Bernstein, A.V.: Manifold learning in data mining tasks. In: Perner, P. (ed.) MLDM 2014. LNCS, vol. 8556, pp. 119–133. Springer, Heidelberg (2014)
53. Kuleshov, A.P., Bernstein, A.V.: Statistical learning on manifold-valued data. In: Perner, P. (ed.) MLDM 2016. LNCS, vol. 9729, pp. 311–325. Springer International Publishing, Switzerland (2016)
54. Stone, C.J.: Optimal rates of convergence for nonparametric estimators. Ann. Stat. **8**, 1348–1360 (1980)
55. Stone, C.J.: Optimal global rates of convergence for nonparametric regression. Ann. Stat. **10**, 1040–1053 (1982)
56. Lee, J.M.: Manifolds and Differential Geometry. Graduate Studies in Mathematics, vol. 107. American Mathematical Society, Providence (2009)
57. Lee, J.M.: Introduction to Smooth Manifolds. Springer, New York (2003)
58. Bernstein, A.V., Kuleshov, A.P.: Tangent bundle manifold learning via Grass-mann & Stiefel eigenmaps. In: arxiv:1212.6031v1 [cs.LG], pp. 1–25 (2012), December 2012
59. Bernstein, A.V., Kuleshov, A.P.: Manifold Learning: generalizing ability and tangent proximity. Int. J. Softw. Inf. **7**(3), 359–390 (2013)
60. Kuleshov, A., Bernstein, A.: Incremental construction of low-dimensional data representations. In: Schwenker, F., Abbas, H.M., El Gayar, N., Trentin, E. (eds.) ANNPR 2016. LNCS, vol. 9896, pp. 55–67. Springer, Cham (2016). doi:10.1007/978-3-319-46182-3_5
61. Golub, G.H., Van Loan, C.F.: Matrix Computation, 3rd edn. Johns Hopkins University Press, Baltimore (1996)
62. Kuleshov, A.P., Bernstein, A.V.: Regression on high-dimensional inputs. In: Workshops Proceedings volume of the IEEE International Conference on Data Mining (ICDM 2016), pp. 732–739. IEEE Computer Society, USA (2016)

63. Burnaev, E., Belyaev, M., Kapushev, E.: Computationally efficient algorithm for Gaussian Processes based regression in case of structured samples. Comput. Math. Math. Phys. **56**(4), 499–513 (2016)
64. Burnaev, E., Panov, M., Zaytsev, A.: Regression on the basis of nonstationary Gaussian processes with Bayesian regularization. J. Commun. Technol. Electron. **61**(6), 661–671 (2016)
65. Burnaev, E., Zaytsev, A.: Surrogate modeling of mutlifidelity data for large samples. J. Commun. Technol. Electron. **60**(12), 1348–1355 (2016)

An Analysis of the Application of Simplified Silhouette to the Evaluation of k-means Clustering Validity

Fei Wang[1,3]($\boxtimes$), Hector-Hugo Franco-Penya[1], John D. Kelleher[1,3], John Pugh[2], and Robert Ross[1,3]

[1] School of Computing, Dublin Institute of Technology, Dublin, Ireland
d13122837@mydit.ie
[2] Nathean Technologies Ltd., Dublin, Ireland
[3] ADAPT Research Centre, Dublin, Ireland

Abstract. This paper analyses the application of Simplified Silhouette to the evaluation of k-means clustering validity and compares it with the k-means Cost Function and the original Silhouette. We conclude that for a given dataset the k-means Cost Function is the most valid and efficient measure in the evaluation of the validity of k-means clustering with the same k value, but that Simplified Silhouette is more suitable than the original Silhouette in the selection of the best result from k-means clustering with different k values.

Keywords: k-means · Simplified Silhouette · Silhouette · Cost Function

1 Introduction

Clustering aims to partition data into homogeneous groups [7,12]. Unlike supervised machine learning methods, clustering does not require external labels as ground truth, but investigates the intrinsic structure and characteristics of data, and partitions data into clusters such that the data in the same cluster are more similar to each other than the data in other clusters. Clustering has been applied in many domains, such as image and text analysis, biology and so on [7], and also noted as an important part of unsupervised learning in many data mining and machine learning text books [8,12].

Our research focuses on the application of clustering to the domain of Business Intelligence. The effective segmentation of customer data is a vital tool for commercial users. For this purpose, two specific characteristics need to be considered in the clustering: (1) there are a large proportion of categorical features in customer data; and (2) users don't have much a priori knowledge about clustering. In our previous research [10], we compared different methods for categorical data clustering, such as 1-of-k coding and k-prototypes. In this paper, we look at k-means clustering, and aim to find the best way to automate the selection of the best clustering result from a set of k-means clusterings with different parameter configurations.

© Springer International Publishing AG 2017
P. Perner (Ed.): MLDM 2017, LNAI 10358, pp. 291–305, 2017.
DOI: 10.1007/978-3-319-62416-7_21

k-means is one of the most widely used clustering algorithms due to its ease of implementation, simplicity, efficiency and empirical success. There are two parameters that need to be set before the start of k-means clustering - the number of clusters k and the initial centroids. Given a fixed parameter configuration, k-means will output a fixed clustering result. However, because different parameter configurations usually lead to different clustering results, a single k-means clustering cannot guarantee the best clustering result. The common way to implement k-means is to run it multiple times with different parameter configurations and select the best one from all the clustering results. The process to find the best result is normally based on the evaluation of the clustering validity, that is, the goodness or quality of the clustering result for a dataset [12].

In this paper, we mainly analyse the application of an internal measure for evaluating the clustering validity - Simplified Silhouette - and compare it with other related measures in k-means clustering. We start with a brief introduction to the background of the evaluation of k-means clustering validity in Sect. 2, followed by a theoretical analysis in Sect. 3. In Sect. 4, we outline the design for our empirical analysis, and then in Sect. 5 present and analyse the experimental results. Finally, in Sect. 6 we draw conclusions and outline future work.

2   Background

Three types of measures can be used to empirically evaluate clustering validity [12]: internal, external and relative measures. Internal measures are based on the intrinsic structure and characteristics of the dataset. External measures are based on labelled datasets, and compare the clustering results with the existing labels. Relative measures compare different clusterings generated with the same clustering algorithm but different parameter settings. Because clustering is usually used in situations where there is no labelled data, internal measures are the most generally used measures for the evaluation of clustering validity in practice and therefore our research focus in this paper.

Internal measures are usually some indices designed to show the compactness and separation of data [12]. The compactness means that the data within the same cluster should be close to each other, and the separation means that the data in different clusters should be widely spaced. There are numerous different internal measures for the evaluation of clustering validity. Different measures show these two concepts in different ways.

First of all, for some clustering algorithms like k-means, the design of the algorithm aims to minimise a cost function. Intuitively, the cost function can be considered as an internal measure for evaluating the clustering validity of this specific algorithm. The k-means Cost Function [7] is defined as the sum of all the distances of each point to its cluster centroid. The process of k-means is designed specifically to reduce the Cost Function by centroid shifts and re-assignments of the data to its closest cluster until the Cost Function converges to a minimum (the optimum), so the convergence of the Cost Function is a monotonic process in k-means. Additionally, because the distances of each data point to its centroid have been calculated during the process of k-means, the calculation of the k-means Cost Function is only to sum up these distances, which requires few extra calculations. Therefore, we can consider the k-means Cost Function as the default internal measure for k-means and the clustering result with the smallest Cost Function as the best result or global optimum. However, for the evaluation of the validity of clustering with different k values, using the Cost Function measure is problematic because it tends to reduce as the k value increases and, consequently, the Cost Function measure has an intrinsic bias toward selecting the result with the largest k as the best result [7]. Therefore we have to use other internal measures.

In addition to the kind of internal measures designed specifically for a clustering algorithm like the k-means Cost Function, there are quite a lot of general internal measures that can be applied in the evaluation of the validity of a set of clustering algorithms: the Dunn index [2] adopts the maximum intra-cluster distance and the minimum inter-cluster distance, the Davies-Bouldin (DB) index [1] evaluates the dispersion of data based on the distances between cluster centroids, the C-index [12] takes the sum of a set of the smallest distances as the baseline, and the SD index [5] is defined based on the concepts of the average scattering for clusters and total separation between clusters.

Silhouette analyses the distances of each data point to its own cluster and its closest neighbouring cluster (defined as the average distance of a data point to all the other data points in its own cluster and that to all the data points in the neighbouring cluster nearest to the data point). Different from most other internal measures, Silhouette is not only used for the evaluation of the validity of a full clustering, but also can be used for that of a single cluster or even a single data point to see if it is well clustered. The calculation of Silhouette starts from each data point, and the Silhouette value of a cluster or a full clustering is just the average of point Silhouette values for all the data involved. Regarding our focus on customer segmentation, it is the advantage of Silhouette that it shows if each customer or customer cluster is well segmented.

Compared with that of the k-means Cost Function, the bias of Silhouette in k-means clustering toward selecting the result with the largest k as the best result exists only when the number of clusters is almost as big as the number of data points. In other situations they can be applied to evaluate the validity of k-means clustering with different k values and select the best result, which can be seen in the experimental results in Sect. 5.4.

Silhouette is one of the most popular internal measures for clustering validity evaluation and it is has been shown to be one of the most effective and generally applicable measures, see e.g. [11]. However, when Silhouette is applied in the evaluation of k-means clustering validity, many more extra calculations are required, and the extra calculations increase following a power law corresponding to the size of the dataset, because the calculation of the Silhouette index is based on the full pairwise distance matrix over all data. This is a challenging disadvantage of Silhouette. From this perspective, Silhouette needs to be simplified for k-means to improve its efficiency.

Simplified Silhouette was introduced by Hruschka in [6]. It inherits characteristics from Silhouette and can be used in the evaluation of the validity of a full clustering and also a single cluster or a single data point. However, the distance of a data point to a cluster in Simplified Silhouette is represented with the distance to the cluster centroid instead of the average distance to all (other) data points in the cluster, just as in the k-means Cost Function. However, Simplified Silhouette has not been systematically analysed in terms of an evaluation metric of k-means clustering validity. In this paper, the application of Simplified Silhouette to k-means is analysed and compared to k-means Cost Function and the original Silhouette. The research targets are to solve these two questions:

1 Does Simplified Silhouette or the original Silhouette perform as well as the k-means Cost Function in the evaluation of k-means clustering validity?
2 Does Simplified Silhouette have competitive performances to the original Silhouette in the evaluation of k-means clustering validity?

The next section, presents a theoretical analysis of the mathematical relationships between Simplified Silhouette and the other two internal measures.

3   Theoretical Analysis

3.1   Mathematics Expressions

Let $X = \{X_1, X_2, ..., X_n\}$ be a set of n data points. $X_i \, (1 \leq i \leq n)$ is one of the data points represented as $[x_{i,1}, x_{i,2}, ..., x_{i,m}]$, where m is the number of features. Given X, an integer $k \, (2 \leq k \leq n)$ and k initial centroids in the domain of X, the k-means algorithm aims to find a clustering of X into k clusters such that it minimises the k-means Cost Function, which is defined as the sum of the distances from a data point to the centroid of the cluster it is assigned to:

$$CF(X, C) = \sum_{l=1}^{k} \sum_{i=1}^{n} w_{i,l} d_E(X_i, C_l) \tag{1}$$

where $d_E(\cdot, \cdot)$ is the squared Euclidean distance, $C = \{C_1, C_2, ..., C_k\}$ is a set of cluster centroids after clustering, and $w_{i,l}$ is the indicator function, which equals to 1 when X_i is in C_l and 0 when X_i is not in C_l. As defined, the smaller the cost function is, the better the corresponding k-means clustering result is, so it can be considered as the default internal measure for k-means clustering. However, for the evaluation of the validity of clustering with different k values, using the Cost Function measure is problematic because it tends to reduce as the k value increases. Therefore, we need other general internal measures like Silhouette.

The calculation of Silhouette doesn't use any representative of a cluster (such as the cluster centroids used by the Cost Function), but is based on the full pairwise distance matrix over all data. For a single data point X_i, its Silhouette value $sil(i)$ is calculated as:

$$sil(i) = \frac{b(i) - a(i)}{max\{a(i), b(i)\}} \tag{2}$$

where $a(i)$ is the distance of X_i to its own cluster, which is defined as the average distance of X_i to all the other data points in its own cluster h as[1]:

$$a(i) = \frac{\sum_{\substack{p=1 \\ p \neq i}}^{n} w_{p,h} d_E(X_i, X_p)}{n_h - 1} \tag{3}$$

where n_h is the number of data points in the cluster h. $b(i)$ is the distance of X_i to its closest neighbouring cluster, which is defined as the average distance of X_i to all the data points in its closest neighbouring cluster as:

$$b(i) = \min_{l \neq h} \frac{\sum_{p=1}^{n} w_{p,l} d_E(X_i, X_p)}{n_l} \tag{4}$$

The $sil(i)$ ranges from -1 to 1. When $a(i)$ is much smaller than $b(i)$, which means the distance of the data point to its own cluster is much smaller than that to other clusters, the $sil(i)$ is close to 1 to show this data point is well clustered. In the opposite way, the $sil(i)$ is close to -1 to show it is badly clustered.

The Silhouette value of a whole cluster or a full clustering is defined as the average value of $sil(i)$ across all the data involved, e.g. the Silhouette value for a full clustering Sil is defined as follows:

$$Sil = \frac{1}{n} \sum_{i=1}^{n} sil(i) \tag{5}$$

Therefore, the Silhouette value for a full clustering Sil also ranges from -1, which shows a very bad clustering, to 1, which shows a perfect clustering.

Simplified Silhouette adopts an approach similar to Silhouette, but simplifies the distance of a data point to a cluster from the average distance of X_i to all (other) data points in a cluster to the distance to the centroid of the cluster:

$$a(i)' = d_E(X_i, C_h) \tag{6}$$

$$b(i)' = \min_{l \neq h} d_E(X_i, C_l); \tag{7}$$

And the Simplified Silhouette value for a single data point $ss(i)$ is defined as:

$$ss(i) = \frac{b(i)' - a(i)'}{max\{a(i)', b(i)'\}} \tag{8}$$

Similarly, the Simplified Silhouette value for a full clustering SS is defined as:

$$SS = \frac{1}{n} \sum_{i=1}^{n} ss(i) \tag{9}$$

The Simplified Silhouette value also ranges from -1 to 1. -1 shows a very bad clustering, while 1 shows a perfect clustering.

---

[1] Here we assume that there are at least two different data points in the cluster. Otherwise, the $a(i)$ is set to be 0, and the $sil(i)$ will be 1.

3.2   Theoretical Comparison

In some sense, Simplified Silhouette can be considered as the medium between the k-means Cost Function and Silhouette, because it evaluates the distances of each data point to its own cluster and its closest neighbouring cluster as Silhouette, and adopts the centroids from the k-means Cost Function as the representatives of clusters. In this section, we compare these different internal measures from a mathematical perspective.

Firstly, because at the end of k-means clustering the distances of a data point to its closest neighbouring cluster centroid $b(i)'$ is always greater than or equal to the distance to its own cluster centroid $a(i)'$, $max\{a(i)', b(i)'\}$ in (8) can be simplified to $b(i)'$. The Simplified Silhouette value for a single data point can also be simplified as follows:

$$ss(i) = 1 - \frac{a(i)'}{b(i)'} \tag{10}$$

It can be easily found that $ss(i)$ is always greater than or equal to 0 after k-means, as well as $SS(i)$.

For the comparison with the Cost Function in (1), Simplified Silhouette for all data points in (9) can also be written as:

$$SS = 1 - \sum_{l=1}^{k} \sum_{i=1}^{n} w_{i,l} \left(\frac{1}{nb(i)'} d_E(X_i, C_l) \right) \tag{11}$$

where $\frac{1}{nb(i)'} d_E(X_i, C_l)$ can be considered as the weighted distance of X_i to the centroid of its cluster l, and $\sum_{l=1}^{k} \sum_{i=1}^{n} w_{i,l}(\frac{1}{nb(i)'} d_E(X_i, C_l))$ can also be considered as the weighted Cost Function of k-means. The weight $\frac{1}{nb(i)'}$ is the only difference from Cost Function as (1). With the weight, the distance of X_i to its closest neighbouring cluster is taken into account. Given the same Cost Function, when the weight gets larger, that is, the data points are far from the centroid of its closest neighbouring cluster, the weighted Cost Function gets smaller and Simplified Silhouette gets a larger value that is closer to 1 to present a good cluster. Otherwise, the weighted Cost Function gets larger and Simplified Silhouette gets a smaller value that is closer to 0 to present a bad cluster.

For the comparison with Silhouette, we firstly expand $a(i)$ in (3) by expanding the squared Euclidean distance as follows:

$$a(i) = \frac{\sum_{\substack{p=1 \\ p \neq i}}^{n} w_{p,h}(\sum_{j=1}^{m}(x_{i,j} - x_{p,j})^2)}{n_h - 1} \tag{12}$$

Similarly $a(i)'$ in (6) can be expanded as follows:

$$a(i)' = \frac{\sum_{j=1}^{m}(\sum_{p=1}^{n} w_{p,h}(x_{i,j} - x_{p,j}))^2}{n_h^2} \tag{13}$$

We look into the mathematical relationship between (12) and (13), and get the equality as follows:

$$a(i) = \frac{n_h}{n_h - 1}a(i)' + \frac{\sum_{p=1}^{n}\sum_{q=1}^{n}w_{p,h}w_{q,h}d_E(X_p, Y_q)}{2n_h(n_h - 1)} \tag{14}$$

It is shown that the $a(i)$ adds a weight $\frac{n_h}{n_h-1}$ that is greater than 1 into $a(i)'$, and takes into account another factor - the sum of all the pairwise distances within its cluster with a weight $\frac{1}{2n_h(n_h-1)}$, therefore is always bigger than $a(i)'$.

Similarly, we can re-write $b(i)$ and $b(i)'$ as follows:

$$b(i) = \underset{l \neq h}{minimum} \frac{\sum_{j=1}^{m}\sum_{p=1}^{n}w_{p,l}(x_{i,j} - x_{p,j})^2}{n_l} \tag{15}$$

$$b(i)' = \underset{l \neq h}{minimum} \frac{\sum_{j=1}^{m}(\sum_{p=1}^{n}w_{p,l}(x_{i,j} - x_{p,j}))^2}{n_l^2} \tag{16}$$

where we denote $\frac{\sum_{j=1}^{m}\sum_{p=1}^{n}w_{p,l}(x_{i,j}-x_{p,j})^2}{n_l}$ as $D_E(X_i, l)$, the distance from a data point X_i to a cluster l that it does not belong to based on Silhouette, while $\frac{\sum_{j=1}^{m}(\sum_{p=1}^{n}w_{p,l}(x_{i,j}-x_{p,j}))^2}{n_l^2}$ as $D_E'(X_i, l)$ based on Simplified Silhouette. Then

$$D_E(X_i, l) = D_E'(X_i, l) + \frac{\sum_{p=1}^{n}\sum_{q=1}^{n}w_{p,l}w_{q,l}d_E(X_p, Y_q)}{2n_l^2} \tag{17}$$

It can be found easily that the $b(i)$ in Silhouette also takes into account one more factor than the $b(i)'$ in Simplified Silhouette - the sum of all the pairwise distances within the corresponding cluster with a weight $\frac{1}{2n_l^2}$.

3.3   Complexity Analysis

Finally, we can analyse the complexity of the computation of these measures. From [9], the overall complexity of the computation of Silhouette is estimated as $O(mn^2)$, while that of Simplified Silhouette is estimated as $O(kmn)$. When k is much smaller than n, Silhouette is much more computationally expensive than Simplified Silhouette. In addition, during the process of k-means clustering, the distance of each data point to its cluster centroid has already been calculated in each iteration, which greatly reduces the calculation of both the Cost Function and Simplified Silhouette. Therefore, the Cost Function and Simplified Silhouette are much more efficient in the evaluation of k-means clustering validity.

3.4   Conclusions of Theoretical Analysis

In summary, from the theoretical comparison, we can conclude that Simplified Silhouette is an internal measure with features related with both the k-means Cost Function and the original Silhouette:

1 It considers more than the k-means Cost Function by additionally bringing
  in the distance of each data point to its closest neighbouring cluster;
2 It also Simplifies Silhouette by ignoring within-cluster pairwise distances.

Therefore, we can consider Simplified Silhouette as a variant of Silhouette for
k-means clustering. In the experimental analysis, we will compare the time consumed by different measures, and most importantly, verify the performance of
Simplified Silhouette compared with k-means Cost Function and the original Silhouette so that to find out if these mathematical differences lead to performance
differences.

4   Experimental Design

4.1   Research Targets

The experimental analysis is designed to evaluate the performances of these three
internal measures, the k-means Cost Function, Silhouette and Simplified Silhouette, in the evaluation of k-means clustering validity and to answer specifically
the two research questions proposed at the end of Sect. 2.

For the evaluation of the validity of clustering with the same k value, we
take the k-means Cost Function as the default measure, and aim to find out if
Silhouette or Simplified Silhouette can perform as well as the Cost Function. On
the other hand, for the evaluation of the validity of clustering with different k
values, we evaluate Silhouette and Simplified Silhouette to find out if Simplified
Silhouette has comparative performances to the original Silhouette so that it can
be used safely instead of the original Silhouette.

4.2   Datasets

This experiment adopts four real world datasets and four synthetic datasets.
The real world datasets are all famous numeric datasets from the UC Irvine
Machine Learning Repository (http://archive.ics.uci.edu/ml/): Iris, Glass, Wine
and Yeast. The labels in these datasets are subjectively labelled only for some
specific purposes, so they cannot reflect exactly the intrinsic structure inside the
data or the ground-truth k value. Therefore, we ignore the labels of these four
datasets in the experiment. The other four datasets are generated artificially for
clustering, and hence the labels in them can be used in the evaluation. The first
two synthetic datasets are the Dim032 and Dim064 datasets from [4] with 32
dimensions and 64 dimensions respectively. The other two synthetic datasets are
the S1 and S3 datasets from [3], which have only two dimensions but many more
instances. The clusters in S1 are separated widely from each other, while those in
S3 are more compact. We select these different datasets in order to evaluate the
internal measures in different situations. Detailed information about the datasets
is summarised in Table 1. As discussed above, we only know the desired k values
of the four synthetic datasets.

Table 1. Experiment datasets

No.	Dataset	#Instances	#Dimensions	Desired k
1	Iris	150	4	None
2	Glass	214	9	None
3	Wine	178	13	None
4	Yeast	1484	8	None
5	Dim032	1024	32	16
6	Dim064	1024	64	16
7	S1	5000	2	15
8	S3	5000	2	15

4.3 Experimental Process

As introduced in Sect. 1, the common way to implement k-means is to run it multiple times with different parameter configurations and select the best result. In this paper, for each dataset we run k-means with the k values ranging from 2 to 30, and for each k value, we run it 30 times with the initial centroids randomly selected from the dataset[2]. As each run of k-means usually takes multiple iterations to process, we keep records of all the clustering labels and the cluster centroids of each iteration, and consider the clustering labels of each iteration in each k-means run as a clustering result. Then we calculate the internal measures of all these clustering results. In this way, these measures are based on not only good clustering results after the convergence of the Cost Function, but also the clustering results during the k-means process that are not very good. Based on these different clustering results, the evaluation of our three measures are more comprehensive and reasonable.

In the experimental process, there are some detailed features that are worth mentioning. Firstly, all the data is normalised to z-score [12] before input into k-means clustering to keep the balance among features with different magnitudes. Secondly, we use a different method to deal with empty clusters in the k-means process. The common way to deal with empty clusters is to re-run the k-means by re-selecting the initial centroids. In order to generate different clustering results for the evaluation, the way we adopt in this work is to find the closest data point to the centroid of each empty cluster, and assign the closest data point to the corresponding empty cluster to form a non-empty cluster.

Based on the total inventory of results accumulated, we then make the following four evaluations:

1 An evaluation of the measures in each iteration of each run of k-means;

---

[2] In preparing our experiments we tested two different initialisation methods for k-means, a random initialisation and a well-known algorithm k-means++. However, we found that the initialisation method made no difference in our results so in this paper we just report the results using the random initialisation.

2 An evaluation of the measures in each run of k-means;

3 An evaluation of the measures in the selection of the best result across all the 30 clustering results with each fixed value of k;

4 An evaluation of the measures in the selection of the overall best result from the best results selected for all the 29 k values.

5   Experimental Results

In this section we detail the results of the four evaluations outlined in the last section, and analyse the performances of the three measures - the k-means Cost Function (CF), Silhouette (Sil) and Simplified Silhouette (SS).

5.1   Evaluation in Each Iteration

Firstly, we look at the performances of the three internal measures in each iteration. As discussed in Sect. 2, the Cost Function of k-means is defined as the default measure for k-means and it decreases monotonically during the k-means process. Therefore, the Cost Function value in the next iteration cannot be larger than that in the previous iteration. However, both Silhouette and Simplified Silhouette are designed for general clustering, so they may not represent the validity of k-means exactly. Table 2 shows the number of iterations in which the k-means Cost Function increases of all the iterations for each dataset, and that in which Silhouette or Simplified Silhouette decreases. Note that a smaller value of the k-means Cost Function indicates a better clustering result, while a bigger value of Silhouette or Simplified Silhouette indicates a better clustering result. The percentages with the parentheses around indicate the proportions of these kinds of iterations in corresponding total iteration numbers.

From Table 2, we can see the Cost Function decreases monotonically as expected, but neither Silhouette nor Simplified Silhouette increases monotonically (although both of them increase in most cases). Based on the definition,

Table 2. Evaluation of iterations

Dataset	Total #Iterations	#Iterations with Increasing CF	#Iterations with Decreasing Sil	#Iterations with Decreasing SS
Iris	5685	0	303 (5.33%)	473 (8.32%)
Glass	7548	0	776 (10.28%)	709 (9.39%)
Wine	5588	0	248 (4.44%)	278 (4.97%)
Yeast	23923	0	5211 (21.78%)	3888 (16.25%)
Dim032	3570	0	119 (3.33%)	93 (2.61%)
Dim064	3342	0	70 (2.09%)	21 (0.63%)
S1	17256	0	5832 (33.80%)	6457 (37.42%)
S3	27329	0	7585 (27.75%)	8234 (30.13%)

the clustering result always gets better along iterations of each run of k-means. Therefore the evaluations of k-means clustering validity with both Silhouette and Simplified Silhouette are inaccurate in some iterations, so we can see neither Silhouette nor Simplified Silhouette performs as well as the Cost Function.

Meanwhile, also from Table 2 we see that there is not much difference between the numbers of iterations with decreasing Silhouette and Simplified Silhouette values, which indicates these two measures perform similarly in the evaluation in iterations.

5.2   Evaluation in Each Run of k-means

. As stated in Sect. 5.1, Silhouette and Simplified Silhouette may be inaccurate in the evaluation of clustering validity in individual iterations. Therefore, for Silhouette and Simplified Silhouette the k-means process may be not a monotonically converging process, and in the last iteration of k-means where the minimum of the Cost Function is always found, the Silhouette or Simplified Silhouette value may be not the best value in the k-means process. We get the results just as we expect: for all datasets, there are always clustering with the last Silhouette or Simplified Silhouette smaller than the best value (due to the limitation of space, the details are not included in the paper). Similarly, we can see that neither Silhouette nor Simplified Silhouette performs as well as the Cost Function.

Even though it may not result in the best Silhouette or Simplified Silhouette value, the last iteration is always taken as the end of a k-means clustering based on its definition. Therefore, the result in the last iteration is always taken as the final clustering result of the k-means clustering in further steps of the experiment.

5.3   Evaluation in the Selection of the Best Result from Clustering with Each Fixed Value of k

For each fixed value of k, we compare 30 k-means clustering results and select the best one among them based on our three internal measures. Table 3 shows the number of k values with which the same best result is selected from clustering, based on every pair of two measures or all three measures.

Silhouette and Simplified Silhouette can select the same best result for most k values, but only for a small number of k values, they can select the same best result as the Cost Function. Similarly, we can see that neither of them can perform as well as the Cost Function. Although Silhouette performs a little better than Simplified Silhouette in this case, there is not much difference.

Based on these results as well as the results in above sections, we can conclude that the k-means Cost Function is the only one among these three internal measures that can accurately evaluate k-means clustering validity. Therefore, the best clustering result for each k value is selected based on the k-means Cost Function in further steps of the experiment.

Table 3. Evaluation of the selection of the best result from clustering with each fixed value of k

Dataset	Total #k Values	#k Values - Sil and SS	#k Values - Sil and CF	#k Values - SS and CF	#k Values - All Measures
Iris	29	23	10	9	8
Glass	29	15	9	5	3
Wine	29	18	13	11	9
Yeast	29	12	6	6	4
Dim032	29	16	13	8	8
Dim064	29	19	17	14	12
S1	29	23	7	7	6
S3	29	22	4	3	3

5.4   Evaluation in the Selection of the Overall Best Result

The selection of the overall best result from all the best results selected for each k value is the last step of the experiment. Table 4 shows the k values corresponding to the overall best results selected based on each internal measure for each dataset. It is shown that the k-means Cost Function is problematic in the evaluation of the validity of k-means clustering with different k values, and tends to select the result with the largest k value as the overall best result, therefore as we discussed, it is not suitable for this case.

On the other hand, Silhouette and Simplified Silhouette select the same overall best result for almost all datasets. For three of the four synthetic datasets that are designed for clustering, both measures can select the results with the desired k values. For the dataset Dim064, they also select the result with the same value. It is common to select results with the non-desired k value based on

Table 4. Evaluation of the selection of the overall best result from the best clustering results selected for different k values based on different measures

Dataset	Desired k Value	Corresponding k for CF	Corresponding k for Sil	Corresponding k for SS
Iris	Unknown	29	2	2
Glass	Unknown	30	2	2
Wine	Unknown	29	3	3
Yeast	Unknown	30	7	8
Dim032	16	29	16	16
Dim064	16	30	20	20
S1	15	30	15	15
S3	15	30	15	15

conditions like this because the initial centroids are randomly selected to generate a variety of clustering results[3]. From this perspective, we can conclude that Simplified Silhouette has competitive performance to Silhouette in the selection of the overall best result.

5.5 Evaluation of Correlations Between Internal Measures

We also evaluate the Pearson correlations between Silhouette and Simplified Silhouette. For each dataset and each k value, the distinct pairs of these two measures are extracted from the results. From Figs. 1 and 2, it is shown that there is highly positive correlation between Silhouette and Simplified Silhouette in an overwhelming majority of situations.

Correlations between Sil and SS

K	Iris	Glass	Wine	Yeast	Dim032	Dim064	S1	S3
2	0.993294	0.992193	0.99702	0.943325	-0.12825	0.300981	0.941844	0.971279
3	0.995385	0.923887	0.996318	0.894682	0.778223	0.425153	0.942118	0.955456
4	0.967166	0.544803	0.961703	0.81958	0.738165	0.439857	0.912224	0.991052
5	0.980247	0.948815	0.951209	0.727211	0.613938	0.948636	0.933215	0.978902
6	0.91578	0.893141	0.973678	0.690819	0.849891	0.858353	0.962757	0.97548
7	0.928572	0.957998	0.959664	0.757024	0.85729	0.967433	0.975869	0.97462
8	0.96134	0.943394	0.90954	0.849662	0.91182	0.954233	0.980063	0.960327
9	0.85654	0.903842	0.70277	0.848189	0.910435	0.966105	0.981642	0.98846
10	0.968438	0.913511	0.924346	0.903346	0.916598	0.958255	0.980504	0.985758
11	0.922668	0.870661	0.928736	0.949788	0.968223	0.980685	0.979796	0.98705
12	0.981977	0.943115	0.875796	0.910057	0.959173	0.960018	0.973384	0.991971
13	0.901127	0.879644	0.92124	0.920059	0.980426	0.965945	0.979321	0.98883
14	0.910819	0.865278	0.831538	0.944559	0.966271	0.961901	0.975649	0.988905
15	0.953423	0.89128	0.90954	0.937084	0.971514	0.980536	0.98125	0.986448
16	0.944367	0.862471	0.876525	0.938838	0.958231	0.977597	0.985526	0.989248
17	0.982379	0.918348	0.935388	0.927498	0.968129	0.963791	0.982856	0.99135
18	0.935021	0.861701	0.905783	0.931496	0.962147	0.950856	0.982589	0.991791
19	0.907564	0.844418	0.91705	0.933302	0.920502	0.922081	0.984666	0.988965
20	0.986338	0.847514	0.818287	0.92105	0.977352	0.961491	0.980543	0.990079
21	0.894858	0.926713	0.832134	0.929169	0.946484	0.916429	0.984255	0.989441
22	0.919939	0.92103	0.899059	0.924478	0.900888	0.933575	0.982873	0.979498
23	0.953855	0.868697	0.841784	0.942225	0.956174	0.904049	0.986931	0.98348
24	0.941008	0.936139	0.85987	0.934342	0.920883	0.914759	0.983635	0.980152
25	0.958949	0.881833	0.88931	0.940603	0.925157	0.887353	0.983654	0.982337
26	0.858996	0.869662	0.847474	0.960428	0.934942	0.761293	0.982828	0.98087
27	0.95194	0.914826	0.87712	0.950632	0.920293	0.782278	0.985089	0.98381
28	0.885049	0.917608	0.874958	0.931141	0.932947	0.881241	0.985711	0.976128
29	0.902892	0.918648	0.901555	0.948535	0.898227	0.924445	0.989047	0.983812
30	0.829004	0.902836	0.867191	0.936504	0.906922	0.800268	0.991359	0.977056

P-value for Correlation Testing between Sil and SS

K	Iris	Glass	Wine	Yeast	Dim032	Dim064	S1	S3
2	8.78E-09	2.22E-16	3.44E-10	0	0.941238	0.184916	0	0
3	1.97E-09	3.35E-13	0	0	0.000143	0.027055	-4.44E-16	0
4	0	0.001045	7.76E-10	4.44E-16	0.000308	0.016956	0	0
5	4.44E-16	4.44E-16	0	0	0.00085	0	0	0
6	8.95E-10	0	0	8.88E-16	7.61E-08	3.36E-10	0	0
7	1.91E-14	0	0	0	4.86E-11	0	0	0
8	5.94E-11	0	2.22E-16	0	0	0	0	0
9	2.04E-10	2.22E-16	2.22E-09	0	0	0	0	4.44E-16
10	4.44E-16	0	0	0	0	0	0	0
11	1.57E-13	0	2.22E-16	0	0	0	0	0
12	0	0	8.88E-16	0	0	0	0	0
13	6.82E-14	0	0	0	0	-4.44E-16	0	0
14	1.05E-10	0	0	0	0	0	-4.44E-16	0
15	0	4.44E-16	2.22E-16	0	0	0	0	0
16	4.88E-15	2.22E-16	0	2.22E-16	0	0	0	0
17	0	0	4.44E-16	0	0	0	0	0
18	0	0	0	4.44E-16	4.44E-16	0	0	0
19	8.88E-16	2.22E-16	0	0	0	0	0	0
20	2.22E-16	0	4.44E-16	0	0	-4.44E-16	0	0
21	1.01E-07	0	0	0	0	0	0	0
22	0	0	0	0	0	0	0	2.22E-16
23	1.33E-15	0	0	0	2.22E-16	0	0	0
24	-4.44E-16	0	0	0	0	0	2.22E-16	0
25	0	0	6.66E-15	0	0	0	0	0
26	1.59E-10	0	0	0	0	0	0	0
27	0	0	2.22E-16	0	2.22E-16	0	0	0
28	3.58E-13	0	0	0	0	0	0	4.44E-16
29	1.18E-10	0	0	0	0	0	0	0
30	4.24E-10	0	0	0	0	0	2.22E-16	2.22E-16	4.44E-16

Fig. 1. Correlations between Sil and SS

Fig. 2. P-value for correlation testing between Sil and SS

5.6 Evaluation of Time Consumed

Finally, we compare the time consumed in the calculation of each internal measure. Figure 3 shows the time consumed (with ms as unit) for the datasets S1, which is the dataset with the most instances. It is shown that the time consumed by Silhouette is much more than that by the Cost Function or Simplified Silhouette, and the differences can be orders of magnitude in size. Similar results are found for other datasets. The rough time consumed in calculation may not

---

[3] If other methods like k-means++ are used for selecting the initial centroids, it is very likely to get all the desired k values for all the synthetic datasets.

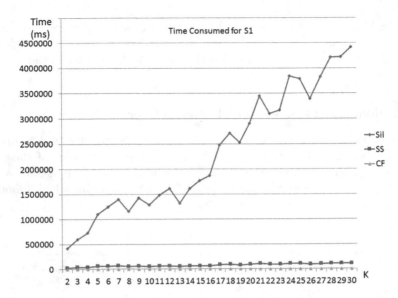

Fig. 3. Time consumed - S1

reflect the genuine efficiency of algorithms exactly, but from the commercial perspective, it is meaningful to notice that the implementation of the Cost Function and Simplified Silhouette is generally much faster than Silhouette.

6 Conclusion

In this paper we have analysed the application of Simplified Silhouette to the evaluation of k-means clustering validity, and compared it with two other internal measures: the k-means Cost Function and the original Silhouette from both theoretical and empirical perspectives.

Theoretically, Simplified Silhouette has a mathematical relationship with both the k-means Cost Function and the original Silhouette. It brings in additionally the distance of each data point to its closest neighbouring cluster to the k-means Cost Function, but simplifies Silhouette by ignoring within-cluster pairwise distances.

Empirically, we can make the following conclusions:

1 Neither Simplified Silhouette nor the original Silhouette can perform as well as the k-means Cost Function in the evaluation of the validity of k-means clustering with the same k value but different initial centroids;
2 Simplified Silhouette has competitive performances to the original Silhouette in the evaluation of k-means validity and is much faster in the calculation;

Therefore, the most suitable method to automate the selection of the best k-means result is using the k-means Cost Function firstly to select the best result for each k value and then using Simplified Silhouette to select the overall best result from the best results for different k values.[4]

Acknowledgement. The authors wish to acknowledge the support of Enterprise Ireland through the Innovation Partnership Programme SmartSeg 2 and the ADAPT Research Centre. The ADAPT Centre for Digital Content Technology is funded under the SFI Research Centres Programme (Grant 13/RC/2106) and is co-funded under the European Regional Development Funds.

References

1. Davies, D.L., Bouldin, D.W.: A cluster separation measure. IEEE Trans. Pattern Anal. Mach. Intell. **2**, 224–227 (1979)
2. Dunn, J.C.: Well-separated clusters and optimal fuzzy partitions. J. Cybern. **4**(1), 95–104 (1974)
3. Fränti, P., Virmajoki, O.: Iterative shrinking method for clustering problems. Pattern Recogn. **39**(5), 761–775 (2006)
4. Franti, P., Virmajoki, O., Hautamaki, V.: Fast agglomerative clustering using a k-nearest neighbor graph. IEEE Trans. Pattern Anal. Mach. Intell. **28**(11), 1875–1881 (2006)
5. Halkidi, M., Vazirgiannis, M., Batistakis, Y.: Quality scheme assessment in the clustering process. In: Zighed, D.A., Komorowski, J., Żytkow, J. (eds.) PKDD 2000. LNCS, vol. 1910, pp. 265–276. Springer, Heidelberg (2000). doi:10.1007/3-540-45372-5_26
6. Hruschka, E.R., de Castro, L.N., Campello, R.J.: Evolutionary algorithms for clustering gene-expression data. In: Fourth IEEE International Conference on Data Mining, ICDM 2004, pp. 403–406. IEEE (2004)
7. Jain, A.K.: Data clustering: 50 years beyond k-means. Pattern Recogn. Lett. **31**(8), 651–666 (2010)
8. Kelleher, J.D., Mac Namee, B., D'Arcy, A.: Fundamentals of Machine Learning for Predictive Data Analytics: Algorithms, Worked Examples, and Case Studies. MIT Press, Cambridge (2015).
9. Vendramin, L., Campello, R.J., Hruschka, E.R.: Relative clustering validity criteria: a comparative overview. Stat. Anal. Data Min. **3**(4), 209–235 (2010)
10. Wang, F., Franco, H., Pugh, J., Ross, R.: Empirical comparative analysis of 1-of-k coding and k-prototypes in categorical clustering (2016)
11. Xiong, H., Li, Z.: Clustering validation measures (2013)
12. Zaki, M.J., Meira Jr., W.: Data Mining and Analysis: Fundamental Concepts and Algorithms. Cambridge University Press, Cambridge (2014)

---

[4] Due to time and resource limitations Simplified Silhouette has not been fully explored in this paper, e.g. the actual industrial datasets are not available. However, this is an attempt to evaluate the internal measures for a specific clustering algorithm. Specific methods should be evaluated, selected and even designed for specific algorithms or conditions, rather than always a same set of general methods for all the situations.

Summarization-Guided Greedy Optimization of Machine Learning Model

Dymitr Ruta$^{(\boxtimes)}$, Ling Cen, and Ernesto Damiani

Emirates ICT Innovation Center (EBTIC), Khalifa University of Science and
Technology, P.O. Box 127788, Abu Dhabi, United Arab Emirates
{dymitr.ruta,cen.ling,ernesto.damiani}@kustar.ac.ae

Abstract. Immense amounts of unstructured data account for up to
90% of all human generated data, yet the attempts to extract significant
value from it with Machine Learning (ML) and Big Data (BD) technologies yield limited successes. We propose a generic approach to deep
data summarization and subsequent automated ML design optimization
to extract maximum predictive value from big data. Knowledge summarization is a central component of the proposed methodology and we
argue that coupled with strictly linear modeling complexity, hierarchical decomposition and optimized model design may define a backbone
of the new platform for automated and scalable construction of robust
ML models. We consider ML build process as data journeys through the
layers of modeling that consistently follow the same patterns of data
summarization and transformation at the subsequent layers of abstraction. In such framework we argue that the robust construction of the ML
model can be achieved through hierarchical greedy optimization of the
links between connected ML model components. We demonstrate several case studies of deep data summarization and automated ML model
design on text, numerical time series and images data. We point out
that application awareness allows to deepen data summarizations while
maintaining or improving its predictive value.

Keywords: Big data · Machine learning · Data summarization ·
Feature selection · Meta-learning · Backward-forward search

1 Introduction

Enormous streams of unstructured data flow every day from countless locations
worldwide, accounting for up to 90% of all human generated data. These data are
collected and stored but the attempts to extract big value or big insights from it
with Machine Learning (ML) and Big Data (BD) technologies yield rather limited successes [1,2], leaving the same question still unanswered: how to monetize
big data? To address this challenge we present a systematic approach to the ML
model design optimization in the wider context of big data processing focused
primarily on extracting and maximizing the predictive value from data. We consider various journeys of data through the layers of machine learning modeling

© Springer International Publishing AG 2017
P. Perner (Ed.): MLDM 2017, LNAI 10358, pp. 306–321, 2017.
DOI: 10.1007/978-3-319-62416-7_22

and argue that all the elements of such modeling from feature selection through model learning, prediction, fusion up to the meta-learning can be considered as instances of the same consistent patterns of data summarization (filtering) and transformation (mapping) at the subsequent layers of abstraction, resembling much acclaimed deep learning methodology to ML [3, 4].

Knowledge summarization becomes a critical component of the proposed ML design methodology. There are many approaches to data summarization depending on the data type, representation and filtering technique [5], yet overall the same objective of maximum data reduction at the maximum retention of the information content is pursued [10, 11]. Beyond data reduction the summaries also catalyze data transformations that project data to the desired representation or stimulate generation of new evidence for predictive models.

We argue that to be effective both filtering and mapping operations on summarizations are required to be at most linearly complex in the number of examples and the dimensionality of the data. We believe this is a necessary condition for any sustainable ML platform to survive the test of time and the scales of big data and propose to solve this problem by means of modular decomposition and a preference towards simple yet robust linear predictors.

Another claim we put forward in this work is that the performance of predictive applications is strongly dependent on how tightly the elements of the model are optimized to each other i.e. how well they exploit each other's prior knowledge about the assumptions, constraints, representations, design flexibility, etc. We claim that on the course of observing various instances of ML model design decisions and the resultant predictive performance it is possible to devise a greedy model composition system to optimize or even fully automate the ML design process [12].

1.1 Knowledge Summarization

Knowledge summarization is a central concept shaping human unprecedented abilities to efficiently perceive, filter, prioritize and learn from limited observations, using limited storage in a limited time, skills that remain unsurpassed by AI systems, despite their huge processing power and storage advantage [13]. Even more surprising is the ease with which human handles different types of data seamlessly merging visual, audio, text and complex numerical and abstract representations to build and store a coherent and synergistic summary of the experiences they can exploit at any time to satisfy individual goals.

Data and knowledge summarization appears to be, therefore, a critical process enabling the efficient extraction, organization and utilization of the surrounding evidence for a particular purpose. Many embodiments of data summarization for different types of data have been analyzed and used for decades [5–11]. Summarization of text typically follows the strategy of selecting the most representative set of sentences [14] or a set of keywords [15], possibly bound by the relationships like in RDF [16]. Other approaches include ontology generation or indexing [5, 17], hierarchical category structures [18] as well as topics and word dictionaries for specific applications and classification or sentiment analysis [9].

It appears that without the specified purpose, generic text summarizations like ontologies offer only limited data reductions compared to the specialized summarization like categories, topics, keywords or sentiment dictionaries [19]. In the subsequent sections we will propose a methodology to deliver keyword-based summarization of the text document data utilizing application context to achieve deeper level of summarization while improving the predictive performance.

Numerical data are typically summarized by a set of parameters of the generative model that can reconstruct the same data structure subject to some bias and variance errors [21,22]. Other approaches involve storing only the most representative set of data points and ignore the rest [23], relating to the problem of outliers detection, or one-class learning [25]. Clustering or unsupervised learning in general represents the essence of numerical summarization [21,22,24]. Nearest mean clustering summarizes the concept of the cluster to just a single point of cluster center [24]. Like for the text, generic summarizations of numerical data like information theoretic data compression is being challenged by the deeper summarizations available when more information exposing the data characteristics or the nature of the application are known. Mixture of Gaussians are good examples of deep summarization observed for clustering or classification if the underlining data are known to have a structure of multiple normal distributions [24], while in [26] even more generic approach to data condensation is proposed. In the subsequent study we propose a further generalization of such numerical data summarization methodology aimed at highly condensed data representations that lead to the same or better classification performance in the supervised learning context.

In the context of generic ML platform development, numerical summarization is the most effective and flexible when applied on the unified data representations. Hierarchical clustering [24], nested set neighborhoods and hypercubes originating from the concept of granular computing, [5] are good examples of the attempts of efficient processing of fused data at different complexity levels thanks to recurrent and modular data representation and fast processing using mostly logical operations, [27]. Another promising approach that shares the generic and decomposable nature of numerical data is through data binarization techniques. Binary representation naturally merges numerical and categorical data, features modularity through hierarchical binary grouping and enables very efficient storage and fast logical processing apparatus. Numerous examples of preferred binary processing of image data [28], text [30] or biological data [29] indicate that binarization enables unified numerical data representation that can be easily applied and controlled at the desired level of granularity.

Due to the limited space we will not elaborate on summarizations along other data representations, just concluding that similar logic could be illustrated across other data types and similarly point at the elementary binary representation that appears to be the most suitable to deliver a unified flexible and scalable summarization that can be seamlessly carried out along the hierarchical data structures and utilized for robust and scalable predictive and knowledge provision services.

1.2   Scalable and Decomposable Machine Learning

Despite the fact that summarizations themselves would deliver immense reduction of the data, their generation and interaction with the raw data streams and predictive models require significant processing cost. To achieve true scalability and to ensure future-proof and real-time applicability, the predictive models are independently required to scale their learning and predicting operations at most linearly in the data size and its dimensionality. It is expected that a combination of the intelligent rapid feature selection techniques [20], coupled with simple probabilistic classification [32], generalized linear regression [33] and approximate hierarchical clustering [34] are able to provide desired linear scalability requirement across large scope of the machine learning methodology.

Further research is required, however, to adapt these linear algorithms to well balanced decomposition methods that would support efficient parallelization at multiple subsequent layers of abstraction. To address this gap, within classification scope, we propose a binary decomposition of the multiple-class classification problem along the output space, that similarly to ECOC [35] returns the binary output code, yet adhering explicitly to the balanced binary hierarchy of the class structure. Such decomposition would admit a much wider plethora of powerful binary classification models to be used as baseline predictors and ensure data-independent parallelization as opposed to data-dependent sequential class decomposition, expensive pairwise classification or under-performing one-class-vs-the-rest decomposition strategies [36]. Other data-independent model decomposition approaches focus on independent processing of different subsets of the overlapping regions of the input space and then merging the results at the output level, or explicit decomposition into specialized models operating on exclusive non-overlapping regions of the input space [37]. We consider the latter approach to offer a true non-redundant and flexible problem decomposition.

Decomposition along the data dimensionality is a scarcely investigated problem as it usually leads to a significant degradation of performance unless a very large set of features is available [38]. However, this is typically true from only global perspective of the data. Locally, in-line with subspace sampling methods [39], some subspaces would actually benefit from degeneration of many redundant dimensions in the classification context. We propose to go a step forward with this approach and explore whether rather than reducing the dimensionality of the local data it could be more beneficial to design complementary small groups of local features and feed them to different base predictors to deliver the outputs at the new converged representation layer at which they can be compared, merged or further combined to deliver maximum predictive performance. Similarly the same logic can be carried out at even higher level of abstraction where the models or applications outputs could be taken as inputs for entirely new meta-learning problems.

2   Automated Model Design Optimization

Observation and careful analysis of the performance of different machine learning model designs leads to a conclusion that the attained predictive performance is strongly dependent on how much individual components of such design are mutually optimized to each other. Optimization is here broadly understood as mutual awareness, exploitation of prior information, modeling assumptions, consistency and fit of data representation etc. To support such conjecture its is easy to notice that all the breakthrough technologies in machine learning like support vector machine, boosting or deep learning networks appear to achieve the significant improvements due to better optimization of all the elements of the ML model design like data, features, learning algorithm, outputs, ensembles etc. Rather than trying to fully optimize ML model towards the absolute global optimum, which might be intractable, we propose to optimize only neighboring elements of the ML design process using specifically designed hierarchical backward-forward greedy search (BFS) algorithm. The method focuses on optimizing the selection of features, their properties and predictors to maximize the predictive performance.

The optimization process starts with the complete feature set, default feature parameters and the prediction models. For every feature or collectively for all the features, the parameters are optimized to maximize the predictive performance. Then the optimized features are subjected to the selection using the same BFS search method. Once the optimal feature subset is found the process steps back to the feature parameter re-optimization and updates feature definitions if they

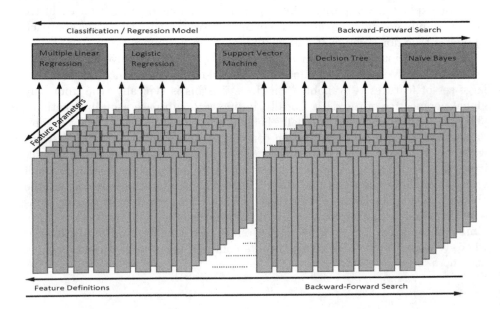

Fig. 1. Greedy ML design optimization with Backward-Forward Search

improve predictive performance. The mutual re-optimization of the selected features and their parameters continues back and forth until no predictive improvement is reported, after which the process proceeds to the selection of the predictor, i.e. the next predictor is now tested with the features optimized to the previous predictor. As before, the backward-forward re-optimization of the feature subset and their parameters are attempted and validated if the performance gain is achieved. This process continues BFS recursively at all hierarchy levels of the ML model design until it is not able to improve predictive performance. The process terminates with all the ML model design optimized at all levels of its hierarchy. An instantiation of such greedy ML design optimization method is depicted in Fig. 1.

3   Case Studies

3.1   Pedestrian Images Summarization and Classification

Automated detection of pedestrians poses significant challenges and is a critical capability in automotive industry that is on the verge of introduction of self-driven vehicles. Such capability requires highly reliable human detection to make rapid decisions based on small and usually poor quality images shot in real-time from a vehicle on the move. While detection accuracy of pedestrians is of prime importance in this context we focus on how we can summarize the original data informing this problem such that we can reduce the data content as much as possible while maintaining its predictive power for this task.

For a demonstration, a sample of 10000 image patches were extracted from a video recorded on vehicle mounted camera and subjected as labeled input examples, such as a few shown in Fig. 2.

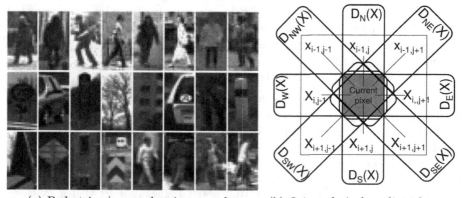

(a) Pedestrian/non-pedestrian samples   (b) Oriented pixel gradient features

Fig. 2. Examples of pedestrian and non-pedestrian images along with the visual definition of oriented pixel difference features.

We then extracted a standard set of differential features formed by aggregation of different directional variants of the neighboring pixel color difference and then aggregated separately for both positive and negative class to generate class-conditional feature templates. The process of classification can be then defined instantly just by measuring the average pixel difference between the new image and class-conditional templates, while all the evidence used for learning from initial training examples is summarized to just a small number of a class templates equal to a pair of images for every feature definition. Some visual depiction of feature families construction is shown in Fig. 2(b), while the selected class-conditional feature templates are presented in Fig. 3.

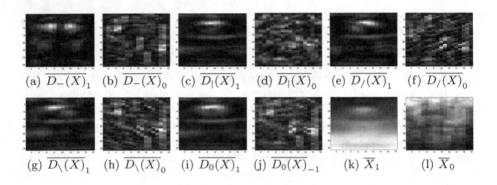

(a) $\overline{D_-(X)}_1$ (b) $\overline{D_-(X)}_0$ (c) $\overline{D_|(X)}_1$ (d) $\overline{D_|(X)}_0$ (e) $\overline{D_/(X)}_1$ (f) $\overline{D_/(X)}_0$

(g) $\overline{D_\backslash(X)}_1$ (h) $\overline{D_\backslash(X)}_0$ (i) $\overline{D_0(X)}_1$ (j) $\overline{D_0(X)}_{-1}$ (k) $\overline{X}_1$ (l) $\overline{X}_0$

Fig. 3. Selected class-conditional feature templates for pedestrian and non-pedestrian

Such simple and deep summarization of input data - training images - was achieved thanks to the context information defining the purpose of the data - detection pedestrians, and frankly all these thousands of images were collected only for this purpose. The actual features associated with the new testing image can be considered as its average Manhattan distance to the corresponding class templates:

$$f_{type}^c(Y) = |D_{type}(Y) - \overline{D_{type}(X)}_c| \tag{1}$$

Given pedestrian detection problem with 2 classes: $(\Omega_1 = 1, \Omega_2 = 0)$ for pedestrian and non-pedestrian respectively, this process returns over 40 features.

The experimental part of this work involved full composition of the robust pedestrian detector model. Given the feature family described above the detection model was exposed to the set of subsequent choices of the feature selection (42 features), classifier selection (logistic regression LR, naive Bayes NB, decision tree DT, linear mixture of Gaussians MG, linear kernel support vector machine SVM), [21,31] and then optionally lightweight classifier combination method (mean, max, product). The model composition was automated to the extend that these choices were evaluated using greedy hierarchical backward-forward search (BFS) as described in Sect. 2. The first step of feature parameters optimization was skipped as the features as defined above were in principle non-parametric.

Hence the design optimization started from optimal feature selection using BFS and evaluated using first classifier i.e. LR. Then using feature subset optimally selected for LR, other classifiers are evaluated with a backward attempts to re-optimize feature selection to improve the performance. Once the top classifier with internally optimized features is found, then the next classifier is attempted to be optimized in the same way with the exception that the performance is now evaluated using mean combiner together with the top classifier optimized in the first stage. This process traverses through all the combinations of combiners and classifiers with re-optimized features until the process terminates with the top design, for which no available design choice improves the prediction performance. Every new choice at the current design layer triggers a step back to see if better choices at the previous layer could further improve the performance by better fitting the design layers below. Such recurrent modular optimization strategy seemingly leads to lengthy deep back re-optimization paths. In practice, however, the process proceeds very fast through the feedback loops as each re-optimization at the specific design layer starts from optimal initial choices fitted to the previous design choice at a higher design layer and are usually near optimal for the new higher-level design choice. What it means, is that for instance feature set optimized for LR classifier is also near-optimal for other classifiers and requires usually only minor adjustments to re-optimize it for the new classifier.

The presented ML design optimization process resulted with MG classifier selected as individual best and the mean-aggregated ensemble of MG, NB and LR classifiers in the combined design version. The Receiver operating characteristic (ROC) curve achieved over the validation set is presented in Fig. 4. Our optimized detector reached the classification performance in excess of 95%. The combined model's ROC curve is smoother and reaches 100% true positive rate

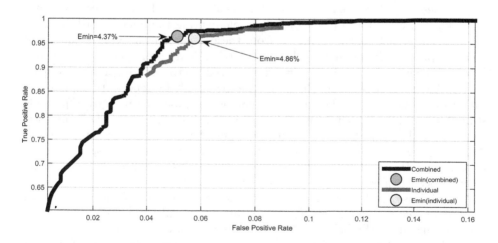

Fig. 4. Pedestrian detection ROC curves for individual and combined models.

at a relatively low false positive rate, reflecting more stable predictions and improved control between the cost and performance.

From the summarization point of view we managed to compress the evidence for the purpose of pedestrian detection from thousands of images to just tens of pedestrian class templates reducing it by more than 99%. On the experimental testing we observed that equally well performing pedestrian detection model is achieved when measuring the distance only to the positive class template (pedestrian) instead of comparing both, hence in-fact twice deeper image summarization level is achieved than initially anticipated.

3.2 Extraction of Fire Incident Risks from Case Text Reports

One of the embodiment of big data is countless text documents, potentially carrying invaluable insights just about anything. A related challenge was formulated by Polish State Fire Department which provided over 50000 written text reports summarizing every fire incident around the city of Warsaw from the last 15 years in an attempt to predict and extract key risk factors of civilian and fire-fighter injuries and casualties.

As an alternative to solving this problem through natural language processing (NLP) one simple strategy was to represent text documents as a collection of unique word incidence features and use them to differentiate between binary injury target on the course of a standard classification process. However, the result of such strategy is a high-dimensional dataset with tens of thousands of very sparse feature-words most of which occur very rarely. To address this issue as in the previous case study we proposes an automated solution for fire injuries prediction that utilizes deep data summarization coupled with self-organizing Machine Learning design.

In the first instance we focused on summarization of over 50000 typed text reports describing the fire incident, each carrying the label of whether any injury or casualty occurred as a result or not. Implementing the binary word-presence representation of features extracted from these reports resulted in a matrix of 50000 reports each represented by over 12000 binary word-presence features. The objective for summarization was now to reduce this set of features as much as possible without reducing the predictive power of the reports to explain the target injuries class variable. Just like before we considered a number of possible simple classifiers that include the Naive Bayes (NB), Decision Tree (DT), Logistic Regression (LR) and Support Vector Machine (SVM), extensively reported in [21, 31]. It was very obvious, however, that given the binary and sparse nature of features NB classifier appears the most suited due to the probabilistic nature of word occurrences and rapid logical algebra used for processing.

In this case the feature selection process was preceded by the reduction of redundant features in two stages. In the first stage all the feature-words that never occurred for the positive class were immediately eliminated reducing the original dataset by over half. In the second stage we considered pairs of feature-words A–B and eliminated a feature A if its true positive set with the target is a subset of true positive set of feature B, while its false positive set fully covers the false positive set of feature B. This step basically eliminates the feature due to the fact that there is a universally better feature in the set i.e. a feature that always makes more accurate detection of the positive class while making less mistakes. This additional elimination not only reduces the feature set to by additional 30% but also ensures that the remaining features are highly diverse and complementary i.e. tend to detect different examples of the positive class, which is a highly desirable property of the robust feature set.

Independently to data reductions through redundancy filtering, we applied further summarization through BFS feature selection forming the first stage of the automated ML model design with the evaluation criterion defined generically by the area under the curve (AUC) of the ROC curve built from comparison of the classifier outputs against the binary target class variable. As before the search proceeded from the selection of about 250 remaining binary feature-words used with NB classifier and then subsequently with DT, LR and SVM classifiers with backward re-optimization of the features sets if they improve performance.

This automated ML design optimization process terminated after about 15 minutes returning the optimal model configuration with NB classifier and 67 binary-features corresponding to the set of keywords that are the most explanative about injuries during fire incidents. The final model achieved average AUC score in excess of 96%.

To further evaluate the quality of the greedy BFS search method we have compared the evolution of the AUC scores of the BFS search against other search models applied for feature selection with the NB classifier fixed. The standard alternatives of random subset, incremental single best, greedy forward and cumulative sparse count were all run and the AUC score evolution recorded throughout the search. The corresponding performance curves are presented in Fig. 5(a). The Greedy BFS search clearly outperforms the other alternatives, yet the simple cumulative sparse count search that replaces the model performance

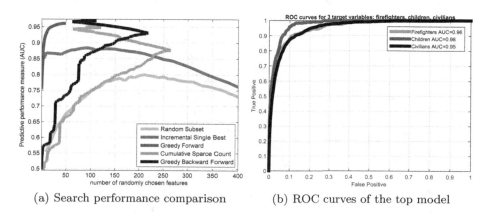

(a) Search performance comparison   (b) ROC curves of the top model

Fig. 5. Performance comparison of several feature selection search algorithms and the ROC curves of the top model

with a simple count of positive features performs almost as good while completes in seconds instead of minutes. The ROC curves of the top model separated into predictions against firefighter, children and civilian injuries target variables are shown in Fig. 5(b)

This case study was yet another instantiation of the proposed automated hierarchically optimized ML model design that utilized deep summarization of initial 50000 text reports down to 67 binary feature-keywords to achieve very high predictive performance of fire incidence injuries detection. Extracted feature-keywords pointed directly at the key factors or injury risk during fire incidents that may be used to design preventive actions and potentially save lives.

3.3   Predicting Methane Concentration Warnings in Coal Mines

Coal mining operation continuously balances the trade-off between the mining productivity and the risk of hazards like methane explosion. Dangerous methane concentration is normally a result of increased cutter loader workload and leads to a costly operation shutdown until the increased concentrations abate. We propose a simple yet very robust methane warning prediction model that can forecast imminent high methane concentrations at least 3 minutes in advance, thereby giving enough notice to slow the mining operation, prevent methane warning and avoid costly shutdowns.

The data and the problem specification come from one of the Polish coal mines, scheme of which is presented in Fig. 6. The shaft covers a system of ventilated corridors equipped with 28 sensors measuring air flow, temperatures, humidity, air pressure, methane concentrations and cutter loader characteristics

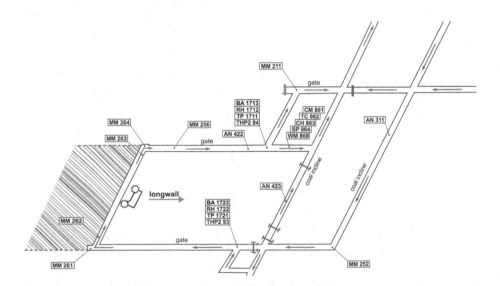

Fig. 6. Mining process scheme

including speed, consumed current and driving direction. The mining process is represented by the cutter loader moving and extracting the coal along the corridor longwalls as shown in Fig. 6.

The input data of 51700 cases each represented by 28 sensor series of 600s each amounted to almost 1 billion numerical measures. The target class variables were defined as binary alert presence at sensors MM263, MM264 and MM256 visible on the scheme in Fig. 6 occurring in the period between 3 to 6 minutes after the end of the individual case time slot. After a brief experimentations guided by the average performance of the several standard predictors including Multiple Linear Regression (MLR), Naïve Bayes (NB), Support Vector Machine (SVM), Decision Tree (DT) and Logistic Regression (LR) several consistent conclusions emerged very quickly:

1. 1-min aggregation of sensor series using simple average (mean) yields the best predictive results
2. For all 28 sensors the most recent section of the input 10-min slot appears the most informative about the target variables
3. Target variables' own histories appear to be by far the best predictors of their futures

As in previous case studies we applied our automated ML design framework based on the hierarchical backward-forward search. Since there was a clear and consistent indication that 1-minute quantization of the sensor data appears optimal, we left this choice fixed and applied BFS to optimize all other design choices of the historical depth, feature and predictor selection. Enumerating by $fname_i^j$ the ith numbered sensor feature built on j last minutes of the data from the 28 sensor variables from Fig. 6, i.e. the top feature selection solutions for the 3 sensor target variables resulted with the following design solutions:

- target MM_{263}: features set $\{TP_{1721}^1, TP_{1711}^1, MM_{263}^1, MM_{256}^1\}$, predictor: LR
- target MM_{264}: features set $\{RH_{1722}^1, MM_{264}^1, CR_{863}^1, P_{864}^1\}$, predictor: LR
- target MM_{256}: features set $\{TP_{1711}^1, MM_{256}^1, P_{864}^1, MM_{256}^1\}$, predictor: LR

First target sensor (MM_{263} was enriched by the temperatures in the same corridor further along the airflow. Second target sensor (MM_{264}) was enriched by the humidity in the corridor preceding the sensor location along the airflow and the methane drainage pressure and its flange pressure difference, while the third sensor (MM_{256}) added a combination of the temperature and the methane drainage pipeline pressure as well as the current of the hydraulic pump engine in the cutter loader.

Interestingly for all three prediction tasks logistic regression has been chosen as the top model beating much more complex models like support vector machines.

The ROC curves along with the AUC scores for the three top prediction models composed by our automated ML design system over the training set are presented in Fig. 7.

What is again notable here is that original sensors' time series carrying billions of data points, were summarized for the purpose of methane concentration

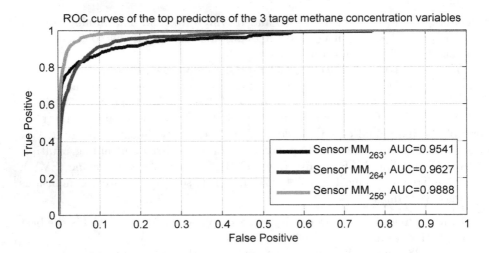

Fig. 7. ROC curves and AUC scores of the top training set predictors

warnings prediction down to a tiny fraction of its original size. Moreover, efficient optimization of the ML design with our hierarchical BFS method delivered very simple, highly optimized and robust predictor.

4   Conclusion

We have sketched the general framework for designing highly scalable and decomposable Machine Learning models that would stand the test of time through efficient processing of summarized big data and successful extraction of its predictive value with highly optimized automated predictive model design. We have highlighted that data summarization is at the center of such framework as it not only enables scalable processing but also catalyzes the essential knowledge generation and storage. We have drafted several conceptual methodologies for flexible summarization that allows to merge several data types together and more crucially is capable to represent the data summary hierarchically at the desired level of abstraction. We have also concluded that linear scalability of the modeling processes in supervised learning is easy to obtain with simple base learning algorithms further stimulated by model decomposition, while the complexity of the model built is shifted towards intelligent data preprocessing, filtering, aggregation and selection of highly compact feature subsets.

Combined with hierarchically summarized data, linearly scalable models could contribute to the breakthrough in ML that is already in making thanks to related advances in deep learning, and fully enable the power of massively parallel big data processing infrastructure. We have also illustrated how to progress towards the long-standing AI vision of automatically composed and self configurable robust ML models deployed along the course of several presented case studies of emerging applications. We argue that mutual optimization of decomposable ML design components is the key to both designing a powerful predictor and understanding why it works for particular applications and data sets.

We have presented several case studies that support various aspects of the proposed framework for automated generation and optimization of ML model design. In particular, we illustrated how much deeper we can push data summarization if the detailed prior knowledge of the model or the application context is available. We demonstrated the methods to achieve deep summarization of text documents that do not affect or even improve its predictive value for classification tasks. We have shown how tens of thousands of images can be compressed to just several templates if the only purpose they serve is to detect specific patterns. We have demonstrated how much the knowledge of the temporal process very rich in high frequency numerical sensors can be reduced to aggregated out states that match the resolution of the target variables at the time that matters for predictions. What is critical, however, is that beyond these deep data reductions our automated ML design framework consistently returns a highly accurate and robust predictive model constructed at the near-real time. We consider this work to be a working progress and continue to pursue many open challenges dedicated to further automation of the ML model design process that enables and efficiently exploits big data.

References

1. Yu, P., McLaughlin, J., Levy, M., Data, B.: A big disappointment for scoring consumer credit risk. NCLC report (2014)
2. Sicular, S.: Big Data is Falling into the Through of Disillusionment. Gartner Blog Network (2013)
3. Bengio, Y., Goodfellow, I., Courville, A.: Deep Learning. MIT Press, Cambridge (2015)
4. Schmidhuber, J.: Deep learning in neural networks: an overview. Nural Netw. **61**, 85–117 (2015)
5. Zhang, Z., Huang, Z., Zhang, Z.: Knowledge summarization for scalable semantic data processing. J. Comput. Inf. Syst. **6**(12), 3893–3902 (2010)
6. Changsheng, X., Maddage, M.C., Xi, S.: Automatic music classification and summarization. IEEE Trans. Speech Audio **13**(3), 441–450 (2004)
7. Ekin, A., Tekalp, A.M.: Automatic soccer video analysis and summarization. IEEE Trans. Image Process. **12**(7), 796–807 (2003)
8. Hori, C., Furui, S.: A new approch to automatic speech summarization. IEEE Trans. Multimedia **5**(3), 368–378 (2003)
9. Yang, C., Junsong, Y., Jiebo, L.: Towards scalable summarization of consumer videos via sparse dictionary selection. IEEE Trans. Multimedia **14**(1), 66–75 (2012)
10. Chandola, V., Kumar, V.: Summarization - compressing data into an informative representation. In: Proceedings 5th IEEE International Conference on Data Mining, New Orleans (2005)
11. Hahn, U., Mani, I.: The challenges of automatic summarization. Computer **33**(11), 29–36 (2000)
12. Bishop, C.M.: Model-based machine learning. Philos. Trans. R. Soc. A (2013)

13. Langley, P.: Artificial intelligence and cognitive systems. AISB Q. **133**, 1–4 (2012)

14. Goldstein, J., Kantrowitz, M., Mittal, V., Carbonell, J.: Summarizing text documents: sentence selection and evaluation metrices. In: Proceedings 22nd International ACM SIGIR Conference on Research and Development in Information Retrieval, pp 121–128 (1999)

15. Steinberger, J., Jezek, K.: Evaluation measures for text summarization. Comput. Inf. **28**, 1001–1026 (2009)

16. d'Acierno, A., Moscato, V., Persia, F., Picariello, A., Penta, A.: Semantic summarization of web documents. In: Proceedings IEEE 4th International Conference on Semantic Computingng, pp: 430–435 (2010)

17. Verma, R., Chen, P., Lu, W.: A semantic free-text summarization system using ontology knowledge. IEEE Trans. Inf. Technol. Biomed. **5**(4), 261–270 (2007)

18. Li, T., Zhu, S., Ogihara, M.: Hierarchical document classification using automatically generated hierarchy. J. Intell. Inf. Syst. **29**(2), 211–230 (2007)

19. Vohra, S.M., Teraiya, J.B.: A comparative study of sentiment analysis techniques. J. Inf. Knowl. Res. Comput. Eng. **2**(2), 313–317 (2013)

20. Saeys, Y., Inza, I., Larranaga, P.: A review of feature selection techniques in bioinformatics. Bioinformatics **19**(23), 2507–2517 (2007)

21. Duda, R.O., Hart, P.E., Stork, D.G.: Pattern Classification. Wiley, New York (2001)

22. Bishop, C.M.: Pattern Recognition and Machine Learning. Springer, New York (2006)

23. Pan, F., Wang, W., Tung, K.H., Yang, J.: Finding Representative Set from Massive Data. Springer, New York (2006)

24. Aggarwal, C.C., Reddy, C.K.: Data Clustering: Algorithms and Applications. CRC Press, Boca Raton (2014)

25. Roth, V.: Outlier detection with one-class kernel fisher discriminant. Adv. Neural Inf. Process. Syst. **17**, 1169–1176 (2004)

26. Mitra, P.: Density-based multiscale data condensation. IEEE Trans. Pattern Anal. Mach. Intell. **24**(6), 734–747 (2002)

27. Yang, P., Li, J.-S., Huang, Y.-X.: HDD: a hypercube division-based algorithm for discretisation. Int. J. Sys. Sci. **42**(4), 557–566 (2010)

28. Sauvola, J., Pietikainen, M.: Adaptive document image binarization. Pattern Recogn. **33**, 225–236 (2000)

29. Zhou, X., Wang, X., Dougherty, E.R.: Binarization of microarray data on the basis of a mixture model. Mol. Cancer Ther. **2**(7), 679–684 (2003)

30. Tomas, J., Cascuberta, F.: Binary feature classification for word disambiguation in statistical machine translation. In: Proceedings 2nd International Workshop on Pattern Recognition in Information Systems (2002)

31. Mitchell, T.: Generative and discriminative classifiers: naive bayes and logistic regression. In: Machine Learning. McGraw Hill (2010)

32. Ng, A.Y., Jordan, M.I.: On discriminative vs. generative classifiers: a comparison of logistic regression and naive Bayes. In: NIPS 14 (2002)

33. Friedman, J., Hastie, T., Tibshirani, R.: Regularization paths for generalized linear models via coordinate descent. J. Stat. Softw. **33**(1), 1–22 (2010)

34. Sugiyama, M., Yamamoto, A.: A fast and flexible clustering algorithm using binary discretization. In: 11th IEEE International Conference on Data Mining, pp. 1213–1217 (2011)

35. Dieterich, T.G., Bakiri, G.: Solving multiclass learning problems via error-correcting output codes. J. Artif. Intell. Res. **2**, 263–286 (1995)

36. Aly, M.: Survey on multiclass classification methods. Caltech Technical report (2005)
37. Rokach, L.: Pattern Classification Using Ensemble Methods. World Scientific, River Edge (2010)
38. Maimon, O., Rokach, L.: Improving supervised learning by feature decomposition. In: Eiter, T., Schewe, K.-D. (eds.) FoIKS 2002. LNCS, vol. 2284, pp. 178–196. Springer, Heidelberg (2002). doi:10.1007/3-540-45758-5_12
39. Drineas, P., Mahoney, M.W., Muthukrishnan, S., Sampling, S.: Approximation, relative-error matrix: column-row-based methods. In: Proceedings 14th Annual Symposium on Algorithms, pp. 304–314 (2006)

Clustering Aided Support Vector Machines

Goce Ristanoski[1], Rahul Soni[1]($\boxtimes$), Sutharshan Rajasegarar[2]($\boxtimes$),
James Bailey[3]($\boxtimes$), and Christopher Leckie[1,3]($\boxtimes$)

[1] Data61, CSIRO, Melbourne, Australia
Goce.Ristanoski@data61.csiro.au, sn.rahul99@gmail.com
[2] School of Information Technology, Deakin University,
Waurn Ponds, Australia
sutharshan.rajasegarar@deakin.edu.au
[3] Department of Computing and Information Systems,
The University of Melbourne, Melbourne, Australia
{baileyj,caleckie}@unimelb.edu.au

Abstract. Support Vector Machines (SVMs) have proven to be an effective approach to learning a classifier from complex datasets. However, highly nonhomogeneous data distributions can pose a challenge for SVMs when the underlying dataset comprises clusters of instances with varying mixtures of class labels. To address this challenge we propose a novel approach, called a cluster-supported Support Vector Machine, in which information derived from clustering can be incorporated directly into the SVM learning process. We provide a theoretical derivation to show that when the total empirical loss is expressed in terms of the combined quadratic empirical loss from each cluster, we can still find a formulation of the optimisation problem that is a convex quadratic programming problem. We discuss the scenarios where this type of model would be beneficial, and present empirical evidence that demonstrates the improved accuracy of our combined model.

1 Introduction

Support Vector Machines, built on the concept of margin theory, have become one of the most widely used supervised learning algorithms. SVMs try to maximize the minimum of the geometric margin i.e., the distance between the decision boundary and the instances closest to it. These instances, called support vectors, are typically small in number compared to the total number of instances, and thus a small number of inner products are required to make a prediction on unseen instances. SVMs can be used for both classification and regression, although the work presented in this paper focuses on improving the classification task.

In practice, the distribution of instances from each class can be highly nonhomogeneous. The density of instances can vary throughout the feature space, corresponding to clusters of instances. Moreover, the distribution of labels can vary from one cluster to another, so that one cluster can be rich in instances belonging to one class, while another cluster may contain a majority of instances from a different class. A challenge for SVMs in this context is that while the chosen

© Springer International Publishing AG 2017
P. Perner (Ed.): MLDM 2017, LNAI 10358, pp. 322–334, 2017.
DOI: 10.1007/978-3-319-62416-7_23

support vectors may reflect the overall distribution of the dataset, but may not be representative of the distribution in each cluster.

Clustering algorithms have been used as a tool in many data mining tasks, for the purpose of understanding the grouping that exists among the samples in a dataset. Our work focuses on how the combined use of clustering algorithms and SVMs can be used to improve the accuracy of classification on highly nonhomogeneous datasets. Past research has shown that the straightforward approach of dividing the data according to the cluster labelling, and then learning a separate model for each cluster can yield improvements in accuracy. However, our aim is to find a way that these two approaches can be more closely integrated. For this reason, we look for a way to incorporate some of the knowledge from the clustering into the learning process of the SVM.

In particular, we aim to modify the loss function of the SVM to reflect the accuracy within each cluster, so that we minimise the variation in accuracy between clusters. While this provides a way to improve the overall accuracy of classification, the underlying optimisation problem of SVM learning is no longer a quadratic programming problem. However, we provide a theoretical proof that the resulting loss can be upper bounded by a linear combination of simpler expressions, which yields a convex quadratic optimisation problem for training our cluster-supported SVM classifier. In this way, we can use clustering to provide an efficient way of modelling local nonhomogeneity in the underlying data distribution when constructing the SVM.

The contributions of the work presented in this paper are:

– We investigate how SVM models perform in relation to the knowledge provided by clustering algorithms, that is, the clusters that can be assigned to the data.
– We design a way to use the clustering assigned to a given dataset in the learning process of SVM.
– We perform a detailed empirical analysis that gives clear insight of the relation of the type of data and the performance of SVM with incorporated clustering knowledge.

The remainder of the paper is organised as follows: Sect. 2 describes the Related work in the area of SVM and clustering for classification. We then present the proposed algorithm in Sect. 3. Section 4 presents the datasets used in the empirical evaluation and the research findings. We conclude into a summary the contributions of the work shown in this paper in Sect. 5.

2   Related Work

The success of Support Vector Machines (SVMs) in the tasks of classification and regression has led to an increased use of this method over different types of analysis. From time series analysis [1], graph matching [10] and dimensionality reduction [11], SVMs have proven to be more than just a powerful classification tool. Practical application has also motivated research into SVM that speed-up the learning time [2,4] and are capable of dealing with large data sets [3].

With the emergence of Big Data, processing large data sets has become a greater challenge than before. A version of SVM that screens the data for features that are irrelevant or with very little relevance has been proposed by [8]. A pre-processing stage of looking into samples which are not contributing to the classifier has been suggested by [5]. Related to that, investigating the option of reducing the number of learned support vectors via multiple kernels in order to allow for faster testing is a useful approach [12]. Additional optimisation of the mean and variance alongside the empirical loss has been introduced by [31]. Even consideration of the underlying hardware architecture and parallel processing has provided a contribution in the area of SVM optimization [9]. With modifications such as using Dual Coordinate Ascent [7] and optimising a specific area of the AUC curve [6], the list of benefits SVMs can provide keeps on increasing.

Research into clustering algorithms has witnessed a substantial increase in interest over the recent years as well. With application areas varying from text processing [20], graph clustering [16], time series data [13,19] and even discrimination aware learning [14], clustering algorithms can provide contributions in the area of classification by itself, and in combination with other learning models. Just like SVMs, clustering algorithms have been modified to cope with high dimensional data in a fast and efficient manner [17]. An important task is also to find clusterings that are easy to interpret [15], in particular when we can have several clusterings over the same dataset [18].

An important research area in which both clustering and SVM methods have been applied is discrimination aware learning. With other methods being investigated [21–23], it is imperative to consider this aspect when testing new methods. Our work not only considers this aspect, but also empirically evaluates the performance of the new method in relation to discrimination aware learning.

SVMs and clustering algorithms have been successfully applied in order to deliver better results. Approaches such as simple clustering of samples and training an SVM model on each cluster [30], to extracting only the most relevant data per cluster [25,26,28,29], have been shown to be beneficial particularly in the case of large datasets [24,27,32]. A modification of the SVM learning process itself, rather than the data, is a more challenging process. Our work aims to investigate this direction, and show that the benefits can be significant over a wide range of data types.

3   Cluster Supported SVM

In this section we define the background of standard SVM and formulate the quadratic objective function for our clustering supported SVM.

3.1   Standard SVM

Let $\mathcal{X} \in \mathbb{R}^d$ be the instance space and $\mathcal{Y} = \{+1, -1\}$ the label set. Let $\mathcal{D}$ be an unknown (underlying) distribution over $\mathcal{X} \times \mathcal{Y}$. Assuming an independent and identically (i.i.d) distributed training set of size m:

$$S = \{(\boldsymbol{x}_1, y_1), (\boldsymbol{x}_2, y_2), \ldots, (\boldsymbol{x}_m, y_m)\},$$

the goal of SVM is to learn a function that can be used to predict the labels for future unseen instances.

For SVMs, f is regarded as a linear model, i.e., $f(\boldsymbol{x}) = \boldsymbol{w}^\top \phi(\boldsymbol{x})$ where $\boldsymbol{w}$ is a linear predictor and $\phi(\boldsymbol{x})$ is a feature mapping of $\boldsymbol{x}$ induced by a kernel k, i.e., $k(\boldsymbol{x}_i, \boldsymbol{x}_j) = \phi(\boldsymbol{x}_i)^\top \phi(\boldsymbol{x}_j)$. The margin of instance $(\boldsymbol{x}_i, y_i)$ is formulated as

$$\gamma_i = y_i \boldsymbol{w}^\top \phi(\boldsymbol{x}_i), \forall i = 1, \ldots, m. \tag{1}$$

It has been shown that in separable cases where the training examples can be separated with the zero error, SVM with hard-margin (or Hard-margin SVM),

$$\min_{\boldsymbol{w}} \frac{1}{2} \boldsymbol{w}^\top \boldsymbol{w}$$
$$\text{s.t.} \ \ y_i \boldsymbol{w}^\top \phi(\boldsymbol{x}_i) \geq 1, \ i = 1, \ldots, m,$$

is regarded as the maximization of the minimum margin $\{\min\{\gamma_i\}_{i=1}^m\}$.

In non-separable cases where the training examples cannot be separated with the zero error, SVM with soft-margin (or Soft-margin SVM) has been developed,

$$\min_{\boldsymbol{w}, \boldsymbol{\xi}} \frac{1}{2} \boldsymbol{w}^\top \boldsymbol{w} + C \sum_{i=1}^m \xi_i$$
$$\text{s.t.} \ \ y_i \boldsymbol{w}^\top \phi(\boldsymbol{x}_i) \geq 1 - \xi_i, \tag{2}$$
$$\xi_i \geq 0, \ i = 1, \ldots, m.$$

where $\boldsymbol{\xi} = [\xi_1, \ldots, \xi_m]^\top$ measure the losses of instances, and C is a trade-off parameter. There exists a constant $\bar{C}$ such that (2) can be equivalently reformulated as,

$$\max_{\boldsymbol{w}} \ \gamma_0 - \bar{C} \sum_{i=1}^m \xi_i$$
$$\text{s.t.} \ \ \gamma_i \geq \gamma_0 - \xi_i,$$
$$\xi_i \geq 0, \ i = 1, \ldots, m,$$

where γ_0 is a relaxed minimum margin, and $\bar{C}$ is the trade-off parameter. Note that γ_0 indeed characterizes the top-p minimum margin; hence, SVMs (with both hard-margin and soft-margin) consider only a single-point margin and have not exploited the whole margin distribution.

3.2 Cluster Supported SVM Formulation

Our work originates from the idea of using additional optimisation via the introduction of a quadratic mean in the empirical loss, applied on some form of grouping in the data [21]. We formulate a mathematical model that takes the root mean squared value of the quadratic empirical loss per group into account, and use clustering methods as the basis of our grouping strategy. We call this model **CluSVM** which is achieved by modifying the optimization framework proposed by [31]. As the model introduces the margin mean and margin variance, we derive our formulas with these components as well, which can be omitted for a version without their optimization by setting them to zero.

Also since it is known that soft margin SVM outperforms hard margin SVM, we consider the soft margin case in the rest of the derivation to follow. CluSVM is henceforth referred to as soft margin CluSVM.

We define total empirical loss R_{Qemp} in terms of the Root mean square error (RMSE) value of the quadratic empirical loss per group as follows:

$$R_{Qemp}(\boldsymbol{w}) = \sqrt{\frac{1}{k}\sum_{i=1}^{k} f_i^2(\boldsymbol{w})}$$

$$f_i(\boldsymbol{w}) = \frac{1}{n_i}\sum_{j\in n_i} l(\boldsymbol{w}, \boldsymbol{x}_j, y_j) \tag{3}$$

$$l(\boldsymbol{w}, \boldsymbol{x}_j, y_j) = \frac{1}{2}(\boldsymbol{w}^\top \phi(\boldsymbol{x}_j) - y_j)^2$$

$$\sum_{i=1}^{k} n_i = m$$

Hence the optimization problem can be written in terms of the RMSE as follows:

$$\min_{\boldsymbol{w},\boldsymbol{\xi}} \ \frac{1}{2}\boldsymbol{w}^\top\boldsymbol{w} + R_{Qemp}(\boldsymbol{w}) + \lambda_1\hat{\gamma} - \lambda_2\bar{\gamma} + C\sum_{i=1}^{m}\xi_i \tag{4}$$
$$\text{s.t.} \ \ y_i\boldsymbol{w}^\top\phi(\boldsymbol{x}_i) \geq 1 - \xi_i,$$
$$\xi_i \geq 0, \ i = 1,\ldots,m.$$

where λ_1 and λ_2 are the parameters for trading-off the margin variance $\hat{\gamma}$, margin mean $\bar{\gamma}$ and model complexity. By substituting (3) in (4) we get:

$$\min_{\boldsymbol{w},\boldsymbol{\xi}} \ \frac{1}{2}\boldsymbol{w}^\top\boldsymbol{w} + \sqrt{\frac{1}{k}\sum_{i=1}^{k}\left\{\frac{1}{n_i}\sum_{j\in n_i}\frac{1}{2}(\boldsymbol{w}^\top\phi(\boldsymbol{x}_j) - y_j)^2\right\}^2}$$
$$+ \frac{2\lambda_1}{m^2}(m\boldsymbol{w}^\top\boldsymbol{X}\boldsymbol{X}^\top\boldsymbol{w} - \boldsymbol{w}^\top\boldsymbol{X}\boldsymbol{y}\boldsymbol{y}^\top\boldsymbol{X}^\top\boldsymbol{w}) \tag{5}$$
$$- \frac{\lambda_2}{m}(\boldsymbol{X}\boldsymbol{y})^\top\boldsymbol{w} + C\sum_{i=1}^{m}\xi_i$$
$$\text{s.t.} \ \ y_i\boldsymbol{w}^\top\phi(\boldsymbol{x}_i) \geq 1 - \xi_i,$$
$$\xi_i \geq 0, \ i = 1,\ldots,m.$$

The RMSE value of the quadratic empirical loss function per group as defined in (3), (5) turns out to be a fourth order function w.r.t the margin parameter $\boldsymbol{w}$, which is a tedious formulation to fit in Lagrange KKT conditions. To overcome this, we can reformulate it as a quadratic loss w.r.t $\boldsymbol{w}$. The quadratic empirical loss function w.r.t $\{f_i(\boldsymbol{w}) : f_i(\boldsymbol{w}) \geq 0 \ \forall \ i = 1, .., k\}$ is upper bounded by its linear combination as follows:

$$
R_{Qemp}(\boldsymbol{w}) = \sqrt{\frac{1}{k} \sum_{i=1}^{k} f_i^2(\boldsymbol{w})}
$$

$$
= \sqrt{\sum_{i=1}^{k} \left(\frac{f_i(\boldsymbol{w})}{\sqrt{k}} \right)^2} \approx \sum_{i=1}^{k} \frac{f_i(\boldsymbol{w})}{\sqrt{k}} \tag{6}
$$

Since $R_{Qemp}(\boldsymbol{w})$ is upper bounded by the R.H.S of (6), a minimizer of the R.H.S of (6) with respect to parameters $a_i \geq 0$, $i = 1, .., k$ also leads to the minimization of $R_{emp}(\boldsymbol{w})$. This means that $R_{Qemp}(\boldsymbol{w})$ is modelled in terms of the margin parameter $\boldsymbol{\alpha}$ as follows:

$$
R_{Qemp}(\boldsymbol{w}) = \sqrt{\frac{1}{k} \sum_{i=1}^{k} f_i^2(\boldsymbol{w})}
$$

$$
= \sqrt{\sum_{i=1}^{k} \left(\frac{f_i(\boldsymbol{w})}{\sqrt{k}} \right)^2} \approx \sum_{i=1}^{k} \frac{f_i(\boldsymbol{w})}{\sqrt{k}} \tag{7}
$$

$$
R_{Qemp}(\boldsymbol{w}) \approx \sum_{i=1}^{k} \frac{1}{n_i \sqrt{k}} \sum_{j \in n_i} l(\boldsymbol{w}, \boldsymbol{x}_j, y_j)
$$

The optimal solution form of $\boldsymbol{w}$ leads to the following representation of quadratic loss $l(\boldsymbol{w}, \boldsymbol{x}_j, y_j)$ in terms of the inner product of the kernel matrix $\boldsymbol{G}$ and solution vector $\boldsymbol{\alpha}$:

$$
l(\boldsymbol{w}, \boldsymbol{x}_j, y_j) = \frac{1}{2} (\boldsymbol{w}^\top \phi(\boldsymbol{x}_j) - y_j)^2
$$

$$
= \frac{1}{2} \{ \boldsymbol{w}^\top \phi(\boldsymbol{x}_j) \}^2 - \boldsymbol{w}^\top \phi(\boldsymbol{x}_j) y_j + \frac{1}{2}
$$

$$
= \frac{1}{2} \boldsymbol{w}^\top \phi(\boldsymbol{x}_j) \phi(\boldsymbol{x}_j)^\top \boldsymbol{w} - \boldsymbol{w}^\top \phi(\boldsymbol{x}_j) y_j + \frac{1}{2} \tag{8}
$$

$$
= \frac{1}{2} \boldsymbol{\alpha}^\top \boldsymbol{X}^\top \phi(\boldsymbol{x}_j) \phi(\boldsymbol{x}_j)^\top \boldsymbol{X} \boldsymbol{\alpha}
$$

$$
- \boldsymbol{\alpha}^\top \boldsymbol{X}^\top \phi(\boldsymbol{x}_j) y_j + \frac{1}{2}
$$

$$
= \frac{1}{2} \boldsymbol{\alpha}^\top \boldsymbol{G}_{:j} \boldsymbol{G}_{:j}^\top \boldsymbol{\alpha} - \boldsymbol{\alpha}^\top \boldsymbol{G}_{:j} y_j + \frac{1}{2}
$$

where $G_{\cdot i}$ denotes the i-th column of G as before.

4 Experiments and Results

Evaluation of the performance of the newly proposed CluSVM focuses on determining what features of the dataset and the clustering contribute to the actual change in performance. For that purpose we generated a series of datasets in which we introduced a range for some of the features that would describe the performance of the model in relation to the nature of the data. This generation of datasets allows for more indepth analysis of how the clusters structure, proximity and densitiy are relevant to the final model performance.

4.1 Dataset Description

We generated a total of 54 synthetic datasets, each with 7 numeric attributes in the range of 0–600, and 5000 samples per dataset. For half of the datasets, we generated clusters in such a way that 2 clusters would not overlap in any of the 7 features. This was accomplished by setting the cluster boundary at half of the minimal allowed distance between two cluster centres, and generating all the data for a given cluster within the cluster boundary. The other 27 datasets were generated in a similar manner, except that for each feature per cluster, the cluster boundary for that feature was set to 2/3 of the allowed minimal distance between cluster centres. This means that in some of the feature spaces, it is possible to have an overlap between 2 clusters, but that overlap will not be a region with the greatest number of samples for each of the clusters overlapping. Furthermore, 75% of the samples were randomly assigned around the cluster centre within the cluster boundary, and the remaining 25% of the samples were assigned closer to the cluster centres at a maximum distance of one half of the cluster boundary, for more concentrated cluster centre areas.

The clusters and cluster labels in the datasets were generated in the range of the following attributes:

- Positive sample rate - the ratio of positive vs negative samples in real life datasets is rarely 1:1. Quite often we are concerned with situations where the class of interest is the less present class in the dataset. As mentioned in the related work section of the paper, research in determining the effect of unbalanced datasets is of significance in the data mining area. For that purpose, we generate datasets where the rate of positive samples in the dataset is 0.5, 0.4 and 0.25 (a rate of 0.5 means that 50% of the samples are from the class).
- Number of clusters - we aim to investigate how the number of clusters contributes to the prediction accuracy. Our initial assumption was that with fewer clusters, the difference in performance between our method and the standard method would not be as significant as when the number of clusters was higher. We also aimed to investigate whether a large number of samples in the dataset results in lower accuracy on its own. For that purpose, we generated datasets with 5, 10 and 15 clusters within the same feature space.

– Cluster homogeneity - the cluster label will correspond to the most frequently occurring class of samples in that cluster. Clusters that only have samples from one class would be ideal, but this is far from expected. By varying the presence of the most frequent class, we can investigate if considering the grouping strategy will still provide a contribution when the cluster homogeneity is lower. The cluster homogeneity was set to 0.7, 0.8 and 0.9 (0.9 corresponds to 90% of the samples in that cluster belonging to the same class, which is the class assigned to the cluster itself).

By varying each of these three attributes, each with three different values, we end up with 27 different combinations corresponding to each of the two groups of 27 datasets generated.

The metrics used to evaluate the performance of the new algorithm are Accuracy (number of correctly classified samples/number of samples), as well as Recall and F-score for the positive samples: Recall=True Positive/(True Positive+False Negative), F-score=2*Precision*Recall/(Precision+Recall), Precision=True Positive/(True Positive+ False Positive). The Recall and F-score indicate how many of the samples with positive class labels have been classified correctly, which is significant in particular for the changes we investigate in the positive sample rate.

The Wilcoxon signed rank test (WSRT p-value) is used for the Accuracy, Recall and F-score values between our method and each of the other baseline methods are used to determine whether the changes are significant enough. We compared our method CluSVM with naive Bayes (NB), Polynomial Kernel SVM (SVM), Normalized Polynomial Kernel SVM (SVM NPK), and Multilayer Perceptron ANN (MLP).

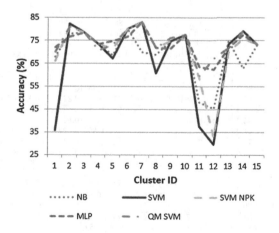

Fig. 1. Individual cluster accuracy for all methods.

4.2   Datasets with Overlapping Cluster Boundary

Presented in Table 1 are the results on the 27 datasets where no clear boundary between the clusters were set, and it was possible for some samples to overlap with others in some of the features. This is why when testing our CluSVM model, we used a KNN algorithm to set the cluster labels and include the clusters overlap information, rather then using our original clustering labels. In a few of the cases, this resulted in 2 close clusters with the same label being labelled as 1 cluster, resulting in a total of 14 clusters, which was still considered to fall in the 15 clusters category.

We can see that in most of the cases, particularly when having 10 and 15 clusters, our methods shows the best results from all metrics: Accuracy, Recall and F-score. MLP is a close runner up, while NB, SVM and even SVM NPK are all outperformed by CluSVM. An interesting observation is that the number of clusters is quite relevant to the performance of all the models: with 5 clusters per dataset the difference in performance between the models is not as great as when we have 10 and 15 clusters. This can be especially seen in the decrease of

Table 1. Performance metrics for datasets with overlapping of cluster boundaries. Our proposed CluSVM shows to have statistically significant improvements compared to NB, SVM and SVM NPK, and still better performance when compared to MLP.

Num. clus.	Cluster hmg.	Accuracy (%) NB	SVM	SVM NPK	MLP	Clu-SVM	Recall NB	SVM	SVM NPK	MLP	Clu-SVM	F-score NB	SVM	SVM NPK	MLP	Clu-SVM
Positive sample rate = 0.5																
5	0.7	60.5	58.3	62.4	**63.1**	62.5	**0.639**	0.598	0.572	0.57	0.564	**0.615**	0.586	0.6	0.604	0.597
	0.8	66.4	64.1	**70.8**	67.5	70.6	0.653	0.615	0.627	**0.701**	0.633	0.659	0.63	0.681	**0.682**	**0.682**
	0.9	69.7	69.3	**78.2**	76.3	77.7	0.694	0.682	0.698	0.725	0.692	0.689	0.683	**0.756**	0.748	0.75
10	0.7	63.3	58.9	67.6	64.1	**68.8**	0.633	0.72	0.68	**0.776**	0.702	0.639	0.632	0.623	0.679	**0.688**
	0.8	71.6	68.8	78.3	**77.8**	78.7	0.694	0.771	0.775	**0.783**	**0.783**	0.707	0.709	0.779	0.777	**0.784**
	0.9	79.1	76.2	88.5	89.4	**89.8**	0.741	0.759	0.883	0.887	**0.889**	0.782	0.763	0.886	0.894	**0.898**
15	0.7	66.9	66.1	69	69.4	**70.1**	0.611	0.605	0.661	0.713	**0.724**	0.64	0.632	0.672	0.692	**0.699**
	0.8	74.9	74.9	78.8	80.1	**80.3**	0.698	0.714	0.792	0.832	**0.834**	0.736	0.74	0.788	0.807	**0.809**
	0.9	83.4	85.1	91	91.7	**92.2**	0.765	0.815	0.925	0.935	**0.944**	0.827	0.851	0.914	0.921	**0.926**
Positive sample rate = 0.4																
5	0.7	73	69.4	72.8	73.1	**74**	0.602	0.477	0.546	0.615	**0.638**	0.636	0.55	0.611	0.615	**0.658**
	0.8	80.4	74.2	80.7	81.1	**81.6**	0.729	0.61	0.694	**0.783**	0.751	0.733	0.636	0.725	**0.754**	0.751
	0.9	84.2	78.1	86.3	86.5	**88.28**	0.772	0.67	0.779	**0.855**	0.824	0.794	0.707	0.818	0.824	**0.847**
10	0.7	68.3	70	74.9	74.5	**75.5**	0.479	0.485	**0.668**	**0.688**	0.687	0.541	0.558	0.68	0.677	**0.684**
	0.8	72.7	75.5	84.2	84.1	**84.6**	0.609	0.584	0.815	0.815	**0.818**	0.635	0.649	0.8	0.799	**0.805**
	0.9	77.1	78.4	91.2	91.7	91.6	0.669	0.642	0.891	**0.894**	**0.894**	0.703	0.707	0.891	**0.897**	0.896
15	0.7	69.4	67.4	71.2	73.6	**74.1**	0.535	0.328	0.522	0.603	**0.643**	0.587	0.45	0.596	0.65	**0.668**
	0.8	74	71.2	78.6	81.7	**82.4**	0.639	0.455	0.621	0.748	**0.753**	0.665	0.561	0.701	0.768	**0.776**
	0.9	78.4	75.6	85.4	90.5	**90.9**	0.722	0.526	0.727	0.864	**0.869**	0.729	0.635	0.801	0.88	**0.885**
Positive sample rate = 0.25																
5	0.7	**82.5**	82.2	80.5	82.2	82	0.569	0.554	0.446	**0.58**	0.5	0.636	0.626	0.552	**0.637**	0.6
	0.8	**86.4**	86.1	85.2	86.3	85.6	**0.644**	0.633	0.607	0.64	0.593	**0.722**	0.715	0.693	0.72	0.694
	0.9	91	**91.1**	89.7	91	89.5	**0.716**	0.709	0.705	0.709	0.673	**0.802**	**0.802**	0.777	0.8	0.765
10	0.7	79.4	75.3	81.3	81.6	**81.7**	0.332	NA	0.445	0.506	**0.51**	0.443	NA	0.541	0.576	**0.579**
	0.8	83.7	84.1	85.7	85.7	**86.1**	0.464	0.506	0.59	0.59	**0.602**	0.577	0.603	0.664	0.664	**0.674**
	0.9	85.1	86.5	89.4	**89.7**	89.6	0.517	0.61	**0.688**	0.68	0.684	0.651	0.708	0.777	**0.78**	0.78
15	0.7	78.3	74.1	80.8	86	**86.4**	0.259	NA	0.409	0.683	**0.707**	0.382	NA	0.525	0.717	**0.729**
	0.8	81.9	80.3	88	90.4	**91.3**	0.39	0.321	0.623	**0.811**	0.527	0.527	0.457	0.738	0.811	**0.828**
	0.9	86.3	85.5	94.2	**95.8**	95.2	0.498	0.478	0.865	**0.948**	0.94	0.646	0.623	0.882	**0.919**	0.908
WSRT p-value		<0.0001			0.008	base	<0.001			0.5	base	<0.0001			0.11	base

the Recall and F-score values for most of the baseline comparison methods as we increase the number of clusters.

The cluster homogeneity behaviour seems to be consistent for all three cases of 5, 10 and 15 clusters: the lower the cluster homogeneity, the lower the performance for all methods. The positive sample rate seems to have a greater impact in the cases when we have a small number of clusters, and seems less relevant when the number of clusters is higher. Figure 1 shows the accuracy per cluster for the case of cluster homogeneity=0.7, positive sample rate=0.4, 15 clusters. The variance in per cluster accuracy is handled by our CluSVM method, and - we can see there is a large difference in the per cluster accuracy for most of the remaining methods.

4.3   Datasets with Clear Cluster Boundary

Table 2 shows the results of testing the 27 datasets with clear cluster boundaries, in which case we used the original cluster labels in our CluSVM. In this case,

Table 2. Performance metrics for datasets without overlapping of cluster boundaries. We can observe similar results as with the case of overlapping boundaries, with our CluSVM showing to outperform other methods in most cases.

Num. clus.	Cluster hmg.	Accuracy (%)					Recall					F-score				
		NB	SVM	SVM NPK	MLP	Clu-SVM	NB	SVM	SVM NPK	MLP	Clu-SVM	NB	SVM	SVM NPK	MLP	Clu-SVM
Positive sample rate = 0.5																
5	0.7	63.1	62.3	64.7	63.2	**65.3**	0.605	0.583	0.564	**0.609**	0.593	0.626	0.613	0.62	0.628	**0.634**
	0.8	71.2	71.4	**73.4**	74.2	73.4	0.687	0.687	0.667	**0.689**	0.671	0.709	0.711	0.719	**0.732**	0.721
	0.9	78	75.8	80.7	**81.7**	80.7	0.742	0.706	0.725	**0.739**	0.731	0.78	0.754	0.798	**0.809**	0.799
10	0.7	63.5	61.8	67.5	67	**68.1**	0.625	0.592	0.625	**0.667**	0.665	0.628	0.605	0.655	0.666	**0.673**
	0.8	71.8	70.5	76.6	77	**78.1**	0.724	0.699	0.742	0.754	**0.76**	0.716	0.7	0.757	0.763	**0.774**
	0.9	80.2	77.8	86.6	88.5	**88.6**	0.835	0.807	0.879	**0.895**	0.887	0.809	0.785	0.868	**0.887**	**0.887**
15	0.7	63.1	59.3	68.1	68.9	**70.1**	0.736	0.726	0.732	0.716	**0.734**	0.665	0.64	0.695	0.696	**0.701**
	0.8	68.7	61.5	79	80.4	**81.6**	0.796	0.738	**0.835**	0.826	**0.835**	0.711	0.65	0.794	0.803	**0.815**
	0.9	76.8	69.4	88.7	90.9	**92.1**	0.862	0.807	0.912	0.921	**0.941**	0.785	0.721	0.888	0.909	**0.921**
Positive sample rate = 0.4																
5	0.7	72.2	69.1	73.7	74	**76.3**	0.575	0.455	0.596	**0.683**	0.681	0.622	0.54	0.643	0.677	**0.696**
	0.8	79	76.2	82	**83**	82.8	0.695	0.622	0.751	**0.776**	0.766	0.724	0.675	0.768	**0.784**	0.78
	0.9	84	80.1	90.7	90.4	**91.4**	0.747	0.683	0.874	0.878	**0.885**	0.796	0.742	0.887	0.885	**0.896**
10	0.7	69.3	66.2	71.3	75.6	**76.2**	0.525	0.369	0.515	0.676	**0.696**	0.58	0.469	0.592	0.691	**0.703**
	0.8	77.5	69.3	82.6	84.4	**84.5**	0.658	0.454	0.742	0.812	**0.817**	0.691	0.531	0.766	0.799	**0.802**
	0.9	82.3	72.9	89.4	90.9	**91.1**	0.711	0.773	0.835	0.859	**0.86**	0.748	0.679	0.854	0.875	**0.877**
15	0.7	65.7	69.9	70.7	74	**75**	0.462	0.503	0.485	0.674	**0.687**	0.516	0.569	0.567	0.672	**0.685**
	0.8	70.3	76	79.8	81.9	**83.9**	0.542	0.59	0.657	0.761	**0.791**	0.595	0.664	0.723	0.772	**0.798**
	0.9	75.5	80.9	87.8	90.4	**91.4**	0.639	0.682	0.804	0.868	**0.886**	0.677	0.742	0.841	0.879	**0.892**
Positive sample rate = 0.25																
5	0.7	82.5	82.8	81.8	81.7	81.9	0.512	0.496	0.46	**0.524**	0.478	0.592	0.589	0.556	0.587	0.572
	0.8	85	84.7	84.9	**85.8**	85.4	0.572	0.544	0.552	**0.589**	0.573	0.648	0.631	0.638	**0.667**	0.654
	0.9	89.5	89.3	89.9	89.8	**90.2**	**0.695**	0.659	0.687	0.687	0.683	0.767	0.754	0.772	0.77	**0.776**
10	0.7	79	73.9	78.2	81.5	**81.8**	0.314	0	0.237	**0.448**	0.444	0.439	NA	0.363	**0.589**	0.561
	0.8	82	72.2	82.2	**85.3**	84.9	0.493	0	0.478	**0.604**	0.579	0.604	NA	0.599	**0.696**	0.681
	0.9	86.4	84.8	87.4	**89.6**	88.6	0.579	0.518	0.615	**0.676**	0.66	0.677	0.628	0.707	**0.763**	0.741
15	0.7	80.2	75.1	83	**85.6**	84.3	0.518			**0.703**	0.615	0.459	NA	0.603	**0.708**	0.661
	0.8	84	83.8	88	89.1	**89.2**	0.483	0.504	0.701	0.735	**0.744**	0.586	0.593	0.732	0.759	**0.763**
	0.9	86.4	84	93.4	**95.4**	95.1	0.61	0.583	0.89	**0.941**	0.917	0.695	0.649	0.873	**0.912**	0.905
WSRT p-value		<0.0001			0.0366	base	<0.0001			0.588	base	<0.0001			0.84	base

the positive sample rate seems to be more relevant, as the sample rate of 0.25 shows that MLP is slightly better than our CluSVM in Accuracy, and has more cases of higher values for Recall and F-Score. Still, our CluSVM seems to be the preferred method when compared in general to all the baseline methods.

The same conclusion that the number of clusters plays a significant role can be derived in this scenario as well - the differences of the Recall and F-score in datasets with 10 and 15 clusters is a clear indicator of the advantages our clustering aided method can deliver.

5   Conclusion

This paper investigates the performance of some of SVM learning methods when clustering is applied on the samples, and suggests a strategy of how to incorporate clustering information directly in the learning process. The CluSVM model with quadratic mean empirical loss has the flexibility to use any type of grouping, and suits the purpose of investigating the performance of this clustering supported model quite well. We have observed an increase in several performance metrics which demonstrates the combination of the two learning models is possible in a more complex, yet still straightforward way.

Future research directions in this are can include other types of grouping strategies, including the adaptation of some form of grouping for datasets that have proven to be more difficult to learn, such as datasets with possible noise, outliers and distribution based grouping.

References

1. Muller, K.R., Smola, A.J., Ratsch, G., Scholkopf, B., Kohlmorgen, J., Vapnik, V.: Using Support Vector Machines for Time Series Prediction Advances in Kernel Methods-Support Vector Learning. MIT Press, Cambridge (1999)
2. Schölkopf, B., Simard, P., Vapnik, V., Smola, A.J.: Improving the accuracy and speed of Support Vector Machines. Adv. Neural Inf. Proc. Syst. **9**, 375–381 (1997)
3. Fan, R.E., Chang, K.W., Hsieh, C.J., Wang, X.R., Lin, C.J.: LIBLINEAR: A library for large linear classification. J. Mach. Learn. Res. **9**, 1871–1874 (2008)
4. Hsieh, C.J., Chang, K.W., Lin, C.J., Keerthi, S.S., Sundararajan, S.: A dual coordinate descent method for large-scale linear SVM. In: Proceedings of the 25th International Conference on Machine learning, pp. 408–415. ACM (2008)
5. Ogawa, K., Suzuki, Y., Takeuchi, I.: Safe screening of non-support vectors in pathwise SVM computation. In: Proceedings of the 26th Annual International Conference on Machine Learning, vol. 3, pp. 1382–1390 (2013)
6. Narasimhan, H., Agarwal, S.: A structural SVM based approach for optimizing partial AUC. In: Proceedings of the 26th Annual International Conference on Machine Learning, vol. 1, pp. 516–524 (2013)
7. Shalev-Shwartz, S., Zhang, T.: Stochastic dual coordinate ascent methods for regularized loss minimization. J. Mach. Learn. Res. **14**, 567–599 (2013)
8. Zhao, Z., Liu, J., Cox, J.: Safe and efficient screening for sparse support vector machine. In: Proceedings of the 20th ACM SIGKDD International Conference on Knowledge Discovery and Data Mining. pp. 542–551. ACM, (2014)

9. Matsushima, S., Vishwanathan, S.V.N., Smola, A.J.: Linear support vector machines via dual cached loops. In: Proceedings of the 18th ACM SIGKDD International Conference on Knowledge Discovery and Data Mining, pp. 177–185. ACM, (2012)

10. Caetano, T.S., McAuley, J.J., Cheng, L., Le, Q.V., Smola, A.J.: Learning graph matching. IEEE Trans. Pattern Anal. Mach. Intell. **31**(6), 1048–1058 (2009)

11. Tao, Q., Chu, D., Wang, J.: Recursive Support Vector Machines for dimensionality reduction. IEEE Trans. Neural Networks **19**(1), 189–193 (2008)

12. Hu, M., Chen, Y., Kwok, J.T.Y.: Building sparse multiple-kernel SVM classifiers. IEEE Trans. Neural Networks **20**(5), 827–839 (2009)

13. Meesrikamolkul, W., Niennattrakul, V., Ratanamahatana, C.A.: Shape-based clustering for time series data. In: Tan, P.-N., Chawla, S., Ho, C.K., Bailey, J. (eds.) PAKDD 2012. LNCS, vol. 7301, pp. 530–541. Springer, Heidelberg (2012). doi:10.1007/978-3-642-30217-6_44

14. Luong, B.T., Ruggieri, S., Turini, F.: k-NN as an implementation of situation testing for discrimination discovery and prevention. In: Proceedings of the 17th ACM SIGKDD International Conference on Knowledge Discovery and Data Mining, pp. 502–510. ACM (2011)

15. Chan, J., Leckie, C., Bailey, J., Ramamohanarao, K.: TRIBAC: Discovering interpretable clusters and latent structures in graphs. In: 2014 IEEE International Conference on Data Mining, pp. 737–742. IEEE (2014)

16. Yang, W., SG, E., Xu, H.: A divide and conquer framework for distributed graph clustering. In: Proceedings of the 32nd International Conference on Machine Learning (ICML-15), pp. 504–513 (2015)

17. Yi, J., Zhang, L., Wang, J., Jin, R., Jain, A.K.: A Single-pass algorithm for efficiently recovering sparse cluster centers of high-dimensional Data. In: Proceedings of the 31st International Conference on Machine Learning, pp. 658–666 (2014)

18. Romano, S., Bailey, J., Nguyen, X.V., Verspoor, K.: Standardized mutual information for clustering comparisons: one step further in adjustment for chance. In: Proceedings of the 31st International Conference on Machine Learning, pp. 1143–1151 (2014)

19. Džeroski, S., Gjorgjioski, V., Slavkov, I., Struyf, J.: Analysis of time series data with predictive clustering trees. In: Džeroski, S., Struyf, J. (eds.) KDID 2006. LNCS, vol. 4747, pp. 63–80. Springer, Heidelberg (2007). doi:10.1007/978-3-540-75549-4_5

20. Yin, J., Wang, J.: A dirichlet multinomial mixture model-based approach for short text clustering. In: Proceedings of the 20th ACM SIGKDD International Conference on Knowledge Discovery and Data Mining, pp. 233–242. ACM, (2014)

21. Liu, W., Chawla, S.: A Quadratic mean based supervised learning model for managing data skewness. In: Proceedings of the 2011 SIAM International Conference on Data Mining, pp. 188–198 (2011)

22. Calders, T., Verwer, S.: Three naive Bayes approaches for discrimination-free classification. Data Min. Knowl. Disc. **21**(2), 277–292 (2010)

23. Kamishima, T., Akaho, S., Asoh, H., Sakuma, J.: Fairness-aware classifier with prejudice remover regularizer. In: Flach, P.A., Bie, T., Cristianini, N. (eds.) ECML PKDD 2012. LNCS, vol. 7524, pp. 35–50. Springer, Heidelberg (2012). doi:10.1007/978-3-642-33486-3_3

24. Yu, H., Yang, J., Han, J.: Classifying large data sets using SVMs with hierarchical clusters. In: Proceedings of the 9th ACM SIGKDD International Conference on Knowledge Discovery and Data Mining, pp. 306–315. ACM, (2003)

25. Li, D.C., Fang, Y.H.: An algorithm to cluster data for efficient classification of support vector machines. Expert Syst. Appl. **34**(3), 2013–2018 (2008)

26. De Almeida, M.B., de Pádua Braga, A., Braga, J.P.: speeding SVMs learning with a priori cluster selection and k-means. In: Proceedings of the 6th Brazilian Symposium on Neural Networks, pp. 162–167. IEEE (2000)
27. Shin, H., Cho, S.: Neighborhood property-ased pattern selection for support vector machines. Neural Comput. **19**(3), 816–855 (2007)
28. Wang, W., Xu, Z.: A heuristic training for support vector regression. Neurocomput. **61**, 259–275 (2004)
29. Guo, G., Zhang, J.S.: Reducing examples to accelerate support vector regression. Pattern Recogn. Lett. **28**(16), 2173–2183 (2007)
30. García-Pedrajas, N.: Constructing ensembles of classifiers by means of weighted instance selection. IEEE Trans. Neural Networks **20**(2), 258–277 (2009)
31. Zhang, T., Zhou, Z.H.: Large margin distribution machine. In: Proceedings of the 20th ACM SIGKDD International Conference on Knowledge Discovery and Data Mining, pp. 313–322. ACM, (2014)
32. Gu, Q., Han, J.: Clustered Support Vector Machines. In: AISTATS, pp. 307–315 (2013)

Mining Player Ranking Dynamics
in Team Sports

Paul Fomenky[1], Alfred Noel[2], and Dan A. Simovici[1($\boxtimes$)]

[1] Computer Science Department, University of Massachusetts Boston,
Boston, USA
{pfomenky,dsim}@cs.umb.edu
[2] Mathematics Department, University of Massachusetts Boston,
Boston, USA,
anoel@math.umb.edu

Abstract. The dynamics of players rankings play an important role in team sports. We use Kendall's τ and Spearman's ρ distances between rankings to study player scoring ranking dynamics in the NBA over the full 2014 regular season. For each team, we study the distances between sequential games, noting the differences between the two distances. Additionally, we define the consistency of teams based on their ranking dynamics. Team consistency and winning percentage are compared. Finally, we use our findings to produce actionable results for sports managers.

1 Introduction

Rankings and their dynamics play an important role in the sports world. In this paper the dynamics of player scoring ranking are studied in an attempt to provide more information about teams. Specifically, player rankings could help describe player quality, dominance, and team management styles. The analysis could provide actionable information to professional team managers about their teams as well as opponents.

To this end, a ranking is viewed as a permutation σ on the set of players. The permutations produced by different rankings are analyzed using Kendall's τ distance and Spearman ρ distance. The rest of this paper is organized as follows: Sect. 2 gives a short review of some recent related works and introduces rank correlations; Sect. 3 describes the data and experiments run; Sect. 4 discusses actionable results and Sect. 5 shows our conclusions.

2 The Metric Space of Rankings

Rankings have been studied in a wide and growing range of fields. Several real world problems boil down to understanding the order in which related objects should be presented. Steck [12], for example, studies rankings to evaluate

© Springer International Publishing AG 2017
P. Perner (Ed.): MLDM 2017, LNAI 10358, pp. 335–344, 2017.
DOI: 10.1007/978-3-319-62416-7_24

movie recommendations. [9] uses the study of rankings for multi object tracking, while [7] uses permutations to study the rankings of distinct lists of objects using prior knowledge. Additionally, rankings have been extensively studied in the context of team sports. Most of the works seeks to either rank players on the different performance parameters [1] or look to predict team ranking [2,5,10]. The current work is different in that we seek to qualify teams by the nature of their rankings. In this work, we focus on the dynamics of the scorer rankings that result from games. Specifically, we seek to quantify the scoring consistency of NBA teams based on their scoring player rankings.

The source of scoring in a team shows its main offensive weapons. Our main contribution is the use of Kendall's τ and Spearman's ρ distance between rankings to understand the scoring dynamics of NBA teams pertaining to:

1. team identity,
2. game outcomes.

To this end, we define four classes of teams and show how opponents can approach facing such teams. Additionally, we define the consistency of a team's scoring, the degree to which it's scoring changes between games.

Rank correlation has received a significant amount of attention including early work by Kendall [6]. An important metric therein introduced is Kendall's τ distance that seeks to measure the degree of agreement between two rankings on a set based on a count of reversed entry pairs.

A *full ranking* L on a set S of cardinality n assigns a position i, where $1 \leq i \leq n$ to each member x_i of the set so that the rank of the item is obtained by $r_L(x_i) = i$. A *partial ranking* assigns such a position to a number $m < n$ of items. A ranking is said to be *linear* when there a no ties between items. That is, there are no two items x_i, x_j such that $r_L(x_i) = r_L(x_j)$. When a full ranking is linear, it induces a permutation on the set S. The permutation generated is linked to the symmetric group S_n [8].

Table 1. Two full rankings

R	1	2	3	4	5	6
R'	6	3	5	1	2	4

In Table 1 we show two permutations, R and R' of the set $\{1, \ldots, 6\}$. If one seeks to estimate the degree to which the rankings R and R' disagree, it makes sense to estimate the number of transformations it takes to get from ranking R to ranking R'. The number of steps it takes to convert one ranking to the other is also the number of pairs i, j such that the rankings R and R' disagree. That is $r_R(x_i) < r_R(x_j)$ but $r_{R'}(x_i) > r_{R'}(x_j)$ or vice versa. One such pair is 1 and 5 with ranks $(1, 5)$ and $(4, 3)$ in R and R' *respectively*. Table 2 shows the 10 pairs over which rankings R and R' disagree. The disagreeing pairs have a score of 1 while the agreeing pairs have a score of 0.

Table 2. Pairs' contributions to the τ distance between R and R'.

Pair	τ score	Pair	τ score	Pair	τ score
$(1,2)$	0	$(2,3)$	1	$(3,5)$	0
$(1,3)$	1	$(2,4)$	0	$(3,6)$	1
$(1,4)$	0	$(2,5)$	1	$(4,5)$	1
$(1,5)$	1	$(2,6)$	1	$(4,6)$	1
$(1,6)$	1	$(3,4)$	0	$(5,6)$	1

Note that, when two rankings are identical, there are 0 pairs over which the rankings disagree; when one ranking is the reverse of the other, there are $\frac{n(n-1)}{2}$ pairs over which the rankings disagree. Therefore dividing the number of pairs over which the rankings disagree by $\frac{n(n-1)}{2}$ places the degree of disagreement between two rankings in the range 0 and 1.

Definition 1. *The Normalized Kendall's τ distance between two rankings R and R' is defined as:*

$$\tau(R, R') = 2 * \frac{d_{R,R'}}{n(n-1)},$$

where $d_{R,R'}$ is the number of pairs x_i, x_j such that $(r_R(x_i) < r_R(x_j) \wedge r_{R'}(x_i) > r_{R'}(x_j)) \vee (r_R(x_i) < r_R(x_j) \wedge r_{R'}(x_i) > r_{R'}(x_j))$.

The normalized Kendall's τ distance between R and R' in Table 1 is 0.667.

Theorem 1. *Kendall's τ distance is a metric*

Proof: To prove the theorem above, it suffices to show three things:

1. $\tau(R, R) = 0$. This follows from the fact $d_{R,R} = 0$.
2. $\tau(R, R') = \tau(R', R)$. This follows from the fact that the number of pairs that differ is identical whichever rank is considered first.
3. $\tau(R, R'') \leq \tau(R, R') + \tau(R', R'')$. This means $d_{R,R''} \leq d_{R,R'} + d_{R',R''}$. When the left hand side is incremented, there exists is a pair x_i, x_j over which rankings R and R'' disagree. Now consider that pair as pertains to the disagreements between R, R' and R', R''. If the rankings R and R' disagree over that pair, then the right hand side of the equation is incremented by $d_{R,R'}$. Note that this implies that R' and R'' are in agreement over that pair. Similarly, if the rankings R and R' agree over the pair, then the right hand side of the equation is incremented by $d_{R',R''}$. Therefore the triangular inequality holds.

An alternative to Kendall's τ metric is Spearman's ρ distance metric between rankings [6]. The ρ metric estimates the magnitude of the positional changes between R and R'.

Definition 2. *Spearman's ρ metric between two rankings R and R' is:*

$$\rho(R, R') = \sum (R(i) - R'(i))^2,$$

where $R(i)$ indicates the item in position i in a ranking R.

In addition to measuring the disagreements between the rankings, ρ assigns a higher weight to the differences between distant rank positions. Intuitively, this metric incorporates the idea that the difference between first and second is less significant than the difference between first and last. The proof of the metricity of ρ is found in [3].

3   Experimental Methods

The dataset consists of extensive data (passes, rebounds, fouls) on all 1228 regular season games of the 2014/15 NBA season. The data was obtained from STAT.com which tracks NBA games with very high granularity. From the raw data scoring information was obtained from each game's box score.

The top seven scorers for each team are selected after each game and ranked. No team has the same top seven scorers every game. For that reason the average top seven scorers over the season are selected. In case where a player is missing from the ranking of a particular game, they are appended to the end of the list. The top seven scorers of each team were selected to capture the contributions of starters as well as non starters. As teams have varying numbers of contributors, capturing seven allows us to get the main scoring signal from every team. The result of the pre-processing is a list of 82 scoring rankings (or permutations) per team; each representing an individual game.

The scoring rankings of NBA teams are examined in two distinct manners:

1. game transitions: the distance between the rankings produced by a team in successive games;
2. full season distances: the distances between all the rankings produced by an individual team are used to define and verify the notion of team consistency.

3.1   Game Transition Distances and the Mean Distance distribution

For each team i we generate a sequence of rankings $\sigma_1^i, \sigma_2^i, \sigma_3^i, \ldots, \sigma_{82}^i$ and a sequence of distances $d_j^i = \tau(\sigma_j^i, \sigma_{j+1}^i)$ for $1 \leq j < 82$. In this section, for computational purposes, we prefer to work with the non-normalized Kendall and Spearman distances.

The transition distances indicate how much change there was in team's scoring from game to game during the season. Tracking this information from game to game gives an indication of how each team's scoring evolves with time. Note that there are several aspects that can lead to game to game variation include opponent, rest and injuries. As a motivating example, Fig. 1 shows the transitional ρ distances of two teams, revealing the different identities of the teams.

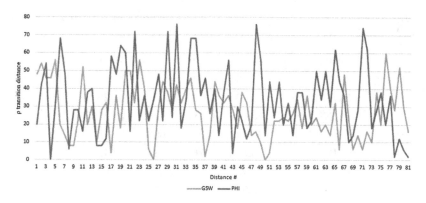

Fig. 1. The ρ transition distances of two NBA teams shows clear difference in "personality". The Warrior show less variation from game to game than 76ers. Could this play a role in team performance?

Information provided by the transition distances of NBA teams can enable team management to evaluate coaching. For less noise for example, 5 game averages in transitional distances could be observed.

Results from Kendall [6], and from Diaconis and Graham [3] show that most of the common distances defined on permutations are in fact asymptotically normal when the permutations are chosen independently and uniformly within the set of permutations S_n. More precisely the following is true [6] Sect. 5, and [3] (Table 1 on p. 264).

Theorem 2. *Let π and μ be permutations chosen independently and uniformly in S_n. Then, as $n \rightarrow \infty$ the expectation of the random variable $\tau(\pi, \sigma)$, $E[\tau(\pi, \sigma)] \rightarrow \frac{1}{4}n^2$ and its variance $Var[\tau(\pi, \sigma)] \rightarrow \frac{1}{36}n^3$. Moreover $\tau(\pi, \sigma)$ standardized by its mean and its standard deviation, is asymptotically normal.*

Consider the mean distance for each team, we can verify that such a random variable has a normal distribution within a 95% confidence interval.

We note that the central limit theorem does not hold in this case since we cannot guarantee independence. Teams play vs. each other, sometimes multiple times! Furthermore, $\bar{X}$ gives a nice summary of the data and should shed more light on future analysis.

For the team i let d_j^i be the τ distance between σ_j^i and σ_{j+1}^i for $1 \leq j \leq 81$. Define $\bar{X}$ as the random variable $\bar{X}(i) = \frac{1}{81} \sum_{j=1}^{81} d_j^i$ for $1 \leq i \leq 30$.

Table 3. Distribution of average transition values

Region	$\bar{X}_i$	p_i	$30p_i$
$y \leq 7.00$	9	0.3594	10.782
$7.00 < y \leq 7.45$	9	0.1765	5.295
$y > 7.45$	12	0.4641	13.923

Table 4. Unnormalized average transition τ and ρ distances of NBA teams during the 2014 NBA season

Team	Mean τ	Mean ρ	Team	Mean τ	Mean ρ
ATL	7.444	27.23	DAL	7.963	38.35
BOS	8.370	35.70	DEN	7.469	31.25
BKN	8.000	40.08	GSW	6.123	24.40
CHA	7.988	34.54	HOU	5.663	31.23
CHI	7.914	35.51	LAC	6.185	50.42
CLE	6.788	34.15	LAL	7.494	41.93
DET	6.543	26.25	MEM	5.951	33.19
IND	9.901	45.78	MIN	6.863	39.98
MIA	7.350	25.01	NOP	7.160	34.74
MIL	8.063	33.88	OKC	6.630	33.93
NYK	6.400	34.05	PHO	8.889	39.23
ORL	7.494	33.63	POR	7.235	40.15
PHI	7.086	30.30	SAC	7.037	36.52
TOR	9.716	52.74	SAN	7.025	32.02
WAS	7.111	33.06	UTA	7.136	36.17

Each team in the dataset had 81 game transitions. Table 4 shows the average transition τ and ρ distance results for the thirty teams in the 2014 NBA season. This variable $\bar{X}$ has a mean of 7.366 and a standard deviation of 0.988. We provide a χ^2-goodness-of-fit test of normality with the level of significance $\alpha = 0.05$ and the test statistic

$$D = \sum_{i=1}^{3} \frac{(\bar{X}_i - 30p_i)^2}{30p_i},$$

where $\bar{X}_i$ is the number of values of $\bar{X}$ that occurs in region i and p_i the probability that the random variable takes values in region i. We collect the needed data in Table 3. Note that all the $30p_i$ values are greater than 5. Therefore D has approximately a χ^2-distribution with 2 degrees of freedom.

The D statistic obtained from Table 3 is

$$D = \frac{(9 - 10.8)^2}{10.8} + \frac{(9 - 5.3)^2}{5.3} + \frac{(12 - 13.9)^2}{13.9} = 3.15$$

We should reject the normality hypothesis if $D \geq \chi^2_{0.95,2} = 5.991$. Thus, the normality is accepted at the 0.05 level of significance.

3.2 Team Consistency

In light of the initial game to game transition data, a more global view of the ranking metrics was sought. We describe the consistency of a team as the variance in the τ (or ρ) distances between the scoring rankings after each game. For each team observed, there are 3,240 such distances. Table 6 shows the relationship between team consistency calculated using both the τ and ρ distance metrics and each team's winning rate (Table 5).

Table 5. Consistency and winning

	τ-consistency	ρ-consistency
Correlation to win rate	−0.4026	−0.3752
P-value	0.0274	0.0410

Results show an inverse correlation between both consistency definitions and team winning rate. This indicates that a team whose rankings vary in a consistent manner is likelier to win during the season. The small difference between the correlation of the consistencies with winning indicate that they capture very similar information about the way teams score. The small *p-values* for the correlation show that the relationship between consistency and winning is very unlikely to be random.

Table 6. Team consistency and winning rate. τ-consistency shows negative correlation with winning of −0.40 while ρ consistency shows correlation of −0.38.

Team	τ-consistency	ρ-consistency	Win rate	Team	τ-consistency	ρ-consistency	Win rate
GSW	8.67	340.76	0.80	PHO	11.74	526.96	0.48
ATL	8.28	347.60	0.69	BOS	11.95	530.37	0.47
CLE	9.12	394.65	0.66	IND	10.96	437.72	0.46
HOU	7.64	288.56	0.66	UTA	10.11	509.72	0.46
LAC	8.96	340.84	0.66	BKN	9.26	392.70	0.46
MEM	8.49	370.98	0.66	MIA	10.15	412.85	0.45
SAN	9.04	388.72	0.65	CHA	9.51	407.91	0.40
POR	10.10	432.50	0.60	DET	9.16	364.43	0.39
CHI	11.12	487.29	0.60	DEN	9.03	380.13	0.37
DAL	9.50	407.24	0.59	SAC	10.40	449.61	0.35
TOR	11.44	527.12	0.57	ORL	9.42	415.79	0.31
WAS	9.04	382.17	0.57	LAL	10.42	441.99	0.26
OKC	9.55	409.77	0.55	PHI	9.74	406.43	0.22
NOP	12.48	558.54	0.52	NYK	10.46	452.33	0.21
MIL	10.00	440.94	0.49	MIN	11.69	516.20	0.20

4   Results and Discussion

The ρ and τ distances capture important if different information about the way team's scorers are ranked. The τ distance seeks to describe the total number of

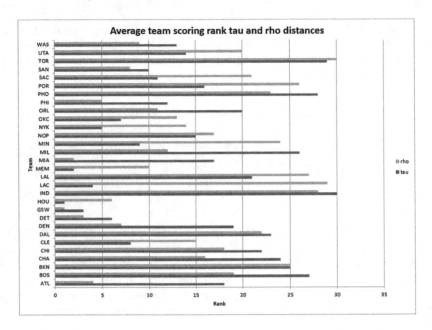

Fig. 2. Average team scoring rank transition τ and ρ distances.

pairs over which the rankings disagree without considering the position of players ranked. The τ distance is valuable in the sense that it counts the variations shown by the group.

In addition to looking at the variations, the ρ metric gives more importance to large swings in player positions between rankings. It correctly sees that if a player goes from sixth in the ranking to the top, then that shows higher volatility than that player moving to fifth or seventh - a distinction the τ distance does not make.

In Fig. 2, the teams are ranked from lowest to highest by both metrics and the differences are shown. In some cases, the two rankings agree for the most part - consider the Toronto Raptors (TOR) and Indiana Pacers (IND) at the bottom of both lists. This indicates that for both of those teams, not only was there significant position changing, but those changes involving large positional swings from game to game. The Raptors and Pacers were unpredictable - possibly a consequence of the their rosters being in development phase [4]. The Golden State Warriors (GSW) showed few positional changes in their rankings as shown in their average τ distance. In fact, their ρ distance indicates that the changes observed were very small - that is between players that were already close in the previous rankings. This shows a stable roster - they went on to win the championship. A third group of teams showed large variation in their rankings by the two metrics. The Miami Heat (MIA) had a low τ but significantly higher ρ positioning, indicating that from game to game they showed few but wild rank positions. From game to game, the Los Angeles Clippers on the other hand showed several tight positional changes.

Our findings are of value to team managers as it gives them an easily understood tool for team evaluation. Our work also has implications in the fantasy sports domain. Team owners would use this information in game to game team selection. Members on LAC would be more trustworthy than their counterparts on MIA. "Sleepers", undervalued players due for big performances could be found on teams like MIA, IND and TOR. The flip side of that coin is the same teams are more likely to provide disappointments.

Team consistency was shown to have a strong inverse correlation with teams' winning percentages. This indicates that teams that play from a stable "script" have a better chance to be successful over the long run. Additionally, we showed that both definitions of consistency captured about the same information. This information is in line with information provided by Glen "Doc" Rivers, an NBA champion coach, during a recent NBA coaches' roundtable [4] on the need for teams to define and understand the rankings they seek to produce on a game to game basis as a first step towards success.

5 Conclusion

In this paper, player scoring dynamics have been used to describe team performance. Starting with Kendall's τ and Spearman's ρ distances, we have described NBA team tendencies including consistent and haphazard performers. We have shown, using the differences in results from Kendall's τ and Spearman's ρ distances, the nature of the dynamics in team scoring. Some teams vary due to clustered changes at the top of their rankings while others show bigger jumps from game to game on average. Using the information from the distances, we have defined the notion of team consistency and shown its strong relationship with team winning percentage. We have shown how understanding the dynamics of player scoring in NBA games is useful for team managers.

Further work includes testing our findings on a larger multi-season dataset. Additionally, we will explore the effect of other rankings produced during NBA game play including rebounds, assists and turnovers on team performance. In [11], the strength of a team is estimated and used to predict game outcomes. We will explore the effectiveness of team consistency in the game prediction process.

Acknowledgement. We would like to thank Charles Rohlf at stats.com for making their NBA dataset available to us. We would also like to thank Marc Pomplun for helpful suggestions.

References

1. Cooper, W.W., Ramón, N., Ruiz, J.L., Sirvent, I.: Avoiding large differences in weights in cross-efficiency evaluations: application to the ranking of basketball players (2011)
2. Dadelo, S., Turskis, Z., Zavadskas, E.K., Dadeliene, R.: Multi-criteria assessment and ranking system of sport team formation based on objective-measured values of criteria set. Expert Syst. Appl. **41**(14), 6106–6113 (2014)

3. Diaconis, P., Graham, R.L.: Spearman's footrule as a measure of disarray. J. Roy. Stat. Soc. Ser. B (Methodol.) **39**, 262–268 (1977)

4. Edition, C.: Open court, October 2006

5. Fogel, F., d'Aspremont, A., Vojnovic, M.: Spectral ranking using seriation. J. Mach. Learn. Res. **17**(88), 1–45 (2016)

6. Gibbons, J.D., Kendall, M.: Rank Correlation Methods. Edward Arnold, London (1990)

7. Huang, J., Guestrin, C.: Riffled independence for ranked data. In: Advances in Neural Information Processing Systems, pp. 799–807 (2009)

8. James, G., Kerber, A.: The Representation Theory of the Symmetric Group. Addison Wesley, Reading (1981)

9. Kondor, R., Howard, A., Jebara, T.: Multi-object tracking with representations of the symmetric group. In: AISTATS, vol. 1, p. 5 (2007)

10. Kvam, P., Sokol, J.S.: A logistic regression/markov chain model for ncaa basketball. Naval Res. Logistics (NrL) **53**(8), 788–803 (2006)

11. Manner, H.: Modeling and forecasting the outcomes of nba basketball games. J. Quant. Anal. Sports **12**(1), 31–41 (2016)

12. Steck, H.: Gaussian ranking by matrix factorization. In: Proceedings of the 9th ACM Conference on Recommender Systems, pp. 115–122. ACM (2015)

ivhd: A Robust Linear-Time and Memory Efficient Method for Visual Exploratory Data Analysis

Witold Dzwine and Rafał Wcisło[(⊠)]

AGH University of Science and Technology, Kraków, Poland
wcislo@agh.edu.pl

Abstract. Data embedding (DE) and graph visualization (GV) methods are very compatible tools used in Exploratory Data Analysis for visualization of complex data such as high-dimensional data and complex networks. However, high computational complexity and memory load of existing DE and GV algorithms, considerably hinders visualization of truly large and big data consisting of as many as $M \sim 10^{6+}$ data objects and $N \sim 10^{3+}$ dimensions. Recently, we have shown that by employing only a small fraction of distances between data objects one can obtain very satisfactory reconstruction of topology of a complex data in 2D in a linear-time $O(M)$. In this paper, we demonstrate the high robustness of our approach. We show that even poor approximations of the nn-nearst neighbor graph, representing high-dimensional data, can yield acceptable data embeddings. Furthermore, some incorrectness in the nearest neighbor list can often be useful to improve the quality of data visualization. This robustness of our DE method, together with its high memory and time efficiency, meets perfectly the requirements of big and distributed data visualization, when finding the accurate nearest neighbor list represents a great computational challenge.

Keywords: Data visualization · Data embedding · Graph visualization · Big data

1 Introduction

Visualization of complex data such as big high-dimensional data and networks, play a very important role in knowledge extraction in various fields of science. It enables to discover instantly some hidden information on mutual relationships between numerous objects both in the global and local scales. For a long time, the graph visualization (GV) methods were developed in parallel and apart to dimensionality reduction and data embedding (DE) schemes. Just recently, with rapidly expanding applications of data science, it appears that many algorithms used by these two data visualization approaches are the same or, at least, very similar [1]. This resemblance between DE and

13th International Conference on Machine Learning and Data Mining MLDM, New York, July 15-20, 2017.

P. Perner (Ed.): MLDM 2017, LNAI 10358, pp. 345–360, 2017.
DOI: 10.1007/978-3-319-62416-7_25

GV can be used in developing more general and efficient tools for Exploratory Data Analysis [2].

We focus here on the visualization of high-dimensional and large datasets $\mathbf{Y}$ consisting of many data objects $\mathbf{y}_i (i = 1, \ldots, M)$ (e.g., feature vectors), which is understood as a transformation $\mathbf{B}$ of $\mathbf{Y}$ into 2D scatter map $\mathbf{X}$, where the data objects $\mathbf{y}_i \in \mathbf{Y}$ are represented by corresponding datapoints $\mathbf{r}_i \in \mathbf{X}$ and $(i = 1, \ldots, M)$. The main goal of $\mathbf{B}$ is to preserve in $\mathbf{X}$ the key topological features of $\mathbf{Y}$, which may carry important information on data separation, clustering, multi-scale structure and local neighborhood. Very intuitive and the most exploited candidates for complex data structures visualization are those based on minimization of a cost function reflecting the dissimilarity between distance matrices $\mathbf{D}$ and $\mathbf{d}$, calculated for data objects $\mathbf{Y}$ and for datapoints $\mathbf{X}$, respectively. These techniques employ the classical multidimensional scaling (MDS) [3, 4], modern stochastic neighbor embedding (SNE) algorithm [5] and their clones [1, 6–10]. However, all of them are rather useless for visualization of truly big data. This is due to their high time complexity and memory load, which are at least $O(M \log M)$ (in respect to a single step of an iterative gradient based minimization method) for the most efficient bh-SNE [7] and sfdp (forceatlas2) [11] DE and GV algorithms, respectively.

In [12, 13], we have shown that the most important structural features of high-dimensional dataset can be preserved in a small set $\boldsymbol{D}_2$ of binary$\{0, 1\}$ distances. We have assumed that the set $\mathbf{E}$ of all distances to the nn nearest neighbors of each $\mathbf{y}_i \in \mathbf{Y}$ are set to 0, while distances to a set $\mathbf{R}$ of selected rn random neighbors are equal to 1. We show in [12, 13] that the values of nn and rn can be small, so we can surmise that the embedding procedure has linear-time $O(M)$ complexity in respect to a single step of a criterion function minimization procedure). Moreover, a list of at most $nn \cdot M$ indices of the nearest neighbors should be stored, instead of $O(M^2)$ floating point distances for classical MDS and t-SNE method [3, 4, 6].

The important contribution of this paper is to present our approach to high-dimensional data visualization in the context of a strong methodological coupling between DE and GV approaches. On the base of rigid graphs theory (e.g. [14]) we present additional arguments on linear-time complexity of our algorithm. We address also an important problem of its application for visualization of truly big and distributed data. It is well known that for big data the search for the nearest neighbors faces a challenging computational problem. To this end, many approximate and fast algorithms for the k-nearest neighbors search were constructed [15], e.g., the most popular locally-sensitive hashing algorithms [16] implemented in Map-Reduce paradigm. All of these algorithms are fast but at rather serious cost of accuracy. In this context, we raise the following question: "To what extent our DE approach is resistant on the false positive nearest neighbors?" which is equivalent to: "Can it be useful for big data visual exploration?".

In the following section we present our DE method, as a method of visualization of nearest neighbor nn-graphs. Then we show and comment the results of tests demonstrating the robustness of our approach in the context of accuracy of nn-graph construction. Next, we discuss the state-of-art related work in domain of big and complex data visualization. Finally, we collect the conclusions.

2 Embedding Algorithm

Data embedding can be understood as a mapping $\mathbf{B}$: $\mathfrak{R}^N \ni \mathbf{Y} \rightarrow \mathbf{X} \in \mathfrak{R}^2$ of N-dimensional *source* data $\mathbf{Y} = \{\mathbf{y}_i = (y_{i1}, \ldots, y_{iN})\}_{i=1,\ldots,M}$ of M vectors (or other data objects for which we can define a dissimilarity measure), where both M and N are huge, into 2D *target* set of datapoints $\mathbf{X} = \{\mathbf{r}_i = (r_{i1}, r_{i2})\}_{i=1,\ldots,M}$. This problem can be also formulated from the point of view of visualization of a weighted, fully connected graph $full\mathbf{G}(\mathbf{V}, \mathbf{E}_{All}, \mathbf{D})$, where vertices $v_i \in \mathbf{V}$ are the indices of the feature vectors $\mathbf{y}_i$, while distance matrix $\mathbf{D} = [d_{ij} = \|\mathbf{y}_i - \mathbf{y}_i\|]_{MxM}$ defines the weights of respective $\mathbf{E}_{All}$ edges. In this case, the goals of 2D visualization from both DE and GV perspective are the same, because $full\mathbf{G}(\mathbf{V}, \mathbf{E}_{All}, \mathbf{D})$ is the full graph and its visualized structure can be entirely defined by $\mathbf{D}$. On the other hand, as comes out from the theory of structural rigidity and theory of rigid graphs [14], the set $\mathbf{Y}$ can be fully represented in d-dimensional space by a set of distances $\mathbf{D}_d \subset \mathbf{D}$ consisting of $L_d \sim M$ distances, and can be defined as a d-rigid graph $r\mathbf{G}(\mathbf{V}, \mathbf{E}_d, \mathbf{D}_d)$. In general, it is possible to find the minimal d-rigid graph of $\mathbf{Y} - rm\mathbf{G}(\mathbf{V}, \mathbf{E}_{dMin}, \mathbf{D}_{dMin}) -$ on M ($\geq d$) vertices in dimension d, where only

$$L_{dMin} = |\mathbf{E}_{dMin}| = d \cdot M - d \cdot (d+1)/2 \tag{1}$$

distances define the number of edges and weights of $rm\mathbf{G}(\mathbf{V}, \mathbf{E}_{dMin}, \mathbf{D}_{dMin})$ and, consequently, the minimal d-dimensional embedding of $\mathbf{Y}$. This time, the graph $r\mathbf{G}(\mathbf{V}, \mathbf{E}_d, \mathbf{D}_d)$ (and $rm\mathbf{G}(\mathbf{V}, \mathbf{E}_{dMin}, \mathbf{D}_{dMin})$) is not fully connected. In 2D, the minimal rigid graphs $rm\mathbf{G}(\mathbf{V}, \mathbf{E}_{2Min}, \mathbf{D}_{2Min})$ representing $\mathbf{X}$ embedding of $\mathbf{Y}$ can be even less connected, and the minimal number of distances defining $\mathbf{X}$ can be as small as $L \sim L_{2Min} = 2 \cdot M - 3$. However, the problem arises with finding a proper sets of "rigid" edges $\mathbf{E}_2$ and respective distances $\mathbf{D}_2$, which should be minimal and satisfactory in terms of both efficiency and quality of DE (and GV).

We have showed in [17] that to assure data (a graph or network) rigidity only a small number of nn the nearest neighbors and rn random neighbors for each $\mathbf{y}_i$ are sufficient (correspondingly, in the case of graphs, we need to know nn connected and random rn disconnected vertices for each v_i [13]). These choice of nn and rn parameters should be estimated on the case by case basis and can vary for various types of datasets and graphs (networks). We have also assumed that to reflect the original N-dimensional topology of $\mathbf{Y}$ and/or visualize the structure of a big graph $r\mathbf{G}(\mathbf{V}, \mathbf{E}_d, \mathbf{D}_d)$ or other complex network (e.g., nn-graph, the road or social networks), one has to define a binary distance between all $\mathbf{y}_i \in \mathbf{Y}$ ($i = 1 \ldots M$) and their nn and rn neighbors (consequently, nn and rn vertices $v_i \in \mathbf{V}$), i.e.

$$\forall \mathbf{y}_i \in \mathbf{Y} \begin{cases} \delta_{ij} = 0 & iff \quad j \in O_{nn}(i) \\ \delta_{ij} = 1 & iff \quad j \in O_{rn}(i) \end{cases} \tag{2}$$

where $O_{nn}(i)$ and $O_{rn}(i)$ ($O_{nn}(i) \cap O_{rn}(i) = \varnothing$) are the sets of data indices (or graph indices) of nn nearest neighbors (connected vertices) and rn random neighbors (disconnected vertices). This choice of distance definition is fully justified for DE case by the "curse of dimensionality" effect in high-dimensional spaces. Namely, the Euclidean distances between feature vectors for high dimensional spaces are in average the same, i.e., the distances of a feature vector to its nearest neighbors do not differ too much from those calculated to other non-neighbor vectors. This lack of distances contrast can completely destroy data separability after **Y** embedding to a lower dimension. This situation is clearly shown in Fig. 1, where the log-lin plot (take into account that y axis is logarithmic!) was made for well structured (10 clear clusters) MNIST dataset (http://yann.lecun.com/exdb/mnist/). All the *rn* random distances are set to 1 because of their similarity in high-dimensional spaces (even as low as 10–15 dimensions [18]). Simultaneously, to increase data contrast, the distances to *nn* nearest neighbors of each y_i were set to 0. The choice of binary metric is also reasonable in the perspective of graph visualization. In the visualized graph the connected neighbors should be as close, while disconnected as far as possible.

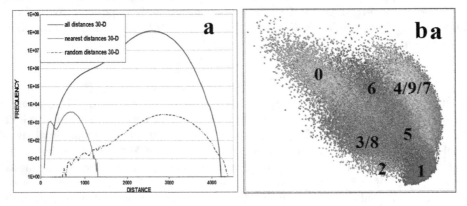

Fig. 1. (a) The plot of envelopes of distance histograms for MNIST dataset ($M = 70000$) transformed by PCA from 784D to 30D space. We display the histograms of both all the distances in the source 30D Y dataset and selected distance subsets: the two nearest (2-*nn*) and one random (1-*rn*) neighbors, respectively. (b) The MDS embedding of MNIST to 2D space, demonstrating the problem of "curse of dimensionality" and low contrast between the two types of distances.

In the case of an unweighted **G**(**V**, **E**) graph or network visualization, we have to construct an augmented graph a**G**(**V**, **E** ∪ **R**), where **R** is the set of edges between randomly selected vertices **V**. Then we need to find its rigid graph ra**G**(**V**, **E**$_2$ ∪ **R**$_2$), where $|\mathbf{E}| \geq |\mathbf{E}_2| = nn$, $|\mathbf{R}| \geq |\mathbf{R}_2| = rn$ and $\mathbf{E}_2 \cap \mathbf{R} = \varnothing$ (*rn* and *nn* are small). We assume that the same number of *rn* random edges incoming from each vertex v_i is specified. Binary distance between vertices is assumed such as in Eq. (2). This structure can be embedded in the same way as **Y**. Resulting 2D embedding can be treated as a visualization of **G**(**V**, **E**) graph. By controlling the values of (*nn, rn*) pair, one can examine various visualization aspects.

Our method of data embedding (and graph visualization) is based on the classical MDS approach. As we postulate at the beginning of this section: each feature vector

$\mathbf{y}_i \in \mathbf{Y}$ (or corresponding graph vertex $v_i \in \mathbf{V}$) is represented by 2D datapoint $\mathbf{r}_i \in \mathbf{X}$. Apart from binary distance set: $\mathbf{D}_2 = \{\delta_{ij} = \{0, 1\}\}$ for the *source* dataset $\mathbf{Y}$, we define corresponding floating-point distance set $\mathbf{d} = \{d_{ij} = sqrt(\|\mathbf{r}_i - \mathbf{r}_j\|)\}_{M \times M}$, for its *target* $\mathbf{X}$ embedding. The mapping $\mathbf{B}: \mathbf{Y} \rightarrow \mathbf{X}$ consists in minimization of $E(\|\mathbf{D}_2 \text{-} \mathbf{d}\|)$ cost function, which represents the mean squared error between $\mathbf{D}_2$ and $\mathbf{d}$. Consequently, we minimize the following cost function:

$$E = \sum_i^N E_i = k_{nn} \cdot \sum_i^N \left(\sum_{j \in \mathbf{O}_{nn}(i)}^{nn} (d_{ij}^n)^2 + c \cdot \sum_{k \in \mathbf{O}_{rn}(i)}^{rn} (1 - d_{ik}^n)^2 \right) \tag{3}$$

We use the force-directed approach as a minimization method [13]. The set of points $\mathbf{X}$ is treated as an ensemble of interacting particles. Initially, the particles are scattered randomly in a 2D computational box. The particles evolve in the Euclidean space with velocities $\mathbf{v}_i^n$ and positions $\mathbf{r}_i^n$ in the subsequent n-th timestep, according to the following discretized Newtonian equations of motion:

$$\mathbf{v}_i^{n+1/2} = \mathbf{v}_i^{n-1/2} + (2k_{nn}\mathbf{f}_i^n - \lambda\mathbf{v}_i^n) \cdot \Delta t,$$
$$\mathbf{v}_i^n = \frac{\mathbf{v}_i^{n+1/2} + \mathbf{v}_i^{n-1/2}}{2}, \ \mathbf{r}_i^{n+1} = \mathbf{r}_i^n + \mathbf{v}_i^{n+1/2} \cdot \Delta t \tag{4}$$

where Δt is the timestep size, k_{nn} and λ are the interparticle and friction force parameters, respectively. The vector $\mathbf{f}_i^n = -\nabla E_i^n$ equal to:

$$\mathbf{f}_i^n = - \sum_{j \in \mathbf{O}_{nn}(i)}^{nn} \mathbf{r}_{ij}^n - c \sum_{k \in \mathbf{O}_{rn}(i)}^{rn} (1 - d_{ik}^n) \cdot \frac{\mathbf{r}_{ik}^n}{d_{ik}^n}, \mathbf{r}_{ik}^n = \mathbf{r}_i^n - \mathbf{r}_k^n \tag{5}$$

is the total force acting on a particle i in the timestep n from neighboring (the nearest and random) particles. Take into account, that the number of forces acting on a particle i can be greater than $nn + rn$, because the particle i for some particle j from $O_{nn}(i)$ or $O_{rn}(i)$ sets may not belong to $O_{nn}(j)$ or $O_{rn}(j)$ sets. We also apply the friction force $\mathbf{f}_D = -\lambda\mathbf{v}_i^n$ to dissipate the energy from the whole particle system. The minimization scheme behaves as the simulated annealing method (SA) at the beginning of simulation and as the gradient descent at its end. This way one can avoid local minima of $E(\ldots)$ reaching a better one, i.e., a closer to the global minimum. The friction force dissipates the energy from the particle system and their motionless positions represent $\mathbf{X}$ embedding of $\mathbf{Y}$ (or graph visualization). Assuming $\Delta t = 1$, only the parameters k_{nn} and λ control the minimization procedure, i.e., a greater value of λ means greater energy dissipation while greater value of k_{nn} pumps more energy to the system. Thus decreasing λ and keeping k_{nn} constant one can obtain better minimum at the cost of longer simulation time. Similarly, for constant λ and greater k_{nn}, more energy will be produced what allows for better inspection of the function domain but extends the minimization process. The most important parameters, which influence the quality of embeddings are nn, rn and c. It is worth to mention that, the value of c from Eqs. (3–5) should be very small ($c = \{0.1; 0,01; 0,001\}$), and should be selected carefully not to overwhelm the influence of the forces coming from the *nn*-nearest neighbors. It can be fitted on the case by case basis and depends on the *nn/rn* ratio. However, as shown in [12, 17], because the method is fast, they can be matched interactively. As show our numerous experiments the "universal" set, i.e., $nn = 3$, $rn = 1$ and $c = 0.1$ (or 0.01), produces embeddings of good quality for the most of data sets (and networks [13]) visualized by the authors.

Below we sum up our previous achievements:

1. In comparison to the other embedding methods such as MDS and SNE clones [1, 3–10], we radically simplified data similarity to the binary {0, 1} measure. This increases the contrast between original (*source*) distances what results in better data separability.
2. We use classical MDS for data embedding (graph visualization) by using only $\sim n_{vi}M$ binary distances mainly responsible for data separability: where $n_{vi} \approx nn + rn$, and *nn* is the number of the *nn* nearest and *rn* random neighbors of y_i for DE or connected and random disconnected vertices v_i in the case of GV. This way we radically decrease the computational complexity and storage load. In comparison to one of the most efficient *bh*-SNE algorithm, we decreased memory load and computational complexity from $O(M \log M)$ to $O(M)$.
3. We have demonstrated that our method is extremely fast and outperforms the other state-of-art DE algorithms, what allows to visualize interactively large datasets consisting of as many as 10^{5+} data objects at the cost of rather small decrease of embedding quality.
4. Moreover, as we show in [12, 13], our method is more general than the state-of-art SNE clones, allowing for graph visualization of big complex networks. Unlike the state-of-art sfdp (forceatlas2) GV algorithm [11], we are unable to visualize large graphs with a few millions vertices and tens million edges (such as social networks) on a regular laptop.

To explore data visually, the calculations of embeddings have to be repeated many times (5–10), to search better the parameter space and to find various visualization modes. To this end, the *nn*-graph representation $nnG(V, E_{nn})$ of a dataset Y, where E_{nn} is the set of edges to the *nn* nearest neighbors of $v_i \in V$, has to be stored on disk (disks) and should have very parsimonious representation to fit in the computer memory. In general, the construction of the *nn*-graph should be highly optimized in terms of storage and time. In our implementation, the DE initialization part, i.e., responsible for calculation of $nnG(V, E_{nn})$, was written in CUDA^{TM} and uses GPU for the nearest neighbors search. However, though this search is performed only once at the beginning of simulation, it takes relatively long time for large data sets. Too long, to use our approach for visualization of truly big and distributed data. That is why, the bottleneck problem of finding in an efficient way *nn*-graph representation of big dataset should be solved. Because of low time and memory complexities of approximate algorithms [15, 16], we propose to use them instead of accurate *k-nn* algorithms. However, we have to check first, how robust is our DE method on possible mistakes in calculation of the nearest neighbors lists. In the following paragraph, we show the results of experiments on some benchmark datasets, assuming that their *nn*-graphs $nnG(V, E_{nn})$ are inaccurate and can include some false positive neighbors.

3 Implementation and Tests

On the base of DE and GV algorithms described above we have developed two GUI interfaces for both DE and GV algorithms, i.e.: **ivhd** interactive **v**isualization of **h**igh-dimensional **d**ata [12, 17] and **ivga** (interactive visualization of **g**raphs) [13], which differ mainly in functionality and the ways of visualization. The two are written in C++, while GUIs were prepared by using OpenGL^{TM} graphics system integrated with the standard Qt^{TM} interface.

To evaluate ivhd robustness we have performed experiments on a few high-dimensional and multi-class datasets (let C is the number of classes). The computations for data and graphs visualizations were performed on the Toshiba Satellite laptop equipped with the Intel® Core (TM) i5 3317U CPU running at 1.7 GHz, 16 GB of operational memory and GeForce GT 630 M graphic board. We have performed the tests by using a few high-dimensional benchmark datasets, which are specified in Table 1.

Table 1. The list of data sets referenced in this paper.

Name	N	M	C	Distance	Description
MNIST	784	70000	10	Euclidean	Well balanced set of grayscale handwritten digit images (http://yann.lecun.com/exdb/mnist/)
NORB	2048	43600	5	Cosine	Small NORB dataset (NYU Object Recognition Benchmark) contains stereo image pairs of 50 uniform-colored toys under 18 azimuths, 9 elevations, and 6 lighting conditions (http://www.cs.nyu.edu/~ylclab/data/norb-v1.0/)
TNG	2000 PCA (30D)	18759	20	Cosine	A balanced collection of documents from 20 various newsgroups. Each vertex stands for a text document represented by bag of words (BOW) feature vector (http://qwone.com/~jason/20Newsgroups/)
Reuters8	2000 PCA (30D)	266931	8	Cosine	Strongly imbalanced text corpus known as RCV1. We used a subset of this repository consisting of 8 clusters (http://about.reuters.com/researchandstandards/corpus/)
Reuters6	2000 PCA (30D)	96117	6	Cosine	the previous dataset without the two largest classes ECAT and C151

The large Reuters (RCV1) dataset and much smaller TNG dataset, both of dimensionality $N > 2000$, we have transformed to 30-dimensional space by using PCA. This procedure has decreased substantially the computational time needed for the initialization phase and also improves considerably the embeddings for these two datasets. They consist of C classes of text documents, which partly overlap each other, and are represented by long and sparse BoW (bag of words) *if-idf* feature vectors. We have remarked that PCA transformation to lower dimension (in this case 30D) reduces the noise by "smoothing" data. The transformations to 100D, 50D, 15D spaces were also examined but they produces distinctly worse embeddings. We retained the high-dimensionality only for MNIST and NORB ($N = 784$ and $N = 2048$, respectively).

To compare data separability and class purity we define the following leave-one-out cf coefficients:

$$cf_{nn} = \frac{\sum_{i=1}^{M} nn(i)}{nn \cdot M} \text{ and } cf = \frac{\sum_{i=1}^{nn_{max}} cf_{nn}}{nn_{max}} \tag{6}$$

where $nn(i)$ means the number of the nearest neighbors from $O_{nn}(i)$, which belong to the same class as y_i. For others classifiers than k-nn, one can expect only better results (greater values) of cf_{nn} and cf. We compute the value of cf for $nn_{max} = 100$ because the smallest class from considered datasets has about 1000 samples. For well separated and pure classes $cf \sim 1$, while for random points from K classes $cf \sim 1/K$.

In Table 2 we show the values of cf_1 and cf_{15} coefficients for the original datasets from Table 1 and for their embeddings $\mathbf{X}$. One can see that these values are smaller for $\mathbf{X}$ than for original data $\mathbf{Y}$. However, as shown in Fig. 2, cf_{nn} for greater nn can be

Table 2. The average values of cf(1) and cf(15) obtained for the source high-dimensional data sets and corresponding 2-D embeddings. The "dis(0, δ)" means that the distances to a random neighbors remain the original ones while "binary(0, 1)" means that it is set to 1. The parameters used for visualization of each dataset are given below their names as triplets (nn, rn, c).

Sort of data	MNIST (2,1.0.01)		NORB (10,1.-0.1)		TNG (5,2,0.1)		Reuters 8 (Reuters 6) (10, 1, 0.1)	
	$cf(1)$	cf (15)	$cf(1)$	cf (15)	$cf(1)$	cf (15)	$cf(1)$	$cf(15)$
Source data cf's	**0.974**	**0.944**	**0.999**	**0.981**	**0.785**	**0.55**	**0.937** **(0.918)**	**0.922** **(0.894)**
Dis (0, δ)	0.89	0.89	0.97	0.96	0.33	0.25	–	–
Binary (0, 1)	0.89	0.89	0.97	0.96	0.33	0.25	–	–
PCA to 30-D (0, δ)	0.88	0.87	0.98	0.95	0.45	0.43	0.840 (0.916)	0.833 (0.862)
PCA to 30-D (0, 1)	0.89	0.88	0.97	0.95	0.45	0.44	0.834 (0.85)	0.828 (0.767)

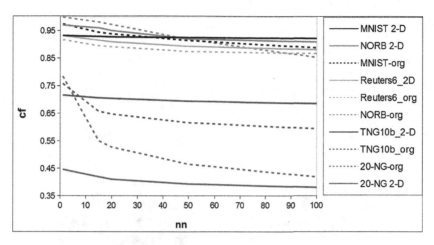

Fig. 2. The plots of the average values of cf_{nn} with nn, for original datasets (dashed) and their respective embeddings (solid). The ordering of plots along the y axis (right hand side) corresponds to their names in the legend. The dataset TNG10b consists of 10 the best separated classes from the original TNG dataset.

smaller for original data than for 2D embeddings, except full TNG dataset. Moreover, the values of cf_{nn} are very stable with nn. It means that though information about local neighborhood is partly lost, our algorithm "smooths" and extracts better the coarse grained data structure. As we show in the following experiments, the datapoints $\mathbf{r}_i$ with the nearest neighbors from various classes concentrates in the center of $\mathbf{X}$ (see Fig. 3a). This mixed cluster is mainly responsible for decreasing the value of cf_{nn}. Meanwhile, the data objects with homogeneous neighborhood separates very well in $\mathbf{X}$, developing compact and dense clusters resulting in a high value of cf_{nn} even for large nn. For datasets, such as TNG, with strongly overlapped clusters and relatively small number of data objects M, the dense and homogeneous clusters are small. Thus the values of cf are the lowest and the result of embedding the poorest. This situation is clearly depicted in Fig. 3.

In this paper we examine how inaccuracies in the lists of the nn-nearest neighbors influence the result of embeddings. To this end, we made the following experiments:

1. We have selected parameters of nn, rn and c giving the highest values of cf coefficient and the best embeddings for each of four datasets from Table 1 (see Table 2).
2. We define a few sets $O_{kk}(i)$ of the extended nearest neighbors lists of each $\mathbf{y}_i$ of various sizes kk where $kk > nn$ ($O_{nn}(i) \subset O_{kk}(i)$).
3. We perform DE experiments assuming that nn inaccurate nearest neighbors are drawn in random from kk elements of $O_{kk}(i)$. For each list we perform five experiments due to the randomness of: the nearest neighbors selected, $O_{rn}(i)$ sets and random initial conditions.
4. The value of rn is the same for each dataset and various kk.

To simulate the situation that some of approximate nearest neighbors can be outliers (and $\delta_{ij} = 1$), we assume that the random neighbor of $\mathbf{y}_i$ can be accidentally drawn

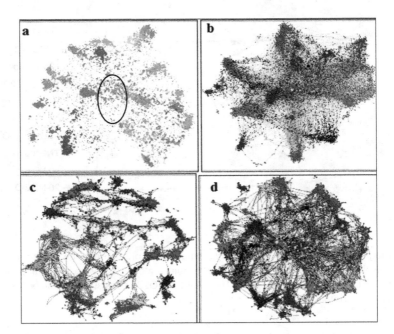

Fig. 3. Visualization of TNG dataset and *nn*-graph of TNG dataset obtained by employing ivhd and ivga interfaces, respectively (*nn* = 4, *rn* = 2 and *c* = 0.1). (a) Data embedding produced by using ivhd (*cf* = 0.43). The cluster of mixed datapoits belonging to various classes is marked by ellipse. (b) Visualization of respective graph (the structures differs due to different initialization in ivhd and ivga). (c) The graph after removal of about 6.5×10^3 with the largest average distance r_{Avg} to the connected neighbors (*cf* = 0.56). (d) The graph after removal of about 6.5×10^3 with the smallest vertex degrees (*cf* = 0.51).

from its nearest neighbor list $O_{kk}(i)$ (i.e., $O_{kk}(i) \cap O_{rn}(i) \neq \mathbf{0}$). However, this may happen few and far between for large M (with probability $\sim (kk\text{-}nn)/M$ for each data object $\mathbf{y}_i$). Of course, the assumption that a random neighbor of $\mathbf{y}_i$ cannot be simultaneously its nearest neighbor, is still hold.

In Fig. 4 we demonstrate the 2D embeddings of MNIST dataset for various values of kk. The first surprise is a very good quality of MNIST visualization for just $nn = 2$ and $rn = 1$ what needs only $L \sim 1.8 \cdot 10^5$ distances, i.e., very close to the minimal number $L_{2Min} = 1.4 \cdot 10^5$ (see Eq. (1)) required for graph rigidity. The influence of inaccurate lists of the nearest neighbors is demonstrated in Figs. 4b, c. It is clearly shown that even for $kk = 50$, the visual representation of the original dataset still reveals important structural features, similar to embeddings with accurate lists of *nn*-nearest neighbors (see Fig. 4a) and to other visualizations of MNIST dataset produced by SNE based methods [1, 5–10]. For example, all the classes are well separated and groups of classes {3, 5, 8} and {4, 7, 9} can be observed.

In Fig. 5 we demonstrate the ivga GV visualization of more dense *nn*-graph $nnG(\mathbf{V}, \mathbf{E}_{nn})$ of MNIST dataset. In the first left panel, we present distribution of vertex degrees, where the highest are located closer to the class centers. Meanwhile, (see Fig. 5b) the datapoints $\mathbf{r}_i$ with the largest average distance to its *nn*-nearest neighbors in $\mathbf{X}$ concentrate

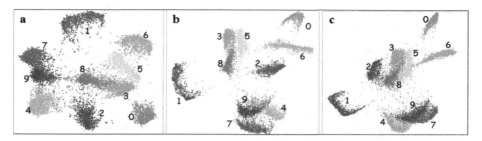

Fig. 4. The embeddings of MNIST dataset obtained by ivhd for $nn = 2$, $rn = 1$ and $c = 0.01$ parameters for various values of kk: (a) $kk = 2$; (b) $kk = 5$; (c) (a) $kk = 50$.

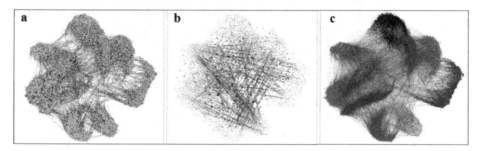

Fig. 5. The nn-graphs for MNIST dataset obtained by ivga for $nn = 5$, $rn = 2$ and $c = 0.05$ parameters where: (a) degrees of vertices; (b) the vertices with the most spread (distant) nearest neighbors; (c) the classes.

in the middle of the whole ensemble. There are the datapoints, for which the corresponding nn-nearest neighbors in **Y** belong to various classes. They are responsible for decrease of the value of cf coefficient. In Fig. 5c one can see the visualization of the full nn-graph with colors representing vertices from various classes. We can see a dense network of edges showing the similarity between data objects from various classes.

The visualization of NORB dataset, shown in Fig. 6, reflects the multi-scale structure of data. The separated clusters {1, 2, 3} represent the stereo images of toys such as: animals, human figures and airplanes, respectively, while two entangled clusters {4, 5} cars and tracks. We can also clearly see the fine data structure as streaks of datapoints, which represent the images of the same toy for various (33) conditions. The Figs. 6a, b, c demonstrate the positive effect of using an inaccurate nearest neighbors lists on data visualization. As shown in Fig. 6a, for $nn = 2$, $rn = 1$, we cannot reconstruct cluster structure of NORB because some images of the same toy create fully connected, very small and very tight "graphlets". This can produce disconnected nn-subgraphs representing separated classes. Then, the lack of contrast between random distances destroys class separability. By assuming very inaccurate lists of the nearest neighbors ($kk \gg nn$) one can disrupt the "graphlets" and create the rigid augmented nn-graph, $annG(\mathbf{V}, \mathbf{E}_{nn} \cup \mathbf{R})$, where $\mathbf{R}_{rn}$ is the set of edges between randomly selected vertices **V**. As shown in Figs. 6c, d, inaccurate neighbor lists can

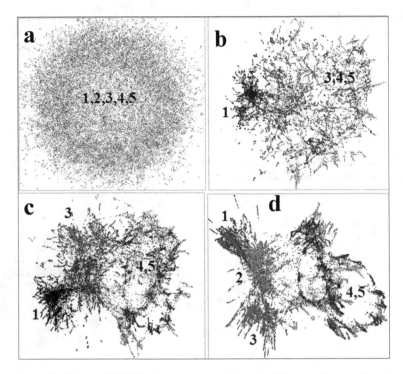

Fig. 6. The embeddings of NORB dataset obtained by ivhd for $nn = 2$, $rn = 1$ and $c = 0.01$ parameters for various values of kk: (a) $kk = 2$; (b) $kk = 5$; (c) $kk = 10$ and for (d) $nn = 10$, $rn = 1$ with $kk = 10$.

produce similar embedding to those obtained for much larger values of nn and accurate lists of nn nearest neighbors.

The embedding of larger Reuters (RCV1) dataset, which consists of eight classes of $2.67 \cdot 10^5$ BOW vectors of articles of various topics, is even harder than TNG. The classes not only can overlap each other but they are also highly imbalanced. Two classes ECAC and C151 represent 64% of the whole datasets while the smallest G154 topic consists of about 1000 samples. Nevertheless, the quality of ivga nn-graph embedding looks quite good (see Fig. 7a). Three classes of vertices ECAC, C151 and C12 are well separated. The edges of corresponding nn-graph clearly show the class relations. However, the rest of dataset with smaller classes is barely visible, overlapped by the C151 class. By using inaccurate lists of nn nearest neighbors ($nn = 10$, $kk = 25$), the overall visualization does not change too much, while the embedding of 6 smallest classes improves a little. As shown in Fig. 7b, the result of embedding of RCV1 without two largest classes, clearly extracts the smallest G154 topic. These observations are also reflected by the values of cf coefficients in Table 3.

One may conclude that, both very stable values of $cf(kk)$ coefficients and presented results of visualization show the high resistance of ivdh on the errors in nn nearest neighbor lists. In the following section ivhd is confronted with the state-of-art embedding methods in terms of quality and efficiency.

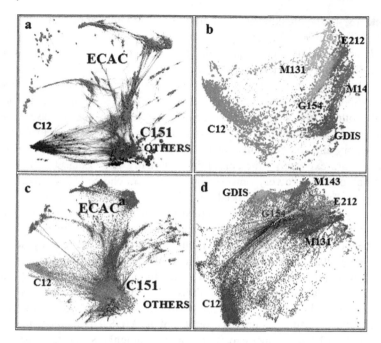

Fig. 7. The embeddings of REUTERS dataset obtained for $nn = 10$, $rn = 1$ and $c = 0.1$, for various values of kk: (a) ivga visualization of 8 clusters and $kk = 10$; (b) ivhd visualization of the smallest 6 clusters with $kk = 10$; (c) ivhd visualization of 8 clusters with $kk = 25$; (d) ivhd visualization of 6 clusters with $kk = 25$. The smallest cluster G154 is clearly visible in Fig. d while it is fully overlapped in Fig. b.

Table 3. The values of cf for original datasets (cf_{org}) and for their ivhd embeddings ($cf(kk)$).

Name	Parameters (nn, rn, c)	cf_{org}	$cf(kk)$ for $kk =$			
			nn	$nn + 3$	$nn + 5$	$nn + 15$
MNIST	(2, 1, 0.01)	**0.94**	0.88 ± 0.01	0.86 ± 0.01	0.84 ± 0.01	0.82 ± 0.01
NORB	(10, 1, 0.1)	**0.92**	0.895 ± 0.005	0.875 ± 0.005	0.865 ± 0.005	0.813 ± 0.004
TNG	(5, 2, 0.1)	**0.61**	0.44 ± 0.005	0.45 ± 0.005	0.44 ± 0.005	0.44 ± 0.005
RCV1 (8 classes)	(10, 1, 0.1)	**0.91**	0.80 ± 0.01	0.80 ± 0.01	0.81 ± 0.01	0.81 ± 0.01
RCV1 (6 classes)	(10, 1, 0.1)	**0.89**	0.91 ± 0.01	0.92 ± 0.01	0.91 ± 0.01	0.91 ± 0.01

4   Related Work

As we have mentioned in the Introduction, in respect of high precision of local neighborhood reconstruction of **Y** in **X**, unbeatable are the algorithms based on the stochastic neighbor embedding (SNE) concept [1, 5–10]. By comparing them to older dimensionality reduction approaches such as MDS and their clones CCA, LLE, CDA,

Isomap, etc. [3, 4], the breakthrough of SNE is in its different approach to construction of the cost function $E(\ldots)$. Namely, instead of minimization of the error between distances **D** in **Y** and distances **d** in **X**, in SNE we minimize the differences between probability distributions of neighbors finding in **Y** and **X**. Therefore, the cost function in SNE [5] and its clones t-SNE [6], bh-SNE [7], ws-SNE [8], q-SNE [9], hierarchical-SNE [10] is defined as the Kullback-Leibler (K-L) divergence, or like in ws-SNE [8], other kind of divergence metrics. Fantastic and very precise 2D embeddings can be found in the webpage [https://lvdmaaten.github.io/tsne/]. However, due to the high memory and computational load – at least $O(M \log M)$ for bh-SNE [7] – these all algorithms are extremely time consuming for really large datasets. Moreover, the SNE based algorithms are very sensitive on parameters choice [1], so matching them interactively is a very awkward procedure.

The duality between DE and the nn-graph visualization was employed in the classical CDA and Isomap [4] dimensionality reduction algorithms. Notwithstanding, all the distances between the vertices in nn-graph should be calculated, by using very demanding computationally Dijkstra's (or Floyd-Warshall) algorithm into the bargain. The linear-time complexity, in turn, were reported in landmark based MDS [3] and fast PCA based Harel data embedding algorithms [13]. However, all of them suffer the "curse of dimensionality" effect, producing very poor embeddings for high-dimensional data. In [12] we described preliminary results of nr-MDS DE method, in which for $y_i(i = 1, \ldots, M)$ all the distances to its nn nearest neighbors are set to 0 while the distances to rn random neighbors remain the original ones (see dis(0, δ) in Table 2). The binary distance defined by Eq. (2) we have applied for the first time for social networks visualization [13]. We have assumed that the distances from a vertex $v_i(i = 1, \ldots, M)$ to connected vertices $nn_i = \deg(v_{ij})$ are set to 0, while to rn randomly selected vertices to 1. Then we have used MDS algorithm for graphs embedding. This method is extremely efficient and amazingly accurate. It allows for reconstructing of global and local topology of big complex networks with a few million edges on a regular laptop. We have developed GUI interface which enable to visualize the networks interactively by manipulating on their structures (adding and removing parts of network, adding vertices, removing edges etc.).

In [17] we have applied the same algorithm for high-dimensional data embedding and nn-graphs visualization. We have confronted it with the state-of-art t-SNE (bh-SNE) method [7] and the recent LargeVis [1] DE algorithm, particularly, in the context of efficiency. The most challenging seemed to be the LargeVis, which is an approximated version of t-SNE algorithm [6] but with linear-time complexity. Similarly to our ivhd algorithm, LargeVis approximates data with a nn-graph. Moreover, it uses a very similar concept we employ in ivga and ivhd, i.e., it tries to keep the nn nearest neighbors of each vertex close to it while others rn randomly chosen vertices far apart. Unlike in our methods, LargeVis visualizes data by employing principled probabilistic model [1], which is an approximation of SNE approach. As shown in [17], ivhd is still more than an order of magnitude faster than both algorithms on tested datasets at the cost of acceptable losses in embedding accuracy. Moreover, it is not so sensitive on parameters choice as SNE based algorithms.

The approximate nearest neighbor search methods based on APQ concept, were used in qSNE method [9] accelerating Dual Tree method in bh-SNE data embedding algorithm. The results of embeddings reported in [9] show high resistance of bh-SNE on this approximation. Similarly in LargeVis by approximating the *nn*-graph prior its embedding both increases the efficiency of DE initialization and does not, in a visible way, deform the embedding effect. However, in this both cases rather small mistakes in the *nn* lists were considered. Meanwhile, for big data the approximate lists can be much more distorted. They can include not only unordered *nn* lists or lists with more distant still close neighbors, but also very distant random neighbors.

5 Concluding Remarks

In this paper we discuss the duality between data embedding and graph visualization, in the context of visual exploration of big high-dimensional data. We surmise that the fully connected and weighed graph, dual to *N*D dataset, can be approximated with a planar rigid graph, which scales linearly with the number of data objects. We assume also that one can do that by selecting a limited (small) number of the edges that represent connections of every vertex to *nn* nearest and *rn* random neighbors. By assuming the binary distance between data objects, which put high contrast between the nearest and random neighbors, and by employing classical MDS algorithm, we are able to observe visually important topological properties of a considered dataset (or its *nn*-graph) in 2D. We showed earlier [12, 17] that our ivhd and ivga methods outperforms the state-of-art DE and GV methods more than one order of magnitude and allow for obtaining high quality embeddings in a linear-time $O(M)$, which reflect multi-scale cluster structure of data. Though we obtain distinctly inferior results than t-SNE in reconstructing of the local neighborhood, in our opinion, the accurate reconstruction of the nearest neighbors is rather a secondary requirement in visual exploration of big data. The accurate *nn* nearest neighbor lists are very hard if impossible to calculate for truly big data. It is mainly, due to the high computational complexity (at least $O(M \log M)$) of the nearest neighbors search. In general, they are not reliable at all because big data undergo dynamic changes and both the measurement errors and the "curse of dimensionality" principle should be taken into account. Furthermore, we show that some incorrectness in the nearest neighbor list can often be useful to improve the quality of data visualization.

Having all of these in mind, we have shown here the robustness of ivhd on serious mistakes in *nn*-neighbor lists for four popular benchmark datasets. This allows for using very approximate and fast methods for the nearest neighbors search such as locality-sensitive hashing techniques [16]. Taking into account that ivhd is extremely fast and sufficiently accurate in preserving data separability in 2D we expect that it is an ideal tool for visualization of truly big data.

Acknowledgments. This research is supported by the Polish National Center of Science (NCN) DEC-2013/09/B/ST6/01549.

References

1. Tang, J., Liu, J., Zhang, M., Mei, Q.: Visualizing large-scale and high-dimensional data. In: Proceedings of the 25th International Conference on World Wide Web, pp. 287–297 (2016)
2. Johnson, W.P., Glenn, J.M.: Making Sense of Data I: A Practical Guide to Exploratory Data Analysis and Data Mining, 2nd edn. (2014)
3. Pawliczek, P., Dzwinel, W., Yuen, D.A.: Visual exploration of data by using multidimensional scaling on multi-core CPU, GPU and MPI cluster. Concurrency Comput. Pract. Experience $26(3)$, 662–682 (2014)
4. van der Maaten, L., Postma, E.O., van den Herik, H.J.: Dimensionality reduction: a comparative review. J. Mach. Learn. Res. 10, 66–71 (2009)
5. Hinton, G.E., Roweis, S.T.: Stochastic neighbor embedding. In: Advances in Neural Information Processing Systems, pp. 833–840 (2002)
6. van der Maaten, L., Hinton, G.: Visualizing data using t-SNE. J. Mach. Learn. Res. 9, 2579–2605 (2011)
7. van der Maaten, L.: Accelerating t-SNE using tree-based algorithms. J. Mach. Learn. Res. 15, 3221–3245 (2014)
8. Zhirong, Y., Peltonen, J., Kaski, S.: Optimization equivalence of divergences improves neighbor embedding. In: Proceedings of the 31st International Conference on Machine Learning, Beijing, China (2014)
9. Ingram, S., Munzner, T.: Dimensionality reduction for documents with nearest neighbor queries. Neurocomputing 150, 557–569 (2015)
10. Pezzotti, N., Höllt, T., Lelieveldt, B., Eisemann,E., Vilanova, A.: Hierarchical stochastic neighbor embedding. Comput. Graph. Forum. $35(3)$, 21–30 (2016)
11. Hu, Y., Lei, S.: Visualizing large graphs. Wiley Interdisc. Rev. Comput. Stat. $7(2)$, 15–136 (2015)
12. Dzwinel, W., Wcisło, R.: Very fast interactive visualization of large sets of high-dimensional data. Procedia Comput. Sci. 51, 572–581 (2015)
13. Dzwinel, W., Wcisło, R., Czech, W.: ivga: a fast force-directed method for interactive visualization of complex networks. J. Comput. Sci. (2016). in print, available on-line
14. Borcea, C., Streinu, I.: The number of embeddings of minimally rigid graphs. Discrete Comput. Geom. $31(2)$, 287–303 (2004)
15. Muja, M., David G.L.: Scalable nearest neighbor algorithms for high dimensional data. IEEE Trans. Pattern Anal. Mach. Intell. $36(11)$, 2227–2240 (2014)
16. Lee, K.M.: Locality-sensitive hashing techniques for nearest neighbor search. Int. J. Fuzzy Logic Intell. Syst. $12(4)$, 300–307 (2012)
17. Dzwinel, W., Wcisło, R., Matwin, S.: ivhd: a fast and simple algorithm for embedding large and high-dimensional data, working version available (2017). www.researchgate.net, doi:10.13140/RG.2.2.28959.15520/1
18. Beyer, K., Goldstein, J., Ramakrishnan, R., Shaft, U.: When is "nearest neighbor" meaningful? In: International Conference on Database Theory, pp. 217–235 (1999)

Personalized Visualization Based upon Wavelet Transform for Interactive Software Customization

Xiaobu Yuan$^{(\boxtimes)}$, Manpreet Kaler, and Vijaya Mulpuri

University of Windsor, Windsor, ON N9B3P4, Canada
xyuan@uwindsor.ca

Abstract. This paper presents an enhancement to a prior work on interactive software customization. By conserving the historical data of belief states collected during a dialogue and analyzing the dataset with wavelet transform, a new approach is developed to categorize users into different clusters so that personalized visualization can be introduced to assist software clients in the process of requirement elicitation. The results of a usability study demonstrate improved efficiency of performance for both experienced and non-experienced users when software visualization is personalized according to users' level of product knowledge and/or skills of software development.

Keywords: Software customization · Software visualization · User clustering · Wavelet transformation

1 Introduction

Software product lines (SPLs) have emerged as a promising new paradigm of software development, whose ultimate goal is to reduce costs, shorten lead-times, improve quality, and enhance flexibility via mass production of customized software products. In a comparable way as software engineering was inspired by the systematic approach of engineering practices to deal with the crisis of software development in 1960s [12], SPL strives to better handle the ever increasing size and complexity of software systems by imitating the production of large amounts of standardized products in industrial manufacturing [11]. In particular, SPL identifies the common and variable system features of a distinct line of software products and represents them with abstract models in its *domain engineering* process [6]. The production of custom-made software is then carried out in the next process of *application engineering* [8].

Despite advancements in the past two decades for the identification and modeling of features in a variety of application domains, noticeable shortage of supporting tools has become a bottleneck that prevents the production of customized software according to the specific needs of software clients [8]. At the current stage of progress in SPL, almost all of the configuration tools available to determine the variable and associated common system features are made

© Springer International Publishing AG 2017
P. Perner (Ed.): MLDM 2017, LNAI 10358, pp. 361–375, 2017.
DOI: 10.1007/978-3-319-62416-7_26

for software developers to act on behalf of software clients [2]. To enable software customization being conducted in a similar manner as customers placing their orders when purchasing consumer products, an approach of interactive software customization with machine-guided requirement elicitation was developed by the authors' research team in [16], which was the very first approach that actually allows the "ordering" of customized software products by means of dialogue in natural language.

This software customization tool has been enhanced with software visualization in [18] to use Petri nets for the illustration of feature selection and the flow of feature evaluation during the interactive process of dialogue-based requirement elicitation. Even though significantly improved efficiency has been achieved with the support of software visualization, a usability study reveals that the use of Petri nets adds additional difficulty for non-experienced users. This observation endorses the claim made by a recent paper that "*current tools lack adequate mechanisms to visualize variability models for different stakeholders, thus making variability harder to manage*" [3]. To address the needs of different software clients, software customization therefore has to determine which system features create value from the view points of stakeholders, who may or may not have knowledge of a software product line and/or skills of software development.

To meet the "*great need for future research to facilitate software customization*" [14], this paper first conducts an analysis of POMDP models with the theory of information space in Sect. 2, and discusses the need of conserving and analyzing the history of belief states for dialogue management. In Sect. 3, a novel approach is developed to use different visualization techniques for different types of users. In the new approach, wavelet transform is applied on the historical data of belief states to help categorizing users into different groups according to the number of sharp variation points. In addition to details of algorithm design, this paper also includes results from a usability study to demonstrate the advantages of personalized visualization in Sect. 4. Finally, Sect. 5 gives conclusions and points out directions for future work.

2 Analysis of POMDP Models with Information Space

For interactive software customization, dialogue management can be based upon different models [18], including those based upon finite state machines, slots, forms, Markov Decision Process (MDP), and Partially Observable Markov Decision Process (POMDP). Among them, the approach based upon POMDP models demonstrates the advantage of handling uncertainty caused by speech act errors [15]. Formally, a POMDP is defined as a tuple $\{S, A_m, T, O, Z, R\}$, where S is a set of states, A_m is a set of actions the system may take, T is the transition model that defines transition probability, O is a set of observations from user's actions, Z is the observation model that defines the observation probability, and R defines the immediate reward $r(s, a_m)$ in expected probability values. POMDP-based dialogue management carries out two tasks. One is to compute or update belief state b, and the other is to find an optimal policy for the selection of actions [10].

Based upon the Bayes filter algorithm, the computation of belief state uses the following equation, where α is a normalizing constant.

$$b_{t+1}(s_{t+1}) = \alpha P(o_{t+1}|s_{t+1}, a_t) \sum P(s_{t+1}|s_t, a_t)b_t(s_t) \tag{1}$$

For the current belief state, Eq. 1 constitutes the flat POMDP model that selects an optimal policy as the maximum of all the expected value function $V^\pi(b)$ with a discounted future reward for a policy π.

$$\pi^* = argmax_\pi E[V^\pi(b)] \tag{2}$$

The factored POMDP model extends the unobserved state to include the user's action model, which is the user's most recent action and relevant dialogue history information from conversation. The extended unobserved state helps to revise Eq. 1 into the following equation to update belief states with more appropriate reward.

$$b'(s'_u, s'_d, a'_u) = \alpha P(o'|a'_u)P(a'_u|s'_u, a_m) \sum P(s'_u|s_u, a_m)$$
$$\sum P(s'_d|a'_u, s_d, a_m) \sum b(s_u, s_d, a_u) \tag{3}$$

The formation of POMDP models follows in part with the theory of information space [9], according to which the only information available to a decision process at stage k of a planning process is the history of all observations $\widetilde{O}_k$ at that stage and the history of all actions $\widetilde{A}_{k-1}$ that have been taken before that stage. Given an initial condition η_0, $\widetilde{O}_k$ and $\widetilde{A}_{k-1}$ are two Cartesian products of observation and action spaces respectively at their corresponding stages.

$$\widetilde{O}_k = O_{k-1} \times O_{k-2} \times O_{k-3} \cdots \times O_0$$
$$\widetilde{A}_{k-1} = A_{k-2} \times A_{k-3} \cdots \times A_0$$

If η_0 belongs to an initial condition space $\mathcal{I}_0$, a history information space $\mathcal{I}_{hist}$ is formed as the union of $\mathcal{I}_0$ and $\mathcal{I}_k = \mathcal{I}_0 \times \widetilde{A}_{k-1} \times \widetilde{O}_k$ for up to the kth stage.

$$\mathcal{I}_{hist} = \mathcal{I}_0 \cup \mathcal{I}_1 \cup \mathcal{I}_2 \cup \cdots \cup \mathcal{I}_k \tag{4}$$

As illustrated in Eq. 5, an information-feedback plan $\pi = (\pi_1, \pi_2, \cdots \pi_k)$ then maps $\mathcal{I}_{hist}$ into a sequence of actions $\mu_1, \mu_2, \cdots \mu_k \in \widetilde{A}_k$, and an optimal plan π^* maximizes a given stage-additive cost function.

$$\pi : \quad \mathcal{I}_{hist} \to \widetilde{A}_k \tag{5}$$

In POMDP models, actions at the previous stage lead to observations probability at the kth stage, which corresponds only to the Cartesian product of $A_{k-1} \times O_k$ in $\mathcal{I}'_k$ as a subset of $\mathcal{I}_k$. This simplification of $\mathcal{I}_{hist}$ to $\mathcal{I}'_k$ leads to a complete loss of history information, including changes in belief states, series of observations, and sequences of actions. It results in the following formula as a much simplified version of Eq. 5.

$$\pi' : \quad \mathcal{I}'_k \to A_k \tag{6}$$

Planning with POMDP models performs much better in real-life applications than all the other approaches as it does not rely on estimated system state, and is able to handle input uncertainty. However, the substitution of $\mathcal{I}_0 \times \widetilde{A}_{k-1} \times \widetilde{O}_k$ with $A_{k-1} \times O_k$ makes it impossible to trace changes in belief states and to retrieve the historical information of observations and actions. In other words, belief state becomes a static probability distribution over the current system state only. In consequence, POMDP-based dialogue management is unable to deal with such uncertainty in belief states that corresponds to uncertainty in either user's actions or the observation of user's actions.

To overcome the shortcomings while retaining advantages of the current POMDP models, a modified planning strategy can be introduced to keep the history of belief states while still using the POMDP models for the selection of actions. As formulated in Eq. 3, the calculation of belief state b_{t+1} relies on both observation o_{t+1} and action a_t. A strategy to use the history of belief states together with $A_{k-1} \times O_k$ can be created by extending Eqs. 6 and 7.

$$\pi_{new} : \quad \mathcal{I}'_{hist} \rightarrow \widetilde{A}_k \tag{7}$$

In the revised mapping, $\mathcal{I}'_{hist}$ is a subset of $\mathcal{I}_{hist}$ as the history of belief states maintains the historical information of both observations and actions. Although information of observations and actions at early stages is not saved explicitly, Eq. 7 reduces the negative effect of Markov assumption and allows POMDP-based dialogue management to plan for actions with not only the current belief state but also belief states before reaching the current state.

3   New Approach of Personalized Visualization

It takes three phases to complete the process of interactive software customization [16,17]. The first phase creates a domain model by initiating an ontology model that consists of functions/qualities, relationships, and rules. The second phase then uses this ontology model as a knowledge base to help identifying required features according to the optional features selected by users via dialogue. The third phase finally generates service descriptions with this subset of system features so that customized software systems can be automatically produced. In support of interactive software customization, this section proposes to support requirement elicitation in the second phase with different styles of software visualization tailored for different types of users.

3.1   Personalized Visualization for Requirements Elicitation

Shown in Fig. 1(a) is the interactive component of a software customization system. In the structure, an ontology knowledge base keeps the ontology model created by domain engineers, which is then used by the dialogue manager to guide the interaction between a user and the system via its input/output controller. During the interaction, the user is presented with variable system features with the support of software visualization for him/her to select or abandon according

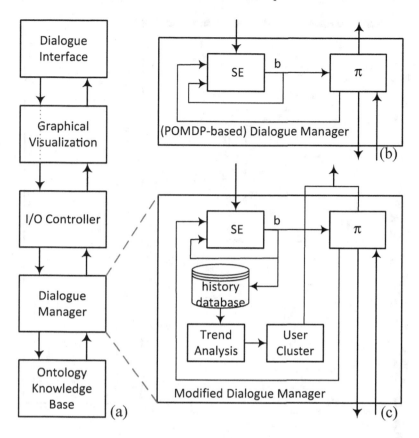

Fig. 1. System structure for the interactive visualization system

to his/her preference. In addition, there are mandatory features that are automatically selected by the system either as common system features or due to its dependency relationship with selected features. They are also presented to the user with explanations.

Figure 1(b) and (c) illustrate the structure of a dialogue manager based upon the original POMDP models (Eq. 6) and the modified version (Eq. 7) respectively. Dialogue management is fundamentally a problem of planning under the influence of uncertainty. The uncertainties that the original POMDP-based approach fails to handle mainly arise from situations in which the user lacks knowledge in the domain or the user's goal cannot be fulfilled due to real-life constraints. By introducing a new branch in the modified structure to keep and analyze the history of belief state, the new approach becomes capable of clustering users into different groups and presenting different styles of visualization accordingly.

In Fig. 1(a), the block of graphical visualization is in charge of software visualization. While user inputs pass through the visualization block directly to I/O controller for language processing, computer-generated utterance is presented to

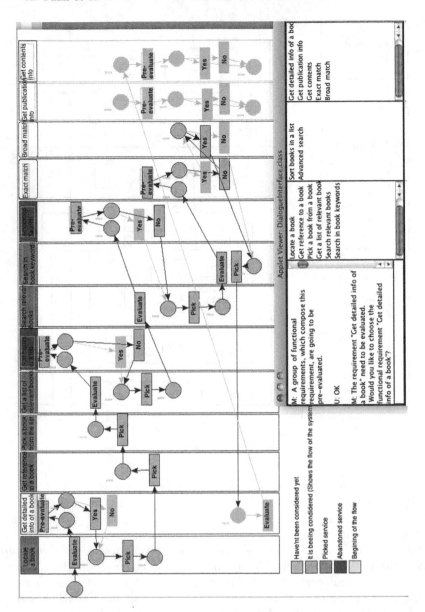

Fig. 2. A two-layer structure of interactive visualization (Color figure online)

the user together with software visualization. By using the modified dialogue manager (Fig. 1(c)), the belief state of each conversation is saved in the database and the historical data is analyzed for user clustering. When utterance is passed over from dialogue manager to the I/O controller together with information about user type, the system directs dialogue via its text-based interface, with software visualization being presented in the background. Figure 2 shows

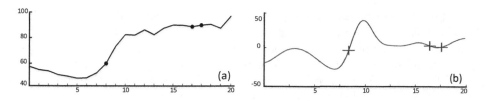

Fig. 3. An illustration of trend changes in belief states and wavelet transform

the layout when the two forms of information are presented together, in which the text-based dialogue interface is in the (right) front for software clients to interact with the system and graphical visualization shows in the back to help them understanding the process and progressing with requirements elicitation.

3.2 Trend Analysis for Personalized Visualization

During requirements elicitation, fluctuation exists in the confidence of belief states [7] as illustrated in Fig. 3(a). Among the abundant techniques for data mining and analysis with or without transformations, wavelet transform captures both frequency information in the transformation space and location information in the chronological order of an original dataset [4]. Given a function $f(x)$, its wavelet transform is defined by the following equation,

$$W_f(a,b) = \frac{1}{\sqrt{a}} \int_{-\infty}^{+\infty} f(x)\varphi(\frac{x-b}{a})dx \qquad a > 0, b \in R \qquad (8)$$

where $\varphi((x-b)/a)$ is called mother wavelet, and a and b are two factors for scaling and (time) shifting respectively. The transformation measures, in a certain direction, the fluctuations of $f(x)$ around a specific time point b on the scale of a. In Eq. 8, $\varphi(x)$ takes the following format when using Mexican Cap wavelet, which yields a maximum of 1 when $x = 0$ and goes to zero when $x \to \infty$.

$$\varphi(x) = (1 - x^2)e^{-\frac{x^2}{2}} \qquad (9)$$

To be applicable to discrete datasets, sampling of wavelet transform has to take place at discrete time, for which Eq. 8 takes a new format as below,

$$W_q(a,t) = \frac{1}{\sqrt{a}} \sum_{n=1}^{N} q(n)\varphi(\frac{n-t}{a}) \qquad (10)$$

where n ranges from 1 to N for the set of data under analysis. In particular, sharp variation points in wavelet transform can be used to identify the locations where the trend of belief states changes [1,5]. As all sharp variation points must satisfy the condition as specified by Eq. 11, they are identified in Fig. 3(b) on the curve for $a = 5$ and the locations are mapped back to Fig. 3(a).

$$W_q(a,t)W_q(a,t+1) < 0 \qquad (11)$$

When the set of data is the history of belief states collected at run time during the dialogue of software customization, the number of sharp variation points tells the frequency of changed preference by a user for optional system features. The overall preference of variable features is expected to remain relatively constant if a user is experienced or change a lot if non-experienced, according to his/her level of knowledge about a software product line and expertise of software development. Therefore, users interacting with the software customization system can be classified into four groups as experts, professionals, amateurs, and novice. With thresholds properly set up, the number of sharp variation points after applying wavelet transform to the historical data of belief states can be used to classify users into different groups.

For the four groups of users, different styles of software visualization can be used to help them with requirements elicitation. In particular, the proposed approach uses Petri nets for experts, directed graphs for professionals, requirement models for amateurs, and block list of requirements for novice users. The four styles of visualization are chosen for the level of details each of them presents and the level of knowledge each requires to understand. For example, experts prefer details and are able to understand more complicated techniques. Consequently, the system chooses to use Petri nets for the presentation of system features with detailed information about parallelism, synchronization, and conflicts. For novice users, however, they usually have neither technical nor business knowledge of software systems. It is more suitable and sufficient to simply provide a simple list of requirements, with different colors to indicate if a system feature is selected, abandoned, or under evaluation.

3.3   An Algorithm of Personalized Visualization

The main difference of the four chosen visualization techniques is the level of details. With a Petri net having the most amount of information in presentation, a directed graph discards the tags. The remaining information of actions is further thrown away in a requirement model, until only titles of requirements are kept in a block list. Illustrated in Fig. 4 is an algorithm for the proposed approach. It carries out requirements elicitation with the support of personalized visualization in three phase respectively for feature preprocessing (`lines 2--10`), interaction with visualization (`lines 12--14`), and after-evaluation processing (`lines 15--22`). To facilitate personalized visualization, the algorithm prepares all information for presentation with a Petri net while allowing the other styles to use only part of the details according the chosen visualization technique.

At the highest level of details, visualization starts with the color of feature titles in white and Petri net in gray. Based on the rules of Petri nets, a transition fires when its associated event occurs. In visualization, it triggers a color change to blue for both the transition node and the input/output arcs linking to the node. As the flow of transition in the Petri net is used to illustrate the process of requirement evaluation, the change of color provides useful information to help experienced clients understanding the progress of requirements elicitation.

```
1    FOR each system feature R to be evaluted
2        IF R is essentail for the system THEN
3            CALL performRequirementSelecting with R
4            SET  R_color = green, R_tag = Pick
5        ELSE IF R is pre-selected THEN
6            CALL performRequirementSelecting with R
7            SET  R_tag = Evaluate
8        ELSE IF R is pre-droped THEN
9            CALL performRequirementDropping with R
10           SET  R_tag = Abandon
11       ELSE
12           CALL determineUtterance
13           CALL personalizeVisualization
14           CALL evaluateRequirement with R
15           SET  R_tag = pre-evaluate
16           IF R is to be selected THEN
17               CALL performRequirementSelecting with R
18               SET  R_color = green, R_tag = Pick
19           ELSE
20               CALL performRequirementDropping with R
21               SET  R_color = red, R_tag = Abandon
22           END IF
23       END IF
24   END FOR
```

Fig. 4. Algorithm for requirements elicitation with personalized visualization

Furthermore, the color of system features in the feature array changes to either green or red to indicate if a feature is selected or abandoned. At the same time, visualization in other styles with the same color change is also updated for use according to need.

While the outer loop of the algorithm goes via all system features one by one, the assessment of R as either a common or variable system feature could fall in one of four cases. In the first case, R is a common feature and therefore is essential to the product. It has to be selected automatically with only the need to inform the clients, without the need of evaluation. The automatic selection of R also results in a change of its title color to green in visualization, and a change of its tag in the Petri net to Pick to indicate that R now becomes a selected feature.

The second case deals with the situation when R is not essential itself, but has to be selected without the clients' evaluation due to such a relationship that makes certain client-selected feature(s) relying on it, certain essential feature(s) decomposing into it, or some essential feature(s) being associated with it while it is a quality. In such a case, R becomes pre-selected but not evaluated. Its tag changes to Evaluate to indicate that this system feature is under evaluation. If, in the third case, a non-essential feature R has a contradiction relationship with certain selected feature(s), it has to be abandoned without the client's evaluation, leading to a change of tag value to Abandon as well. This could also happen when certain abandoned feature(s) relies on, decomposes into, or is associated with R when R is a quality.

The fourth case covers all the remaining features when R is non-essential and has neither been selected nor abandoned yet. In this last case, R is presented via the text-based dialogue interface to the clients, together with an update to software visualization. An action by the user to either select or abandon the

```
 1  performRequirementSelecting with R
 2      CALL selectRequirement with R
 3      IF  R_tag = enabled THEN
 4          SET R_token = out
 5          // move Petri net token to R's output place
 6          SET R.arc_color = blue
 7      ELSE wait until R_tag is enabled and goto 4
 8      CALL preSelectRequirement with all R'
 9          on which R relies
10      CALL preDropRequirement with all R'
11          to which R contradicts
12      CALL preEvaluateRequirement with all R'
13          into which R decomposes
14      IF R is function THEN
15          CALL preEvaluateRequirement with all R'
16              where each R' is a quality and
17              R is associated with it
18      END IF
19
20  performRequirementDropping with R
21      CALL dropRequirement with R
22      IF  R_tag = enabled THEN
23          SET R_color = red, R_token = out
24          SET R.arc_color = blue
25      ELSE wait until R_tag is enabled and goto 23
26      CALL preDropRequirement with all R' relying on R
```

Fig. 5. Two subroutines of the personalized visualization algorithm

feature is visualized correspondingly with a change of title color to either green or red, and a change of tag to either `Pick` or `Abandon`. During the process, subroutine `personalizeVisualization` makes one visualization style active for presentation according to the result after applying wavelet transform to the history of belief states passed over from `determineUtterance`. Afterwards, `evaluateRequirement` returns the flow of control to the text-based interface of interactive requirements elicitation. The other two subroutines are further illustrated in Fig. 5, where the additional subroutines perform the operations as suggested by their names.

During requirements elicitation, the pre-selection of a feature further leads to pre-selection of all the features that it relies on, and pre-dropping of all the features in contradiction with it. Moreover, the action of pre-dropping a feature will result in the labeling of `Evaluate` if not labeled yet, and more pre-dropping of the features relying on it. To ensure completeness and consistency of requirements, if a selected feature becomes abandoned, all the selected features relying on it will also be abandoned. Similarly, if an abandoned feature becomes selected, the selected features that contradict with it will be abandoned and the abandoned features that it relies on will be selected. Details of interactive software customization for the production of custom-made software are available in [16].

4   Experiment and Usability Study

The proposed system has been implemented with Java programming language using Eclipse IDE. In addition, a Java-based framework called "Rakiura JFern"

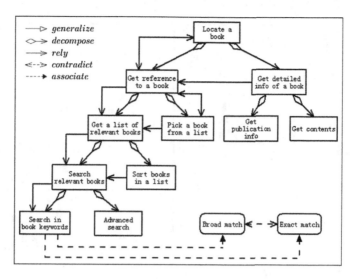

Fig. 6. The requirement model instantiated with book-locating service

is used to prepare graphical presentations for all the four styles of software visualization. After training the system with a set of simulated data, the number of sharp variation points for four different groups of users is set to a range respectively from 0 to 7 for experts, 8 to 14 for professionals, 15 to 21 for amateurs, and 22 or more for novice users.

4.1   Experiment Results

An ontology model has been created to enable the customization of online book shopping systems for experiment. Figure 6 shows the book-locating module of the ontology, with five relationships annotated at the top-left corner. The ontology consists in total of 52 functions, 6 qualities, and 2 soft-goals. In addition, there are 162 relationships in the model, among which 48 are for *decompose*, 102 for *rely*, 6 for *contradict*, and 6 for *associate*. The dialogue interface (in the front of Fig. 2) is further divided into three parts. The utterances generated by the dialogue manager are displayed in the upper left text box. Users can type their response in the lower left text box. Meanwhile, the three columns on the right side list the selected, abandoned, and to-be-evaluated system features respectively.

Figure 2 illustrates the progress of interactive requirements elicitation with Petri nets being the default style of software visualization after reaching the point to evaluate system feature Broad match. The two red-colored titles in the feature array indicate that both Soft books in a list and Advanced search have been abandoned by the user's action in the dialogue interface during evaluation. In a similar way, the green-colored titles indicate that these features have been selected either automatically because they are essential, e.g., Locate a book, or interactively by the user, e.g., Get detailed info of a book. The other three

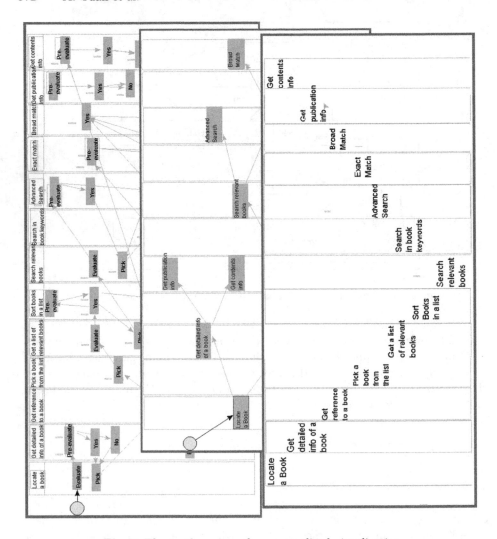

Fig. 7. Three other views for personalized visualization

styles of visualization are shown in Fig. 7, and only one of four styles is active according to the type of users determined by wavelet transform.

In the example, the decomposition relationship between concepts in the ontology model (Fig. 6) results in three features, e.g., Sort books in a list, being pre-evaluated before a decision is made to either select or abandon their preceding features, e.g., Get a list of relevant books. The contradiction relationship between two concepts in the ontology model, on the other hand, results in a re-evaluation of related features to make sure that all the features that rely on the chosen feature are selected and all those contradict to it are abandoned. This re-evaluation traces back all the way to Get detailed info of a book.

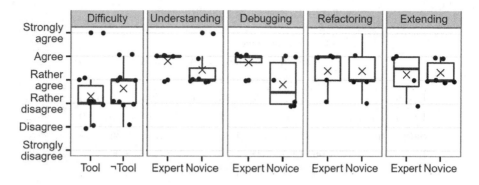

Fig. 8. Results of a usability study on visualization [13]

A change of title color from green back to white signals the new status of this feature, whose color will be updated together with those for the contradicting pair of features when re-evaluation is finished.

4.2   A Usability Study

There is plenty of publications on visualization, but very little about its usability. Shown in Fig. 8 is one of the few that studies the usability of visualization [13]. To evaluate personalized visualization for software customization, a usability study has also been conducted by asking users with one (Petri nets) visualization technique or with all the four different techniques. Users are all students in the university at graduate/undergraduate levels, majoring in areas such as computer science, engineering, mathematics, business, physics, economics, library, and human kinematics. As computer science students have more knowledge and experience of software development, they are placed in a different group than the others when analyzing the results of study. In the study, all users are requested to complete software customization by choosing three out of seven variable features for an online bookstore system.

With only brief description of functionality being given, a successful user is expected to pick only the three but abandon the other four features. For each user who finishes software customization, a score is obtained by calculating the number of correctly selected or abandoned features. An error rate is measured by subtracting the score from a total score of seven, and error is simply the complementary of score. In addition, on-site assessment of time records the duration that a user takes to finish the work. Table 1 provides the results in average (avg) and standard deviation ($stdv$), together with several other values obtained from questionnaires filled in by users about their opinions of easiness, understandability, satisfaction, and improvement.

As shown in Table 1, the use of personalized visualization has brought positive changes for both groups in regard to all attributes, including the opinions about satisfaction and improvement that were different in a prior study when Petri nets were used as the only technique for visualization [18]. If the study in [13]

Table 1. Results of usability study

	Computer science				Others			
	Visualized		Personalized		Visualized		Personalized	
	avg	stdv	avg	stdv	avg	stdv	avg	stdv
Easiness	6.83	1.12	8.44	0.83	6.13	1.29	8.50	1.05
Understandable	6.54	2.20	8.01	1.58	4.36	1.23	6.40	1.90
Satisfaction	7.22	1.27	8.11	1.33	7.44	1.18	8.38	1.05
Improvement	7.65	1.06	8.18	1.23	7.06	1.04	8.25	1.05
Time	8.33	1.94	7.89	1.76	9.19	1.91	8.13	1.13
Score	6.06	0.97	6.59	0.59	4.94	1.26	5.94	0.99
Error rate	0.94	0.97	0.41	0.59	2.06	1.26	1.06	0.99

had used more visualization techniques than dependency graphs alone, more consistent and better results could have been achieved as well. Overall in the study, understandability has been further improved at an impressive rate of 22% for computer science students and at a high rate of 47% for other users. For efficiency, which is an important attribute that indicates the number of accomplished tasks in a time interval, an additional increase of 15% is achieved for computer science students and 35% for other students. It means that, with the support of personalized visualization, all types of users are now able to correctly perform much more tasks than before.

5   Conclusions

Aiming at helping software clients with requirements elicitation during software customization, an approach of personalized software visualization is presented in this paper. In addition to the enhancement of the original dialogue-based interface with live visualization of activity flow and evaluation progress, this work further allows the style of visualization to be changed according to the user's knowledge of software systems and skills of software development. The results of a usability study provide convincing evidence that personalized software visualization improves understandability and increases efficiency for not only experienced but also non-experienced users at different levels. Active investigation is being conducted to include emotion in dialogue-based software customization for further enhancement of the existing system.

References

1. Addison, P.: The Illustrated Wavelet Transform Handbook: Introductory Theory and Applications in Science, Engineering, Medicine and Finance, 2nd edn. CRC Press, Boca Raton (2017)
2. Bhumula, M.: Comparative study and analysis of variability tools. Master's thesis, School of Computing, IT and Engineering, University of East London (2013)

3. Bosch, J., Capilla, R., Hilliard, R.: Trends in systems and software variability. IEEE Softw. **32**(3), 44–51 (2015)
4. Burrus, C., Gopinath, R., Guo, H.: Introduction to Wavelets and Wavelet Transforms. Prentice Hall, Englewood Cliffs (1998)
5. Calderbank, A., Daubechies, I., Sweldens, W., Yeo, B.: Wavelet transforms that map integers to integers. Appl. Comput. Harmonic Anal. **5**(3), 332–369 (1998)
6. Czarnecki, K., Grünbacher, P., Rabiser, R., Schmid, K., Wąsowski, A.: Cool features and tough decisions: a comparison of variability modeling approaches. In: Proceedings of the 6th International Workshop on Variability Modeling of Software-Intensive Systems, pp. 173–182 (2012)
7. Doshia, F., Roy, N.: Spoken language interaction with model uncertainty: an adaptive human-robot interaction system. Connect. Sci. **20**(4), 299–318 (2008)
8. Heradio, R., Perez-Morago, H., Fernandez-Amoros, D., Cabrerizo, F.J., Herrera-Viedma, E.: A science mapping analysis of the literature on software product lines. In: Fujita, H., Guizzi, G. (eds.) SoMeT 2015. CCIS, vol. 532, pp. 242–251. Springer, Cham (2015). doi:10.1007/978-3-319-22689-7_18
9. LaValle, S.: Planning Algorithms. Cambridge University Press, New York (2006)
10. Panella, A., Gmytrasiewicz, P.: Interactive POMDPs with finite-state models of other agents. Auton. Agents Multi-Agent Syst. **31**, 1–44 (2017)
11. Pohl, K., Böckle, G., van der Linden, F.: Software Product Line Engineering: Foundations, Principles and Techniques. Springer, Heidelberg (2005)
12. Randell, B.: Software engineering in 1968. In: Proceedings of the 4th International Conference on Software Engineering, pp. 1–10 (1979)
13. Rentschler, A., Noorshams, Q., Happe, L., Reussner, R.: Interactive visual analytics for efficient maintenance of model transformations. In: Duddy, K., Kappel, G. (eds.) ICMT 2013. LNCS, vol. 7909, pp. 141–157. Springer, Heidelberg (2013). doi:10.1007/978-3-642-38883-5_14
14. Weiss, M., Schweiggert, F.: Opportunities and challenges of software customization. ACEEE Int. J. Inf. Technol. **3**(4), 1–11 (2013)
15. Young, S., Breslin, C., Gašić, M., Henderson, M., Kim, D., Szummer, M., Thomson, B., Tsiakoulis, P., Hancock, E.: Evaluation of statistical POMDP-based dialogue systems in noisy environments. In: Rudnicky, A., Raux, A., Lane, I., Misu, T. (eds.) Situated Dialog in Speech-Based Human-Computer Interaction, pp. 3–14. Springer, Cham (2016)
16. Yuan, X., Zhang, X.: An interactive approach of online software customization via conversational Web agents. In: 2013 IEEE International Conference on Internet of Things, pp. 327–334 (2013)
17. Yuan, X., Zhang, X.: An ontology-based requirement modeling for interactive software customization. In: Proceedings of the 5th IEEE International Model-Driven Requirements Engineering Workshop, pp. 21–30 (2015)
18. Yuan, X., Sadri, V.: An approach of interactive visualization for software customization. In: Proceedings of the 7th IEEE International Conference on Software Engineering and Service Science, pp. 21–24 (2016)

Automatic Detection of Knee Joints and Quantification of Knee Osteoarthritis Severity Using Convolutional Neural Networks

Joseph Antony[1]([✉]), Kevin McGuinness[1], Kieran Moran[1,2], and Noel E. O'Connor[1]

[1] Insight Centre for Data Analytics, Dublin City University, Dublin, Ireland
joseph.antony@insight-centre.org
[2] School of Health and Human Performance, Dublin City University, Dublin, Ireland

Abstract. This paper introduces a new approach to automatically quantify the severity of knee OA using X-ray images. Automatically quantifying knee OA severity involves two steps: first, automatically localizing the knee joints; next, classifying the localized knee joint images. We introduce a new approach to automatically detect the knee joints using a fully convolutional neural network (FCN). We train convolutional neural networks (CNN) from scratch to automatically quantify the knee OA severity optimizing a weighted ratio of two loss functions: categorical cross-entropy and mean-squared loss. This joint training further improves the overall quantification of knee OA severity, with the added benefit of naturally producing simultaneous multi-class classification and regression outputs. Two public datasets are used to evaluate our approach, the Osteoarthritis Initiative (OAI) and the Multicenter Osteoarthritis Study (MOST), with extremely promising results that outperform existing approaches.

Keywords: Knee osteoarthritis · KL grades · Automatic detection · Fully convolutional neural networks · Classification · Regression

1 Introduction

Knee Osteoarthritis (OA) is a debilitating joint disorder that mainly degrades the knee articular cartilage. Clinically, the major pathological features for knee OA include joint space narrowing, osteophytes formation, and sclerosis. Knee OA has a high-incidence among the elderly, obese, and those with a sedentary lifestyle. In its severe stages, it causes excruciating pain and often leads to total joint arthoplasty. Early diagnosis is crucial for clinical treatments and pathology [10,14]. Despite the introduction of several imaging modalities such as MRI, Optical Coherence Tomography and ultrasound for augmented OA diagnosis, radiography (X-ray) has been traditionally preferred, and remains the main accessible tool and "gold standard" for preliminary knee OA diagnosis [10,15,17].

© Springer International Publishing AG 2017
P. Perner (Ed.): MLDM 2017, LNAI 10358, pp. 376–390, 2017.
DOI: 10.1007/978-3-319-62416-7_27

Previous work has approached automatically assessing knee OA severity [14, 17,20] as an image classification problem. In this work, we train CNNs from scratch to automatically quantify knee OA severity using X-ray images. This involves two main steps: (1) automatically detecting and extracting the region of interest (ROI) and localizing the knee joints, (2) classifying the localized knee joints.

We introduce a fully-convolutional neural network (FCN) based method to automatically localize the knee joints. A FCN is an end-to-end network trained to make pixel-wise predictions [9]. Our FCN based method is highly accurate for localizing knee joints and the FCN can easily fit into an end-to-end network trained to quantify knee OA severity.

To automatically classify the localized knee joints we propose two methods: (1) training a CNN from scratch for multi-class classification of knee OA images, and (2) training a CNN to optimize a weighted ratio of two loss functions: categorical cross-entropy for multi-class classification and mean-squared error for regression. We compare the results from these methods to WND-CHARM [15,17] and our previous study [1]. We also compare the classification results to both manual and automatic localization of knee joints.

We propose a novel pipeline to automatically quantify knee OA severity including a FCN for localizing knee joints and a CNN jointly trained for classification and regression of knee joints. The main contributions of this work include the fully-convolutional network (FCN) based method to automatically localize the knee joints, training a network (CNN) from scratch that optimizes a weighted ratio of both categorical cross-entropy for multi-class classification and mean-squared error for regression of knee joints. This multi-objective convolutional learning improves the overall quantification with an added benefit of providing simultaneous multi-class classification and regression outputs.

2 Related Work

Assessing knee OA severity through classification can be achieved by detecting the variations in joint space width and osteophytes formation in the knee joints [10,14,15]. In a recent approach, Yoo et al. used artificial neural networks (ANN) and KNHANES V-1 data, and developed a scoring system to predict radiographic and symptomatic knee OA [20] risks. Shamir et al. used WND-CHARM: a multipurpose bio-medical image classifier [11] to classify knee OA radiographs [16,17] and for early detection of knee OA using computer aided analysis [14]. WND-CHARM uses hand-crafted features extracted from raw images and image transforms [11,16].

Recently, convolutional neural networks (CNNs) have outperformed many methods based on hand-crafted features and they are highly successful in many computer vision tasks such as image recognition, automatic detection and segmentation, content based image retrieval, and video classification. CNNs learn effective feature representations particularly well-suited for fine-grained classification [19] like classification of knee OA images. In our previous study [1], we showed that the off-the-shelf CNNs such as the VGG 16-Layers network [18], the VGG-M-128 network [2], and the BVLC reference CaffeNet [5,6] trained on ImageNet LSVRC dataset [13] can be fine-tuned for classifying knee OA images through transfer learning. We also argued that it is appropriate to assess knee OA severity using a continuous metric like mean-squared error instead of binary or multi-class classification accuracy, and showed that predicting the continuous grades through regression reduces the mean-squared error and in turn improves the overall quantification.

Previously, Shamir et al. [14] proposed template matching to automatically detect and extract the knee joints. This method is slow for large datasets such as OAI, and the accuracy and precision of detecting knee joints is low. In our previous study, we introduced an SVM-based method for automatically detecting the center of knee joints [1] and extract a fixed region with reference to the detected center as the ROI. This method is also not highly accurate and there is a compromise in the aspect ratio of the extracted knee joints that affects the overall quantification.

3   Data

The data used for the experiments and analysis in this study are bilateral PA fixed flexion knee X-ray images. The datasets are from the Osteoarthritis Initiative (OAI) and Multicenter Osteoarthritis Study (MOST) in the University of California, San Francisco, and are standard datasets used in knee osteoarthritis studies.

3.1   Kellgren and Lawrence Grades

This study uses Kellgren and Lawrence (KL) grades as the ground truth to classify the knee OA X-ray images. The KL grading system is still considered the gold standard for initial assessment of knee osteoarthritis severity in radiographs [10–12,15]. It uses five grades to indicate radiographic knee OA severity. 'Grade 0' represents normal, 'Grade 1' doubtful, 'Grade 2' minimal, 'Grade 3' moderate, and 'Grade 4' represents severe. Figure 1 shows the KL grading system.

3.2   OAI and MOST Data Sets

The baseline cohort of the OAI dataset contains MRI and X-ray images of 4,476 participants. From this entire cohort, we selected 4,446 X-ray images based on the availability of KL grades for both knees as per the assessments by Boston University X-ray reading center (BU). In total there are 8,892 knee images and the distribution as per the KL grades is as follows: Grade 0 - 3433, Grade 1 - 1589, Grade 2 - 2353, Grade 3 - 1222, and Grade 4 - 295.

The MOST dataset includes lateral knee radiograph assessments of 3,026 participants. From this, 2,920 radiographs are selected based on the availability of KL grades for both knees as per baseline to 84-month Longitudinal Knee

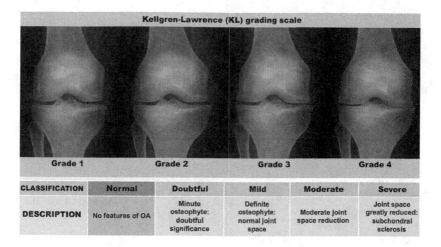

Fig. 1. The KL grading system to assess the severity of knee OA.

Radiograph Assessments. In this dataset there are 5,840 knee images and the distribution as per KL grades is as follows: Grade 0 - 2498, Grade 1 - 1018, Grade 2 - 923, Grade 3 - 971, and Grade 4 - 430.

4   Methods

This section introduces the methodology used for quantifying radiographic knee OA severity. This involves two steps: automatically detecting knee joints using a fully convolutional network (FCN), and simultaneous classification and regression of localized knee images using a convolutional neural network (CNN). Figure 2 shows the complete pipeline used for quantifying knee OA severity.

4.1   Automatically Localizing Knee Joints Using a FCN

Assessment of knee OA severity can be achieved by detecting the variations in joint space width and osteophytes formation in the knee joint [10]. Thus, localizing the knee joints from the X-ray images is an essential pre-processing step before quantifying knee OA severity, and for larger datasets automatic methods are preferable. Figure 3 shows a knee OA radiograph and the knee joints: the region of interest (ROI) for detection. The previous methods for automatically localizing knee joints such as template matching [14] and our own SVM-based method [1] are not very accurate. In this study, we propose a fully convolutional neural network (FCN) based approach to further improve the accuracy and precision of detecting knee joints.

FCN Architecture: Inspired by the success of a fully convolutional neural network (FCN) for semantic segmentation on general images [9], we trained

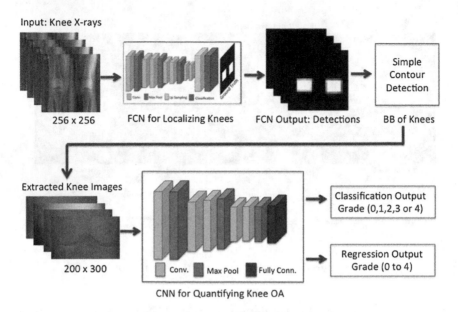

Fig. 2. The pipeline used for quantifying knee OA severity.

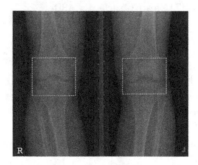

Fig. 3. A knee OA X-ray image with the region of interest: the knee joints.

a FCN to automatically detect the region of interest (ROI): the knee joints from the knee OA radiographs. Our proposed FCN is based on a lightweight architecture and the network parameters are trained from scratch. Figure 4 shows the architecture. After experimentation, we found this architecture to be the best for knee joint detection. The network consists of 4 stages of convolutions with a max-pooling layer after each convolutional stage, and the final stage of convolutions is followed by an up-sampling and a fully-convolutional layer. The first and second stages of convolution use 32 filters, the third stage uses 64 filters, and the fourth stage uses 96 filters. The network uses a uniform $[3 \times 3]$ convolution and $[2 \times 2]$ max pooling. Each convolution layer is followed by a batch normalization and a rectified linear unit activation layer (ReLU). After the final convolution layer, an $[8 \times 8]$ up-sampling is performed as the network uses 3

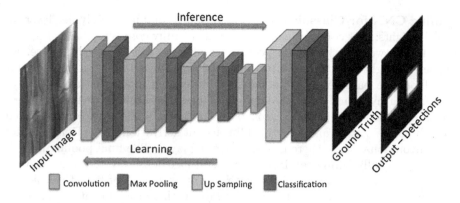

Fig. 4. The fully convolutional network for automatically detecting knee joints.

stages of $[2 \times 2]$ max pooling. The up-sampling is essential for an end-to-end learning by back propagation from the pixel-wise loss and to obtain pixel-dense outputs [9]. The final layer is a fully convolutional layer with a kernel size of $[1 \times 1]$ and uses a sigmoid activation for pixel-based classification. The input to the network is of size $[256 \times 256]$ and the output is of same size.

FCN Training: We trained the network from scratch with training samples of knee OA radiographs from the OAI and MOST datasets. The ground truth for training the network are binary images with masks specifying the ROI: the knee joints. Figure 4 shows an instance of the binary masks: the ground truth. We generated the binary masks from manual annotations of knee OA radiographs using a fast annotation tool that we developed. The network was trained to minimize the total binary cross entropy between the predicted pixels and the ground truth. We used the adaptive moment estimation (Adam) optimizer [7], with default parameters, which we found to give faster convergence than standard SGD.

Extracting Knee Joints: We deduce the bounding boxes of the knee joints using simple contour detection from the output predictions of FCN. We extract the knee joints from knee OA radiographs using the bounding boxes. We upscale the bounding boxes from the output of the FCN that is of size $[256 \times 256]$ to the original size of each knee OA radiograph before we extract the knee joints so that the aspect ratio of the knee joints is preserved.

4.2 Quantifying Knee OA Severity Using CNNs

We investigate the use of CNNs trained from scratch using knee OA data and jointly train networks to minimize the classification and regression losses to further improve the assessment of knee OA severity.

Training CNN for Classification: The network contains mainly five layers of learned weights: four convolutional layers and one fully connected layer. Figure 5 shows the network architecture. As the training data is relatively scarce, we considered a lightweight architecture with minimal layers and the network has 5.4 million free parameters in total. After experimenting with the number of convolutional layers and other parameters, we find this architecture to be the best for classifying knee images. Each convolutional layer in the network is followed by batch normalization and a rectified linear unit activation layer (ReLU). After each convolutional stage there is a max pooling layer. The final pooling layer is followed by a fully connected layer and a softmax dense layer. To avoid over-fitting, we include a drop out layer with a drop out ratio of 0.2 after the last convolutional (conv4) layer and a drop out layer with a drop out ratio of 0.5 after the fully connected layer (fc5). We also apply an L2-norm weight regularization penalty of 0.01 in the last two convolutional layers (conv3 and conv4) and the fully connected layer (fc5). Applying a regularization penalty to other layers increases the training time whilst not introducing significant variation in the learning curves. The network was trained to minimize categorical cross-entropy loss using the Adam optimizer [7]. The inputs to the network are knee images of size [200 × 300]. We chose this size to approximately preserve the aspect ratio based on the mean aspect ratio (1.6) of all the extracted knee joints.

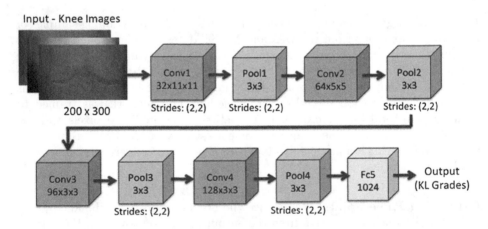

Fig. 5. The network architecture for classifying knee joint images.

Jointly Training CNN for Classification and Regression: In general, assessing knee OA severity is based on the multi-class classification of knee images and assigning KL grade to each distinct category [10,11,14,17]. As the disease is progressive in nature, we argued in our previous paper [1] that assigning a continuous grade (0–4) to knee images through regression is a better approach for quantifying knee OA severity. However, with this approach there is no ground truth of KL grades in a continuous scale to train a network directly for regression output. Therefore, we train networks using multi-objective convolutional learning [8] to optimize a weighted-ratio of two loss functions: categorical cross-entropy and mean-squared error. Mean squared error gives the network information about ordering of grades, and cross entropy gives information about the quantization of grades. Intuitively, optimizing a network with two loss functions provides a stronger error signal and it is a step to improve the overall quantification, considering both classification and regression results. After experimenting, we obtained the final architecture shown in Fig. 6. This network has six layers of learned weights: 5 convolutional layers and a fully connected layer, and approximately 4 million free parameters in total. Each convolutional layer is followed by batch normalization and a rectified linear activation (ReLU) layer. To avoid over-fitting this model, we include drop out ($p = 0.5$) in the fully connected layer (fc5) and L2 weight regularization in the fully connected layer (fc5) and the last stage of convolution layers (Conv3-1 and Conv3-2). We trained the model using stochastic gradient descent with *Nesterov* momentum and a learning rate scheduler. The initial learning rate was set to 0.001, and reduced by a factor of 10 if there is no drop in the validation loss for 4 consecutive epochs.

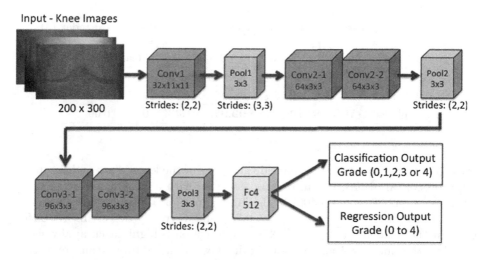

Fig. 6. The network architecture for simultaneous classification and regression.

5   Experiments and Results

5.1   Localizing the Knee Joints Using a FCN

We trained FCNs to automatically localize and extract the knee joints from knee OA X-ray images. We use the well-known Jaccard index to evaluate the detection result. The datasets are split into a training/validation set (70%) and test set (30%). The training and test samples from OAI dataset are 3,146 images and 1,300 images. The training and test samples from MOST dataset are 2,020 images and 900 images. First, we trained the network with training samples from OAI dataset and tested it with OAI and MOST datasets separately. Next, we increased our training samples by including the MOST training set and the test set is a combination of both OAI and MOST test sets. Before settling on the final architecture, we experimented by varying the number of convolution stages, the number of filters and kernel sizes in each convolution layer. The final network (shown in Fig. 4) was trained with the samples from both OAI and MOST datasets.

Evaluation: The automatic detection is evaluated using the well-known Jaccard index i.e. the intersection over Union (IoU) of the automatic detection and the manual annotation of each knee joint. For this evaluation, we manually annotated all the knee joints in both the OAI and MOST datasets using a fast annotation tool that we developed. Table 1 shows the number (percentage) of knee joint correctly detected based on the Jaccard index (J) values greater than 0.25, 0.5 and 0.75 along with the mean and the standard deviation of J. Table 1 also shows detection rates on the OAI and MOST test sets separately.

Table 1. Comparison of automatic detection based on the Jaccard Index (J)

Test data	$J \geq 0.25$	$J \geq 0.5$	$J \geq 0.75$	Mean	Std. dev
OAI	**100%**	**99.9%**	89.2%	0.83	0.06
MOST	99.5%	98.4%	85.0%	0.81	0.09
Combined OAI-MOST	99.9%	**99.9%**	**91.4%**	0.83	0.06

Results: Considering the anatomical variations of the knee joints and the imaging protocol variations, the automatic detection with a FCN is highly accurate with 99.9% (4,396 out of 4,400) of the knee joints for $J \geq 0.5$ and 91.4% (4,020 out of 4,400) of the knee joints for $J \geq 0.75$ being correctly detected. Section 5.3 gives further evidence that the FCN based detection is highly accurate by showing that the quantification results obtained with the automatically extracted knee joints gives results on par with manually segmented knee joints.

5.2   Classification of Knee OA Images Using a CNN

We use the same train-test split for localization and quantification to maintain uniformity in the pipeline and to enable valid comparisons of the results obtained across the various approaches. We include the right-left flip of each knee joint image to increase the training samples and this doubles the total number of training samples available. As an initial approach, we trained networks to classify manually annotated knee joint images. After experimenting, we obtained the final architecture shown in Fig. 5.

Results: we compare the classification results from our network to WND-CHARM, the multipurpose medical image classifier [11,16,17] that gave the previous best results for automatically quantifying knee OA severity. Table 2 shows the multi-class classification accuracy and mean-squared error of our network and WND-CHARM. The results show that our network trained from scratch for classifying knee OA images clearly outperforms WND-CHARM. Also these results show an improvement over our earlier reported methods [1] that used off-the-shelf networks such as VGG nets and the BVLC Reference CaffeNet for classifying knee OA X-ray images through transfer learning. These improvements are due to the lightweight architecture of our network trained from scratch with less (5.4 million) free parameters in comparison to 62 million free parameters of BVLC CaffeNet for the given small amount of training data. The off-the-shelf networks were trained using a large dataset like ImageNet containing millions of images, whereas our dataset contains much fewer ($\sim$10, 000) training samples. We show further improvements in the results for quantifying knee OA severity in the next section.

Table 2. Classification results of our network and WND-CHARM.

Method	Test data	Accuracy	Mean-squared error
Wndchrm	OAI	29.3%	2.496
Wndchrm	MOST	34.8%	2.112
Fine-tuned BVLC CaffeNet	OAI	57.6%	0.836
Our CNN trained from Scratch	OAI & MOST	**60.3%**	0.898

5.3   Jointly Trained CNN for Classification and Regression

The KL grades used to assess knee OA is a discrete scale, but knee OA is progressive in nature. We trained networks to predict the outcomes in a continuous scale (0–4) through regression. Even though we obtained low mean-squared error values for regression, the classification accuracy reduces when the continuous grades are rounded. Next, to obtain a better learning representation we trained networks that learn using a weighted ratio of two loss functions: categorical cross entropy for classification and mean-squared error for regression. We experimented with values from 0.2 to 0.6 for the weight of regression loss and we fixed the weight at 0.5 as this gave the optimal results. Figure 6 shows our network jointly trained for classification and regression of knee images. Figure 7 shows the learning curves of the network trained for joint classification and regression. The learning curves show a decrease in training and validation losses, and also an increase in training and validation accuracies over the training.

Comparing Manual and Automatic Localization: We present the classification and regression results obtained using both the manual and the automatic methods for localizing the knee joints in Tables 3 and 4. From the results, it is evident that the classification and regression of the knee joint images after automatic localization are comparable with the results after manual localization.

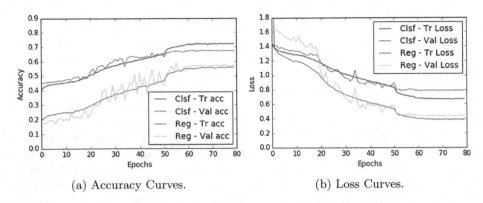

(a) Accuracy Curves.   (b) Loss Curves.

Fig. 7. (a) Training (Tr) and validation (Val) accuracy (acc), (b) Training and validation loss for joint classification (Clsf) and regression (Reg) training.

Table 3. Classification of knee joints after manual and automatic localization.

Method	Classification-Acc	Classification-MSE	Regression-MSE
Manual localization	**63.6%**	**0.706**	**0.503**
Automatic localization	61.9%	0.781	0.541

Table 4. Classification metrics after localizing knee joints.

Grade	Manual localization			Automatic localization		
	Precision	Recall	F_1	Precision	Recall	F_1
0	0.66	0.87	0.75	0.64	0.88	0.74
1	0.39	0.06	0.10	0.33	0.02	0.04
2	0.52	0.60	0.56	0.50	0.57	0.53
3	0.75	0.72	0.73	0.73	0.73	0.73
4	0.78	0.78	0.78	0.75	0.66	0.70
Mean	0.60	0.64	0.59	0.57	0.62	0.56

Comparing Joint Training with Classification Only: From the results shown in Tables 2 and 3, the network trained jointly for classification and regression gives higher multi-class classification accuracy of 63.4% and lower mean-squared error 0.661 in comparison to the previous network trained only for classification with multi-class classification accuracy 60.3% and mean-squared error 0.898. Table 5 shows the precision, recall, F_1 score, and area under curve (AUC) of the network trained jointly for classification and regression and the network trained only for classification. These results show that the network jointly trained for classification and regression learns a better representation in comparison to the previous network trained only for classification.

Table 5. Metrics comparing joint training for classification and regression to network trained for classification only.

Grade	Joint training for Clsf & Reg				Training for only Clsf			
	Precision	Recall	F_1	AUC	Precision	Recall	F_1	AUC
0	0.68	0.80	0.74	0.87	0.63	0.82	0.71	0.83
1	0.32	0.15	0.20	0.71	0.25	0.04	0.06	0.66
2	0.53	0.63	0.58	0.82	0.47	0.57	0.51	0.78
3	0.78	0.74	0.76	0.96	0.76	0.71	0.73	0.94
4	0.81	0.75	0.78	0.99	0.78	0.77	0.77	0.99
Mean	0.61	0.63	0.61	-	0.56	0.60	0.56	-

Error Analysis: From the classification metrics (Table 5), the confusion matrix (Fig. 8) and the receiver operating characteristics (Fig. 9), it is evident that classification of successive grades is challenging, and in particular classification metrics for grade 1 have low values in comparison to the other Grades.

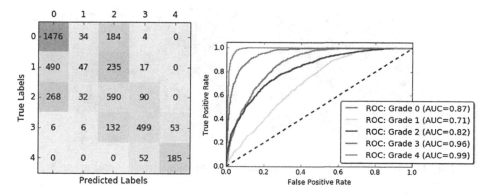

Fig. 8. Confusion matrix.   **Fig. 9.** ROC for joint training.

Figure 10 shows some examples of mis-classifications: grade 1 knee joints predicted as grade 0, 2, and 3. Figure 11 shows the mis-classifications of knee joints categorized as grade 0, 2 and 3 predicted as grade 1. These images show minimal variations in terms of joint space width and osteophytes formation, making them challenging to distinguish. Even for the more serious mis-classifications in Fig. 12, e.g. grade 0 predicted as grade 3 and vice versa, do not show very distinguishable variations.

Even though the KL grades are used for assessing knee OA severity in clinical settings, there has been continued investigation and criticism over the use of KL grades as the individual categories are not equidistant from each other [3,4]. This could be a reason for the low multi-class classification accuracy in the automatic quantification. Using OARSI readings instead of KL grades could

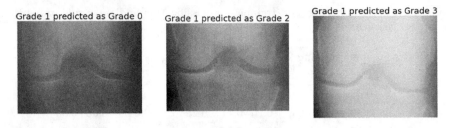

Fig. 10. Mis-classifications: grade 1 joints predicted as grade 0, 2, and 3

Fig. 11. Mis-classifications: other grade knee joints predicted as grade 1

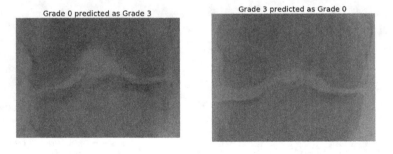

Fig. 12. An instance of more severe mis-classification: grade 0 and grade 3

possibly provide better results for automatic quantification as the knee OA features such as joint space narrowing, osteophytes formation, and sclerosis are separately graded.

6   Conclusion

We proposed new methods to automatically localize knee joints using a fully convolutional network and quantified knee OA severity through a network jointly trained for multi-class classification and regression where both networks were trained from scratch. The FCN based method is highly accurate in comparison to the previous methods. We showed that the classification results obtained with automatically localized knee joints is comparable with the manually segmented knee joints. There is an improvement in the multi-class classification accuracy, precision, recall, and F_1 score of the jointly trained network for classification and regression in comparison to the previous method. The confusion matrix and other metrics show that classifying Knee OA images conditioned on KL grade 1 is challenging due to the small variations, particularly in the consecutive grades from grade 0 to grade 2.

Future work will focus on training an end-to-end network to quantify the knee OA severity integrating the FCN for localization and the CNN for classification. It will be interesting to investigate the human-level accuracy involved in assessing the knee OA severity and comparing this to the automatic quantification methods. This could provide insights to further improve fine-grained classification.

Acknowledgment. This publication has emanated from research conducted with the financial support of Science Foundation Ireland (SFI) under grant numbers SFI/12/RC/2289 and 15/SIRG/3283.

The OAI is a public-private partnership comprised of five contracts (N01-AR-2-2258; N01-AR-2-2259; N01-AR-2-2260; N01-AR-2-2261; N01-AR-2-2262) funded by the National Institutes of Health, a branch of the Department of Health and Human Services, and conducted by the OAI Study Investigators. Private funding partners include Merck Research Laboratories; Novartis Pharmaceuticals Corporation, GlaxoSmithKline; and Pfizer, Inc. Private sector funding for the OAI is managed by the Foundation for the National Institutes of Health.

The MOST is comprised of four cooperative grants (Felson – AG18820; Torner – AG18832; Lewis – AG18947; and Nevitt – AG19069) funded by the National Institutes of Health, a branch of the Department of Health and Human Services, and conducted by MOST study investigators. This manuscript was prepared using MOST data and does not necessarily reflect the opinions or views of MOST investigators.

References

1. Antony, J., McGuinness, K., Connor, N.E., Moran, K.: Quantifying radiographic knee osteoarthritis severity using deep convolutional neural networks. In: Proceedings of the 23rd International Conference on Pattern Recognition. IEEE (2016)
2. Chatfield, K., Simonyan, K., Vedaldi, A., Zisserman, A.: Return of the devil in the details: delving deep into convolutional nets. In: Proceedings of British Machine Vision Conference (2014)
3. Emrani, P.S., Katz, J.N., Kessler, C.L., Reichmann, W.M., Wright, E.A., McAlindon, T.E., Losina, E.: Joint space narrowing and Kellgren-Lawrence progression in knee osteoarthritis: an analytic literature synthesis. Osteoarthr. Cartil. **16**(8), 873–882 (2008)
4. Hart, D., Spector, T.: Kellgren & Lawrence grade 1 osteophytes in the knee-doubtful or definite? Osteoarthr. Cartil. **11**(2), 149–150 (2003)
5. Jia, Y., Shelhamer, E., Donahue, J., Karayev, S., Long, J., Girshick, R., Guadarrama, S., Darrell, T.: Caffe: convolutional architecture for fast feature embedding. In: Proceedings of the ACM International Conference on Multimedia, pp. 675–678 (2014)
6. Karayev, S., Trentacoste, M., Han, H., Agarwala, A., Darrell, T., Hertzmann, A., Winnemoeller, H.: Recognizing image style. arXiv preprint arXiv:1311.3715 (2013)

7. Kingma, D., Ba, J.: Adam: A method for stochastic optimization. arXiv preprint arXiv:1412.6980 (2014)
8. Liu, S., Yang, J., Huang, C., Yang, M.H.: Multi-objective convolutional learning for face labeling. In: Proceedings of the IEEE Conference on Computer Vision and Pattern Recognition, pp. 3451–3459 (2015)
9. Long, J., Shelhamer, E., Darrell, T.: Fully convolutional networks for semantic segmentation. In: Proceedings of the IEEE Conference on Computer Vision and Pattern Recognition, pp. 3431–3440 (2015)
10. Oka, H., Muraki, S., Akune, T., Mabuchi, A., Suzuki, T., Yoshida, H., Yamamoto, S., Nakamura, K., Yoshimura, N., Kawaguchi, H.: Fully automatic quantification of knee osteoarthritis severity on plain radiographs. Osteoarthr. Cartil. **16**(11), 1300–1306 (2008)
11. Orlov, N., Shamir, L., Macura, T., Johnston, J., Eckley, D.M., Goldberg, I.G.: WND-CHARM: multi-purpose image classification using compound image transforms. Pattern Recogn. Lett. **29**(11), 1684–1693 (2008)
12. Park, H.J., Kim, S.S., Lee, S.Y., Park, N.H., Park, J.Y., Choi, Y.J., Jeon, H.J.: A practical MRI grading system for osteoarthritis of the knee: association with Kellgren-Lawrence radiographic scores. Eur. J. Radiol. **82**(1), 112–117 (2013)
13. Russakovsky, O., Deng, J., Su, H., Krause, J., Satheesh, S., Ma, S., Huang, Z., Karpathy, A., Khosla, A., Bernstein, M., et al.: ImageNet large scale visual recognition challenge. Int. J. Comput. Vis. **115**(3), 211–252 (2015)
14. Shamir, L., Ling, S.M., Scott, W., Hochberg, M., Ferrucci, L., Goldberg, I.G.: Early detection of radiographic knee osteoarthritis using computer-aided analysis. Osteoarthr. Cartil. **17**(10), 1307–1312 (2009)
15. Shamir, L., Ling, S.M., Scott, W.W., Bos, A., Orlov, N., Macura, T.J., Eckley, D.M., Ferrucci, L., Goldberg, I.G.: Knee X-ray image analysis method for automated detection of osteoarthritis. IEEE Trans. Biomed. Eng. **56**(2), 407–415 (2009)
16. Shamir, L., Orlov, N., Eckley, D.M., Macura, T., Johnston, J., Goldberg, I.: WND-CHARM: multi-purpose image classifier. Astrophysics Source Code Library (2013)
17. Shamir, L., Orlov, N., Eckley, D.M., Macura, T., Johnston, J., Goldberg, I.G.: Wndchrm-an open source utility for biological image analysis. Source Code Biol. Med. **3**(1), 13 (2008)
18. Simonyan, K., Zisserman, A.: Very deep convolutional networks for large-scale image recognition. arXiv preprint arXiv:1409.1556 (2014)
19. Yang, S.: Feature engineering in fine-grained image classification. Ph.D. thesis, University of Washington (2013)
20. Yoo, T.K., Kim, D.W., Choi, S.B., Park, J.S.: Simple scoring system and artificial neural network for knee osteoarthritis risk prediction: a cross-sectional study. PLoS ONE **11**(2), e0148724 (2016)

High Accuracy Predictive Modelling for Customer Churn Prediction in Telecom Industry

R. Prashanth[✉], K. Deepak, and Amit Kumar Meher

Flytxt Mobile Solutions Pvt. Ltd.,
Carnival Technopark (Formerly Leela Infopark), Technopark,
Trivandrum 695581, Kerala, India
{prashanth.ravindran,deepak.k,amit.meher}@flytxt.com

Abstract. Churn prediction is an important factor to consider for Customer Relationship Management (CRM). In this study, statistical and data mining techniques were used for churn prediction. We use linear (logistic regression) and non-linear techniques of Random Forest and Deep Learning architectures including Deep Neural Network, Deep Belief Networks and Recurrent Neural Networks for prediction. This is the first time that a comparative study of conventional machine learning methods with deep learning techniques have been carried out for churn prediction. It is observed that non-linear models performed the best. Such predictive models have the potential to be used in the telecom industry for making better decisions and customer management.

Keywords: Churn prediction · Deep learning · Data mining · Statistical testing

1 Introduction

Customer retention is one of the important factors in the telecom industry with regard to Customer Relationship Management (CRM). Due to saturated markets and intensive competition, most companies realize that their existing customers is their most valuable asset, and it is more beneficial to keep and satisfy existing customers than to constantly attract new customers who are characterized by a high attrition rate [1–4]. The need to identify customers that are most prone to switching, is of high priority. It has been shown that a small change in retention rate can result in significant changes in the revenue generation. It is crucial to develop an effective and accurate customer churn model in order to efficiently manage customer relationship.

Data mining techniques carries huge potential in churn prediction [5–8]. A data mining process can be used to describe (i.e. discover interesting patterns or relationships in the data), and predict (i.e. predict or classify the behaviour of the model based on the input data). Over the time, various predictive modelling techniques have been attempted and used, ranging from logistic regression

© Springer International Publishing AG 2017
P. Perner (Ed.): MLDM 2017, LNAI 10358, pp. 391–402, 2017.
DOI: 10.1007/978-3-319-62416-7_28

to more complex ensemble based methods. These data mining techniques can effectively assist in selecting customers who are most prone to churn. [1] used hierarchical multiple kernel support vector machine for churn prediction. [2] used a dataset containing 2000 subscribers and observed that a combination of logistic regression and perceptron algorithm gave the best results. [3] used a dataset of 3333 customers and used neural networks, SVMs and Bayesian networks for predictive modelling. [4] used a dataset involving 7190 customers for training a predictive model. [5] used artificial neural networks (ANN) on a dataset of 159 samples to get the best results. [6] use AntMiner+ (a technique involving Ant Colony Optimization) and Active learning based approach (ALBA) with support vector machine for churn prediction. [7] used a dataset containing 51306 subscriber information and used hybrid artificial neural networks for analysis. [8] made a comparative study of different machine learning problems on a churn dataset containing 5000 samples and observed that SVM with polynomial kernel using AdaBoost gave the best performance. Most of these studies used a smaller dataset to make predictive models. In recent years, deep learning has become one of the biggest trends due to its huge potential in big data processing [9]. In contrast to most conventional machine learning methods, which are considered using shallow-structured learning architectures, deep learning refers to machine learning techniques that use supervised and/or unsupervised strategies to automatically learn hierarchical representations in deep architectures. In this study, for the first time, we perform a comparative study of conventional techniques with deep learning techniques for churn prediction using a large dataset.

2   Materials and Methods

A flow chart of the process followed is as shown in the Fig. 1. Description of the process followed is explained in following subsections.

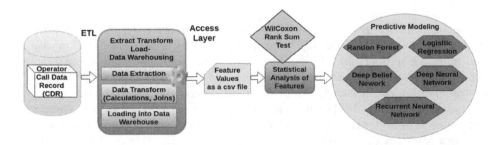

Fig. 1. Flowchart of the process involved

2.1   Data Extraction

For the study, we use a raw data provided by a major Asian telecom service provider for analysis. Raw data is in the form of Call Data Record (CDR) which contains telecommunication transaction data such as metadata of the data fields that describes specific instance of the data describing the duration of the call, start and end time of call, mobile originated call (outgoing call), mobile terminated call (incoming call) and session download (total volume of downloaded data in a data session).

Extract Transform Load (ETL) is used to migrate data from one database to another database to form data marts. Extract is the process of reading the CDR files from database and storing in the big data platform. Transformation is the process of converting the data from its previous form of raw unstructured data into form of structured data by defining parameter specification using rules, lookup tables or combining with other data. *From domain experts, a total of 36 features were chosen to be extracted from the huge set of CDR files.* Load is the process of writing data into the target database which can be accessed for further analysis for developing predictive models.

2.2   Dataset Details

The extracted data volume had information (36 features) of 337817 subscribers out of which 64599 were churners and the rest were non-churners. A brief description of these features is given below:

1. **Days_Last_Usage:** Days since last used
2. **Days_Last_Recharge:** Days since last recharged
3. **Average_Recharge_Count:** No. of recharges made in last 3 months
4. **Last_Recharge_MRP:** Last recharge value
5. **Average_Recharge_MRP:** Average Maximum Retail Price (MRP) of recharge in last 3 months
6. **AON:** Age on network
7. **Voice_OG_Call_Days_M1:** No. of days with outgoing calls in month 1
8. **Voice_OG_Call_Days_M2:** No. of days with outgoing calls in month 2
9. **Voice_OG_Call_Days_M3:** No. of days with outgoing calls in month 3
10. **Voice_IC_Call_Days_M1:** No. of days with incoming calls in month 1
11. **Voice_IC_Call_Days_M2:** No. of days with incoming calls in month 2
12. **Voice_IC_Call_Days_M3:** No. of days with incoming calls in month 3
13. **OG_MOU_M1:** Total outgoing minutes of usage in month 1
14. **OG_MOU_M2:** Total outgoing minutes of usage in month 2
15. **OG_MOU_M3:** Total outgoing minutes of usage in month 3
16. **Total_Decrement_M1:** Total balance decrement in month 1
17. **Total_Decrement_M2:** Total balance decrement in month 2
18. **Total_Decrement_M3:** Total balance decrement in month 3
19. **On_Net_Share_M1:** On-net share in month 1
20. **On_Net_Share_M2:** On-net share in month 2
21. **On_Net_Share_M3:** On-net share in month 3
22. **STD_Contribution_M1:** STD contribution in month 1
23. **STD_Contribution_M2:** STD contribution in month 2
24. **STD_Contribution_M3:** STD contribution in month 3

25. Max_Days_Between_Recharge_M1: Maximum days between successive recharges for month 1

26. Max_Days_Between_Recharge_M2: Maximum days between successive recharges for month 2

27. Max_Days_Between_Recharge_M3: Maximum days between successive recharges for month 3

28. Max_Days_Between_Recharge.3_Months: Maximum days between successive recharges for 3 months

29. Median_Delay_Between _Recharge_M1: Median days between successive recharges for month 1

30. Median_Delay_Between_Recharge_M2: Median days between successive recharges for month 2

31. Median_Delay_Between_Recharge_M3: Median days between successive recharges for month 3

32. Median_Delay_Between_Recharge.3_Months: Median days between successive recharges for 3 months

33. Max_Consistent_Non_Usage_Days_M1: Maximum days between successive usages for month 1

34. Max_Consistent_Non_Usage_Days_M2: Maximum days between successive usages for month 2

35. Max_Consistent_Non_Usage_Days_M3: Maximum days between successive usages for month 3

36. Max_Consistent_Non_Usage_Days.3_Months: Maximum days between successive usages for 3 months

Table 1 shows the list of features along with z-statistic and p-value from two-sided Wilcoxon rank sum test.

2.3 Predictive Modelling Methodologies

Various techniques have been used and compared for the study including general linear model (logistic regression) [8], Random forest [10], and deep learning techniques of Deep Neural Networks [9,11], Deep Belief Networks [12] and Recurrent Neural Networks [13]. A brief description of the methods is given below. The parameters for the classifiers were estimated using grid search and cross validation using about 10% the data to eliminate the chances of overfitting during predictive modelling.

Logistic Regression. Logistic regression attempts to fit a logistic model to describe the relationship between a dichotomous variable and a set of independent variables [8]. For model fitting, regression coefficients are estimated using maximum likelihood estimation. Regularization is applied to reduce the chances of over-fitting and to obtain the best possible model. The elastic net penalty is used for parameter regularization which gives rise to two tunable parameters in the logistic model, namely, alpha and lambda. Lambda controls the degree of

Table 1. Study of statistical significance of features

Features	Z-stat	p-value*
Voice_OG_Call_Days_M3	95	≈0
Voice_IC_Call_Days_M3	87.69	≈0
OG_MOU_M3	82.95	≈0
Total_Decrement_M1	78.37	≈0
Voice_OG_Call_Days_M2	70.05	≈0
Voice_IC_Call_Days_M2	67.46	≈0
Voice_OG_Call_Days_M1	64.71	≈0
Voice_IC_Call_Days_M1	62.39	≈0
Average_Recharge_Count	62.09	≈0
OG_MOU_M2	61.58	≈0
Total_Decrement_M2	59.69	≈0
OG_MOU_M1	58.24	≈0
Total_Decrement_M3	56.41	≈0
On_Net_Share_M3	50.57	≈0
Max_Days_Between_Recharge.3_Months	47.15	≈0
Max_Days_Between_Recharge_M2	46.35	≈0
Max_Days_Between_Recharge_M1	46.25	≈0
Median_Delay_Between_Recharge_M2	45.96	≈0
Median_Delay_Between_Recharge_M3	45.75	≈0
Median_Delay_Between_Recharge_M1	45.73	≈0
Median_Delay_Between_Recharge.3_Months	45.55	≈0
Last_Recharge_MRP	44.25	≈0
Average_Recharge_MRP	43.49	≈0
Max_Days_Between_Recharge_M3	42.75	≈0
On_Net_Share_M2	42.48	≈0
On_Net_Share_M1	42.08	≈0
AON	30.95	≈0
STD_Contribution_M3	20.98	≈0
STD_Contribution_M1	17.41	≈0
STD_Contribution_M2	17.15	≈0
Max_Consistent_Non_Usage_Days_M3	−32.19	≈0
Max_Consistent_Non_Usage_Days_M1	−32.22	≈0
Max_Consistent_Non_Usage_Days_M2	−34.52	≈0
Days_Last_Recharge	−36.21	≈0
Max_Consistent_Non_Usage_Days.3_Months	−46.62	≈0
Days_Last_Usage	−90.13	≈0

*≈0 indicates a value ≪0.

regularization (0 for no regularization and infinity for ignoring all variables) and alpha controls the degree of mix of $l1$ and $l2$ regularizations (0 for only $l2$ and 1 for only $l1$). The optimization equation is as given below.

$$\underset{w}{\text{minimize}} \quad \sum L(w^T x, y) + \lambda h(w) \tag{1}$$

where $L(w^T x, y)$ is the logistic loss and $h(w)$ is the elastic net regularization penalty which is the weighted sum of the $l1$ and $l2$ norms of the coefficients vector.

These models are linear in nature. Due to which they have the limitation that they cannot handle non-linearities in the data. Here, we also use non-linear techniques that are briefly described below.

Random Forest. Random forest is an ensemble technique which is based on bootstrap aggregation or bagging and random selection of features in learning [10]. It fits a number of decision trees on various sub-samples of the dataset and then takes into consideration the decision from all the trees. The sub-sample size is always the same as the original input sample size but the samples are drawn with replacement. This process improves predictive accuracy and controls over-fitting. The final outcome is made by taking the majority vote of the predictions from all the individual trees. Split at a node is based on the Gini impurity criterion which evaluates the quality of a split for a feature. The Gini impurity value used for splitting is computed using the below formula

$$GI = 1 - \sum_i p(i)^2 \tag{2}$$

where $p(i)$ is the fraction of items with class i that reach the node.

Along with classification, Random forest also provides feature importance estimates based on out-of-bag error computation. The importance of feature is computed by permuting its value in the training data and the difference in out-of-bag error is calculated. Higher the difference, higher the importance of the feature.

Deep Neural Networks. Artificial neural networks with multiple hidden layers can be used to accomplish deep learning tasks [11]. The architecture of a deep neural network with 2 hidden layers is as shown in Fig. 2. The loss function is in the form of cross-entropy (log-loss) which is minimized using stochastic gradient descent computed via backpropagation [14].

The initialization scheme used in the network is the uniform adaptive which is an optimized initialization based on the size of the network. The activation function is the rectified linear, or rectifier that has shown high performance on complex tasks and is a more biologically accurate model of neuron activations [14]. The rectifier activation function is given by $f(x) = max(0, x)$.

The loss function $L(W, B|j)$ is minimized using stochastic gradient descent (SGD) with the gradient $\Delta L(W, B|j)$ computed via backpropagation [14]. SGD involves the following update equations for all weights $w \in W$ and biases $b \in B$.

$$w^* = w - \alpha \frac{\delta L(W, B|j)}{\delta w}$$
$$b^* = b - \alpha \frac{\delta L(W, B|j)}{\delta b} \tag{3}$$

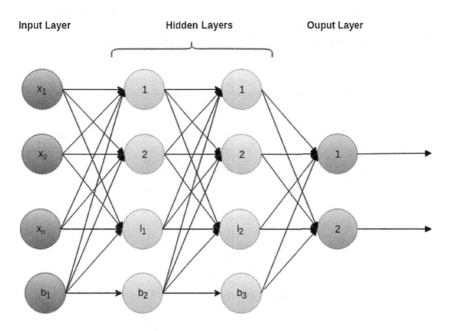

Fig. 2. Basic architecture of the deep neural network used

Deep Belief Networks. Deep Belief Networks is a deep learning architecture formed by stacking Restricted Boltzmann Machines (RBM), which is a genera-tive stochastic neural network that can learn a probability distribution over its set of inputs, and trained in a greedy manner [12,15]. RBMs consist of hidden, visible and bias layer. Here, the connections between the visible and hidden layers are undirected and fully connected. In the training process, there is an unsuper-vised layer-wise pre-training followed by supervised fine-tuning via supervised gradient descent as in deep multilayer perceptron. The joint distribution between observed vector x and the k hidden layers $(h^1, h^2, .., h^k)$ as follows:

$$P(x, h^1, h^2, .., h^k) = \left(\prod_{l=0}^{k-2} P(h^l|h^{l+1}) \right) P(h^{k-1}, h^k) \tag{4}$$

where $x = h^0$, $P(h^l - 1|h^l)$ is a conditional distribution for the visible units conditioned on the hidden units of the RBM at level k, and $P(h^{k-1}, h^k)$ is the visible-hidden joint distribution in the top-level RBM.

Recurrent Neural Networks. Recurrent neural network (RNN) are class of neural networks with loops in them, allowing information to persist. In the analysis, Long Short Term Memory (LSTM) networks which are a special kind of RNN, that are capable of learning long-term dependencies [13]. The basic architecture of an LSTM is shown in Fig. 3. For training, gradient descent via technique called backpropagation through time, can be used to change each weight in proportion to the derivative of the error with respect to that weight, provided the non-linear activation functions are differentiable.

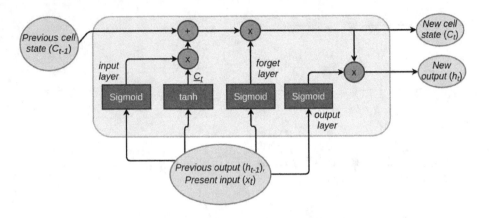

Fig. 3. Basic architecture of an LSTM network

2.4   Parameter Selection and Performance Metrics

The parameters of the classifiers were computed based on cross-validation on about 10% of the data. After parameter selection, predictive models were trained and evaluated using the complete data through 5-fold cross validation. Performance metrics used to evaluate the classifiers are Accuracy, Sensitivity, Specificity and Area Under the ROC curve (AUC).

3   Results and Discussion

Figure 4 shows the variation of cross-validation errors (log-loss) for different combinations of parameters for each classifier. Different sets of values (grid search) for the tunable parameters were tested to find the best set of classifier parameters. Table 2 shows the best parameters obtained for different classifiers. The figure shows the log-loss errors obtained during parameter optimization. The error values are sorted in the ascending order for the sake of visualizing the variation of errors when the parameters are changing. The y-axis represents an iteration. For instance, in GLM, there are two tunable parameters, namely alpha and lambda. The grid values for alpha and lambda parameters were [0.01, 0.3, 0.5] and [1e–5, 1e–6, 1e–7, 1e–8], respectively. The first point, in the plot corresponding to GLM, the y-axis represents a log-loss of 0.295304 and x-axis represents an iteration corresponding to alpha = 0.3 and lambda = 1e−5. Best parameters for other classifiers were similarly chosen based on the log loss errors. Class balancing was taken into account for predictive modelling.

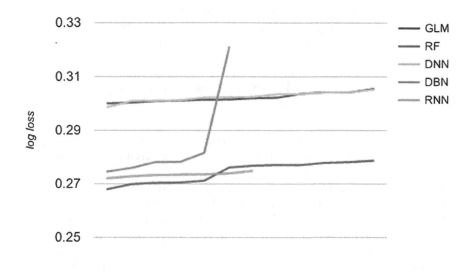

Fig. 4. Plot of cross-validation errors for various classifiers for parameter selection. GLM, RF, DNN, DBN and RNN represent General Linear Model (logistic regression), Random Forest, Deep Neural Networks, Deep Belief Networks and Recurrent Neural Networks, respectively.

Table 3 shows the performances obtained for various classifiers. It is observed that the non-linear techniques performed better than the linear one with respect to all performance metrics used. All the non-linear techniques almost gave a

Table 2. Parameters used for various classifiers as obtained via cross-validation

Method	Parameters
General Linear Model	alpha = 0.3, lambda = 1e−05
Random forest	n_tress = 100, max_depth = 15
Deep Neural Networks	activation = 'Rectifier', hidden = [32, 32], epochs = 10.4, l1 = 0.001
Deep Belief Networks	Hidden = 100, learn_rates = 0.1, learn_rate_decays = 0.5, epochs = 50
Recurrent Neural Network (LSTM)	Output_dim = 100, init = 'glorot_uniform', inner_init = 'orthogonal', forget_bias_init = 'one', activation = 'tanh', inner_activation = 'hard_sigmoid'

Table 3. Performance measures for various classifiers used

Metrics	GLM	RF	DNN	DBN	RNN
Accuracy	87.01	88.36	87.55	87.79	87.75
AUC	90.48	92.81	91.93	92.33	92.48
Specificity	93.86	94.82	93.36	93.12	92.95
Sensitivity	58.04	61.06	62.97	64.77	65.76

where GLM, RF, DNN, DBN and RNN stands
for general linear model (logistic regression),
Random Forest, Deep Neural Networks, Deep
Belief Networks and Recurrent Neural Network, respectively.

comparable performance. In terms of accuracy, AUC and Specificity, Random
Forest gave the best performance, and in terms of Sensitivity, Recurrent Neural
Networks gave the best performance.

Among the deep learning techniques, RNNs performed better than the DBN
and DNN. This may be due to the fact that RNNs can use their internal memory
to process arbitrary sequences of inputs. LSTM, the RNN architecture used
in the study, which is augmented by recurrent gates called forget gates can
prevent backpropagated errors from vanishing or exploding. Instead, error flows
backwards through unlimited numbers of virtual layers in LSTM RNNs unfolded
in space. That is, LSTM can learn "Very Deep Learning" task that require
memories of events that happened many discrete time steps ago. DBNs on the
other hand, might have worked better than the DNN due to the pre-training
method that better initializes the weights of the network.

It is to be noted that the classification problem is challenging as observed
from the sensitivity obtained from the logistic model. This sensitivity is greatly
increased using non-linear techniques, especially from the Recurrent Neural
Networks.

Figure 5 shows the plot of feature importance obtained by Random Forest
technique. It is observed that the features corresponding to days since last usage,
days since last recharge, outgoing calls and incoming call days were coming
among the most important features, and features such as STD contribution
were the least important. This feature importance estimate correlates well with
the domain experts opinion.

In this work, although only one dataset was used for predictive modelling,
we have used and compared multiple methods including the methods used in the
state-of-the-art studies for predictive modelling. A possible future work to the
present study is to carry out modelling using the approaches as discussed in the
article, with other datasets.

This paper presents a variety of methods for predictive modelling for customer churn. But it is to be noted that these predictive modelling methodologies are not specific for the customer churn prediction, and could be applied to
other predictive problems as well in the similar manner. The parameters for the
classifiers have to be appropriately estimated in that case.

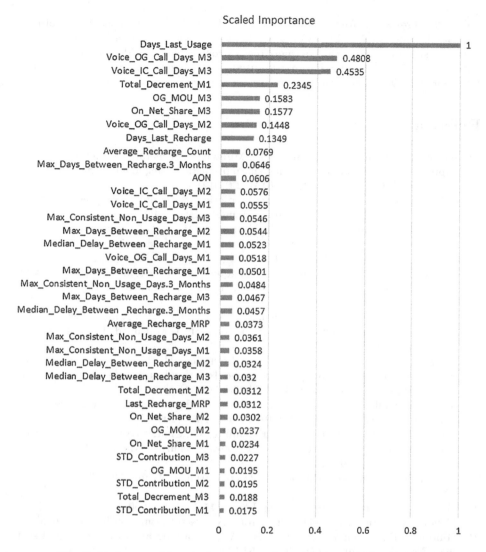

Fig. 5. Estimate of feature importance from Random Forest method

4 Conclusion

This study carried out predictive modelling for customer churn prediction which is crucial factor in customer relationship management. Machine learning techniques used in the study include general linear modelling, Random forest, and deep learning. The non-linear techniques performed better than the linear one. Among the non-linear techniques of Random forest, and deep learning, both gave comparable performance. These models have the potential to be used in telecom industry for making better decisions for churn management.

Acknowledgement. We thank our colleagues Jobin Wilson and Jisha Nelliat of Fly-txt Mobile Solution Pvt Ltd who provided insight and helped in simulations that greatly assisted this research. We thank the telecom service provider for providing the data. We would also like to show our gratitude to Professor Santanu Chaudhury, IIT Delhi for sharing his pearls of wisdom with us during the course of this research.

References

1. Chen, Z.Y., Zhen-Yu, C., Zhi-Ping, F., Minghe, S.: A hierarchical multiple kernel support vector machine for customer churn prediction using longitudinal behavioral data. Eur. J. Oper. Res. **223**(2), 461–472 (2012)
2. Olle, G., Georges, O.: A hybrid churn prediction model in mobile telecommunication industry. Int. J. e-Educ. e-Bus. e-Manag. e-Learn. **4**(1), 55 (2014)
3. Brandusoiu, I., Ionut, B., Gavril, T., Horia, B.: Methods for churn prediction in the pre-paid mobile telecommunications industry. In: International Conference on Communications (COMM) (2016)
4. Lu, N., Ning, L., Hua, L., Jie, L., Guangquan, Z.: A customer churn prediction model in telecom industry using boosting. IEEE Trans. Ind. Inf. **10**(2), 1659–1665 (2014)
5. Ismail, M.R., Awang, M.K., Rahman, M.N.A., Mokhairi, M.: A multi-layer perceptron approach for customer churn prediction. Int. J. Multimed. Ubiquitous Eng. **10**(7), 213–222 (2015)
6. Verbeke, W., Wouter, V., David, M., Christophe, M., Bart, B.: Building comprehensible customer churn prediction models with advanced rule induction techniques. Expert Syst. Appl. **38**(3), 2354–2364 (2011)
7. Tsai, C.F., Chih-Fong, T., Yu-Hsin, L.: Customer churn prediction by hybrid neural networks. Expert Syst. Appl. **36**(10), 12547–12553 (2009)
8. Vafeiadis, T., Diamantaras, K.I., Sarigiannidis, G., Chatzisavvas, K.C.: A comparison of machine learning techniques for customer churn prediction. Simul. Model. Pract. Theor. **55**, 1–9 (2015)
9. LeCun, Y., Bengio, Y., Hinton, G.: Deep learning. Nature **521**(7553), 436–444 (2015)
10. Breiman, L.: Random forests. Mach. Learn. **45**(1), 5–32 (2001)
11. Candel, A., Parmar, v., LeDell, E., Arora, A.: Deep learning with H2O (2015)
12. Hinton, G.E., Osindero, S., Teh, W.Y.: A fast learning algorithm for deep belief nets. Neural Comput. **18**(7), 1527–1554 (2006)
13. Schmidhuber, J.: Deep learning in neural networks: an overview. Neural Netw. **61**, 85–117 (2015)
14. LeCun, Y., Bottou, L., Orr, G.B., Müller, K.-R.: Efficient BackProp. In: Orr, G.B., Müller, K.-R. (eds.) Neural Networks: Tricks of the Trade. LNCS, vol. 1524, pp. 9–50. Springer, Heidelberg (1998). doi:10.1007/3-540-49430-8_2
15. Hinton, G.E., Salakhutdinov, R.R.: Reducing the dimensionality of data with neural networks. Science **313**(5786), 504–507 (2006)

You Are What You Tweet:
A New Hybrid Model for Sentiment Analysis

Arthur Huang[✉], David Ebert, and Parker Rider

Department of Engineering and Computer Science, Tarleton State University,
Stephenville, TX 76402, USA
ahuang@tarleton.edu

Abstract. The rise of social media has provided new opportunities to study human emotions through self-reported information such as text, emojis/emoticons, and geo-locations. Research has shown that hybrid models which integrate lexicons and machine learning methods can improve the accuracy of sentiment prediction. We propose the Normalized Difference Sentiment Index (NDSI) to identify frequently-occurring words that are predictive of positive or negative sentiments. Furthermore, we propose *e-senti*, a new hybrid model which combines 3 attributes (lexicons, a new NDSI word rank list, and tweet features) into a random forest classifier. We contribute to the methodology of sentiment analysis by introducing a model that is easy to implement, efficient, and accurate. We compare four widely used lexicons and find the AFINN lexicon most effective and efficient for our model. We test the *e-senti* model based on the sentiment140 data and tweets from Los Angeles County, California. Our results show that the maximum accuracy for the sentiment140 data is 86.1% and for our Los Angeles County data is 74.6%, outperforming most existing methods. Our future work will link the geo-tagged sentiment data to land use data to reveal how emotions and the built environment are connected.

1 Introduction

Emotions permeate our daily activities. Emotions often happen in the context of social and natural environment and can affect our behavior, daily decisions, and various forms of interactions with others [8]. Therefore, assessing emotions is critical for understanding human nature and the human-environment relationship. Traditionally, emotions were analyzed by neuro-psychologists and neuro-psychiatrists through facial expressions of speech [41], physiological signals (such as heart-rate [39] and functional MRI images [32]). While these studies are valuable in revealing the fundamental mechanisms of emotional activation and expression, they face the challenges of limited sample sizes and relatively high cost of experiment. The fast development of social media and mobile devices has brought much attention to large-scale user-generated data from social media platforms. Social media generates large volumes of real-time social signals which can offer new insight on human behavior and emotions.

© Springer International Publishing AG 2017
P. Perner (Ed.): MLDM 2017, LNAI 10358, pp. 403–416, 2017.
DOI: 10.1007/978-3-319-62416-7_29

There has been a growing interest in exploring text mining methods to assess emotions based on social media data. This research contributes to the methodology of sentiment analysis by proposing a new hybrid model *e-senti* to efficiently identify the sentiments of large volumes of tweets. We propose a new Normalized Difference Sentiment Index (NDSI) to identify words highly predictive of sentiments and incorporate it into our *e-senti* hybrid model. Our *e-senti* model combines lexicons, a new NDSI word rank list, and tweet features into a random forest classifier. We test this model based on the sentiment140 data and geo-tagged tweets from Los Angeles County, California.

The rest of this paper is organized as follows: we begin with a summary of related work in sentiment analysis. Following a description of our data set, we propose a framework for collecting, cleaning, analyzing, and assessing sentiments from social media. We then describe our hybrid *e-senti* model. Finally, we analyze the results, compare our results with other models, and conclude this paper.

2   Related Work

Sentiment analysis refers to the study of individuals' emotions, opinions, and attitudes toward services, products, and topics by detecting, extracting, and analyzing information from user-shared content [22,25]. The basic tasks of sentiment analysis include polarity classification and agreement detection [6]. Polarity classification refers to classifying a single issue as a positive or negative sentiment [25,40]. Agreement detection assigns emotions on a continuum between positive and negative [7]. The techniques for sentiment classification can be categorized into three types: lexicon-based approach, machine learning approach, and hybrid approach [20].

The lexicon-based approach uses a previously-defined list of words where each word is associated with a sentiment score. The simplest approach to classifying sentiments is to count positive and negative words from a lexicon. This method has been used in sentence and aspect-level sentiment classification [12]. Lexicon-based classifiers are easy to implement since they does not require a training set. However, it is often difficult to find a robust lexicon-based dictionary fit for all contexts [2]. Moreover, the lexicon-based approach requires additional work to treat objective tweets [37]. Some previous research removed objective tweets which did not express positive or negative sentiment [1,4], while other research indicated that certain objective texts could be used to determine subjective feelings. Xiang et al. [44] showed that the sentiment analysis results can be further improved after classifying tweets into groups based on topics and performing analysis on each group independently. Cambria et al. [7] indicated that the lexicon-based approach was more effective in classifying sentiments when combined with other approaches such as part-of-speech tagging and trained classifiers.

The machine learning approach predicts sentiments by training and testing data sets. This approach is advantageous in adapting models for specific purposes and contexts [2]. When the labeled training classifiers are available, the supervised method can be used to learn the various features of an existing text which can be applied to analyzing a new text. When the labeled training classifiers are unavailable, an alternative method called *semi-supervised* or *distant supervision* uses the semantic orientation (SO) of the text to determine the thresholds of sentiments. Examples of the SO include hashtags and emoticons. While such symbols are not always reliable indicators of sentiments for individual sentences, previous research has successfully used the semi-supervised method to train classifiers [37]. Some representative machine learning methods for sentiment analysis include: Bayesian networks, naive Bayes (NB), support vector machine (SVM), maximum entropy (ME), neural networks, and random forests [17,18]. In addition, Galavotti et al. [13] proposed a simplified variant of χ^2 statistics as the feature selection technique which was further used to distinguish sentiment words [45].

The hybrid approach combines the machine learning approach and lexicon-based approach to take advantage of existing public resources for initial sentiment detection and the ability to create trained models for specific purposes [2]. Mudinas et al. [23] proposed a hybrid approach *pSenti* which aimed to attain high stability from a carefully designed lexicon and high accuracy from a supervised learning algorithm. Ghiassi et al. [14] proposed a hybrid approach using n-grams and statistical analysis to develop a Twitter-specific lexicon, and found that it produced higher accuracy than traditional SVM method. Poria et al. [27] developed a hybrid approach comprised of an event concept extraction algorithm, common-sense computing, and the extreme learning machine (ELM), and achieved a higher accuracy than the state-of-art approach in predicting the sentiments of sentences with conjunctions and comparisons. Prabowo and Thelwall [29] developed new hybrid classifiers based on several existing classifiers (general inquirer-based classifier, rule-based classifier, and statistics-based classifiers) and achieved an accuracy of 90% on some small data sets. Xia et al. [43] integrated ensemble feature sets and machine learning models (NB, ME, and SVM) for sentiment classification and tested it on some existing smaller data sets with an accuracy of above 70%. One common issue with the above research is that the sizes of the data set tested tend to be limited. There is a need to test these algorithms on bigger and more complex data sets [31].

3   Methodology

We advance the methodology of sentiment analysis by proposing the Normalized Difference Sentiment Index (NDSI). The goal is to identify frequently-occurring words that are predictive of positive or negative sentiments. We then propose a new hybrid model *e-senti* which integrates lexicons, a new NDSI word rank list, and tweet features into a random forest classifier. One of our contributions is to explicitly use emojis/emoticons as indicators of emotions for training our data (also known as distant supervision [15] and semi-supervised learning [11]). We provide an actionable framework which incorporates data collection, pre-processing, analysis, and evaluation for high volumes of tweets.

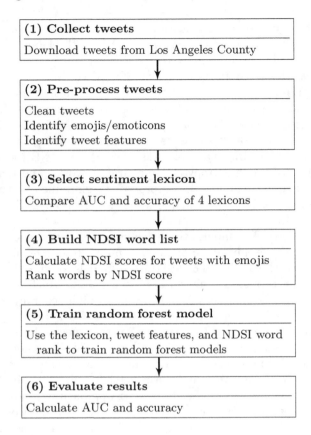

Fig. 1. Analysis framework for sentiment analysis.

Our analysis framework is shown in Fig. 1. We start by collecting geo-tagged English tweets from Los Angeles County, California using the Twitter streaming API via the streamR package [3]. Our corpus consists of 4.65 million geo-tagged tweets downloaded continuously during August 2016. Figure 2 shows the distribution of the geo-tagged tweets in our study area.

We pre-process tweets by removing all blank and non-English tweets. Each tweet is cleaned by setting all text to lowercase and removing punctuation. Additionally, four tweet features are used for sentiment analysis: (1) a count of the username mentions, (2) a count of the URL links, (3) a count of the hashtags, and (4) a dummy variable indicating whether a tweet is a reply. We hypothesize that these features may enhance our model's performance as they may represent emotions (e.g., `#happy, #awful`, etc.).

As we lack labeled (happy or sad) training data, the tweets' emojis and emoticons are used to form a semi-supervised training data set [10] because they indicate emotions. We select tweets with the common emojis and emoticons shown in Fig. 3. In total we identified 214,077 tweets with happy emojis/emoticons and 176,039 tweets with sad emojis/emoticons.

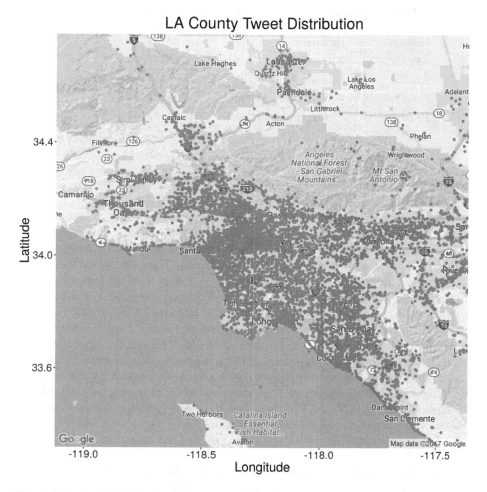

Fig. 2. Distribution of 4.65 million geo-tagged English tweets from Los Angeles County area, California (08/2016).

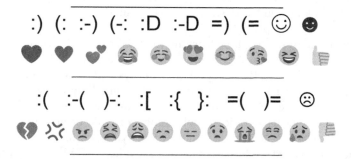

Fig. 3. Common emojis and emoticons used to create a semi-supervised training set.

Table 1. Publicly available sentiment lexicons for comparison.

Lexicon	Reference	Words	Range
AFINN	Nielsen [24]	2,477	−5 to 5
ANEW	Bradley and Lang [5]	13,915	−4.80 to 2.46
OpinionFinder	Wilson et al. [42]	6,884	−1 or +1
EmoLex	Mohammad and Turney [21]	14,182	−1 or +1

For our *e-senti* model, it is necessary to identify an appropriate sentiment lexicon prior to creating a hybrid classifier. We compare the accuracy and efficiency of 4 widely used lexicons (Table 1). The AFINN lexicon [24] consists of 2,477 words, each assigned an integer sentiment score from −5 (sad) to +5 (happy). The Affective Norms for English Words (ANEW) lexicon [5] consists of 13,915 words, each assigned a decimal score between 1 and 9. To distinguish sad words from happy words, we subtract the mean score from each word's score, so that negative values indicate sadness and positive values indicate happiness. The OpinionFinder lexicon [42] consists of 6,884 words, and the NRC Word-Emotion Association lexicon (also known as EmoLex) [21] includes 14,182 words. For both lexicons, positive and negative words are scored as 1 and −1 respectively.

We apply the four lexicons to a subset of tweets with emojis/emoticons. We assign a score to each tweet by summing the sentiment scores of words contained in each lexicon. To account for negation, we assign the opposite value to word pairs that begin with a negation. For example, the word *happy* receives a score of 3 based on the AFINN lexicon, whereas the word pair *not happy* scores −3 because of the negation (based on a list of words such as *no*, *not*, *none*, *neither*, *nothing*, and *never* etc. from the Cambridge Online Dictionary [30] and the verbs negated with n't [9]).

We evaluate the lexicons in two ways. First, we calculate AUC, the area under the receiver operating characteristic curve. The receiver operating characteristic (ROC) curve is a graphical representation of true positive classification rate versus false positive classification rate. For our balanced, two-class classification, a perfect classifier will have an AUC of 1, while random guessing will have an AUC near 0.5. In addition to AUC, we also calculated the accuracy [19] attained by each classifier at its optimal cutoff.

In sentiment analysis, machine learning models commonly use unigrams (one word), bigrams (two consecutive words), and n-grams. Previous research has found unigrams to be effective in sentiment analysis [26]. While most unigram-based classifiers use words that occur frequently, some frequently-occurring unigrams may not be as efficient in predicting sentiments as others. For example, in our tweets with emojis/emoticons, while the word *last* occurs more frequently than the word *sick*, *sick* is more helpful in identifying sentiments. Therefore, we propose the Normalized Difference Sentiment Index (NDSI) to identify words (unigrams) that occur frequently and are predictive of positive or negative sentiment based on a subset of tweets with emojis/emoticons.

After stemming each word using Porter's stemming algorithm [28], an NDSI score was calculated for each word occurring in at least 0.01% of tweets in the training data used to build the classifier. For a word w, the NDSI score is the absolute value of the difference between the number of times occurring in happy tweets and sad tweets:

$$|n(w|\text{happy}) - n(w|\text{sad})|$$

where $n(w|\text{happy})$ indicates the number of times that word w occurs in tweets with a happy emoji/emoticon, and $n(w|\text{sad})$ denotes the number of times that word w occurs in tweets with a sad emoji/emoticon. Scores were normalized by dividing the highest NDSI score, yielding values from 0 to 1, with 0 indicating a word occurs equally in happy and sad tweets and 1 denoting a word that only occurs frequently in either happy or sad tweets. This index is easy to calculate and identifies words highly indicative of happiness/sadness. Words are ranked by NDSI scores.

Next, we use the following attributes to create a random forest classifier: (1) tweet features (counts of the number of hashtags, mentions, URL's, and a dummy variable indicating whether the tweet is a reply), (2) the count of NDSI-ranked unigram words, and (3) lexicon sentiment scores. Random forests are an ensemble machine learning method in which a multitude of decision trees are constructed with each tree predicting happiness or sadness. Random forest models are robust and can handle both categorical and numerical data [38]. The output is a value between 0 and 1, where 0 indicates that all decision trees predict sadness and 1 denotes that all decision tress predict happiness. Prior research in sentiment analysis has shown that random forest models outperformed naive Bayes, max entropy, and boosted trees in sentiment analysis [16].

4   Results

4.1   Lexicon Selection

Table 2 compares the AUC and accuracy of the lexicons applied to our tweets with emojis/emoticons. The AFINN lexicon produced the highest accuracy and similar AUC to the ANEW lexicon. The AFINN lexicon has 2,477 words and ANEW has 13,915 words. The AFFIN lexicon also incorporates on-line slang (such as lol) which fits the context of social media. Therefore, we choose the AFINN lexicon for our model because it is smaller and more efficient.

Table 2. The AUC and accuracy of lexicons based on tweets with emojis/emoticons in our Los Angles County data.

Lexicon	AUC	Accuracy
AFINN	0.690	64.3%
ANEW	0.696	63.8%
OpinionFinder	0.647	60.9%
EmoLex	0.644	60.7%

4.2   NDSI Scores

We randomly select 20,000 tweets with happy emojis/emoticons and 20,000 tweets with sad emojis/emotions from our data to calculate NDSI scores for the words in tweets and to create the random forest classifier. Table 3 shows a sample of words from our training data ranked by NDSI scores. Words like "love", "thank", and "birthday" occur frequently in happy tweets and words like "want" appear frequently in sad tweets. Thus, they all have relatively high NDSI scores. Words like "human" and "film" occur less frequently but are still good indicators of either happiness or sadness.

To determine an appropriate number of words to train the random forest classifier, we test different numbers of words (100 to 1000) based on our emoji/emoticon training data, and compare the AUC and accuracy of the test results. Figure 4 shows that the accuracy increases by 3% to 5% as the number of words expands from 100 to 300. Yet when the number of words rises beyond 600, both accuracy and AUC improves by less than 0.5% while the computing time goes up by about 10% for each increase of 100 words. Therefore, we decide to include 600 words with the highest NDSI scores in our random forest classifier.

Table 3. Normalized Difference Sentiment Index (NDSI) word rank list.

Rank	Word w	$n(w\vert\text{happy})$	$n(w\vert\text{sad})$	NDSI(w)
1	love	1794	426	1.00
2	thank	1407	189	0.89
3	happi	1159	135	0.75
4	birthday	880	103	0.57
5	want	194	723	0.39
⋮	⋮	⋮	⋮	⋮
597	human	17	6	0.01
598	film	18	7	0.01
599	rose	18	7	0.01
600	ball	9	20	0.01

4.3   Model Results

From 40,000 randomly selected tweets, we use 70% of the data to train each classifier and the other 30% for testing. Table 4 shows the AUC and accuracy of each *e-senti* model using different combinations of attributes. We also test a model using only the AFINN lexicon for comparison. We find that all attributes (AFINN lexicon, tweet features, and NDSI unigrams) increase the model's accuracy and AUC. The combination of all 3 attributes achieves the highest AUC and accuracy. In order to further validate this strongest model, we randomly partition the data into 10 test sets of 4,000 tweets in order to perform 10-fold cross validation. For each of the ten test sets of 4,000 tweets, a model is trained over the remaining 36,000 tweets. We record the AUC and accuracy of each model. The average AUC of the 10 models is 0.814, and the average accuracy is 74.6%.

We further test our model using the sentiment140 data from [15]. This data set contains 177 negative and 182 positive tweets, and has been widely used in the literature to compare different models [33]. Table 5 shows the results of the *e-senti* models based on the sentiment140 data. All 3 attributes improve the accuracy, and the *e-senti* model with all 3 attributes yields the highest AUC of 0.926 and an accuracy of 86.1%. Table 6 compares our *e-senti* model's result with the existing approaches based on the sentiment140 data. [15] reported an accuracy of 83.0%. [36] used the label propagation method and attained an accuracy of 84.7%. [34] proposed the interpolation method for naive Bayes classifier and produced the best accuracy of 81.3%. Our *e-senti* model outperforms most of the existing approaches and achieves comparable performance to [35] which used the sentiment-topic features (86.3% accuracy).

The histograms in Fig. 5 show the output of our *e-senti* model based on the Los Angeles County test data and sentiment140 data. Sad tweets are shown in blue and happy tweets are shown in red. Purple indicates the overlapping

Table 4. Results of the *e-senti* model using different combinations of attributes based on Los Angeles County data.

Model	AUC	Accuracy
AFINN + tweet features + NDSI unigrams	**0.814**	**74.6%**
NDSI unigrams + tweet features	0.806	73.5%
AFINN + NDSI unigrams	0.794	73.8%
NDSI unigrams	0.786	72.4%
AFINN + tweet features	0.744	69.0%
Tweet features	0.675	64.7%
AFINN lexicon	0.641	64.1%

Table 5. Results of the *e-senti* model using different combinations of attributes based on sentiment140 data.

Model	AUC	Accuracy
AFINN + tweet features + NDSI unigrams	**0.930**	**86.1%**
AFINN + NDSI unigrams	0.908	82.4%
AFINN + tweet features	0.889	78.7%
NDSI unigrams + tweet features	0.853	78.7%
NDSI unigrams	0.861	78.7%
AFINN lexicon	0.855	76.8%
Tweet features	0.574	55.6%

Table 6. Classification accuracy based on sentiment140 data.

Model	Accuracy
Saif et al. [35]	86.3%
e-senti	**86.1%**
Speriosu et al. [36]	84.7%
Go et al. [15]	83.0%
Saif et al. [34]	81.3%

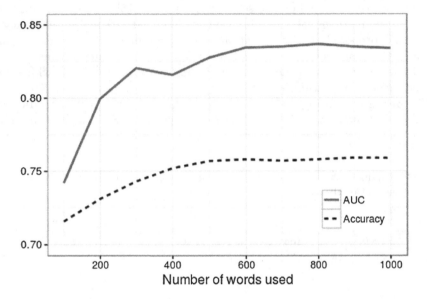

Fig. 4. The AUC and accuracy of the *e-senti* model using different numbers of NDSI unigrams.

of happy and sad tweets (i.e., the tweets are not distinguished by the model). These figures demonstrate that our *e-senti* model effectively separates happy tweets from sad tweets for both data sets.

4.4   Discussion

By proposing and testing a new hybrid model *e-senti*, we incorporate key attributes for classifying tweet sentiments into a random forest classifier. In comparing existing lexicons, we find that the AFINN lexicon outperforms other lexicons because it accounts for varying levels of happiness and sadness. Moreover, it incorporates slang and non-standard English words common to social media. In fact, as a standalone predictor, a simple summation of AFINN scores performs moderately well (64.1% accuracy for our Los Angeles County data). The results are further improved (69.0% accuracy) when the AFINN lexicon is

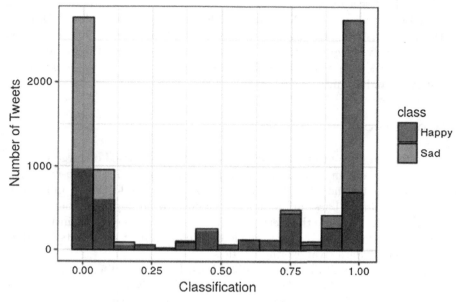

(a) Los Angeles County emoticon/emoji tweets

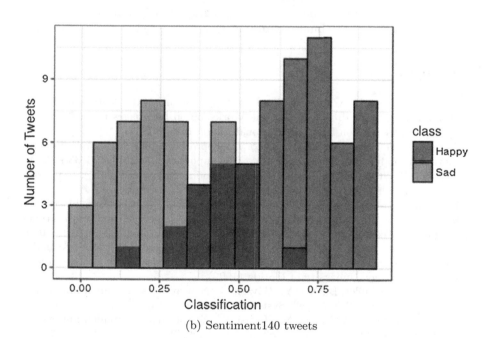

(b) Sentiment140 tweets

Fig. 5. Random forest classification results of our *e-senti* model. Red indicates happy tweets, blue indicates sad tweets, and purple represents the overlapping of happy and sad tweets. (Color figure online)

combined with tweet features such as hashtags, URL's, and username mentions. Furthermore, we propose the Normalized Difference Sentiment Index (NDSI) for identifying words that are highly representative of emotions. This new index can be effectively integrated with other attributes to enhance the model's performance (74.6% accuracy). Our final model with all 3 attributes achieves the highest AUC and accuracy.

5   Conclusions

There has been a growing effort to investigate sentiments based on large volumes of social media data. In this paper, we propose a new hybrid model *e-senti* that combines attributes including the AFINN lexicon, tweet features, and a new NDSI word rank list. These attributes are all effective in improving our model's accuracy. We contribute to the methodology of sentiment analysis by introducing a model that is easy to implement, efficient, and accurate. Our results based on the sentiment140 data rival existing models.

Our future work will take several directions. First, we will examine the sentiments of tweets based on different topics. Second, we will investigate bigrams and other machine learning algorithms including support vector machines and maximum entropy. Third, we will link the geo-tagged sentiment data to land use and infrastructure data to understand the geographic distribution of emotions and how emotions and the built environment are connected.

References

1. Agarwal, A., Xie, B., Vovsha, I., Rambow, O., Passonneau, R.: Sentiment analysis of Twitter data. In: Proceedings of the Workshop on Languages in Social Media, pp. 30–38. Association for Computational Linguistics (2011)
2. Alessia, D., Ferri, F., Grifoni, P., Guzzo, T.: Approaches, tools and applications for sentiment analysis implementation. Int. J. Comput. Appl. **125**(3), 26–33 (2015)
3. Barbera, P.: StreamR: access to Twitter streaming API via R (2014). R package version 0.2.1. https://CRAN.R-project.org/package=streamR
4. Barbosa, L., Feng, J.: Robust sentiment detection on Twitter from biased and noisy data. In: Proceedings of the 23rd International Conference on Computational Linguistics: Posters, pp. 36–44. Association for Computational Linguistics (2010)
5. Bradley, M.M., Lang, P.J.: Affective norms for English words (ANEW): instruction manual and affective ratings. Technical report C-1, The Center for Research in Psychophysiology, University of Florida (1999)
6. Cambria, E., Schuller, B., Xia, Y., Havasi, C.: New avenues in opinion mining and sentiment analysis. IEEE Intell. Syst. **28**(2), 15–21 (2013)
7. Cambria, E., Schuller, B., Xia, Y., Havasi, C.: New avenues in opinion mining and sentiment analysis. IEEE Intell. Syst. **2**, 15–21 (2013)
8. Chanel, G., Kronegg, J., Grandjean, D., Pun, T.: Emotion assessment: arousal evaluation using EEG's and peripheral physiological signals. In: Gunsel, B., Jain, A.K., Tekalp, A.M., Sankur, B. (eds.) MRCS 2006. LNCS, vol. 4105, pp. 530–537. Springer, Heidelberg (2006). doi:10.1007/11848035_70

9. Dadvar, M., Hauff, C., de Jong, F.: Scope of negation detection in sentiment analysis. In: Proceedings of the Dutch-Belgian Information Retrieval Workshop, Amsterdam, pp. 16–20 (2011)
10. Davidov, D., Tsur, O., Rappoport, A.: Enhanced sentiment learning using Twitter hashtags and smileys. In: Proceedings of the 23rd International Conference on Computational Linguistics: Posters, pp. 241–249. Association for Computational Linguistics (2010)
11. Davidov, D., Tsur, O., Rappoport, A.: Semi-supervised recognition of sarcastic sentences in Twitter and Amazon. In: Proceedings of the Fourteenth Conference on Computational Natural Language Learning, pp. 107–116. Association for Computational Linguistics (2010)
12. Ding, X., Liu, B., Yu, P.S.: A holistic lexicon-based approach to opinion mining. In: Proceedings of the 2008 International Conference on Web Search and Data Mining, pp. 231–240. ACM (2008)
13. Galavotti, L., Nardi, V.J., Sebastiani, F., Simi, M.: Feature selection and negative evidence in automated text categorization. In: Proceedings of the 4th European Conference on Research and Advanced Technology for Digital Libraries (ECDL 2000) (2000)
14. Ghiassi, M., Skinner, J., Zimbra, D.: Twitter brand sentiment analysis: a hybrid system using n-gram analysis and dynamic artificial neural network. Expert Syst. Appl. **40**(16), 6266–6282 (2013)
15. Go, A., Bhayani, R., Huang, L.: Twitter sentiment classification using distant supervision. CS224N Project Report, Stanford, 1, 12 (2009)
16. Gupte, A., Joshi, S., Gadgul, P., Kadam, A.: Comparative study of classification algorithms used in sentiment analysis. Int. J. Comput. Sci. Inf. Technol. **5**(5), 6261–6264 (2014)
17. Lima, A.C.E., de Castro, L.N., Corchado, J.M.: A polarity analysis framework for Twitter messages. Appl. Math. Comput. **270**, 756–767 (2015)
18. Lin, C., He, Y.: Joint sentiment/topic model for sentiment analysis. In: Proceedings of the 18th ACM Conference on Information and Knowledge Management, pp. 375–384. ACM (2009)
19. Manning, C.D., Schütze, H.: Foundations of Statistical Natural Language Processing, vol. 999. MIT Press, Cambridge (1999)
20. Maynard, D., Funk, A.: Automatic detection of political opinions in tweets. In: García-Castro, R., Fensel, D., Antoniou, G. (eds.) ESWC 2011. LNCS, vol. 7117, pp. 88–99. Springer, Heidelberg (2012). doi:10.1007/978-3-642-25953-1_8
21. Mohammad, S., Turney, P.: Crowdsourcing a word-emotion association lexicon. Comput. Intell. **29**(3), 436–465 (2013)
22. Mohammad, S.M.: Sentiment analysis: detecting valence, emotions, and other affectual states from text. In: Emotion Measurement (2015)
23. Mudinas, A., Zhang, D., Levene, M.: Combining lexicon and learning based approaches for concept-level sentiment analysis. In: Proceedings of the 1st International Workshop on Issues of Sentiment Discovery and Opinion Mining, p. 5 (2012)
24. Nielsen, F.: A new anew: evaluation of a word list for sentiment analysis in microblogs. arXiv:1103.2903 (2011)
25. Pang, B., Lee, L.: Opinion mining and sentiment analysis. Found. Trends Inf. Retr. **2**(1–2), 1–135 (2008)
26. Pang, B., Lee, L., Vaithyanathan, S.: Thumbs up? Sentiment classification using machine learning techniques. In: Proceedings of the ACL-02 Conference on Empirical Methods in Natural Language Processing, vol. 10, pp. 79–86 (2002)

27. Poria, S., Cambria, E., Winterstein, G., Huang, G.B.: Sentic patterns: dependency-based rules for concept-level sentiment analysis. Knowl. Based Syst. **69**, 45–63 (2014)
28. Porter, M.F.: An algorithm for suffix stripping. Program **14**(3), 130–137 (1980)
29. Prabowo, R., Thelwall, M.: Sentiment analysis: a combined approach. J. Informetr. **3**(2), 143–157 (2009)
30. Cambridge University Press: Cambridge online dictionary. Accessed 1 Mar 2017
31. Ravi, K., Ravi, V.: A survey on opinion mining and sentiment analysis: tasks, approaches and applications. Knowl. Based Syst. **89**, 14–46 (2015)
32. Sabatinelli, D., Keil, A., Frank, D.W., Lang, P.J.: Emotional perception: correspondence of early and late event-related potentials with cortical and subcortical functional MRI. Biol. Psychol. **92**(3), 513–519 (2013)
33. Saif, H., Fernandez, M., He, Y., Alani, H.: Alleviating data sparsity for Twitter sentiment analysis. In: 1st Interantional Workshop on Emotion and Sentiment in Social and Expressive Media: Approaches and Perspectives from AI (ESSEM 2013) (2013)
34. Saif, H., He, Y., Alani, H.: Semantic smoothing for twitter sentiment analysis. In: Proceeding of the 10th International Semantic Web Conference (ISWC) (2011)
35. Saif, H., He, Y., Alani, H.: Evaluation datasets for Twitter sentiment analysis: a survey and a new dataset, the STS-Gold. In: CEUR Workshop Proceedings, vol. 838 (2012)
36. Speriosu, M., Sudan, N., Upadhyay, S., Baldridge, J.: Twitter polarity classification with label propagation over lexical links and the follower graph. In: Proceedings of the First workshop on Unsupervised Learning in NLP, pp. 53–63 (2011)
37. Taboada, M., Brooke, J., Tofiloski, M., Voll, K., Stede, M.: Lexicon-based methods for sentiment analysis. Comput. Linguist. **37**(2), 267–307 (2011)
38. Tan, P.N., Steinbach, M., Kumar, V., et al.: Introduction to Data Mining, vol. 1. Pearson Addison Wesley, Boston (2006)
39. Thayer, J.F., Åhs, F., Fredrikson, M., Sollers, J.J., Wager, T.D.: A meta-analysis of heart rate variability and neuroimaging studies: implications for heart rate variability as a marker of stress and health. Neurosci. Biobehav. Rev. **36**(2), 747–756 (2012)
40. Turney, P.D.: Thumbs up or thumbs down? Semantic orientation applied to unsupervised classification of reviews. In: Proceedings of the 40th Annual Meeting on Association for Computational Linguistics, pp. 417–424 (2002)
41. Valstar, M.F., Mehu, M., Jiang, B., Pantic, M., Scherer, K.: Meta-analysis of the first facial expression recognition challenge. IEEE Trans. Syst. Man Cybern. Part B (Cybern.) **42**(4), 966–979 (2012)
42. Wilson, T., Wiebe, J., Hoffmann, P.: Recognizing contextual polarity in phrase-level sentiment analysis. In: Proceedings of the Conference on Human Language Technology and Empirical Methods in Natural Language Processing, pp. 347–354. Association for Computational Linguistics (2005)
43. Xia, R., Zong, C., Li, S.: Ensemble of feature sets and classification algorithms for sentiment classification. Inf. Sci. **181**(6), 1138–1152 (2011)
44. Xiang, B., Zhou, L., Reuters, T.: Improving Twitter sentiment analysis with topic-based mixture modeling and semi-supervised training. In: ACL, Maryland, pp. 434–439 (2014)
45. Zhou, H., Chen, L., Shi, F., Huang, D.: Learning bilingual sentiment word embeddings for cross-language sentiment classification. In: Proceedings of the 53rd Annual Meeting of the Association for Computational Linguistics and the 7th International Joint Conference on Natural Language Processing, pp. 430–440 (2015)

Mining Frequent Closed Set Distinguishing One Dataset from Another from a Viewpoint of Structural Index

Yoshiaki Okubo and Makoto Haraguchi[(⊠)]

Graduate School of Information Science and Technology,
Hokkaido University, N-14 W-9, Sapporo 060-0814, Japan
mh@ist.hokudai.ac.jp

Abstract. The variety of concept's specialization can be an index of how the concept is significant. From this viewpoint, given two incident relations as datasets, we consider formal concepts with many frequent subconcepts in one dataset, while those have few frequent subconcepts in another dataset. Instead of calculating the number of frequent subconcepts directly, we introduce a structural index that approximates the depth complexity of join semilattice of frequent concepts, and consider an anti-monotonic constraint for one dataset and a monotonic constraint for another one. Based on these two constraints, we develop a procedure to search for "emerging concepts" with respect to the structural index. Although it is generally a hard task to compute the structural index, the index we choose is known as efficient for large sparse data. The experimental results show the effectiveness of proposed method, involving some interesting output concepts contrasting two datasets.

Keywords: Formal concepts · Emerging concepts · Complexity of concept lattice

1 Introduction - The Problem to Be Solved

Starting with the study of frequent itemset mining [1], various classes of frequent itemsets have been studied so as to restrict our attention to those that are worth discovering. For this aim, some interestingness measures [8] and several classes of frequent itemsets as in [9,10] are considered. A statistical correlation [11] and a set theoretic correlation [12] are also introduced to obtain itemsets of correlated items. On the other hand, even non-frequent itemsets, referred to as "Rare Concept" [13,24], have been discussed in terms of "Formal Concept" [16]. Both of the frequentness and rareness are naturally related to the problem of emerging patterns [14,15] and contrast set mining [11]. In these studies, under two or more datasets (transaction databases), frequencies of itemsets for two or more datasets are compared and contrasted. More precisely, itemsets with relatively higher frequency in one dataset and relatively lower frequency at another

© Springer International Publishing AG 2017
P. Perner (Ed.): MLDM 2017, LNAI 10358, pp. 417–430, 2017.
DOI: 10.1007/978-3-319-62416-7_30

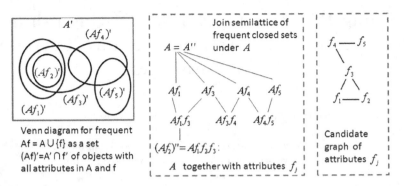

Fig. 1. An example of join semilattice of frequent closed attribute set A is shown in the middle, where, for closed sets A_1 and A_2 of attributes, A_2 is said more general than A_1 and placed higher than A_1 in the figure if the corresponding object set of A_2, denoted by A_2', includes A_1' for A_1. The relationship among A_j' of objects is depicted by Venn diagram at the left. A'' is called a closure to obtain closed attribute set corresponding to the set A' of objects. We will give its short explanation in Sect. 2.

set are regarded. We can say that those itemsets distinguish one dataset from another from the viewpoint of frequency.

This paper similarly uses such a strategy of "Compare and Contrast" to get characteristic itemsets, where we compare rather structural indices of itemsets while requiring that they must be frequent at both datasets.

(FC: Frequentness Condition). The target itemsets must be frequent at the two, and we compare structural indices of itemsets. In case of "Emerging Pattern", itemsets with low frequencies that become to have higher frequencies after we change datasets are preferred. In our case, every non-frequent itemset is disregarded by the reason that it is too much individual.

According to the studies of "Formal Concept" [16–18], we identify itemsets whenever they have the same object sets, and consider only frequent closed itemsets (intents of formal concepts). Then we pay our attention to a degree of how frequent closed itemsets are derived from a closed itemset A. In what follows, we use the term "attribute" and "closed (attribute) set" rather than "item" and "closed itemset", respectively.

As is well known, the derivation of closed set A to more specific ones is well described by drawing formal concept lattice [16] of closed sets. We illustrate such a lattice by the middle figure in Fig. 1, where every non-frequent closed sets are removed in the figure, resulting join semilattice with A as the most general one. Then, each derivation of A is realized by a path from A to its more specific frequent closed set $Af_1f_2f_3 = A \cup \{f_1, f_2, f_3\}$ for instance. We consider in this paper that, as the path length is longer, the conceptual structure about A is more rich. Suppose a closed set does not have such a rich structural property at one dataset, while it has at another dataset. Then the closed set could show some characteristic feature contrasting two datasets from a structural viewpoint. In Sect. 7, we present an actual example of such a concept which has a simple structure consisting of just two specific concepts in one dataset, while in another a complex structure with ten specific concepts.

We thus adopt a path length as a structural index of join semilattice, and make the following constraints in addition to the frequentness constraint.

(DCT: Depth Complexity at Target Dataset). The path length of join semilattice of frequent closed set below A, constructed at one dataset referred to as a target t, must be greater than a parameter κ_t.

(DCB: Depth Complexity at Base Dataset). On the other hand, at another dataset referred to as base b, the maximum path length of join semilattice of frequent closed set below A must not exceed a given parameter κ_b.

To check the above depth complexity for various frequent closed sets, a naive strategy is to enumerate every possible closed sets by using fast enumerator as LCM [2] and build the join semilattice of all frequent closed sets. However, this strategy allows useless generation of closed sets. For instance, whenever we know that A violate (DCT), we can reject any B more specific than A, independently of the corresponding structure of A at base. In other words, we need not generate frequent closed set under A at base.

Another possible strategy to cope with our problem is to construct "borders" [14] along which the truth of the conditions changes, where our border notion will be addressed from the structural property, while the original notion of borders is based on the frequentness and non-frequentness only. Instead of border notion, we introduce "Covering Condition" that allows us to output general closed set which covers several closed set meeting (FC), (DCT) and (DCB). As the (DCT) is anti-monotonic and (DCB) is monotonic, we can stop the generation of closed set whenever tentative closed set with (DCT) turns out to satisfy (DCB). All the maximal solutions are represented by that general solution.

However there remains the problem of how we compute such a path length as a structural index. We here propose to use an approximated index based on clique sizes. Again in Fig. 1, the rightmost figure shows how a candidate attributes that can be added to the present closed itemset A is correlated, where attributes f, g are adjacent iff they can be added to A simultaneously without violating frequentness. An undirected graph thus defined is called a candidate graph, and we use the maximum clique size as an approximated structural index. In case of Fig. 1, the maximum clique size is just the longest path length in join semilattice. Strictly speaking, the maximum clique size is an upper bound of path length. Thus we use clique size as an approximated structural index.

The problem of finding maximum clique size is however NP-hard [23]. On the other hand, some fast methods for large sparse graphs have been extensively studies [19,20]. The study [22] shows that the effectiveness of clique algorithm is practically efficient for large sparse graphs, and the study [21] proves that the problem of maximum clique finding is polynomial solvable provided the maximum degree of vertices is bounded. As we see in Sect. 7, the graphs constructed from our datasets (newspaper articles) is actually sparse enough to make our algorithm run fast for datasets with about 10 thousand attributes.

This paper is organized as follows. In Sect. 2, some basic notions about closed attribute sets are presented. (FC) condition is discussed in Sect. 3. Some elemental property of structural index is presented in Sect. 4. Based on these consideration, Sect. 5 gives our problem specification solved by a procedure presented in Sect. 6. The performance of procedure is experimentally analyzed in Sect. 7. Section 8 summarizes this paper.

2   Preliminary Definitions

Let $\mathcal{F}$ and $\mathcal{O}$ be a set of objects (newspaper articles in our examples) and a set of attributes (terms extracted from the articles). Each attribute f is supposed to have its object set $f' \subseteq \mathcal{O}$. f' is exactly the set of index terms in case of newspaper articles. For attribute set A, B, $A \cup B$ is abbreviated to AB. We also write Af to denote $A \cup \{f\}$ for an attribute f. The closure A'' of A is formed by adding every f implied by A to A. where we say that A implies f if $A' \subseteq f'$. A is closed if $A'' = A$. The support $sup(A)$ is simply $|A'|/|\mathcal{O}|$. Then, given a minimum support parameter, $minsup$, we say that A is frequent if $sup(A) \geq minsup$. A straightforward but important facts we note here are given as follows:

$$sup(A) = sup(A'').\ \ A \subseteq B \Rightarrow B' \subseteq A'\ .A' = \bigcap_{f \in A} f'.$$
$$\text{So, } (AB)' = A' \cap B' \text{ and } (Afg)' = (Af)' \cap (Ag)'.$$

The object set A' is also written as $ext_\sigma(A)$ when we need to designate a dataset referred by $\sigma \in \{b, t\}$, where b and t are abbreviation of "base" and "target", respectively. Similar notations with subscript σ as $sup_\sigma(A)$ are used throughout this paper. A toy example of base and target in the form of standard incident relation [16] (boolean data matrix) is illustrated in Fig. 2.

3   Frequentness Condition and Candidate Graphs

We regard only common attributes in $\mathcal{F} = \mathcal{F}_b \cap \mathcal{F}_t$. Also, we use the same minimum support parameter, minsup for the two datasets. Under this assumption, a closed set A must satisfy

(CF) A is frequent at b, and the corresponding closed set $b(A)$ at b must be also frequent.

As the two datasets are different, a closed set A at t is not necessarily closed at b. So we correspond A to a closed set $b(A)$ at base. Although there exist several ways to make such a correspondence, we define $b(A) = \{f \in \mathcal{F} \mid ext_b(A) \subseteq ext_b(f)\}$. In case Fig. 2, $\{f_2\}$ is closed at t, while it is not closed at b as f_2 implies f_0 at b. As a result, $b(\{f_2\}) = \{f_0, f_2\}$.

According to the standard frequent set mining method (see [18] for instance), a notion of candidates is normally used to expand a closed set to a larger one. For a closed set A at σ, an attribute $f \notin A$ is called a candidate of A at σ if Af is also frequent at σ. The set of all candidates is written as $Cand_\sigma(A)$. In case of

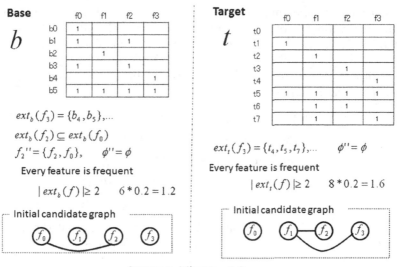

Common Minsup = 0.2

Fig. 2. A toy example of base and target with four common attributes f_0, f_1, f_2, f_3. The greatest closed sets are both empty set $\phi = \phi''$ in this case. Note that ϕ'' is not empty iff there exists some attribute with all objects. So, in this example, ϕ at base is also a closed set corresponding to ϕ at target. Also initial candidate graph with edges (f_i, f_j) meaning that f_i and f_j co-occur $(sup_\sigma(f_i f_j) \geq minsup)$ is drawn.

target domain t in Fig. 2, $\{f_2\}$ is closed and has f_1 as its unique candidate. The other attributes f_k disappear whenever we choose $\{f_2\}$ as the present closed set.

Then, for a closed set B at σ, we can define a candidate graph $G_\sigma(B) = (Cand_\sigma(B), E_\sigma(B))$, showing the joint frequentness (co-occurrence) of attributes together with B. The vertex set is just the $Cand_\sigma(B)$. The adjacency relation $E_\sigma(B)$ is just defined as $(f, g) \in E_\sigma(B)$ iff Bfg is frequent at σ. In Fig. 2, we draw only initial candidate graphs for the most general closed set ϕ'' at t. We make such new candidate graphs for both datasets whenever a new closed set $(Af)''$ at t is created from a tentative closed set A. Then, from the anti-monotonicity of frequentness, both of the vertices and the edges are reduced as we add a candidate attribute g to the present closed set B.

$$Cand_\sigma((Bg)'') \subseteq Cand_\sigma(B), \quad E_\sigma((Bg)'') \subseteq E_\sigma(B) \tag{1}$$

4   A Structural Index

Now let us return to our problem of defining indices for structural complexity of join semilattice of frequent closed sets under a closed set B. As the argument in this section does not depend on the choice of σ, we omit σ from our notations. Firstly we note a fact that the length of path between frequent closed sets B

and C is at most the maximum clique size of $Cand(B)$ if B is more general. For the terminology about maximum or maximal cliques, refer [19–21] for instances.

Let $B = B_0 \subset B_1 \subset \cdots \subset B_\ell = C$ be a chain of frequent closed sets with $B_{j+1} = (B_j \cup \{b_j\})''$. As $Bb_1b_2 \cdots b_\ell \subseteq C$, b_i and b_j are jointly frequent together with B. Any paired b_i and b_j are thus adjacent in $Cand(B)$. That is, $\{b_1, b_2, ..., b_\ell\}$ form a clique of $Cand(B)$.

$$\text{path length } l \text{ from } B$$
$$\leq \text{ the maximum clique size of } Cand(B), \tag{2}$$
$$\text{the maximum clique size of } Cand((Bg)'')$$
$$\leq \text{ the maximum clique size of} Cand(B)$$
$$\text{for any } g \in Cand(B) \tag{3}$$

The inequality (3) is directly followed by (1), and means that the maximum clique size is decreasing as we add candidates. This achieves a safe pruning rule introduced in Sect. 5.

Figure 1 shows a case in which the equality holds for (3). However in general, the equality may not hold. Such a case occurs when distinct candidates f and g are equivalent together with a closed set. For instances in Fig. 1, suppose furthermore $(Af_2)' = (Af_1)' \cap (Af_3)'$. Then f_2 implies both f_1 and f_3 under A, and conversely, f_1 together with f_3 implies f_2 under A. So, $(Af_2)''$ and $(Af_1f_3)''$ are the same closed set. This reduces the path length from the root A to $(Af_2)''$, and we get the actual path length 2 which is less than the clique size 3.

It is also possible to calculate the actual path length from A by computing equivalence classes of candidates until we reach to leaf closed sets in the semilattice. This is however the same task as building the semilattice. As the present closed set A is general, and as the actual depth from A is more deep, the computational costs for that construction is not cheap one. For this reason, we use the maximal clique size as a kind of approximated depth complexity of join semilattice of A.

5   The Problem Specification Using the Structural Indices for Two Domains

Now this section provides a specification of our problem for finding closed sets that distinguish the target from the base based on structural indices.

(Potential solution). Given b and t, a closed attribute set A at t is called a solution if it satisfies the (DC) constraint in addition to (FC) in Sect. 3.

(Depth Complexity Constraint (DC)). consisting of (DCT) and (DCB).
(DCT) The maximum clique size of $Cand_t(A) > \kappa_t$, and
(DCB) the maximum clique size of $Cand_b(b(A)) \leq \kappa_b$,
where κ_σ is a parameter binding the approximated depth complexity.

Note that, from (3), (DCT) is an anti-monotonic constraint meaning that closed set A satisfies (DCT) whenever its super set (more specific) B does (DCT).

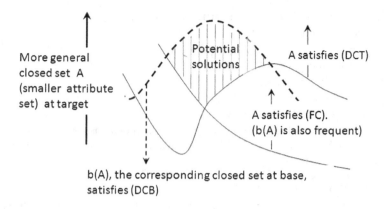

More general
closed set A
(smaller attribute
set) at target

Potential
solutions

A satisfies (DCT)

A satisfies (FC).
(b(A) is also frequent)

b(A), the corresponding closed set at base,
satisfies (DCB)

Fig. 3. The area of potential solutions, segmented by (FC), (DCB) and (DCT).

Hence, just like standard frequent itemset miner [1, 2] enjoying anti-monotonicity of frequentness, we can safely reject B and its any super set whenever B fails in (DCT) test. (DCB) is, on the other hand, a monotonic constraint. We utilize (DCB) as a stopping condition while keeping the condition that the set of output solutions covers every maximal potential solution satisfying (FC) and (DC). Refer to Fig. 3 illustrating the area of potential solutions.

(Maximal Solution). A maximal solution is a potential solution that is maximal among potential ones.

(Covering Condition for Output Solutions). We design our algorithm so as to output potential solutions such that, for any maximal solution M, there exists some output solution A more general than M ($A \subseteq M$).

By the Covering Condition, we can stop a process of expanding tentative closed set A whenever A achieves (DCB) while keeping (DCT). The detailed design ensuring this property is given in the next section.

6   Algorithm

Firstly, we present a procedure checking if every clique size is less than or equal to a parameter κ binding clique sizes. It tries to enumerate maximal cliques by expanding tentative cliques, and stops the expansion process whenever their sizes exceed κ. It is also equipped with a branch-and-bound pruning that cuts off possible clique expansion whenever any expansion, including maximal ones, is proved not to exceed κ by estimating clique sizes. The procedure shown by Fig. 4 is based on the maximal clique generator in [20] with a branch-and-bound pruning technique. So refer to [20] about maximal clique enumeration and [21] for the branch-and-bound control for details. We apply it to the base candidate graphs, given $\kappa = \kappa_b$, expecting true answer. For the target candidate graphs, we use the same procedure under $\kappa = \kappa_t$ to check if the answer is false.

```
boolean MC(G = (V, E)) = {  return Expand(φ, V, V);  }
//Q: tentative clique,  SUBG = ∩ Γ(q): candidate set
                                q∈Q
//CAND is the set of candidates not yet tried
boolean Expand(Q, SUBG, CAND) {// CAND ⊆ SUBG
  if (|Q| > κ) return false;
  if (SUBG = φ) {return true; };//Q is maximal
  u = arg  max   |CAND ∩ Γ(v)|;
        v∈SUBG
    // candidate adjacent to u need not be expanded (useless branch)
  SC = CAND − Γ(u): selectable candidate which is never useless branch
  for (each q ∈ SC) {
    Qnew = Q ∪ {q}; SUBGnew = SUBG ∩ Γ(q);
    CANDnew = CAND ∩ Γ(q);
    if (|Qnew| + |SUBGnew| ≤ κ) continue;// branch-and-bound control
    if (Expand(Qnew, SUBGnew, CANDnew) is false) return false ;
    CAND = CAND − {q};//trial for q ends
  };
  return true;
}
```

Fig. 4. Structural index checker, given κ, where $\Gamma(v)$ for a vertex $v \in V$ is the set of adjacent vertices to v in the input graph $G = (V, E)$.

Now, we are ready to present our main procedure searching for frequent closed sets satisfying (FC) and (DC). The basic strategy is to generate closed set A at the target t, where closed set $b(A)$ at b must be also frequent. So, given a tentative frequent closed A satisfying both (FC) and (DCT), a attribute f to be added to A is restricted. It must be a common candidate in $Cand_t(A) \cap Cand_b(b(A))$. Any other attribute need not be regarded at least for generating possible closed sets. It should be noted here that non-common attribute is still a member of $Cand_t(A)$ or $Cand_b(b(A))$ for evaluating structural indices. Then, our procedure traces a path of search tree in a depth-first manner. The search node on the search path has $A, Cand_t(A)$ and $Cand_b(b(A))$. The path is a sequence of $\phi'' = A_0 \subset A_1 \subset \cdots A_\ell = A$ such that $A_{k+1} = (A_k f_k)''$ at t for some $f_k \in Cand_t(A) \cap Cand_b(b(A))$. After checking (DCT) condition, new search node A_{k+1} is actually generated. So, we can stop expansion and output A_{k+1} whenever it satisfies (DCB) even when common candidates still remain according to the Covering Condition for output solutions. The standard technique to prevent the procedure from duplicated enumeration of closed sets is also equipped with it. In a word, a linear ordering over attributes is initially fixed, and the paths are tried according to the lexicographic order $\prec$ based on the attribute ordering. Then, for a tentative closed set A_k, it suffices to consider only a common attribute f such that $A \prec f$ to get next closure $A_{k+1} = (A_k f)''$ at t. In addition, we can reject f whenever $A_k f$ implies some common g such that $g \prec f$. The algorithm is shown in Fig. 5.

Before closing this section, we note the validness of our procedure w.r.t. the Covering Condition. Suppose A_{max} is maximal potential solution. Then, there

```
void searchMain() {
    A_t^{(0)} = cl_t(φ); A_b^{(0)} = cl_b(A_t^{(0)});
    make n(φ) = (A_b^{(0)}, Cand_b(A_b^{(0)}), A_t^{(0)}, Cand_t(A_t^{(0)}));
    if (csize(Cand_t(A_t^{(0)})) ≤ κ_t) return (fail); // no solution
    if (csize(Cand_b(A_b^{(0)})) ≤ κ_b) {print cl_t(φ); return (success);};
    search(n(φ));
}
void search(n(P))  {// n(P) = (B, Cand_b(B), A, Cand_t(A)), cl_t(P) = A, B = cl_b(A)
                    // csize(Cand_t(A)) > κ_t is already verified when invoked
    if (csize(Cand_b(B)) ≤ κ_b) {print A; return (success);};// Covering Condition
    CC(A) = Cand_b(B) ∩ Cand_t(A);
    if (CC(A) = φ) ⇒ return (fail);
    for (each f_s ∈ CC(A) s.t. P ≺ f_s) { // f_s is selected to expand in the order of ≺
        if (Af_s implies some g ∈ CC(A) such that g ≺ f_s) continue;
                        // if f_s together with A implies prior g then skip
        compute A_new = cl_t(Af_s) and Cand_t(A_new)
            // g ∈ Cand_t(A)\{f_s} is added to A_new when Af_s implies g at t
            // g not implied by Af_s is registered to Cand_t(A_new) when Af_s g is frequent at t
        if (csize(Cand_t(A_new) ≤ κ_t)) continue; // violation of (DCT)
        compute B_new = cl_b(A_new) and Cand_b(B_new);
            // similar construction as A_new and Cand_t(A_new)
        make the next node n(Pf_s) = (B_new, Cand_b(B_new), A_new, Cand_t(A_new));
        search(n(Pf_s));
    };
}
```

Fig. 5. The pseudo-code of main procedure, where $cl_σ(A)$ denotes the closure A'' at $σ$, and $csize(G) ≤ κ$ returns true iff all cliques in G have their sizes $≤ κ$.

exists the leftmost generator P_{max} of A_{max} according to $≺$, where "generator" here means that $P''_{max} = A_{max}$ at t. As A_{max} satisfies (FC), (DCT) and (DCB), and the (DCB) condition is monotonic, let P_{cc} be the first initial segment of P_{max} meeting (DCB). Clearly P''_{cc} at t satisfies (FC) and (DCT) as they are anti-monotonic. Hence the closure of P_{cc} is an output covering A_{max}.

7   Experimental Results

In this section, we present our experimental results to verify usefulness of our method.

We have implemented our system in JAVA and executed it on a PC with Intel® Core™-i3 M380 (2.53 GHz) CPU and 4 GB main memory.

7.1   Datasets

We have prepared a pair of document datasets each of which is a collection of news articles released by The Mainichi, a daily newspaper company in JAPAN[1].

---

[1] We have collected them from a CD-ROM edition of the newspapers.

The two datasets are referred to as **Before** and **After**. We assume **Before** and **After** to be the base and the target, respectively, in order to observe differences between two periods. **Before** consists of the articles in "Hokkaido" edition issued in the period from July to September in 2006[2]. On the other hand, **After** is the collection of those issued in the following period from October to December.

For each article, a set of index terms extracted from the original document is provided by The Mainichi. In our datasets, therefore, each article is represented as the set of such index terms. As a pre-process, we have removed terms appeared in a very few or too many articles. Under several thresholds, we have obtained datasets with various numbers of terms. Table 1 summarizes scales of the datasets.

Table 1. Scales of datasets

Name of DB	# of doc	# of terms in DB pair		
Before	1603	2682	3427	5663
After	1588			

7.2   Example of Emerging Concept

We present here an example of emerging concept actually extracted.

Intent in Target: { "Education", "Hokkaido Government Board of Education"}. For the dataset pair **Before-After**, under the parameter setting of 0.003, 1 and 4, we have extracted a concept whose intent is

{Education,   Hokkaido_Government_Board_of_Education(HGBE)}

in the target dataset as an emerging concept. It has the extent with 8 objects (articles) in the target and the extent of the corresponding concept in the base also consists of 8 objects.

Roughly speaking, the articles associated with the concept in the base are almost concerned with a fundamental policy of high school consolidations in Hokkaido made by HGBE. A union had a mass meeting to discuss the policy, a town had a meeting to continue a school there and a reporter summarized current situations in several areas and expected influences.

On the other hand, in the period of the target, the articles associated with the concept in the target includes various educational topics with which HGBE is concerned. We find, for example, articles about inadequate class plannings, an inauguration speech by the new chairman of the government board of education, an integrated education from primary school through high school, some dishonest accountings by a town board of education as well as high school consolidations.

---

[2] "Hokkaido" is the northernmost prefecture in Japan.

The conceptual structures rooted with the extracted emerging concept in the base and the target are illustrated in Fig. 6[3]. As is easily observed, the concept in the target has a structure which is more complex than that in the base.

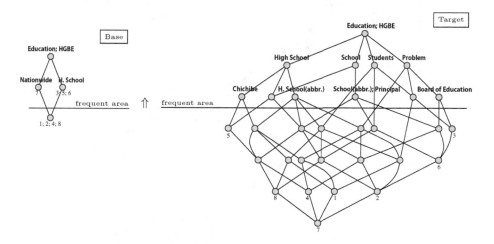

Fig. 6. Conceptual structures rooted with emerging concept in base and target

7.3 Computational Performance

We present here computational performance of our mining system.

To the best of our knowledge, there exists no system which can detect our emerging concepts directory. However, if some additional efforts are acceptable for us, we can simply obtain emerging concepts with the help of any formal concept miner or closed itemset miner. That is, we can first enumerate all frequent concepts in each of both base and target databases, and then try to find concepts which can become emerging ones in our sense. Since it has been well known that several closed itemset mining algorithms, e.g., [2,4,5], can extract frequent concepts very efficiently, one might claim that such a naïve approach would be sufficient for our purpose. In order to make a comparison between this kind of naïve method and ours, we observe numbers of solutions and computation times by each method under several minimum support thresholds. Concretely speaking, as an efficient closed itemset miner, we have made choice of LCM [2] which is the winner of the IEEE FIMI Workshop in 2004 [3] and still provides some key principles in efficient computation for recent pattern mining problems [6,7].

Figure 7 shows numbers of solutions and computation times by both systems. Note here that numbers of solutions by LCM are those of concepts actually enumerated in the naïve approach and computation times are those taken for the enumeration process. If we try to detect our emerging concepts by the naïve

---

[3] The lattices have been drawn by `Graphviz` (http://www.graphviz.org) via `FcaStone` (http://fcastone.sourceforge.net).

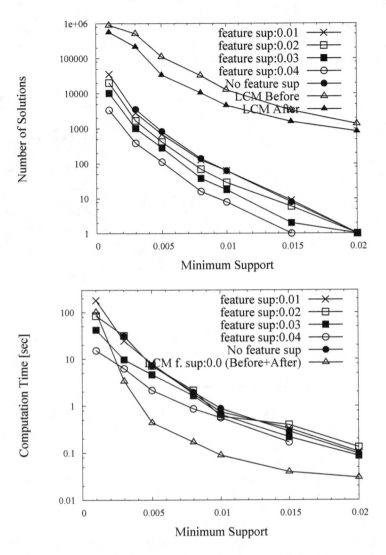

Fig. 7. Number of solutions and computation times

approach, we have to individually extract a huge number of candidate concepts in the base and the target, respectively. As a post-process, then, for each extracted concept in the target, we need to identify the sub-lattice rooted with the concept whose corresponding sub-lattice in the base is much simpler. As can be observed in the lower figure, the concept enumeration can be efficiently performed with less that 10.0 s in most cases. However, since the numbers of enumerated concepts are considerably larger than those of actual solutions detected by our system, we can easily guess that the post-process in the naïve approach would be very time-consuming. As the results, our system would be much more efficient than the naïve method.

8   Concluding Remarks and Discussion

This paper presents a depth-first search procedure for detecting frequent closed sets satisfying structural constraints based on clique sizes.

As a structural complexity, this paper considered an approximated depth complexity. That is, clique sizes work as upper bound of path lengths of join semilattice of frequent closed sets. On the other hand, it is interesting to investigate "width complexity" meaning the variation of direct child concepts. A procedure with such a width complexity is now planned based on the calculation of equivalence classes of candidate sets.

References

1. Agrawal, R., Srikant, R.: Fast algorithms for mining association rules. In: Proceedings of the 20th International Conference on Very Large Databases (VLDB 1994), pp. 487–499 (1994)
2. Uno, T., Kiyomi, M., Arimura, H.: LCM ver. 2: efficient mining algorithm for frequent/closed/maximal itemsets. In: Proceedings of IEEE ICDM 2004 Workshop (FIMI 2004) (2004). http://sunsite.informatik.rwth-aachen.de/Publications/CEUR-WS//Vol-126
3. Workshop on Frequent Itemset Mining Implementations (FIMI 2004) (2004). http://fimi.ua.ac.be/fimi04/
4. Zaki, M.J., Hsiao, C.: CHARM: an efficient algorithm for closed itemset mining. In: Proceedings of the 2002 SIAM International Conference on Data Mining (SDM 2002), pp. 457–453 (2002)
5. Wang, J., Han, J., Pei, J.: CLOSET+: searching for the best strategies for mining frequent closed itemsets. In: Proceedings of the 9th ACM SIGKDD International Conference on Knowledge Discovery and Data Mining (KDD 2003), pp. 236–245 (2003)
6. Leroy, V., Kirchgessner, M., Termier, A., Amer-Yahia, S.: TopPI: an efficient algorithm for item-centric mining. Inf. Syst. **64**, 104–118 (2017). Elsevier
7. Zida, S., Furnier-Viger, P., Lin, J.C., Wu, C., Tseng, V.S.: EFIM: a fast and memory efficient algorithm for high-utility itemset mining. Knowl. Inf. Syst. **51**, 595–625 (2016). Online First Articles, Springer
8. Geng, L., Hamilton, H.J.: Interestingness measures for data mining: a survey. ACM Comput. Surv. **38**(3), Article 9 (2006)
9. Han, J., Cheng, H., Xin, D., Yan, X.: Frequent pattern mining - current status and future directions. Data Mining Knowl. Disc. **15**(1), 55–86 (2007). Springer
10. Zhu, F., Yan, X., Han, J., Yu, P.S., Cheng, H.: Mining colossal frequent patterns by core pattern fusion. In: Proceedings of the 23rd IEEE International Conference on Data Engineering (ICDE 2007), pp. 706–715 (2007)
11. Bay, S.D., Pazzani, M.J.: Detecting group differences: mining contrast sets. Data Mining Knowl. Disc. **5**(3), 213–246 (2001). Kluwer Academic Publishers
12. Omiecinski, E.R.: Alternative interest measures for mining associations in databases. IEEE Trans. Knowl. Data Eng. **15**(1), 57–69 (2003)
13. Szathmary, L., Napoli, A., Valtchev, P.: Towards rare itemset mining. In: Proceedings of the 19th IEEE International Conference on Tools with Artificial Intelligence (ICTAI 2007), pp. 305–312 (2007)

14. Dong, G., Li, J.: Mining border descriptions of emerging patterns from dataset pairs. Knowl. Inf. Syst. **8**(2), 178–202 (2005). Springer
15. Li, J., Dong, G., Ramamohanarao, K.: Making use of the most expressive jumping emerging patterns for classification. Knowl. Inf. Syst. **3**(2), 131–145 (2001). Springer
16. Ganter, B., Wille, R.: Formal Concept Analysis: Mathematical Foundations. Springer, Heidelberg (1999)
17. Pasquier, N., Bastide, Y., Taouil, R., Lakhal, L.: Efficient mining of association rules using closed itemset lattices. Inf. Syst. **24**(1), 25–46 (1999). Elsevier
18. Vychodil, V.: A new algorithm for computing formal concepts. In: Proceedings of The 19th European Meeting on Cybernetics and Systems Research, pp. 15–21 (2008)
19. Bron, C., Kerbosch, J.: Algorithm 457 - finding all cliques of an undirected graph. Commun. ACM **16**(9), 575–577 (1973)
20. Tomita, E., Tanaka, A., Takahashi, H.: The worst-case time complexity for generating all maximal cliques and computational experiments. Theor. Comput. Sci. **363**(1), 28–42 (2006). Elsevier
21. Tomita, E., Nakanishi, H.: Polynomial-time solvability of the maximum clique problem. In: Computing and Computational Intelligence, pp. 203–208. World Scientific and Engineering Academy and Society (2009)
22. Eppstein, D., Strash, D.: Listing all maximal cliques in large sparse real-world graphs. In: Pardalos, P.M., Rebennack, S. (eds.) SEA 2011. LNCS, vol. 6630, pp. 364–375. Springer, Heidelberg (2011). doi:10.1007/978-3-642-20662-7_31
23. Garey, M.R., Johnson, D.S.: Computers and Intractability: A Guide to the Theory of NPCompleteness. W.H. Freeman and Company, New York (1979)
24. Okubo, Y., Haraguchi, M.: An algorithm for extracting rare concepts with concise intents. In: Kwuida, L., Sertkaya, B. (eds.) ICFCA 2010. LNCS (LNAI), vol. 5986, pp. 145–160. Springer, Heidelberg (2010). doi:10.1007/978-3-642-11928-6_11

Methods of Hyperparameter Estimation in Time-Varying Regression Models with Application to Dynamic Style Analysis of Investment Portfolios

Olga Krasotkina[1(✉)], Vadim Mottl[2], Michael Markov[3],
Elena Chernousova[4], and Dmitry Malakhov[1]

[1] Moscow State University, Leninskie Gory, Moscow, Russia
o.v.krasotkina@yandex.ru
[2] Computing Center of the Russian Academy of Sciences, Moscow, Russia
[3] Markov Processes International, Summit, NJ, USA
[4] Moscow Institute of Physics and Technology, Moscow, Russia

Abstract. The problem of estimating time-varying regression inevitably concerns the necessity to choose the appropriate level of model volatility – ranging from the full stationarity of instant regression models to their absolute independence of each other. In the stationary case the number of regression coefficients, constituting the model parameter to be estimated, equals that of regressors, whereas the absence of any smoothness assumptions augments the dimension of the unknown vector by the factor of the time-series length. We consider here a family of continuously nested a priori probability distributions matching the specificity of time-varying data models, in which the dimension of the parameter is fixed, but the freedom of its values is softly constrained by a family of continuously nested a priori probability distributions, which contains a number of hyperparameters. The aim of this paper is threefold. First, in accordance with the specificity of the time-varying regression, we modify three commonly adopted methods of estimating hyperparameters in data models, namely, Leave-One-Out Cross Validation, Evidence Maximization and Hypothetical Cross Validation. Second, we experimentally compare these methods on both simulated and real-world data. Third, on the basis of the proposed technique we develop a new approach to the problem of detecting the hidden dynamics of an investment portfolio in respect to certain market or economic factors.

Keywords: Time-varying regression · Leave-one-out · Evidence maximization · Akaike information criterion · Style analysis of investment portfolios

1 Introduction

There exists a wide class of signal analysis problems in which it is required, for the given signal on the axis of a discrete argument $t = 1, \ldots, T$ (usually time), to estimate the sequence of hidden values of a sufficiently smoothly changing parameter

P. Perner (Ed.): MLDM 2017, LNAI 10358, pp. 431–450, 2017.
DOI: 10.1007/978-3-319-62416-7_31

$\boldsymbol{\beta} = \left(\boldsymbol{\beta}_t = (\beta_{t,1}, \ldots, \beta_{t,n}) \in \mathbb{R}^n, t = 1, \ldots, T \right) \in \mathbb{R}^{nT}$ considered as a nonstationary local model of the observed signal $\mathbf{y} = (y_t, t = 1, \ldots, T)$ at all the time points.

We study here the problem of time-varying regression estimation which appears in many applications and is an important particular case of nonstationary signal analysis. In this particular problem, the observed vector signal to be analyzed

$$(\mathbf{y}, \mathbf{x}) = ((y_t, \boldsymbol{x}_t), \; t = 1, \ldots, T) \tag{1}$$

consists of two components, namely, the vector sequence $\boldsymbol{x}_t \in \mathbb{R}^n$ and the scalar sequence $y_t \in \mathbb{R}$. It is assumed that, at each point of observation, the values y_t are conditionally independent normally distributed random variables with different mathematical expectations being linear functions of the respective known vector $\boldsymbol{x}_t$ and the same variance α:

$$y_t = \sum\nolimits_{i=1}^{n} \beta_{t,i} x_{t,i} + e_t = \boldsymbol{\beta}_t^T \boldsymbol{x}_t + e_t, \; \boldsymbol{\beta}_t \in \mathbb{R}^n, \; e_t \sim \mathcal{N}(e_t \mid 0, \alpha), \; E(e_t e_s) = \alpha, \; t = s, \; = 0, \; t \neq s. \tag{2}$$

The regression model $\boldsymbol{\beta}_t \in \mathbb{R}^n$ is assumed to be changing gradually over the observation interval $\boldsymbol{\beta} = (\boldsymbol{\beta}_t \in \mathbb{R}^n, t = 1, \ldots, T)$.

In this scenario, the dimension of the parameter vector is fixed $\boldsymbol{\beta} \in \mathbb{R}^{nT}$ and n times exceeds the number of observations T. This fact makes stable estimation of $\boldsymbol{\beta}$ practically impossible. As a regularization factor, it is commonly adopted to assume that the sequence of regression coefficients to be estimated is a hidden Markov random process [8–10]

$$\boldsymbol{\beta}_t = \mathbf{V}_t \boldsymbol{\beta}_{t-1} + \boldsymbol{\xi}_t, \; \boldsymbol{\xi}_t \sim \mathcal{N}(\boldsymbol{\xi}_t \mid 0, \lambda), \; E(\boldsymbol{\xi}_t) = 0, \; E(\boldsymbol{\xi}_t \boldsymbol{\xi}_s) = \begin{cases} \lambda, & t = s, \\ \lambda = 0, & t \neq s, \end{cases} \tag{3}$$

where the choice of the fixed sequence of nondegenerate matrices $\mathbf{V}_t$ defines the desired a priori model of the hidden dynamics of the regression coefficients, that starts from an unknown first value $\boldsymbol{\beta}_1 \sim \mathcal{N}(\boldsymbol{\beta}_1 \mid \mathbf{0}, \rho\mathbf{I}), \; \rho \to \infty$.

Actually, the assumption (3) is a Hidden Markov Model (HMM) of the sequence of unknown regression coefficients, which is a continuous generalization of the classical HMM originally designed for discrete data [1]. The HMM defines a family of continuously nested a priori probability distributions, in which the dimension of the general vector parameter is fixed $\boldsymbol{\beta} \in \mathbb{R}^{nT}$, but the freedom of its values is softly constrained by the variance of the inner noise λ. The observation noise variance $\alpha > 0$ (2) and the inner noise variance $\lambda > 0$ (3) are the two hyperparameters, whose values specify a particular signal model within the parametric family of them. In the stationary case $\alpha/\lambda \to \infty$, the number of regression coefficients, constituting the model parameter to be estimated, equals that of regressors n, whereas the absence of any smoothness assumptions $\alpha/\lambda \to 0$ augments the dimension of the unknown vector nT by the factor of the time-series length T.

There are many advantages of using models with varying coefficients. Varying coefficient models inherit simplicity and easy interpretation of the traditional linear models in contrast to nonparametric models [2, 3]. Nonparametric methods fail when they are applied to high- dimensional spaces. Varying coefficient models allow for avoiding the curse of dimensionality. Despite the popularity of varying coefficient models, most of them are yet intrinsically nonparametric and based on the so called kernel smoothing technique [4]. The hidden nonparametric structure of varying coefficients leads to a number of problems with choosing the kernel function and estimating the kernel smoothing hyperparameter. Three general approaches are cross-validation, information criterions and evidence maximization. Cross-validation is computationally expensive for smoothing kernel models, whereas information criterions and evidence require some approximation and asymptotic analysis. Fan and Zang [5] proposed a bandwidth selector that minimizes the evidence approximation criterion. An alternative approach is based on a penalized sum of squared residuals. In the classical nonparametric regression, Härdle et al. [6] prove that a penalized sum of squared residuals is asymptotically equivalent to cross-validation. The penalized least squares bandwidth selection is computationally less expensive than cross-validation. A technique was elaborated recently by Mammen and Park [7] for smooth backfitting estimators of the additive regression model.

In this paper, we consider the HMM as generating process for regression coefficients $\boldsymbol{\beta} = (\boldsymbol{\beta}_t,\ t = 1,\ldots,N)$ discussed earlier in the statistical literature during, at least, the last twenty years [8–10]. However, the cited papers considered primarily the problem of estimating the hidden sequence of regression coefficients, whereas the challenge of adjusting the hyperparameters to the observed time series $(\mathbf{y},\mathbf{x}) = ((y_t,\boldsymbol{x}_t),\ t = 1,\ldots,T)$ remained touched only perfunctorily. This approach does not exploit the nonparametric models at all and gives strictly optimal procedures for all hyperparameters estimation methods.

This paper is meant to fill in this gap. In Sect. 2, we recall the Bayesian approach to parameter estimation as applied to time-varying regression. In Sect. 3, we adapt three well-known methods of estimating hyperparameters to the specificity of choosing the most relevant level of time volatility in time-varying regression models – Leave-One-Out Cross Validation, Evidence Maximization and Hypothetical Cross Validation (a continuous generalization of the Akaike Information Criterion). Some modifications of these three methods are considered, respectively, in Subsects. 3.1, 3.2 and 3.3. In Sect. 4 we consider the problem of Dynamic Style Analysis of investment portfolios as a glowing class of practical problems concerned with the necessity of estimating a time-varying regression model. Finally, in Sect. 5, the results of an experimental study of hyperparameter estimation methods for dynamic portfolio data are presented.

2   The Bayesian Approach to Time-Varying Regression Estimation

We shall consider here as random only the observations $\mathbf{y} = (y_t,\ t = 1,\dots,T) \in \mathbb{R}^T$, and treat the sequence of regressors $\mathbf{x} = (x_t,\ t = 1,\dots,T) \in \mathbb{R}^{nT}$ as fixed (1). Then, the observation model (2) will produce the parametric family of conditional normal probability densities

$$\Phi(\mathbf{y}|\mathbf{x}, \boldsymbol{\beta}, \alpha) = \mathcal{N}(\mathbf{y}|\mathbf{x}^T\boldsymbol{\beta}, \alpha\mathbf{I}) = \frac{1}{\alpha^{T/2}(2\pi)^{T/2}} \exp\left\{-\frac{1}{2\alpha}(\mathbf{y} - \mathbf{X}^T\boldsymbol{\beta})^T(\mathbf{y} - \mathbf{X}^T\boldsymbol{\beta})\right\}, \quad (4)$$

where $\mathbf{X} = \begin{pmatrix} x_1 & 0 & \cdots & 0 \\ 0 & x_1 & \cdots & 0 \\ \vdots & \vdots & \ddots & \vdots \\ 0 & 0 & \cdots & x_T \end{pmatrix}$ $(nT \times T)$, $\mathbf{y} = \begin{pmatrix} y_1 \\ y_2 \\ \vdots \\ y_T \end{pmatrix} \in \mathbb{R}^T$, with mathematical expecta-

tion $\mathbf{X}^T\boldsymbol{\beta} = (x_t^T\beta_t,\ t = 1,\dots,T) \in \mathbb{R}^T$ and covariance matrix $\alpha\mathbf{I}(T \times T)$.

In accordance with the Markov assumption on the hidden sequence of regression coefficients $\boldsymbol{\beta} = (\beta_t \in \mathbb{R}^n, t = 1,\dots,T)$ (3), its a priori probability density is also normal:

$$\Psi(\boldsymbol{\beta}|\lambda) = \mathcal{N}(\boldsymbol{\beta}|\mathbf{0},\ \lambda\mathbf{B}_\rho^{-1}) = \frac{1}{\left|\lambda\mathbf{B}_\rho^{-1}\right|^{1/2}(2\pi)^{Tn/2}} \exp\left\{-\frac{1}{2}\boldsymbol{\beta}^T(\lambda^{-1}\mathbf{B}_\rho)\boldsymbol{\beta}\right\} =$$

$$\frac{1}{\rho^{n/2}\lambda^{n/2}(2\pi)^{n/2}} \exp\left(-\frac{1}{2\rho\lambda}\beta_1^T\beta_1\right) \prod_{t=2}^{T} \frac{1}{\lambda^{n/2}(2\pi)^{n/2}} \exp\left(-\frac{1}{2\lambda}(\beta_t - \mathbf{V}_t\beta_{t-1})^T(\beta_t - \mathbf{V}_t\beta_{t-1})\right). \quad (5)$$

The covariance matrix $\lambda\mathbf{B}_\rho^{-1}$ is inversion of the matrix $\lambda^{-1}\mathbf{B}_\rho$, which, in its turn, is easy to show to be block three-diagonal:

$$\frac{1}{\lambda}\mathbf{B}_\rho = \frac{1}{\lambda} \begin{pmatrix} \mathbf{V}_2^T\mathbf{V}_2 + (1/\rho)\mathbf{I} & -\mathbf{V}_2^T & 0 & \cdots & 0 & 0 \\ -\mathbf{V}_2 & \mathbf{V}_3^T\mathbf{V}_3 + \mathbf{I} & -\mathbf{V}_3^T & \cdots & 0 & 0 \\ 0 & -\mathbf{V}_3 & \mathbf{V}_4^T\mathbf{V}_4 + \mathbf{I} & \cdots & 0 & 0 \\ \vdots & \vdots & \vdots & \ddots & \vdots & \vdots \\ 0 & 0 & 0 & -\mathbf{V}_{N-1} & \mathbf{V}_N^T\mathbf{V}_N + \mathbf{I} & -\mathbf{V}_N^T \\ 0 & 0 & 0 & 0 & -\mathbf{V}_N & \mathbf{I} \end{pmatrix}, \left|\mathbf{B}_{\rho,\lambda}\right| = \left(\frac{1}{\rho}\right)^n. \quad (6)$$

Given the input signal $(\mathbf{y}, \mathbf{x}) = ((y_t, x_t),\ t = 1,\dots,T)$, the a posteriori probability density of the hidden sequence of regression coefficients $\boldsymbol{\beta} = (\beta_t \in \mathbb{R}^n, t = 1,\dots,T) \in \mathbb{R}^{nT}$ remains normal:

$$P(\boldsymbol{\beta}|\mathbf{y}, \mathbf{x}, \alpha, \lambda) = \mathcal{N}(\boldsymbol{\beta}|\hat{\boldsymbol{\beta}}_{\alpha/\lambda}, \hat{\mathbf{G}}_{\alpha,\lambda}) = \frac{1}{\left|\mathbf{G}_{\alpha,\lambda}\right|^{1/2}(2\pi)^{nN}} \exp\left\{-\frac{1}{2}(\boldsymbol{\beta} - \hat{\boldsymbol{\beta}}_{\alpha/\lambda})^T\hat{\mathbf{G}}_{\alpha,\lambda}^{-1}(\boldsymbol{\beta} - \hat{\boldsymbol{\beta}}_{\alpha/\lambda})\right\}. \quad (7)$$

It is easy to show that the a posteriori mathematical expectation $\hat{\boldsymbol{\beta}}_{\alpha,\lambda} = (\hat{\beta}_{1,\alpha,\lambda}, \dots, \hat{\beta}_{T,\alpha/\lambda}) \in \mathbb{R}^{nT}$, $\hat{\boldsymbol{\beta}}_{t,\alpha/\lambda} \in \mathbb{R}^{n}$, is nothing else than the Bayesian estimate of the hidden sequence of regression coefficients

$$
\begin{aligned}
\hat{\boldsymbol{\beta}}_{\alpha/\lambda} &= \arg\min\big((1/\lambda)\boldsymbol{\beta}^{T}\mathbf{B}_{\rho}\boldsymbol{\beta} + (1/\alpha)(\mathbf{y} - \mathbf{X}^{T}\boldsymbol{\beta})^{T}(\mathbf{y} - \mathbf{X}^{T}\boldsymbol{\beta})\big) \\
&= \arg\min\big((\alpha/\lambda)\boldsymbol{\beta}^{T}\mathbf{B}_{\rho}\boldsymbol{\beta} + (\mathbf{y} - \mathbf{X}^{T}\boldsymbol{\beta})^{T}(\mathbf{y} - \mathbf{X}^{T}\boldsymbol{\beta})\big), \quad \big(\mathbf{X}\mathbf{X}^{T} + (\alpha/\lambda)\mathbf{B}_{\rho}\big)\hat{\boldsymbol{\beta}}_{\alpha/\lambda} = \mathbf{X}\mathbf{y},
\end{aligned}
\tag{8}
$$

and the a posteriori covariance matrix has the form

$$
\begin{aligned}
\mathbf{G}_{\alpha,\lambda} &= \big((1/\alpha)\mathbf{X}\mathbf{X}^{T} + (1/\lambda)\mathbf{B}_{\rho}\big)^{-1} = \alpha\big(\mathbf{X}\mathbf{X}^{T} + (\alpha/\lambda)\mathbf{B}_{\rho}\big)^{-1} \\
&= \lambda\big((\lambda/\alpha)\mathbf{X}\mathbf{X}^{T} + \mathbf{B}_{\rho}\big)^{-1}(nT \times nT).
\end{aligned}
\tag{9}
$$

Despite the fact that the system of linear equations (8) may be of a huge dimensionality $(nT \times nT)$, it is computationally very easy to solve it without direct inverting its $(nT \times nT)$ matrix $\hat{\boldsymbol{\beta}}_{\alpha/\lambda} = \big(\mathbf{X}\mathbf{X}^{T} + (\alpha/\lambda)\mathbf{B}_{\rho}\big)^{-1}\mathbf{X}\mathbf{y}$, because this matrix is block three-diagonal with relatively small blocks $(n \times n)$ like (6). This fact allows for the double sweep method of solving the system, or, what is completely the same, for application of the quadratic dynamic programming procedure, which, in its turn, is completely equivalent to the Kalman filter and smoother. It is shown in [9–11] that a slight modification of the backward path of the Kalman smoother sequentially yields the optimal vectors of regression coefficients in the inverse order $\hat{\boldsymbol{\beta}}_{\alpha/\lambda} = (\hat{\boldsymbol{\beta}}_{T,\alpha/\lambda},$ $hat\beta_{T-1,\alpha/\lambda}, \dots, \hat{\boldsymbol{\beta}}_{1,\alpha/\lambda})$.

The computational complexity of such algorithms is proportional to $n^{3}T$, i.e., is linear with respect to the length T of the time series.

The result of training (8), i.e. the Bayesian estimate of the hidden sequence of regression coefficients, will drastically depend on the hyperparameters of the time-varying regression model – the observation noise variance α (4) and the variance of the inner noise λ in the hidden Markov model (5). These two hyperparameters jointly determine the time-volatility of the regression coefficients, ranging from the full stationarity of instant regression models if $\alpha/\lambda \to \infty$, and to their absolute independence of each other if $\alpha/\lambda \to 0$.

3 Methods of Estimating Hyperparameters of the Time-Varying Regression Model

3.1 Leave-One-Out Cross Validation

Cross validation is a statistical method for evaluation and comparison of learning methods, in particular, different values of hyperparameters within the same method, by dividing the data set into two segments – one used to train the model and the other used to validate it [12, 13]. Currently, cross-validation is widely accepted in data mining and machine learning community, and serves as a standard procedure for performance estimation and model (hyperparameter) selection.

The basic form of cross-validation is k-fold cross-validation, in which the data is first partitioned into k equally sized segments or folds. Subsequently, k iterations of training and validation are performed such that, within each iteration, a different fold of the data is held-out for validation while the remaining $k - 1$ folds are used for learning.

Leave-one-out cross validation (LOO) is a particular case of k-fold cross-validation, where k equals the number of instances in the data. In other words, nearly all the data except for a single observation are used in each iteration for training, and the model is tested on that single observation. An accuracy estimate obtained using LOO is known to be almost unbiased [14]. If T is the size of the training set, the LOO test of model accuracy requires, in the general case, just T runs of the training algorithm.

As applied to the time-varying regression model with some tentative ratio of hyperparameters α/λ, the LOO indicator is the sum

$$LOO_{\alpha/\lambda} = (1/T) \sum_{t=1}^{T} (y_t - \mathbf{x}_t^T \hat{\boldsymbol{\beta}}_{t,\,\alpha/\lambda}^{(t)})^2, \tag{10}$$

where $\hat{\boldsymbol{\beta}}_{\alpha/\lambda}^{(t)} = (\hat{\boldsymbol{\beta}}_{1,\alpha/\lambda}^{(t)}, \ldots, \hat{\boldsymbol{\beta}}_{t,\alpha/\lambda}^{(t)}, \ldots, \hat{\boldsymbol{\beta}}_{T,\alpha/\lambda}^{(t)})$ is the sequence of regression coefficients obtained from the full time series (1) by deleting (leaving one out) the respective element

$$(\mathbf{y}, \mathbf{x})^{(t)} = ((y_1, \mathbf{x}_1), \ldots, (y_{t-1}, \mathbf{x}_{t-1}), (?, ?), (y_{t+1}, \mathbf{x}_{t+1}), \ldots, (y_T, \mathbf{x}_T)). \tag{11}$$

Generally speaking, the leave-one out estimate $\hat{\boldsymbol{\beta}}_{\alpha/\lambda}^{(t)}$ is nothing else than the solution of the slightly modified system of linear equations (8)

$$\left(\left(\mathbf{X}\mathbf{X}^T - \mathbf{X}_0^{(t)} \mathbf{X}_0^{(t)T} \right) + (\alpha/\lambda) \mathbf{B}_\rho \right) \hat{\boldsymbol{\beta}}_{\alpha/\lambda}^{(t)} = (\mathbf{X} - \mathbf{X}_0^{(t)}) \mathbf{y}, \tag{12}$$

where $\mathbf{X}_0^{(t)}$ is $(nT \times T)$ matrix that, as distinct from the block diagonal matrix $\mathbf{X}$ of the same size (4), has only one non-zero $(n \times 1)$ block $\mathbf{x}_t$ on the main diagonal.

The following theorem shows that there is no need to repeatedly inverse matrix $\left(\left(\mathbf{X}\mathbf{X}^T - \mathbf{X}_0^{(t)} \mathbf{X}_0^{(t)T} \right) + (\alpha/\lambda) \mathbf{B}_\rho \right) (nT \times nT)$ in (12) for different left-out observations (t), because only the diagonal blocks $\mathbf{D}_{tt}(n \times n)$ of its full inversion are required.

Theorem. The LOO indicator (10) can be represented as

$$LOO_{\alpha/\lambda} = \frac{1}{T} \sum_{t=1}^{T} \frac{(y_t - \mathbf{x}_t^T \hat{\boldsymbol{\beta}}_{t,\alpha/\lambda})^2}{(1 - \mathbf{x}_t^T \mathbf{D}_{tt} \mathbf{x}_t)^2}. \tag{13}$$

Proof is based on Woodbury's formula of matrix identity.

In accordance with this Theorem, it is enough to solve the block three-diagonal system of linear equations (8) only once without any deletions. The matrices $\mathbf{D}_{tt}$ will be computed in the inverse order at the backward path of the Kalman smoother along with the full estimates $\hat{\boldsymbol{\beta}}_{\alpha/\lambda} = (\hat{\boldsymbol{\beta}}_{T,\alpha/\lambda}, \dots, \hat{\boldsymbol{\beta}}_{1,\alpha/\lambda})$. The detailed algorithm is to be found in [9, 10].

Thus, the LOO indicator can be easily computed for each ratio of hyperparameters $\gamma = \alpha/\lambda$ (10)

$$(\hat{\alpha}/\hat{\lambda}) = \arg \min LOO_{\alpha/\lambda}, \cdots \hat{\alpha} = \min LOO_{\alpha/\lambda}. \tag{14}$$

The less $LOO_{\alpha/\lambda}$ the higher adequacy of the model with the given value of the volatility parameter λ to the observed time series $((y_1, \boldsymbol{x}_1), \dots, (y_T, \boldsymbol{x}_T))$. The value

$$(\hat{\alpha}/\hat{\lambda}) = \arg \max PR^2(\alpha/\lambda), \ PR^2(\alpha/\lambda) = \left(1 - LOO_{\alpha/\lambda}/D(\mathbf{y})\right), \ D(\mathbf{y})$$
$$= (1/T) \sum_{t=1}^{T} (y_t)^2, \tag{15}$$

where $PR^2(\alpha/\lambda)$ is the so-called *Predicted R squared* criterion, should be taken as the smoothing parameter recommended for the given time series.

However, finding the estimates $\hat{\alpha}$ and $\hat{\lambda}$ of both hyperparameters is possible only by search on a sufficiently tense discrete set of ratios $\gamma = \alpha/\lambda$.

3.2 Marginal Likelihood Maximization

The principle of marginal likelihood maximization allows for finding an appropriate combination of both hyperparameters (α, λ) as result of an iterative procedure without computationally expensive search on a discrete grid.

Let the sequence of regressors $\mathbf{x} = (\boldsymbol{x}_t, \ t = 1, \dots, T) \in \mathbb{R}^{nT}$ in (1) be fixed. If $\Phi(\mathbf{y}|\mathbf{x}, \boldsymbol{\beta}, \alpha)$ (4) is the parametric family of conditional probability densities over all the feasible realizations of the observable signal, and $\Psi(\boldsymbol{\beta}|\lambda)$ (5) is the assumed parametric family of a priori densities over all the possible sequences of regression coefficient vectors, then the continuous mixture

$$F(\mathbf{y}|\mathbf{x}, \alpha, \lambda) = \int_{\mathbb{R}^{nT}} \Phi(\mathbf{y}|\mathbf{x}, \boldsymbol{\beta}, \alpha) \Psi(\boldsymbol{\beta}|\lambda) d\boldsymbol{\beta} \tag{16}$$

has the sense of the likelihood function over the range of hyperparameter combinations (α, λ). It is commonly adopted to say of this function as Marginal Likelihood or Evidence Function [15, 16]. Maximization of the marginal likelihood, which is completely defined by the given data set $(\mathbf{y}, \mathbf{x})$, is a popular way of choosing appropriate values of hyperparameters [17]:

$$(\hat{\alpha}, \hat{\lambda}) = \arg \max F(\mathbf{y}|\mathbf{x}, \alpha, \lambda). \tag{17}$$

This way is especially attractive in our particular case of time-varying regression, since both mixed and mixing distributions are normal in the integral (16), $\Phi(\mathbf{y}|\mathbf{x}, \boldsymbol{\beta}, \alpha) = \mathcal{N}(\mathbf{y}|\mathbf{x}^T\boldsymbol{\beta}, \alpha\mathbf{I})$ (4), $\Psi(\boldsymbol{\beta}|\lambda) = \mathcal{N}(\boldsymbol{\beta}|\mathbf{0}, \lambda\mathbf{B}_\rho^{-1})$ (5). Therefore, the random vector $\mathbf{y} = \mathbf{X}^T\boldsymbol{\beta} + \boldsymbol{\xi}$ (16) remains to be a normal variable, whose mathematical expectation and covariance matrix are, respectively, $E(\mathbf{y}) = \mathbf{0} \in \mathbb{R}^T$ and $Cov(\mathbf{y}) = \lambda \mathbf{X}^T \mathbf{B}_\rho^{-1}\mathbf{X} + \alpha \mathbf{I} \, (T \times T)$. So,

$$F(\mathbf{y} \,|\, \mathbf{x}, \alpha, \lambda) = \mathcal{N}\left(\mathbf{y} | \mathbf{0}, \, (\lambda \mathbf{X}^T \mathbf{B}_\rho^{-1}\mathbf{X} + \alpha \mathbf{I})\right)$$

$$= \frac{\left|\lambda \mathbf{X}^T \mathbf{B}_\rho^{-1}\mathbf{X} + \alpha \mathbf{I}\right|^{-1/2}}{(2\pi)^{T/2}} \exp\left\{-\frac{1}{2}\mathbf{y}^T\left(\lambda \mathbf{X}^T \mathbf{B}_\rho^{-1}\mathbf{X} + \alpha \mathbf{I}\right)^{-1}\mathbf{y}\right\},$$

$$\ln F(\mathbf{y}|\mathbf{x}, \alpha, \lambda) = const - \frac{1}{2}\ln\left|\lambda \mathbf{X}^T \mathbf{B}_\rho^{-1}\mathbf{X} + \alpha \mathbf{I}\right| - \frac{1}{2}\mathbf{y}^T\left(\lambda \mathbf{X}^T \mathbf{B}_\rho^{-1}\mathbf{X} + \alpha \mathbf{I}\right)^{-1}\mathbf{y}$$

$$= const + \frac{1}{2}\left[T\ln\frac{\alpha}{\lambda} - T\ln\alpha - \ln\left|\mathbf{X}^T \mathbf{B}_\rho^{-1}\mathbf{X} + \frac{\alpha}{\lambda}\mathbf{I}\right| - \frac{1}{\alpha}(\mathbf{y} - \mathbf{X}^T \hat{\boldsymbol{\beta}}_{\alpha/\lambda})^T(\mathbf{y} - \mathbf{X}^T \hat{\boldsymbol{\beta}}_{\alpha/\lambda})\right], \tag{18}$$

where $\gamma = \alpha/\lambda > 0$ is smoothness parameter.

It can be shown that the logarithmic marginal likelihood is unimodal and differentiable with respect to both hyperparameters α and $\gamma = \alpha/\lambda$. Differentiation of the logarithmic marginal likelihood (18) with respect to α immediately yields the best observation noise variance as function of smoothness $\gamma = \alpha/\lambda$:

$$\alpha = (1/T)(\mathbf{y} - \mathbf{X}^T \hat{\boldsymbol{\beta}}_{\alpha/\lambda})^T(\mathbf{y} - \mathbf{X}^T \hat{\boldsymbol{\beta}}_{\alpha/\lambda}). \tag{19}$$

Substitution of this function back into (18) turns the problem of maximizing the marginal likelihood into that of finding the maximum point of a unimodal function of one variable:

$$\left\{T\ln\gamma - T\ln\left[(1/T)(\mathbf{y} - \mathbf{X}^T \hat{\boldsymbol{\beta}}_\gamma)^T(\mathbf{y} - \mathbf{X}^T \hat{\boldsymbol{\beta}}_\gamma)\right] - \ln\left|\mathbf{X}^T \mathbf{B}_\rho^{-1}\mathbf{X} + \gamma \mathbf{I}\right|\right\} \to \max(\gamma).$$

To simplify the numerical solution of the latter optimization problem, we represent the logarithmic determinant as function of eigenvalues $(\mu_1, \ldots, \mu_T)$ of the positive definite matrix $\mathbf{X}^T \mathbf{B}_\rho^{-1}\mathbf{X}(T \times T)$, which is completely defined by the data set. It is enough to compute them only once, because the eigenvalues of all the matrices $\mathbf{X}^T \mathbf{B}_\rho^{-1}\mathbf{X} + \gamma \mathbf{I}$ are simple functions of them $\mu_i + \gamma$, $i = 1, \ldots, T$. Since the determinant of any matrix is product of its eigenvalues, we have

$$\ln\left|\mathbf{X}^T \mathbf{B}_\rho^{-1}\mathbf{X} + \gamma \mathbf{I}\right| = \ln\prod_{i=1}^{T}(\mu_i + \gamma) = \sum_{i=1}^{T}\ln(\mu_i + \gamma).$$

The marginal likelihood estimate of the smoothness parameter is the maximum point

$$\hat{\gamma} = \arg\max\left\{T\ln\gamma - \ln(\mathbf{y} - \mathbf{X}^T \hat{\boldsymbol{\beta}}_\gamma)^T(\mathbf{y} - \mathbf{X}^T \hat{\boldsymbol{\beta}}_\gamma) - \sum_{i=1}^{T}\ln(\mu_i + \gamma)\right\},$$

which can be found by any appropriate method, for instance, by the golden section procedure. The estimates of the original hyperparameters immediately follow from (19):

$$\hat{\alpha} = \frac{1}{T}(\mathbf{y} - \mathbf{X}^T \hat{\boldsymbol{\beta}}_{\hat{\gamma}})^T(\mathbf{y} - \mathbf{X}^T \hat{\boldsymbol{\beta}}_{\hat{\gamma}}), \cdots \hat{\lambda} = \frac{\hat{\alpha}}{\hat{\gamma}}.$$

An advantage of the marginal likelihood maximization method is the fact that it yields both hyperparameters $(\hat{\alpha}, \hat{\lambda})$ without discrete search.

3.3 Hypothetical Cross Validation: Continuous Generalization of the Akaike Information Criterion

The traditional Akaike Information Criterion (AIC) [17, 18] is adopted in data analysis as a simple and effective means of adjusting the most adequate model to the available data set among a discrete succession of nested parametric model classes.

Let the given data set $\mathbf{y} = (y_t,\ t = 1, \ldots, T)$ be considered as a sample of random variables whose actual joint distribution obeys some unknown density $\Phi^*(\mathbf{y})$, whereas the observer assumes a parametric family $\Phi(\mathbf{y}|\boldsymbol{\beta})$, $\boldsymbol{\beta} \in \mathbb{R}^m$. It is a typical case that the parameter dimension m is certainly too large for both "actual" density $\Phi^*(\mathbf{y})$ and size T of the sample, what makes senseless the maximum-likelihood estimate

$$\hat{\boldsymbol{\beta}} = \arg\max \Phi(\mathbf{y}|\boldsymbol{\beta}). \tag{20}$$

The assumption that underlies the traditional AIC is that elements of $\boldsymbol{\beta}$ are naturally ordered by their decreasing "importance" $(\beta_1, \beta_2, \ldots, \beta_m)$. The idea is to truncate the parameter vector by putting $\beta_i = 0$ for $n < i \le m$:

$$\boldsymbol{\beta} = (\boldsymbol{\beta}_n, \boldsymbol{\beta}_{m-n}), \ \boldsymbol{\beta}_n \in \mathbb{R}^n, \ \boldsymbol{\beta}_{m-n} = \mathbf{0} \in \mathbb{R}^{m-n}.$$

So, the density family $\Phi(\mathbf{y}|\boldsymbol{\beta})$ turns into a succession of nested families

$$\Phi(\mathbf{y}|\boldsymbol{\beta} = (\boldsymbol{\beta}_n, \mathbf{0})),\ \mathbb{R}^{n_{min}} \subset \cdots \subset \mathbb{R}^{n_{max}}.$$

The classical AIC is a criterion of choosing the dimension as the most appropriate level of model complexity $\hat{n}(\mathbf{y}) = \text{argmax}_n[\ln \Phi(\mathbf{y}|(\boldsymbol{\beta}_n(\mathbf{y}), \mathbf{0})) - n]$ instead of the plain likelihood maximization (20). However, this formula was designed under the assumption that $\nabla^2_{\boldsymbol{\beta}_n \boldsymbol{\beta}_n} \ln \Phi(\mathbf{y}|(\boldsymbol{\beta}_n, \mathbf{0}))$ is a full rank matrix at the point of the maximum likelihood, and, so, the estimate $\hat{\boldsymbol{\beta}}_n(\mathbf{y})$ is unique. To cover the most general case, the penalty n should be replaced by the rank of this matrix:

$$\hat{n}(\mathbf{y}) = argmax_n \left\{ \ln \Phi\left(\mathbf{y}|\left(\hat{\boldsymbol{\beta}}_n(\mathbf{y}), \mathbf{0}\right)\right) - Rank\left[\nabla^2_{\boldsymbol{\beta}_n \boldsymbol{\beta}_n} \ln \Phi\left(\mathbf{y}|\left(\hat{\boldsymbol{\beta}}_n(\mathbf{y}), \mathbf{0}\right)\right)\right] \right\}. \tag{21}$$

The main idea underlying the AIC is the view of the maximum point of Kulback similarity between the model and universe

$$n^* = \arg\max_n \int \left[\ln \Phi(\mathbf{y}|(\boldsymbol{\beta}_n^*, \mathbf{0})) \right] \Phi^*(\mathbf{y}) d\mathbf{y} \tag{22}$$

just as the desired dimension under the assumption that $\Phi^*(\mathbf{y}) = \Phi(\mathbf{y}|(\boldsymbol{\beta}_{n^*}^*, \mathbf{0}))$ with some value $(\boldsymbol{\beta}_n^*, \mathbf{0})$ cut out from the unknown $\boldsymbol{\beta}^* = (\boldsymbol{\beta}_1^*, \ldots, \boldsymbol{\beta}_m^*)$.

One of the first applications of AIC was modeling of a nonstationary signal on the discrete time axis by dividing the time interval into an unknown number n of blocks and adjusting a locally stationary autoregression model of a fixed order k to each of them [19].

After Akaike's pioneering paper [18], numerous modifications of the information-based parsimony principle in model building were proposed [20], among which the Bayesian Information Criterion (BIC) [17] has found the most wide adoption.

However, all the known model selection criteria are aimed at the problem of choosing the most appropriate model within a succession of rigidly nested model classes. The search for ways of generalizing the classical AIC [21] was prompted just by the needs of the time-varying regression problem considered in this paper. The family of continuously nested a priori probability distributions, that matches the specificity of the time-varying data model, is defined by the hidden Markov model (3) and its interaction with the observation model (2). In this family, the dimension of the time-varying parameter $\boldsymbol{\beta} = (\boldsymbol{\beta}_t \in \mathbb{R}^n, \ t = 1, \ldots, T) \in \mathbb{R}^{nT}$ is fixed, but its time-volatility is softly constrained by the ratio of two hyperparameters $\gamma = \alpha/\lambda$.

Let the sequence of regressors $\mathbf{x} = (x_t \in \mathbb{R}^n, \ t = 1, \ldots, T) \in \mathbb{R}^{nT}$ be fixed. With the purpose of extending the computationally perfect Akaike's principle onto the case of data models with continuously changing effective dimension of unknown parameter, just as in [21], we consider in this paper the parametric model of the unknown universe $F^*(\mathbf{y}|\mathbf{x})$ of observable sequences $\mathbf{y} = (y_t, \ t = 1, \ldots, T)$ as a continuous mixture of conditional densities from the given family $\Phi(\mathbf{y}|\mathbf{x}, \boldsymbol{\beta}, \hat{\alpha})$ (4), $\boldsymbol{\beta} \in \mathbb{R}^{nT}$, with some "actual" unknown mixing density $\Psi^*(\boldsymbol{\beta})$:

$$F^*(\mathbf{y}|\mathbf{x}) = \int_{\mathbb{R}^{nT}} \Phi(\mathbf{y}|\mathbf{x}, \boldsymbol{\beta}, \hat{\alpha}) \Psi^*(\boldsymbol{\beta}) d\boldsymbol{\beta}, \tag{23}$$

where the observation noise variance $\hat{\alpha}$ in (4) is assumed to be known.

The idea of hypothetical cross validation is based on the hypothesis about two independent realizations of the observable signal $\mathbf{y}$ and $\tilde{\mathbf{y}}$ produced by the same known random mechanism $\Phi(\mathbf{y}|\mathbf{x}, \boldsymbol{\beta}, \hat{\alpha})$. It is assumed that one of them, let it be $\tilde{\mathbf{y}}$, is used for estimating the hidden random parameter with some tentative variance λ of the inner noise in the hidden Markov model (5), and

$$\hat{\boldsymbol{\beta}}_\gamma(\tilde{\mathbf{y}}) = \left(\mathbf{X}\mathbf{X}^T + (\hat{\alpha}/\lambda)\mathbf{B}_\rho\right)^{-1}\mathbf{X}\tilde{\mathbf{y}},$$

is the resulting random estimate (8).

It appears natural to choose the ratio $\gamma = \hat{\alpha}/\lambda$, which maximizes the Kullback similarity between, on the one hand, the density $\Phi(\mathbf{y}|\mathbf{x}, \boldsymbol{\beta}, \hat{\alpha})$ and, on the other hand, the estimated density $\Phi\left(\mathbf{y}|\mathbf{x}, \hat{\boldsymbol{\beta}}_\gamma(\tilde{\mathbf{y}}), \hat{\alpha}\right)$, both defined in the same space $\mathbb{R}^N$, like in the Akaike Information Criterion (22):

$$\int_{\mathbb{R}^T} \left[\ln \Phi\left(\mathbf{y}|\mathbf{x}, \hat{\boldsymbol{\beta}}_{\hat{\alpha}/\lambda}(\tilde{\mathbf{y}}), \hat{\alpha}\right)\right] \Phi(\mathbf{y}|\mathbf{x}, \boldsymbol{\beta}, \hat{\alpha})d\mathbf{y} \rightarrow \max(\lambda).$$

Since both mental "training data" $\tilde{\mathbf{y}}$ and initial parameter $\boldsymbol{\beta}$ are also random here, the double mathematical expectation over these variables should be maximized:

$$\int_{\mathbb{R}^{nT}} \left\{\int_{\mathbb{R}^T} \left(\int_{\mathbb{R}^T} \left[\ln \Phi\left(\mathbf{y}|\mathbf{x}, \hat{\boldsymbol{\beta}}_{\hat{\alpha}/\lambda}(\tilde{\mathbf{y}}), \hat{\alpha}\right)\right] \Phi(\mathbf{y}|\mathbf{x}, \boldsymbol{\beta}, \hat{\alpha})d\mathbf{y}\right) \Phi(\tilde{\mathbf{y}}|\mathbf{x}, \boldsymbol{\beta}, \hat{\alpha})d\tilde{\mathbf{y}}\right\} \Psi^*(\boldsymbol{\beta})d\boldsymbol{\beta} \rightarrow \max(\lambda). \quad (24)$$

This criterion expresses the idea of hypothetical cross validation.

Theorem ([21]). Under the assumptions (4) and (5)

$$\int_{\mathbb{R}^T} \left(\int_{\mathbb{R}^T} \left[\ln \Phi\left(\mathbf{y}|\mathbf{x}, \hat{\boldsymbol{\beta}}_{\hat{\alpha}/\lambda}(\tilde{\mathbf{y}}), \hat{\alpha}\right)\right] \Phi(\mathbf{y}|\mathbf{x}, \boldsymbol{\beta}, \hat{\alpha})d\mathbf{y}\right) \Phi(\tilde{\mathbf{y}}|\mathbf{x}, \boldsymbol{\beta}, \hat{\alpha})d\tilde{\mathbf{y}}$$
$$= \int_{\mathbb{R}^T} \left(\ln \Phi\left(\mathbf{y}|\mathbf{x}, \hat{\boldsymbol{\beta}}_{\hat{\alpha}/\lambda}(\mathbf{y}), \hat{\alpha}\right)\right) \Phi(\mathbf{y}|\mathbf{x}, \boldsymbol{\beta}, \hat{\alpha})d\mathbf{y} - Tr\left\{\mathbf{X}\mathbf{X}^T \left(\mathbf{X}\mathbf{X}^T + (\hat{\alpha}/\lambda)\mathbf{B}_{\rho,\lambda}\right)^{-1}\right\}.$$

Thus, criterion (24) has now the form

$$\int_{\mathbb{R}^{nT}} \left\{\int_{\mathbb{R}^T} \left(\ln \Phi\left(\mathbf{y}|\mathbf{x}, \hat{\boldsymbol{\beta}}_{\hat{\alpha}/\lambda}(\mathbf{y}), \hat{\alpha}\right)\right) \Phi(\mathbf{y}|\mathbf{x}, \boldsymbol{\beta}, \hat{\alpha})d\mathbf{y}\right\} \Psi^*(\boldsymbol{\beta})d\boldsymbol{\beta} =$$
$$\int_{\mathbb{R}^T} \left(\ln \Phi\left(\mathbf{y}|\mathbf{x}, \hat{\boldsymbol{\beta}}_{\hat{\alpha}/\lambda}(\mathbf{y}), \hat{\alpha}\right)\right) F^*(\mathbf{y}|\mathbf{x})d\mathbf{y} - Tr\left\{\mathbf{X}\mathbf{X}^T \left(\mathbf{X}\mathbf{X}^T + (\hat{\alpha}/\lambda)\mathbf{B}_\rho\right)^{-1}\right\} \rightarrow \max(\lambda), \quad (25)$$

but it is impossible to use it, as before, since the data source distribution $F^*(\mathbf{y}|\mathbf{x})$ remains unknown since we don't know the density $\Psi^*(\boldsymbol{\beta})$ in (23), which is a "secret" of the nature.

As an additional heuristics, we propose to ignore the mathematical expectation in (25), i.e. consider the only available signal $\mathbf{y} = (y_t, t = 1, \ldots, T)$ as its unbiased estimate:

$$J(\lambda|\mathbf{y}, \mathbf{x}, \hat{\alpha}) = \ln \Phi\left(\mathbf{y}|\mathbf{x}, \hat{\beta}_{\hat{\alpha}/\lambda}(\mathbf{y}), \hat{\alpha}\right) - Tr\left\{\mathbf{X}\mathbf{X}^T\left(\mathbf{X}\mathbf{X}^T + (\hat{\alpha}/\lambda)\mathbf{B}_\rho\right)^{-1}\right\} \to \max(\lambda). \quad (26)$$

Such a heuristics is especially natural if the hidden Markov process (3) is ergodic.

Thus, in accordance with (4), to find the most appropriate value of time-volatility parameter $\hat{\lambda}$, it is enough to solve the one-dimensional optimization problem:

$$\hat{\lambda} = \arg\min_{\lambda}\left\{(1/2\hat{\alpha})\left(\mathbf{y} - \mathbf{X}^T\hat{\beta}_{\hat{\alpha}/\lambda}\right)^T\left(\mathbf{y} - \mathbf{X}^T\hat{\beta}_{\hat{\alpha}/\lambda}\right) + Tr\left[\mathbf{X}\mathbf{X}^T\left(\mathbf{X}\mathbf{X}^T + (\hat{\alpha}/\lambda)\mathbf{B}_\rho\right)^{-1}\right]\right\}. \quad (27)$$

However, as distinct from the methods considered in precedent Sects. 3.1 and 3.2, the hypothetical cross validation yields an estimate $\hat{\lambda}$ of only the inner noise variance in the a priori hidden Markov model (3), whereas the observation noise variance $\hat{\alpha}$ in (4) is to be estimated by some other method.

4   Dynamic Style Analysis of Investment Portfolios

A typical example of a time-varying regression estimation problem is that of determining major market factors affecting performance of an investment portfolio. We consider here a dynamic generalization of this problem, the original static formulation of which belongs to William Sharpe, 1990 Nobel Prize winner in Economics [22].

4.1   Returns Based Style Analysis by William Sharpe

Let the capital of an investment company be fully invested in assets of n kinds in the proportion $\boldsymbol{\beta} = (\beta_0, \beta_1, \ldots, \beta_n)$, $\sum_{i=0}^{n} \beta_i = 1$, $\beta_i \geq 0$. The set of assets or asset classes for which $\beta_i > 0$ is called the company's portfolio, and vector $\boldsymbol{\beta} = (\beta_1, \ldots, \beta_n)$ represents the capital share within it.

The notation β_0 is reserved here for a short-term instrument, such as bank deposit in an interest bearing account, often referred to as a risk-free asset.

The non-negativity constraints $\beta_i \geq 0$ express the assumption that assets can be purchased only using inner capital without borrowing money or assets from outside sources, or, as it is commonly adopted to say in portfolio management, without holding short (negative) positions – most mutual funds are not allowed to do this. Investment companies which take negative (short) positions $\beta_i < 0$ that correspond to borrowed assets are called hedge funds.

Let $z_{beg,i}$ and $z_{end,i}$ be the prices of assets at the beginning and the end of some time interval called the holding period, respectively, $z_{beg,p}$ and $z_{end,p}$ will be the cost of the portfolio as a whole. The ratio $y = (z_{end,p} - z_{beg,p})/z_{beg,p}$ is referred to as the portfolio return for the holding period, and $x_i = (z_{end,i} - z_{beg,i})/z_{beg,i}$ are returns of assets. If asset classes are considered, their expected returns x_i are usually estimated by special analytical companies as return indices.

In Sharpe's model, the periodic return of a portfolio is equal to the linear combination of periodic returns of the assets or classes of assets with coefficients having the meaning of the portfolio's shares invested in each of them at the beginning of the period under the assumption that the entire budget is fully spent on the investment:

$$y = a + x + \varepsilon = a + \sum_{i=0}^{n} \beta_i \tilde{x}_i + \varepsilon = a + \boldsymbol{\beta}^T \tilde{x} + \varepsilon, \text{ where } \varepsilon \text{ is random noise.} \quad (28)$$

Portfolio managers are, as a rule, very secretive about what they buy and sell, and the fractional structure of the portfolio $\boldsymbol{\beta} = (\beta_0, \beta_1, \ldots, \beta_n)$ is typically concealed from the public. Such information, having been recovered, would be of great interest for those who monitor the portfolio. In particular, it would provide investors in this portfolio with an early warning.

For consecutive time moments $t = 1, 2, 3, \ldots$ making the succession of holding periods, for instance, stock exchange workdays, months, and quarters, the periodic returns of both portfolio and assets form a number of time series, respectively, y_t and $x_{i,t}$. These values can be considered as known, because any investment company must regularly publish the return of its portfolio, and returns of assets classes are regularly published in both on-line and printed financial media. In the simplest case of daily periodicity, the return of each single asset can be immediately computed from the known changes in its price. This reasoning underlies Sharpe's principle of estimating the portfolio structure, which leads to the quadratic programming problem:

$$\begin{cases} (\hat{a}, \hat{\beta}_0, \hat{\beta}_1, \ldots, \hat{\beta}^{(n)}) : \sum_{t=0}^{N} \left(y_t - a - \sum_{i=0}^{n} \beta_i x_{i,t} \right)^2 \to \min, \\ \beta_i \geq 0, \sum_{i=0}^{n} \beta_i = 1. \end{cases} \quad (29)$$

An actual portfolio may contain hundreds and thousands of kinds of assets. Sharpe proposed approximation of the resulting portfolio return, being a linear combination of such an immense set of asset returns, by a small number of market indices each of which represents a certain investment style. The principle of approximation (29), having been used by Sharpe for just this purpose, is commonly adopted in the modern investment analysis as Returns Based Style Analysis (RBSA) [22].

An essential limitation of the model (28) is the unlikely assumption on the invariability of the portfolio structure during the entire holding period. To overcome this drawback, Sharpe used a moving window of some preset length [22], but this technique remains based on the forced assumption that the structure of the portfolio is constant inside the window.

4.2 Dynamic Returns Based Style Analysis

In contrast to Sharpe's static model (28), we consider here a model in which the fractional structure of the portfolio is changing in time. Let $t = 1, 2, \ldots, T$ be a sequence of holding periods, for instance, stock exchange workdays, weeks, months, quarters or years, and

$$\boldsymbol{\beta} = (\boldsymbol{\beta}_t, \ t = 1, \ldots, T), \ \boldsymbol{\beta}_t = (\beta_{0,t}, \beta_{1,t}, \ldots, \beta_{n,t}), \ \sum_{i=0}^{n} \beta_{i,t} = 1, \qquad (30)$$

be the respective sequence of the portfolio's fractional structures at the end of each period.

The new dynamic model of periodic portfolio returns can be written as follows:

$$y_t \cong \sum_{i=0}^{n} \beta_{i,t} x_{i,t} = \boldsymbol{\beta}_t^T \boldsymbol{x}_t. \qquad (31)$$

It is required to estimate the vectors of time-varying fractional asset weights $\boldsymbol{\beta}_t = (\beta_{0,t}, \beta_{1,t}, \ldots, \beta_{nt}) \in \mathbb{R}^{n+1}$ (31) in contrast to time-invariant vector $\boldsymbol{\beta} = (\beta_0, \beta_1, \ldots, \beta_n)$ in the static model (28).

Equation (31) represents the time series of periodic returns on the portfolio under analysis y_t as time-varying regression of the time-varying returns of assets $x_{i,t}$ $i = 0, 1, \ldots, n, t = 1, \ldots, T$. The key element of the proposed Dynamic Style Analysis (DSA) is the treatment of fractional asset weights $\boldsymbol{\beta}_t = (\beta_t^{(1)}, \ldots, \beta_t^{(n)})$ as a hidden process assumed a priori to possess the Markov property (5).

In these terms, the solution of the DSA problem estimates the unknown time-varying fractional structure of the portfolio $\boldsymbol{\beta}_t = (\beta_t^{(1)}, \ldots, \beta_t^{(n)})$ from its input-output observations

$$((y_t, \boldsymbol{x}_t), i = 1, \ldots, n, t = 1, \ldots, N).$$

Thus, we arrive at the problem of time-varying regression estimation (5), in which choosing the appropriate value of time volatility parameter λ is especially important. Small value of the variance λ in the Markov assumption (3) is expected to constrain the time volatility of the hidden portfolio structure in a more delicate manner than the moving-widow technique. This expectation allows us to omit the non-negativity restriction $\beta_{i,t} \geq 0$ which, in the static case $\beta_i \geq 0$ (29), essentially regularized Sharpe's technique. This means that our analysis will be applicable to hedge funds which take negative (short) positions $\beta_{i,t} < 0$, which correspond to borrowed assets.

5   Experimental Study of Hyperparameter Estimation Methods for Dynamic Portfolio Data

5.1   Experiments on Simulated Data

5.1.1   Semi-simulated Ground-Truth Data Set

The idea of ground-truth experiments on simulated data is to comparatively test the three methods of choosing hyperparameters (α, λ) when the random vector signal to be analyzed $(\boldsymbol{y}, \boldsymbol{x}) = ((y_t, \boldsymbol{x}_t), t = 1, \ldots, T)$ (1) is formed by the known sequence of regression coefficients (2). The comparison criterion is discrepancy between the known $\boldsymbol{\beta} = (\boldsymbol{\beta}_t \in \mathbb{R}^n, t = 1, \ldots, T)$ and estimated $\hat{\boldsymbol{\beta}}_{\alpha,\lambda} = (\hat{\boldsymbol{\beta}}_{\alpha,\lambda,t} \in \mathbb{R}^n, t = 1, \ldots, T)$ sequences.

In accordance with the Bayesian approach to time-varying regression estimation, we consider as random only the hidden coefficients $\boldsymbol{\beta} = (\boldsymbol{\beta}_t \in \mathbb{R}^n, t = 1, \ldots, T)$ and the observations $\mathbf{y} = (y_t, t = 1, \ldots, T) \in \mathbb{R}^T$, whereas the sequence of regressors $\mathbf{x} = (x_t, t = 1, \ldots, T) \in \mathbb{R}^{nT}$ is treated as fixed.

We call this group of our experiments semi-simulated because the regressors are not randomly generated but taken from the real universe of stocks, which circulate in the stock market, and their characteristics.

5.1.2 Fama-French Three-Factor Model

The backbone of the Returns Based Style Analysis (RBSA) is the Capital Asset Pricing Model (CAPM), development of which was started by William Sharpe as the primary contributor [23]. In CAPM, a single index x is used as a proxy to represent the equilibrium return of the market (28). As the next step, this index was extended in RBSA to allow for multiple market proxy indices $x = \sum_{i=0}^{n} \beta_i \tilde{x}_i$.

As an extension of CAPM, the Fama-French three factor model [24] uses, in addition to the traditional market excess return $x = x_1$, also two extra factors: x_2- size premium factor SMB (small minus big) based on company size, and x_3 - the value premium factor HML (high book-to-market ratio minus low book-to-market ratio).

The SMB and HML factors are constructed using the 6 value-weight portfolios formed on size and book-to-market: Large Value, Large Growth, Large Neutral, Small Value Small Growth and Small Neutral. Thus, SMB factor is designed to account for the historical tendency for the stocks of firms with smaller market capitalizations to outperform the stocks of firms with larger capitalizations. It is calculated as the simple average of the returns on the three Small portfolios above minus the average of the returns on the three "large" ones.

The value premium factor HML is the average of the returns on the two Value portfolios above ("High" book-to-market) minus the average of the returns on the two Growth portfolios ("Low" book-to-market).

In our experiments, we used the three-dimensional sequence of Fama-French Indices for 200 holding periods $(\mathbf{x}_t = (x_{1,t}, x_{2,t}, x_{3,t}), t = 1, \ldots, T = 200)$, in which $x_{1,t}$ is the market index of a conventional portfolio, and two latter ones $(x_{2,t}, x_{3,t})$ are, respectively, the SMB and HML indices.

5.1.3 Experimental Set-Up

The experiments are built by the following scheme:

- simulating two sets of 200 random time series $\mathbf{y} = (y_t, t = 1, \ldots, T)$ each with two known relatively smooth sequences of three-dimensional regression coefficients $(\boldsymbol{\beta}_1^{*1}, \ldots, \boldsymbol{\beta}_T^{*1})$ and $(\boldsymbol{\beta}_1^{*2}, \ldots, \boldsymbol{\beta}_T^{*2})$ having essentially different styles of oscillation;
- inferring the Bayesian estimates $(\hat{\boldsymbol{\beta}}_{\alpha, \lambda, 1}, \ldots, \hat{\boldsymbol{\beta}}_{\alpha, \lambda, T})$ from each of 200×2 realization $(\mathbf{x}, \mathbf{y})$ by the Bayesian criterion (8) with choosing the time-volatility parameter λ, for each of them, via three methods: Leave-one-out cross validation (Sect. 3.1), Maximum likelihood maximization (Sect. 3.2) and Hypothetical cross validation (Sect. 3.3);
- comparing the six averaged estimation errors.

One of the sequences of ground-truth regression coefficients $(\boldsymbol{\beta}_1^{*1}, \ldots, \boldsymbol{\beta}_T^{*1})$ was built in full accordance with the theoretically assumed normal Markov model (3). On the contrary, the second sequence $(\boldsymbol{\beta}_1^{*2}, \ldots, \boldsymbol{\beta}_T^{*2})$ was formed by three sinusoidal functions of time $\beta_{i,t}^{*2} = 4\sin\left((2\pi/T)t + (2\pi/3)(i-1)\right)$ mutually shifted by phase. For the normal Markov model we vary the time-volatility parameter λ from full stationarity of the instant models to complete independence them from each other. For sinusoid model we change the frequency $1/T$ of regression coefficients.

All the 400 time series $((y_1, \boldsymbol{x}_1), \ldots, (y_T, \boldsymbol{x}_T))$ had the length $T = 200$ and were simulated with the same sequence of Fama-French indices $\mathbf{x} = (\boldsymbol{x}_t, t = 1, \ldots, T)$. The random output values y_t (2) were generated with 10% noise variance $\alpha = 0.1\left((1/T)\sum_{t=1}^{T} \boldsymbol{x}_t^T \boldsymbol{\beta}_t^*\right)$ depending on the respective ground-truth sequence of regression coefficients.

5.1.4   Result of Experiments

For each of time series, the two estimates of regression coefficients $(\hat{\boldsymbol{\beta}}_{\alpha,\lambda,1}, \ldots, \hat{\boldsymbol{\beta}}_{\alpha,\lambda,T})$ were compared with the respective ground-truth model by the criterion of relative mean deflection of the estimate from the model

$$\varepsilon = \sum_{t=1}^{T} (\hat{\boldsymbol{\beta}}_{\alpha,\hat{\lambda},t} - \boldsymbol{\beta}_t^*)^T (\hat{\boldsymbol{\beta}}_{\alpha,\hat{\lambda},t} - \boldsymbol{\beta}_t^*) \Big/ \sum_{t=1}^{T} (\boldsymbol{\beta}_t^*)^T \boldsymbol{\beta}_t^*.$$

We estimate also the smoothness power $\gamma = \hat{\alpha}/\hat{\lambda}$ in each experiment.
We obtained the following results:

Time variability estimation criterion	$\lambda = 10$			$\lambda = 0.1$			$\lambda = 0.001$		
	ε	$E[\gamma]$	$E[\gamma^2]$	ε	$E[\gamma]$	$E[\gamma^2]$	ε	$E[\gamma]$	$E[\gamma^2]$
Leave-one-out cross validation	0.0398	852.7	363.6	0.0261	744.3	205.2	0.0071	9066	2953.9
Marginal likelihood maximization	0.0414	787.9	294.4	0.0235	408.7	185.1	0.0060	10^5	0.32
Hypothetical cross validation	0.0388	656.0	268.8	0.0248	469.7	178.9	0.0048	9932	211

Time variability estimation criterion	$2T$			T			$T/4$		
	ε	$E[\gamma]$	$E[\gamma^2]$	ε	$E[\gamma]$	$E[\gamma^2]$	ε	$E[\gamma]$	$E[\gamma^2]$
Leave-one-out cross validation	0.0114	846.1	274.7	0.0239	298.9	205.2	0.0961	20.9	5.6
Marginal likelihood maximization	0.0168	371.0	294.4	0.0235	127.3	23.4	0.1117	18.7	4.9
Hypothetical cross validation	0.0115	942.0	156.8	0.0248	262.8	33.18	0.1228	25.6	6.5

The results show that, despite the fact that the Leave-One-Out cross validation, Marginal Likelihood Maximization and Hypothetical Cross Validation are essentially different by the principle of adjusting a model to the given data set within a succession of nested model classes, the final accuracy of estimating the hidden time-varying regression coefficients remains practically the same. Despite the accuracy is close for all methods, the Leave-One-Out cross validation leads in most case to the more smooth models than the other methods and the smoothness power γ has bigger variance across the experiments.

5.2   Experiment on Real-World Data: Bridgewater All Weather Hedge Fund

Bridgewater Associates founded in 1975 is the biggest hedge fund firm in the world with assets over \$150B [25]. It also manages the two world largest hedge funds: Pure Alpha and All Weather with over \$60B in assets each as of 2016. The All Weather fund, launched in 1996, as the name suggests was intended to perform well in any market "weather" conditions. It may invest in a broad range of global assets: stocks, bonds including high yield, inflation linked bonds, emerging market bonds and equities. In managing the All Weather, Bridgewater pioneered an approach that was later called "risk parity" where portfolio allocations to the above assets is intended to balance their risk levels [26] and leverage is used to attain the desired level of risk. Such allocations are adjusted over time as the volatility (risk) of underlying assets is changing. A strategy like this, during a market crisis, would dynamically scale down both its leverage and allocations to risky assets such as equities. During the financial crisis of 2008–2009 the All Weather fund was one of the handful of hedge funds that exhibited positive returns. It is worth noting that the fund doesn't publish its positions, so the exact dynamics of its allocations is generally hidden from investors. When commenting on the crisis period, the fund management confirmed that they switched the portfolio to a "safe" mode with limited allocation to risky assets and no leverage. Uncovering hidden dynamic of the fund allocations is the focus of our analysis below. What makes this analysis difficult is that the fund provides only monthly performance data, even though the trades may occur intra-month [27].

The data set we analyze $((y_t, x_t), t = 1, \ldots, T)$ (1) consists of $T = 72$ monthly returns $(y_t \in \mathbb{R}, t = 1, \ldots, T)$ and the same number of market index vectors $(x_t \in \mathbb{R}^{n+1}, t = 1, \ldots, T)$, $x_t = (x_{0,t}, x_{i,t}, i = 1, \ldots, n)$, $n = 6$, where the chosen indices are denoted as $x_{i,t}$ whereas notation $x_{0,t}$ is reserved for cash. The aim of the analysis is estimating $n + 1$ time varying coefficients $\beta_t = (\beta_{0,t}, \beta_{i,t}, i = 1, \ldots, n)$, in accordance with the condition that coincides in the main with the time-varying regression problem (8)

$$\hat{\beta}_t = (\hat{\beta}_{0,t}, \hat{\beta}_{i,t}, i = 1, \ldots, n) = \arg\min J(\beta_t^0, \beta_t^1, \ldots, \beta_t^n, t = 1, \ldots, T) =$$

$$\frac{1}{\alpha} \sum_{t=1}^{T} \left\{ y_t - \beta_t^0 x_t^0 - \sum_{i=1}^{n} \beta_t^i x_t^i \right\}^2 + \sum_{t=2}^{T} \left\{ \frac{1}{\lambda_0}(\beta_t^0 - \beta_{t-1}^0)^2 + \sum_{i=1}^{n} \frac{1}{\lambda_i}(\beta_t^i - \beta_{t-1}^i)^2 \right\}, \quad \sum_{i=0}^{n} \beta_t^i = 1. \tag{32}$$

There are, however, two distinctions. First, the additional budget constraint like in (29) means that the capital is fully invested in assets. Second, the possibility to consider different volatilities λ_i at different indices x_t^i is stipulated.

The experience shows that, when working with real portfolio data, it is reasonable to assume the time volatility of β-coefficients at each index x_t^i proportional to its time variance $\lambda_i = \lambda g_i^{-1} = \lambda\left((1/T)\sum_{t=1}^{T}(x_t^i - \bar{x}^i)^2\right)^{-1}$, where $\lambda > 0$ is the general volatility to be adjusted as in Sect. 3. All three method hyperparameter estimation from Sect. 3 suggest the same overall volatility λ and have led to the following β-coefficients:

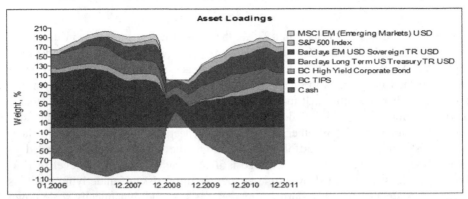

Created with MPI Analytics

As we see from the picture before summer 2008 despite the depression gauge was moving higher the portfolio was not changed from to the All Weather mix. That all changed when the U.S. government allowed Lehman Brothers to fail and the fund quickly shifted its portfolio into the "safe" mode. Over the final months of 2008 and into 2009 the credit intermediation system was not functioning normally and the portfolio was remained in the "safe" mode. In early May 2009 it became clear that the Federal Reserve has the ability to create and spend enough money to offset the deflationary depression, and the fund gradually shifted its 100% Safe portfolio to the All Weather and Safe mix. This trick allows Bridgewater to save the money its clients form crisis in late 2008.

6 Conclusions

The proposed technique has made it possible to fundamentally improve the existing Returns Based Style Analysis methodology currently employed in Finance. The proposed method that implements existing quadratic optimization algorithms is very practical and can be executed on a personal computer. For the very first time, it provides a framework for truly dynamic analysis of hedge funds, resulting in increased transparency and better due diligence which is of crucial importance for financial institutions today.

Acknowledgement. We would like to acknowledge support from grants of the Russian Foundation for Basic Research 17-07-00436 and 17-07-00993.

References

1. Rabiner, L.R.: A tutorial on hidden Markov models and selected applications to speech recognition. Proc. IEEE **77**(2), 257–286 (1989)

2. Hastie, T., Tibshirani, R.: Varying-coefficient models. J. R. Stat. Soc. Ser. B (Methodol.) **55**, 757–796 (1993)
3. Zhao, Z., et al.: Parametric and nonparametric models and methods in financial econometrics. Stat. Surv. **2**, 1–42 (2008)
4. Park, B.U., Mammen, E., Lee, Y.K., Lee, E.R.: Varying coefficient regression models: a review and new developments. Int. Stat. Rev. **83**(1), 36–64 (2015)
5. Fan, J., Zhang, W.: Statistical methods with varying coefficient models. Stat. Interface **1**, 179–195 (2008)
6. Härdle, W., Hall, P., Marron, J.S.: How far are automatically chosen regression smoothing parameters from their optimum? J. Am. Stat. Assoc. **83**, 86–101 (1998)
7. Mammen, E., Park, B.U.: Bandwidth selection for smooth backfitting in additive models. Ann. Stat. **33**, 1260–1294 (2005)
8. Kalaba, R., Tesfatsion, L.: Time-varying linear regression via flexible least squares. Int. J. Comput. Math. Appl. **17**, 1215–1245 (1989)
9. Markov, M., Krasotkina, O., Mottl, V., Muchnik, I.: Time-varying regression model with unknown time-volatility for nonstationary signal analysis. In: Proceedings of the 8th IASTED International Conference on Signal and Image Processing, Honolulu, 14–16 August 2006, paper 534-196 (2006)
10. Markov, M., Muchnik, I., Mottl, V., Krasotkina, O.: Dynamic analysis of hedge funds. In: Proceedings of the 8th IASTED International Conference on Financial Engineering and Applications, Massachusetts, 9–11 October 2006, paper 546-028. MIT, Cambridge (2006)
11. Mottl, V.V., Muchnik, I.B.: Bellman functions on trees for segmentation, generalized smoothing, matching and multi-alignment in massive data sets. DIMACS Technical report 98-15, February 1998. Rutgers University, the State University of New Jersey, 63 p. (1998)
12. Larson, S.: The shrinkage of the coefficient of multiple correlation. J. Educ. Psychol. **22**(1), 45–55 (1931)
13. Stone, M.: Cross-validatory choice and assessment of statistical predictions. J. R. Stat. Soc. **36**(2), 111–147 (1974)
14. Efron, B.: Estimating the error rate of a prediction rule: improvement on cross-validation. J. Am. Stat. Assoc. **78**, 316–331 (1983)
15. MacKay, D.J.C.: Hyperparameters: optimize, or integrate out? In: Heidbreder, G.R. (ed.) Maximum Entropy and Bayesian Methods. Fundamental Theories of Physics, vol. 62, pp. 43–59. Springer, Dordrecht (1993)
16. Li, C., Mao, Y., Zhang, R., Huai, J.: On hyper-parameter estimation in empirical Bayes: a revisit of the MacKay algorithm. In: Proceedings of the Thirty-Second Conference on Uncertainty in Artificial Intelligence, Jersey City, 25–29 June 2016, pp. 477–486 (2016)
17. Bishop, C.: Pattern Recognition and Machine Learning. Springer, New York (2006)
18. Akaike, H.: A new look at the statistical model identification. IEEE Trans. Autom. Control **19**, 716–723 (1974)
19. Kitagawa, G., Akaike, H.: A procedure for the modeling of no-stationary time series. Ann. Inst. Stat. Math. Part B **30**, 351–363 (1987)
20. Rodrigues, C.C.: The ABC of model selection: AIC, BIC and new CIC. In: AIP Conference Proceedings, vol. 803, 23 November 2005, pp. 80–87 (2005)
21. Ezhova, E., Mottl, V., Krasotkina, O.: Estimation of time-varying linear regression with unknown time-volatility via continuous generalization of the Akaike Information Criterion. In: Proceedings of World Academy of Science, Engineering and Technology, March 2009, vol. 51, pp. 144–150 (2009)
22. Sharpe, W.F.: Asset allocation: management style and performance measurement. J. Portf. Manag. **7**–19 (1992)

23. Sharpe, W.F.: Capital asset prices: a theory of market equilibrium under conditions of risk. J. Financ. **19**(3), 425–442 (1964)
24. Fama, E.F., French, K.R.: The cross-section of expected stock returns. J. Financ. **47**(2), 427–465 (1992)
25. Wieczner, J.: Ray Dalio's McDonald's-Inspired Hedge Fund Is Crushing His, Flagship Fund, 07 July 2016. http://fortune.com/2016/07/07/bridgewater-hedge-fund-ray-dalio/
26. https://en.wikipedia.org/wiki/Risk_parity
27. http://www.eurekahedge.com

Author Index

Printed in the United States
By Bookmasters